Herbs, Botanicals, and Teas

FUNCTIONAL FOODS AND NUTRACEUTICALS SERIES

Series Editor
G. Mazza, Ph.D.

Senior Research Scientist and Head
Food Research Program
Pacific Agri-Food Research Centre
Agriculture and Agri-Foods Canada
Summerland, British Columbia

**Functional Foods: Biochemical and Processing Aspects
Volume 1**
Edited by G. Mazza, Ph.D.

Herbs, Botanicals and Teas
Edited by G. Mazza, Ph.D., and B.D. Oomah, Ph.D.

**Functional Foods: Biochemical and Processing Aspects
Volume 2**
Edited by John Shi, Ph.D., G. Mazza, Ph.D., and Marc Le Maguer, Ph.D.

Methods of Analysis for Functional Foods and Nutraceuticals
Edited by W. Jeffrey Hurst, Ph.D.

FUNCTIONAL FOODS AND NUTRACEUTICALS SERIES

Herbs, Botanicals, and Teas

EDITED BY

G. Mazza, Ph.D.
Agriculture and Agri-Food Canada
Pacific Agri-Food Research Centre
Summerland, BC, Canada

Functional Foods Alberta Centre of Excellence
University of Alberta
Edmonton, AB, Canada

B.D. Oomah, Ph.D.
Agriculture and Agri-Food Canada
Pacific Agri-Food Research Centre
Summerland, BC, Canada

CRC Press
Taylor & Francis Group
Boca Raton London New York

CRC Press is an imprint of the
Taylor & Francis Group, an **informa** business

CRC Press
Taylor & Francis Group
6000 Broken Sound Parkway NW, Suite 300
Boca Raton, FL 33487-2742

First issued in paperback 2019

© 1998 by Taylor & Francis Group, LLC
CRC Press is an imprint of Taylor & Francis Group, an Informa business

No claim to original U.S. Government works

ISBN-13: 978-1-56676-851-1 (hbk)
ISBN-13: 978-0-367-39852-1 (pbk)

**Visit the Taylor & Francis Web site at
http://www.taylorandfrancis.com**

**and the CRC Press Web site at
http://www.crcpress.com**

Library of Congress Cataloging-in-Publication Data

Main Entry under title:
Functional Foods & Neutraceuticals Series: Herbs, Botanicals and Teas

Library of Congress Card Number 00-102584

Table of Contents

Series Editor's Preface

THE Functional Foods and Nutraceuticals Series was developed to serve all those involved with and interested in foods and/or food components that provide health benefits beyond those that can be attributed to basic nutritional functions. It offers a comprehensive treatment of the emerging science and technology of functional foods and nutraceuticals shown to play a role in preventing or delaying the onset of diseases, especially chronic diseases. Books in the Series cover a wide range of developments in chemistry, biochemistry, pharmacology, epidemiology and engineering of products from plant and animal sources, results of animal and clinical trials and regulatory, standardization and quality control issues. At least one volume will deal with methods of analysis of functional foods and nutraceuticals.

In developing the series, the Series Editor and Technomic Publishing Company, Inc., recognized the need for assembling, reviewing, condensing and disseminating the rapidly accumulating information on food and health issues to food scientists and technologists, nutritionists, public health professionals, regulators, entrepreneurs and sophisticated consumers. The Series was launched at a time when the diet-health paradigm of food being the source of essential nutrients to sustain life and growth was beginning to change into one in which foods are also called on to provide protection against diseases. The paradigm shift is occurring as consumer interest in diet and health is at an all-time high, and it is due in part to the growing body of epidemiological evidence associating a diet rich in fruits and vegetables with the reduced risk of certain types of cancer and other chronic diseases.

Health professionals are gradually recognizing the role of phytochemicals in health enhancement, and currently there is a frenzy of epidemiological animal and human studies. Emerging results indicate that specific phytochemicals may prevent chronic diseases, aid in the management of symptoms of chronic disorders, improve immune response and reduce negative effects of aging. Food components such as anthocyanins, catechins, lycopene, lutein, allicin, isoflavones, resveratrol, lignans, isothiocyanates, phytosterols, omega-3 polyunsaturated fatty acids, saponins and complex polysaccharides have been shown to modify metabolic processes and influence disease risks. Elucidating the chemistry, biology, pharmacology and the many potential health effects of the countless bioactives present in plant and animal products, developing sound scientific data to support health claims and developing food products that address the needs of an increasingly health conscious consumer interested in self-medication offer tremendous challenges and opportunities for the scientific community and the food industry worldwide.

The aim of this book series is to bring together, in a timely manner, reliable information that will serve the needs of food, nutrition and health practitioners, provide reference to material that may otherwise be difficult to locate and provide a starting point for further research.

G. MAZZA
Series Editor

Preface

CONSUMERS, especially baby boomers, who have grown disillusioned with modern medicine, pharmaceuticals and the healthcare system and who want more control over their health, are driving the ever-growing demand for functional foods and nutraceuticals. As a result, there has been an explosion of interest in information on herbs, botanicals and teas in recent years. It has become clear that a book on the current state of knowledge of these products is needed. *Herbs, Botanicals and Teas,* written by leading researchers presently contributing to this field, provides the latest scientific and technical information on the chemical, pharmacological, epidemiological and clinical aspects of garlic, ginseng, Echinacea, ginger, fenugreek, St. John's wort, Ginkgo biloba, goldenseal, saw palmetto, valerian, evening primrose, licorice, bilberries and blueberries, and green and black teas. The book contains 13 chapters, 10 on herbs and botanicals, one on teas, one on international regulations and the last chapter on quality assurance and control for the herbal and tea industry.

The content of a typical chapter on a given product includes an introduction, sections on chemical composition and chemistry, physiological actions of the product, clinical or therapeutic applications, conclusions, and references. Whenever possible, results of chemical, biological and/or clinical studies are presented in tables or figures. The chapter on ginseng, for example, presents and discusses the latest research results on its chemical and pharmacological properties. Physiological actions addressed include endocrine activity, neurotransmitter activity, neurophysiological responses, antioxidant activity and ginseng's effects on the immune system. Similarly, the chapter on evening primrose reviews the current knowledge of evening primrose oil (EPO) and its major constituent gamma-linolenic acid (GLA). It also contains an in-depth discussion of the result of recent research on the metabolism of GLA, nutritional relevance of EPO, the role of EPO as a functional food and therapeutic applications of EPO in cardiovascular disease, diabetic neuropathy, gastrointestinal, gynecological, and neurological disorders, rheumatoid arthritis, viral infection and immunological disorders.

The distinction between herbs, botanicals and teas may not always be apparent in this book. The reason for this seeming lack of clarity is that these terms are defined in several ways depending on the context in which the words are used. The term herbs generally refers to plants with leaves or stems that are used for medicine, seasoning, food or perfume. Here it is defined as a crude extract, or a product, produced from an edible plant and sold in pill, powder and other medicinal forms and demonstrated to have a physiological benefit or provide protection against disease. A botanical is traditionally a drug made from roots, leaves, flowers or other parts of plants. It can also imply herbal preparation, herbal tea, herbal mixture or medicinal herb. Here it is synonymous with nutraceutical. The term teas refers to black and green teas made from the leaves of *Theae nigrae folium*.

With over 2,000 scientific references, this book provides our readers (food scientists, nutritionists, biochemists, food technologists, toxicologists, chemists, molecular biologists and public health professionals) with a comprehensive and up-to-date publication on the chemistry, pharmacology and clinical applications of the major herbs, botanicals and teas.

We express our sincere thanks and appreciation to all the contributors who by freely and willingly giving their knowledge and expertise have made this book possible. Our gratitude is also extended to colleagues who have reviewed various chapters and the editorial staff and publishers at Technomic Publishing Company, Inc. for their contribution in bringing this work to publication. Many thanks are also extended to Linda Kerr, Paul Ferguson, Rachel Mazza and Michael Weis who helped with the preparation of portions of the book.

We hope that this book will serve to further stimulate the development of functional foods and nutraceuticals and provide consumers worldwide with products that prevent diseases and help them maintain healthier lives throughout the new millennium.

G. MAZZA
B.D. OOMAH

List of Contributors

ATHAR, M.
Department of Dermatology
College of Physicians and Surgeons
Columbia University
630 West 168th Street
New York, NY 10032 USA

BAISSAC, Y.
Laboratoire de Recherche sur les
   Substances Naturelles Végétales
(UPRES EA 1677)
Université Montpellier II
34095 Montpellier, France

BALENTINE, D. A.
Lipton
800 Sylvan Avenue
Englewood Cliffs, NJ 07632 USA

BAUER, R.
Institut für Pharmazeutische
   Biologie
Heinrich-Heine-Universität
   Düsseldorf
Universitätsstr. 1
D-40225 Düsseldorf, Germany

BICKERS, D. R.
Department of Dermatology
College of Physicians and Surgeons
Columbia University
630 West 168th Street
New York, NY 10032 USA

BROADHURST, C. L.
Environmental Chemistry
   Laboratory
U.S. Department of Agriculture
Agricultural Research Service
Building 012, BARC-West
Beltsville, MD 20705-2350 USA

CAMIRE, M. E.
Department of Food Science and
   Human Nutrition
University of Maine
5736 Holmes Hall
Orono, ME 04469-5736 USA

DIAMOND, S.
Flora Manufacturing and
   Distribution Ltd.
7400 Fraser Park Drive
Burnaby, BC V5J 5B9 Canada

FEDEC, P.
POS Pilot Plant Corporation
118 Veterinary Road
Saskatoon, SK S7N 2R4 Canada

HEGARTY, P. V.
Institute for Food Laws and
  Regulations
165 C National Food Safety and
  Toxicology Center
Michigan State University
East Lansing, MI 48824 USA

HOLUB, B. J.
Department of Human Biology and
  Nutritional Sciences
University of Guelph
Guelph, ON N1G 2W1 Canada

KIKUZAKI, H.
Department of Food and Nutrition
Faculty of Human Life Science
Osaka City University
3-138, Sugimoto 3-CHOME
Sumiyoshi, Osaka, 558 Japan

KITTS, D. D.
Food, Nutrition and Health Program
Faculty of Agricultural Sciences
6650 North West Marine Drive
University of British Columbia
Vancouver, BC V6T 1Z4 Canada

KOLODZIEJCZYK, P. P.
POS Pilot Plant Corporation
118 Veterinary Road
Saskatoon, SK S7N 2R4 Canada

MAZZA, G.
Food Research Program
Agriculture and Agri-Food Canada
Pacific Agri-Food Research Centre
Summerland, BC V0H 1Z0 Canada

NAGPURKAR, A.
Department of Human Biology and
  Nutritional Sciences
University of Guelph,
Guelph, ON N1G 2W1 Canada

OOMAH, B. D.
Food Research Program
Agriculture and Agri-Food Canada
Pacific Agri-Food Research Centre
Summerland, BC V0H 1Z0 Canada

PAETAU-ROBINSON, I.
Lipton
800 Sylvan Avenue
Englewood Cliffs, NJ 07632 USA

PESCHELL, J.
Department of Human Biology and
  Nutritional Sciences
University of Guelph
Guelph, ON N1G 2W1 Canada

PETIT, P.
Laboratoire de Pharmacologie
(UPRES EA 1677)
Faculté de Médecine,
Université Montpellier I
34060 Montpellier, France

RIBES, G.
UMR 9921, CNRS
1919 Route de Monde
34293 Montpellier, France

SAUVAIRE, Y.
Laboratoire de Recherche sur les
  Substances Naturelles Végétales
(UPRES EA 1677)
Université Montpellier II
34095 Montpellier, France

TOWERS, G. H. N.
Department of Botany
University of British Columbia
Vancouver, BC V6T 1Z4 Canada

WINTHER, M.
QuantaNova Canada Ltd.
PO Box 818
Kentville, NS B4N 4H8 Canada

WANG, Z. Y.
42 Bartha Avenue
Edison, NJ 08817 USA

CHAPTER 1

Garlic Constituents and Disease Prevention

A. NAGPURKAR
J. PESCHELL
B. J. HOLUB

1. INTRODUCTION

In recent years, sources of disease-modifying foods and their functional components have attracted attention in nutrition and clinical research. Garlic appears to be a food item that contributes multiple constituents that can potentially benefit human health. Papers, journals and the media have been reporting on various medical findings related to the effects of garlic. Over 2,000 papers have been written about garlic or its constituents (Lawson, 1998a). As a consequence, some health organizations have targeted garlic as a prime candidate for the development of low-cost "functional foods" and "nutraceuticals" that help to reduce risk factors associated with cardiovascular disease and cancer (Block, 1992). Garlic is fast becoming one of the most significant nutraceuticals of our time.

1.1. BACKGROUND INFORMATION

For centuries, garlic has been thought of as a magical healing plant. Its use as a medicinal agent was recorded in Sanskrit documents over 5,000 years ago. Garlic was used by Egyptian slaves during the building of the pyramids and by the Romans who ate it to strengthen themselves during battles. Hippocrates, considered to be the "father of medicine," and Dioscorides, the "father of pharmacy," recommended garlic for several conditions such as infection and blood flow problems. In both world wars, garlic was used to treat wounds when other antibiotics were not available (Fenwick and Hanley, 1985a; Lawson, 1998a). Medicinally, garlic has been used in numerous disorders including

1

atherosclerosis, hypertension, colds, headaches, worms and tumors (Fenwick and Hanley, 1985a; Murry, 1995).

Botanically, garlic is known as *Allium sativum* L. (family Liliaceae). The exact origin of the genus name is unknown, nevertheless, a relation to the Latin word *olere* is often made, meaning, "to smell." The general characteristics of this genus are plants that are herbaceous perennials and usually form bulbs. Garlic grows to approximately 30–90 cm in height in well-fertilized, sandy, loamy soil in warm sunny locations during spring and summer. Leeks, onions and shallots are all part of the *Allium* genus along with garlic (Murry, 1995; Hahn, 1996).

In 1997, the global production of garlic was 12 million metric tons, with China being the leading producer (8.8 million metric tons), followed by South Korea and India (0.5 million metric tons each). The United States (252,000 metric tons) and Argentina (90,000 metric tons) are the major producers in North and South America, respectively (Food and Agriculture Organization of the United Nations, 1997). The highest producer of garlic per hectare of the above-mentioned countries is the United States (16,816 kg/ha), whereas the largest producer per hectare in the world is Egypt (22,729 kg/ha) with an overall production yield of 159,000 metric tons. In Canada, the commercial production of garlic is a relatively new industry, with Ontario producing 2,220 metric tons during 1998 (87% of the Canadian production) (Ontario Garlic Growers Association, 1998).

In 1994, the U.S. consumption of garlic, primarily as food, was 199 million kg, while in Canada, the 1998 consumption was 10.9 million kg. The U.S. herbal supplement industry is a fast-growing industry with sales doubling from 1994 to 1997. The herbal industry in 1997 had sales of $3.65 billion, of which garlic supplements accounted for $200 million (Monmaney and Roan, 1998). The two major garlic supplement producers are Lichtwer Pharma GmbH, in Germany, whose products are sold under the brand name Kwai, and Wakunaga Pharmaceutical Co. of Japan, whose products are sold under the brand name Kyolic.

Although garlic has been used for centuries in herbal medicine, it is only during the last 15–20 years that some of the health claims associated with garlic (e.g., lowering of blood cholesterol and blood pressure, and its anticancer properties) have been tested rigorously for legitimate scientific merit (Han et al., 1995; Lawson, 1998a). With over 2,000 studies, garlic can be considered one of the most researched medicinal plants. Several comprehensive reviews have been published on garlic (Block, 1985; Fenwick and Hanley, 1985a, 1985b, 1985c; Kendler, 1987; Kleijnen et al., 1989; Block, 1992; Reuter and Sendl, 1994; Agarwal, 1996; Koch and Lawson, 1996; Lawson, 1998a), yet the complex biological and pharmacological actions of garlic and its constituents are still not completely understood. The pace of the biological studies has far exceeded the chemical and analytical studies required for identifying the constituents and active components in the various garlic preparations used in the *in vitro* and *in vivo* studies (Lawson, 1993).

2. CHEMICAL COMPOSITION AND CHEMISTRY

2.1. COMPOSITION

Garlic is composed mainly of water (56–68%), followed by carbohydrates (26–30%). The most significant components, medicinally, are the organo-sulfur-containing compounds (11–35 mg/g fresh garlic). Garlic contains nearly three times as much sulfur-containing compounds (per g fresh weight) as broccoli, onions, apricots or cauliflower (Lawson, 1996). The investigation of garlic usually concerns the sulfur-containing compounds, possibly due to their presence in garlic in unusually high amounts or to the pharmacological activities attributed to various sulfur-containing compounds (e.g., penicillin, probucol).

Garlic also contains various other compounds such as saponins, vitamins (ascorbic acid 30 mg/100 g fresh weight, vitamin E 9.4 µg/g), minerals (selenium 0.014 mg/100 g, chromium 0.05 mg/100 g) plus others (Fenwick and Hanley, 1985b; U.S. Department of Agriculture, 1998).

2.2. CHEMISTRY

2.2.1. γ-Glutamyl-Cysteines and Cysteine Sulfoxides

The mature, intact garlic clove contains mainly cysteine sulfoxides such as alliin (5–14 mg/g fresh garlic), followed by methiin and isoalliin that are formed from the γ-glutamyl-cysteines (γ-glutamyl-S-*trans*-1-propenylcys-teine, γ-glutamyl-S-allylcysteine and γ-glutamyl-S-methylcysteine) (Figure 1.1). During wintering and sprouting, the γ-glutamyl-cysteines are hydrolyzed to form cysteine sulfoxides by increasing levels of γ-glutamyl-transpeptidase. This process also occurs during storage and occurs more rapidly at cooler temperatures. The cysteine sulfoxides (8–19 mg/g fresh weight) and γ-gluta-myl-cysteines (5–16 mg/g) account for approximately 82% of the total sulfur in fresh garlic (Lawson, 1993). When garlic is cut, crushed or chewed, the enzyme alliinase is released, converting the cysteine sulfoxides into the thiosul-finates. The bulb contains most of the cysteine sulfoxides (alliin), approxi-mately 85%, whereas the leaves (12%) and the roots (2%) contain smaller amounts (Lawson, 1998a).

2.2.2. Alliinase

Alliinases or alliin lyases (EC 4.4.1.4) are pyridoxal 5′-phosphate dependent a,b-eliminating lyases that catalyze the conversion of the cysteine sulfoxides

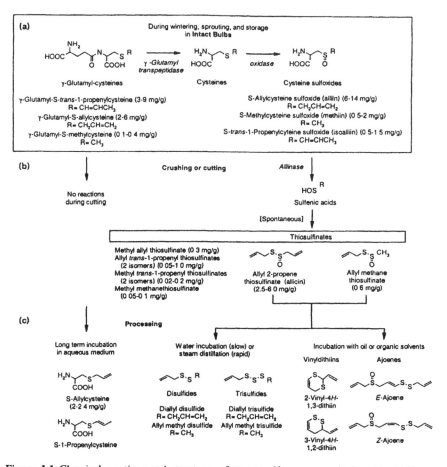

Figure 1.1 Chemical reactions and structures of organosulfur compounds found in (a) intact cloves, (b) upon crushing and (c) through processing [amounts (mg/g) are approximate levels in fresh garlic]. Adapted from Lawson (1996, 1998a).

to the biologically active dialkyl thiosulfinates via sulfenic acid intermediates (Block, 1992). Alliinase is localized to a few vascular bundle sheath cells around the veins or phloem (Ellmore and Feldberg, 1994), whereas alliin and the other cysteine sulfoxides are found in the mesophyll cells. Two different alliinase activities have recently been found in garlic. One is specific for alliin and isoalliin, the other methiin. The former has a pH optimum of 4.5 and cleaves 97% of its substrate within 0.5 min at 23°C, the latter has a pH optimum of 6.5, cleaving 97% of its substrate within 5 min (Block, 1992). Alliinase is temperature and pH dependent and can be irreversibly deactivated at pH 1.5–3.0. This enzyme is approximately 10 times more abundant in the cloves than in the leaves and accounts for at least 10% of the total protein in the cloves (Ellmore and Feldberg, 1994).

2.2.3. Thiosulfinates

The thiosulfinates (2–9 mg/g fresh crushed garlic) are reactive, volatile, odor producing substances formed enzymatically when garlic is cut, crushed or chewed. Allicin (allyl 2-propenethiosulfinate) is the most abundant thiosulfinate (70%) formed via alliinase reactions, with allyl methanethiosulfinate being the second most abundant (18%). Various other thiosulfinates are formed in low concentrations. Conversion of the cysteine sulfoxides to thiosulfinates via alliinase occurs rapidly, within 0.2–0.5 min at room temperature for allicin and 1.5–5 min for allyl methyl thiosulfinates. The stability of thiosulfinates is dependent on solvent, temperature, concentration and purity. Allicin is soluble in organic solvents, especially polar solvents, but it is less soluble in water. The half-life of pure allicin in water and 1 mM citric acid is 30 and 60 days, respectively. Without a solvent, the half-life of allicin decreases to 16 hours (Lawson, 1993). Thiosulfinates in garlic homogenates are less stable than their pure forms, possibly due to water-soluble substances in the garlic homogenates. Refrigeration greatly increases the stability, but the only long-term storage option is freezing at −70°C (Lawson, 1996).

The thiosulfinates undergo various transformations depending on temperature, pH and solvent conditions, to form more stable compounds such as di- and tri-sulfides, allylsulfides, vinyl dithiins, ajoenes and mercaptocysteines. Incubation of allicin or allyl methane thiosulfinate in low-polarity solvents or without solvents produces mainly vinyl dithiins followed by ajoenes in smaller amounts. Incubation in alcohol gives variable results forming ajoenes, diallyl trisulfide or vinyl dithiins, depending on whether pure allicin or cloves are incubated. Steam distilled garlic oil consists of diallyl disulfide, diallyl trisulfide and mono- to hexasulfides (Lawson, 1993). It should be noted that intact garlic cloves are relatively odorless and do not contain allicin and allicin-generated compounds, although they do contain cysteine sulfoxides and γ-glutamyl-cysteines, precursors of allicin.

2.3. VARIATION IN ORGANOSULFUR COMPOUNDS IN GARLIC

The generation of allicin and other organosulfur compounds varies depending on soil conditions, climate, garlic variety, harvest dates and post-harvest handling. According to Lawson (1998a), the variation among garlic strains typically found in grocery stores (in California) differs 1.8–2.7-fold in alliin and 1.5–4.2-fold in γ-glutamylcysteines. The addition of ammonium sulfate to the soil, as compared to ammonium nitrate, causes a proportional increase of allicin-releasing potential in garlic bulbs (Reuter and Sendl, 1994; Kosian, 1998).

3. COMMERCIAL GARLIC PREPARATIONS

Several types of garlic preparations are commercially available (see Table 1.1). These products are produced under a variety of conditions such as low-temperature drying, steam distillation, and long-term incubation in various mediums. Most commercial products on the market either generate allicin or contain allicin-derived compounds. Analysis of garlic supplements by HPLC for the estimation of allicin in some of the commercial and noncommercial products are shown in Figure 1.2.

3.1. ALLICIN-DERIVED PRODUCTS

Fresh garlic and powders that have been prepared by careful drying at low temperatures (< 60°C) are capable of producing allicin. These garlic supplements contain the enzyme alliinase and its substrate alliin. The potency of these supplements is usually expressed as "allicin-releasing potential" or "allicin yield," the amount of allicin generated from alliin when the tablet or capsule is exposed to an aqueous medium. The most popular forms of garlic supplements are made from garlic powders and are sold as either enteric-coated tablets (e.g., Kwai) or nonenteric-coated capsules/tablets. Most of the garlic tablets are standardized for their allicin yield. This ensures reproducible dosages of the supplement. The variation in the allicin-releasing potential of the major brands ranges from 0.1 to 8.9 mg/g powder (Han et al., 1995). Drying at higher temperatures often results in irreversible loss of activity of the enzyme alliinase, thereby preventing allicin generation. Garlic powder as a spice may or may not contain active alliinase, depending on drying conditions.

Some products do not generate allicin but contain allicin-derived compounds such as garlic macerates in oil and garlic oil products. These products are often not standardized for organosulfur compounds and contain varying amounts of allicin-derived compounds such as polysulfides, ajoenes and vinyl dithiins.

3.2. NON-ALLICIN-DERIVED PRODUCTS

Although most garlic products rely on allicin-derived compounds for their therapeutic effect, some products do not produce any significant amounts of either allicin or allicin-derived compounds. Aged garlic extracts (e.g., Kyolic) contain γ-glutamylcysteine-derived compounds such as S-allylcysteine (SAC) and small amounts of S-allylmercaptocysteine. Several prod-

TABLE 1.1. Commercially Available Products and Their Possible Active Principles.

Products	Processing	Possible Active Principle(s)	Notes
Fresh garlic	None	Allicin and allicin-derived sulfur compounds generated *in vivo*	Heating may cause loss of allinase activity (>60°C)
Garlic powder tablets (e.g., Kwai, etc.)	Drying/grinding	Allicin and allicin-derived sulfur compounds generated *in vivo*	May be enteric or nonenteric coated and/or standardized for alliin content/allicin-generating potential
Oil-macerated garlic Garlic oil	Incubation in oil Steam distillation	Vinyl dithiins, ajoenes, allyl sulfides Allyl di- and trisulfides	Not often found commercially in North America
Aged garlic extract (e.g., Kyolic, etc.)	Incubation in ethanol (>1 yr.)	S-allylcysteine (SAC) S-allylmercaptocysteine	Contains no allicin or allicin-derived compounds

7

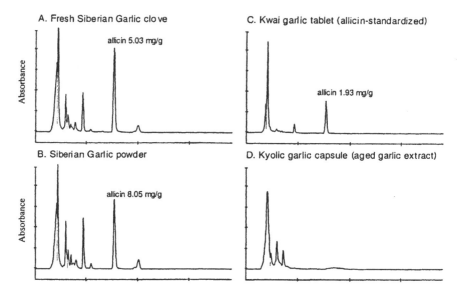

Figure 1.2 HPLC of allicin from (A) Ontario-grown Siberian garlic bulb, (B) dried powder derived from Ontario-grown Siberian garlic bulb, (C) Kwai garlic tablet (Lichtwer Pharma GmbH, Germany) and (D) Kyolic garlic capsule supplement (Wakunaga of America Co , Ltd., USA). Allicin was determined by HPLC using a C-18 reverse-phase column with a 50 50 water·methanol (v/v) eluant, at a flow rate of 1 mL/min (Lawson et al., 1991).

ucts are known to contain a combination of allicin-derived and non-allicin-derived compounds (e.g., Quintessence).

4. METABOLIC FATE OF GARLIC-DERIVED ORGANOSULFUR COMPOUNDS

Many biological studies have shown physiological effects with a variety of garlic preparations, ranging from fresh and powdered garlic to garlic oils, containing various active components. However, it has not been possible to clearly identify the organosulfur metabolites that may be responsible for the physiological effects of garlic. This is primarily due to the fact that the metabolic fate of allicin, generally regarded to be the most important biologically active compound derived from garlic, is not well understood (Lawson and Wang, 1993; Reuter and Sendl, 1994; Freeman and Kodera, 1995; Lawson and Block, 1997; Freeman and Kodera, 1997; Lawson, 1998a).

An abbreviated version of the potential pathways for the metabolism of certain predominant organosulfur compounds derived from garlic is shown in Figure 1.3. This summarizes the events that occur when fresh garlic or commercial garlic preparations are consumed, starting from the chewing/biting and swallowing of garlic, followed by digestion in the stomach and finally moving

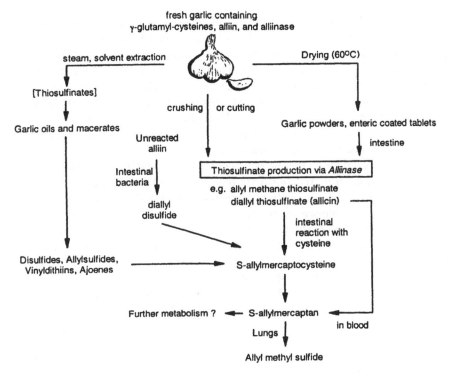

Figure 1.3 Abbreviated scheme for the possible pathways involved in the metabolism of garlic and garlic-derived compounds.

to the intestine. If fresh garlic or garlic powder capsules, which are not enteric-coated, are consumed, alliin would be converted to allicin in the stomach. This would occur only if the acidity of the stomach content is above its normal pH of 1–3, otherwise, the enzyme alliinase would be irreversibly deactivated. Consequently, this often results in inefficient and highly variable generation of allicin required for the physiological effects attributed to garlic. For this reason, it is often recommended that garlic preparations that are enteric-coated be taken. This allows the garlic to withstand the acidic environment of the stomach, and upon reaching the intestine, where the pH is neutral, extensive conversion of alliin to allicin takes place. It has been suggested that alliin that has survived the acidic environment in the stomach could potentially be converted to physiologically-active diallyl disulfide in the intestine (Block, 1992).

At present, the role of allicin in garlic research is controversial (Freeman and Kodera, 1995; Lawson and Block, 1997; Freeman and Kodera, 1997). Although allicin and allicin-derived organosulfur compounds are regarded by most investigators to be responsible for the various physiological effects of garlic, allicin does not remain in the blood for longer than 5 minutes after the consumption of fresh garlic or commercial garlic preparations (Freeman and Kodera, 1995; Lawson, 1998a). Recent studies have shown that allicin is very unstable in the blood and is metabolized rapidly to form allyl mercaptan, which can potentially be metabolized to further metabolites (Lawson, 1998a).

A recent study has shown the presence of garlic metabolites (e.g., allyl methyl sulfide, dimethyl sulfide and acetone) in the human breath 30 h after garlic consumption (Taucher et al., 1996; Lawson, 1998a). It has been suggested that these metabolites may be associated with the various physiological effects associated with garlic (Lawson, 1998a).

Another potential metabolic pathway for allicin and its transformation compounds is through its reaction with the amino acid cysteine (in the intestine) to form S-allylmercaptocysteine, which is converted to allyl mercaptan. *In vitro* studies have shown physiological activity associated with allyl mercaptan and S-allylmercaptocysteine, which are potential precursors of organosulfur-containing metabolites (Wattenberg et al., 1989; Lee et al., 1994; Gebhardt and Beck, 1996; Pinto et al., 1997; Sigounas et al., 1997a, 1997b). Recent human studies have shown the formation of increased levels of acetone in the breath after consumption of garlic, suggested to be indicative of enhanced metabolism of blood lipids (Taucher et al., 1996).

5. HEALTH BENEFITS

5.1. CARDIOVASCULAR EFFECTS

5.1.1. Cholesterol Lowering

Since 1900, approximately 166 clinical studies have focused on the potential hypocholesterolemic effect of garlic in humans (Reuter and Sendl, 1994; Han et al., 1995; Lawson, 1998a). Most of these studies have used garlic preparations that have either allicin-releasing potential or have allicin-derived organosulfur compounds as their main ingredient (fresh garlic, garlic powders, garlic oils, garlic oil macerates). A combination of a commercial garlic preparation along with an omega-3 fish oil concentrate has recently been reported to offer reductions in cholesterol plus triglyceride levels in moderately hypercholesterolemic subjects (Adler and Holub, 1997). These studies have reported, on average, a lowering of total serum cholesterol by about 10–12%. Two recent meta-analyses of the primary clinical trials have also confirmed the cholesterol lowering effect (~10%) in hypercholesterolemic subjects (Warshafsky et al., 1993; Silagy and Neil, 1994a). A German study found that the cholesterol-lowering effect with garlic powder (900 mg/day) was as effective as the drug bezafibrate (Holzgartner et al., 1992).

The early introduction of garlic supplements with proven clinical efficacy to individuals with normal or slightly elevated cholesterol levels can potentially reduce the progression of their cholesterol values to levels that would require the eventual intervention with beneficial but costly pharmaceutical therapies. However, the new generation of cholesterol-lowering drugs, hepatic HMG-CoA reductase inhibitors (statins), can provide cholesterol reductions of 20–25%. The cost (per 1% reduction of cholesterol) of putting a patient on established pharmaceutical therapies is considerably more than that offered by alternative, clinically proven garlic supplements.

Although it is not yet known which of the compounds in garlic are responsible for the blood cholesterol-lowering effect, there is considerable evidence from animal and other *in vitro* studies to suggest that allicin or allicin-generated organosulfur compounds are responsible for the effect (Gebhardt et al., 1994; Yeh and Yeh, 1994; Gebhardt and Beck, 1996). Tissue culture studies have shown allyl disulfide and allyl mercaptan to inhibit HMG-CoA reductase, the rate-limiting hepatic enzyme for the biosynthesis of cholesterol in the body. The mechanism of this reaction is thought to involve the formation of sulfide bridges within the HMG-CoA reductase enzyme by interaction with the allicin-derived compounds, rendering the enzyme inactive (Gebhardt et al., 1994).

Although there is considerable evidence based on published studies (Kleijnen et al., 1989; Warshafsky et al., 1993; Silagy and Neil, 1994a; Agarwal, 1996; Reuter et al., 1996) in support of the cholesterol-lowering effect of garlic, there is some growing concern about the usefulness of garlic as a cholesterol-lowering supplement. This concern has arisen from recent human clinical studies that have shown no cholesterol-lowering effect (Simons et al., 1995; Berthold et al., 1998; McCrindle et al., 1998; Isaacsohn et al., 1998) and one meta-analysis (Neil et al., 1996) carried out in the UK that found no statistically significant cholesterol lowering due to garlic consumption in human subjects. These studies have created uncertainty about the therapeutic benefits of garlic supplements for lowering cholesterol (Berthold and Sudhop, 1998). It should be pointed out, however, that most of the recent negative results from studies that have used Kwai garlic supplements may be due to a change in the composition of these supplements as suggested by Lawson (1998b).

There are various potential reasons for the lack of effect in the above mentioned studies. One hypothesis may be that (at present) most garlic supplements are sold as dried powder and do not contain preformed allicin, the thiosulfinate that is supposedly required for the cholesterol-lowering effect. Although it is true that the allicin-releasing potential is determined under simulated conditions that are likely to exist in the intestine where the enteric-coated garlic supplements are supposed to disintegrate and generate allicin, conditions in the intestine may vary from one individual to another. Likewise, there may be other factors present in the intestine that may affect the allicin-releasing potential. This could result in varying amounts of allicin and allicin-derived compounds produced between individuals resulting in varying therapeutic effects, eventually affecting the statistical significance.

Another possibility for lack of effect may be related to dietary factors. Several recent studies that have shown a lack of effect with garlic supplementation used a diet restriction regime (NCEP step I/II diets); in contrast, studies that have shown an effect usually had no dietary restrictions.

5.1.2. Blood Pressure Lowering

The effect of garlic preparations on blood pressure has been studied since 1921. A meta-analysis of controlled human trials using garlic preparations (powders) concluded that garlic significantly (although very modestly) lowers systolic and diastolic blood pressures, although the reductions were generally greater for systolic pressure (Silagy and Neil, 1994b). Although several animal and human studies have demonstrated a moderate hypotensive effect, the precise active principal(s) present in garlic preparations responsible for blood pressure lowering are not known. Sendl and colleagues (1992) have shown that the aqueous extract of garlic leaves, containing mainly glutamyl peptides, exhibits greater inhibitory activity (compared to the bulb) toward the angiotensin-converting enzyme (an important pathway in blood pressure regulation). More recently, it has been shown that aqueous extracts of garlic increase the activity of nitric oxide synthase, an enzyme that produces nitric oxide, which is associated with a lowering of blood pressure (Das et al., 1995a, 1995b; Pedraza-Chaverri et al., 1998). As with aforementioned blood cholesterol trials, not all garlic studies have shown blood pressure reducing effects (Silagy and Neil, 1994b).

5.1.3. Anti-Thrombotic Activity

One of the therapeutic effects of garlic known since 1978 is its potential effect on blood coagulation and platelet aggregation (Lawson et al., 1992; Han et al., 1995), two important factors involved in cardiovascular disease and its associated thrombogenicity. It has been shown through several *in vitro* and *ex vivo* studies (animals and humans) that certain garlic constituents affect platelet aggregation (Makheja and Bailey, 1990; Venton et al., 1991; Lawson et al., 1992; Han et al., 1995). Adenosine, which is present in garlic cloves, is known to inhibit platelet aggregation, but this inhibition was seen only in platelet-rich plasma and was absent in whole blood (Han et al., 1995). Thus, the effect with adenosine may be of questionable physiological relevance (Lawson et al., 1992). Perhaps the most interesting characteristic is the ability of garlic constituents to inhibit the action of most platelet agonists (Venton et al., 1991). A number of garlic preparations (ether extracted garlic oil, steam distilled garlic oil, fresh garlic, garlic powder, aqueous garlic extract, synthetic and natural allicin, ajoenes) have been used to study the effects of garlic on platelet aggregation (Lawson et al., 1992). The authors have compared the effects of these commercial preparations and identified the important active principles as allicin, ajoenes, vinyl dithiins and diallyl trisulfide. Allicin and diallyl trisulfide were found to be the most active garlic-derived sulfur compounds in inhibiting platelet aggregation *in vitro*. The precise mechanisms for the inhibition of platelet aggregation by garlic-derived compounds are not known but likely include multiple effects such as the inhibition of eicosanoid formation from arachidonic acid (Mohammad and Woodward, 1986; Apitz-Castro et al., 1986, 1991, 1992; Ali et al., 1990; Makheja and Bailey, 1990; Lawson et al., 1992; Srivastava and Tyagi, 1993; Ali and Thomson, 1995; Batirel et al., 1996).

Garlic is also known to have an effect on blood coagulation and fibrinolytic activity (breakup of the fibrin clot into fibrin degradation products). Deficient fibrinolytic activity due to increased plasminogen activator inhibitor type-1 levels or reduced levels of tissue-type plasminogen activator predisposes patients to thrombotic events. Human studies have shown that garlic consumption leads to increased fibrinolytic activity, increased blood coagulation time and decreased fibrinogen levels (Han et al., 1995). A recent study in Germany showed that patients taking a garlic powder supplement had more flexibility in their blood vessels that may provide a cardioprotective effect (Breithaupt-Grogler et al., 1997).

5.2. ANTICANCER PROPERTIES

A role for garlic in the prevention of cancer comes from epidemiological studies that have shown that consumption of *Allium* vegetables such as garlic resulted in a significant reduction in the development of stomach cancer (Lau et al., 1990; Han et al., 1995; Milner, 1996; Lawson, 1998a). An epidemiological study comparing two counties in China determined the incidence of stomach cancer in a region of people consuming 20 g fresh garlic per day was 8% of that where people ate less than 1 g per day (Mei et al., 1982). Members of the stomach-cancer-prone population were shown to have higher levels of nitrites in their gastric juices. Garlic has been shown to have an inhibitory effect on nitroso compound formation *in vivo* by decreasing the amount of nitrate-reducing bacteria in the stomach (Mei et al., 1989). A survey conducted by Hopkins and Williams (1981) found 246 risk factors associated with cardiovascular disease, of which low intake of garlic was a factor. A more recent prospective study in the United States, the "Iowa Women's Health Study" (41,837 women) (Steinmetz et al., 1994), determined an inverse relationship between colon cancer risk and garlic consumption, with an age and energy adjusted relative risk of 0.68.

Studies have indicated that an average intake of 4–6 g of garlic per day is inversely correlated with the incidence of gastrointestinal cancers (Lawson, 1998a). Although several epidemiological studies have proposed an anticancer role for garlic, three epidemiological studies from the Netherlands have found no association between garlic supplement consumption and the prevention of breast, colon and lung cancers (Dorant et al., 1994, 1995, 1996).

Numerous *in vitro* and *in vivo* studies have determined that various constituents of garlic inhibit tumor growth and possess other anticancer properties (Lau et al., 1990; Brady et al., 1991; Han et al., 1995; Milner, 1996; Schaffer et al., 1996; Srivastava et al., 1997; Pinto et al., 1997; Riggs et al., 1997; Hu and Singh, 1997; Sakamoto et al., 1997; Lawson, 1998a). The allyl-containing organosulfides such as diallyl di- and trisulfides, allyl methyl di- and trisulfides, allylmercaptan, and S-allylcysteine as well as fresh garlic extract have inhibitory effects on the formation of tumors induced by various initiators (Dorant et al., 1993; Heber, 1997). Mice treated with thiosulfinates or freshly ground garlic showed no tumor formation after being injected with ascites tumor cells. Furthermore, injections directly into existing tumors stopped their growth (Fujiwara and Natata, 1967). Studies with aged garlic extract fed to rats at 2–4% of the diet decreased carcinogen-induced tumors and DNA adducts. The active principle is unknown, but it has been postulated that S-allylcysteine may be the effective compound (Liu and Milner, 1990). Garlic-derived organosulfides have also been shown to inhibit the conjugation of carcinogens with mammary cell DNA to form DNA adducts. Changes in the DNA adducts serve as an early indicator of the alterations in the initiation phase of carcinogenesis. The anticarcinogenic action of these organosulfides may be caused by a direct inhibition of tumor cell metabolism (cytochrome P-450), inhibition of initiation/promotion through carcinogen detoxification via increased glutathione S-transferase or through modulation of host immune response (Lau et al., 1990; Brady et al., 1991; Srivastava et al., 1997; Pinto et al., 1997).

5.3. ANTIOXIDANT EFFECTS

Garlic has antioxidant effects. Aged garlic extract has been shown to inhibit lipid peroxidation in rat liver microsomes in a dose-dependent manner (Horie et al., 1992). S-allylcysteine and S-allylmercaptocysteine also have radical scavenging activities (Imai et al., 1994). One human clinical trial with garlic supplementation (Kwai, 600 mg/d) resulted in less susceptibility to copper-induced oxidation in low-density lipoprotein (Phelps and Harris, 1993). Allyl mercaptan generated by blood cells also appears to have an antioxidant effect *in vivo* (Lawson and Wang, 1993). The antioxidant effect is useful in cancer prevention as well as in the prevention of cardiovascular disease. Garlic also prevents the oxidation of low-density lipoprotein (LDL) to the atherogenic oxidized LDL species in human subjects (Phelps and Harris, 1993).

5.4. ANTIMICROBIAL EFFECTS

In addition to cardiovascular and anticancer effects, which are well documented in peer-reviewed journals, there are other known physiological effects of garlic. The antibacterial and antifungal activities (against a variety of Gram-negative and Gram-positive microorganisms) of garlic have been attributed to allicin. The antibiotic activity of 1 mg of allicin has been equated to that of 15 IU of penicillin (Han et al., 1995). Ajoene, an organosulfur compound derived from garlic, has been shown to increase the antimalarial activity of chloroquine against the parasite *P. berghei* (Perez et al., 1994). Recent studies have also demonstrated an inhibitory effect by aqueous garlic extracts on *Helicobacter pylori*, a bacterium implicated in the etiology of stomach cancer (Cellini et al., 1996; Sivam et al., 1997).

6. ADVERSE EFFECTS

Garlic is widely used as a food and is not considered dangerous, in fact it is "Generally Regarded as Safe" (GRAS) by the FDA. However, various negative effects can occur after garlic consumption that are generally nonlife-threatening but are discomforting and embarrassing for some individuals. Odoriferous garlic breath and perspiration often occur after garlic consumption due mainly to the allyl sulfide compounds (allyl methyl sulfide, disulfide, diallyl sulfide, disulfide and 2-propenethiol, which are expelled from the blood via the lungs or by sweat) (Block, 1992). Garlic may also cause an acid reflux effect that is due to transient lower-esophageal sphincter relaxation (Koch, 1996). A few individuals may experience allergic-type reactions (e.g., dermatitis, asthma) to fresh garlic or diallyl disulfide (Asero et al., 1998; Jappe et al., 1999).

On the other hand, a controversy exists about possible adverse effects of allicin According to some researchers, synthetic allicin (0.2 mg/mL), when tested *in vitro*, is considered toxic based on their observations of the oxidation of blood hemoglobin (Freeman and Kodera, 1995). The LD_{50} of allicin in mice is 60 mg/kg intravenously and 120 mg/kg subcutaneously (Block, 1992). Aqueous extracts of allicin have also been shown to have adverse effects in mice at 300–600 mg/kg/d (Fehri et al., 1991). These dosages are several hundredfold above any normal dose that would generally be ingested by humans. Moreover, garlic only produces on average 2–6 mg allicin/g, which at normal garlic consumption is well below the toxic dose.

In general, based on numerous human clinical trials and studies, fresh garlic ingestion at 10 g per day (one to two cloves or 4–6 g garlic powder) is considered safe when taken with a meal (Koch, 1996). For all intents and purposes, garlic is considered safe and nontoxic.

7. CONCLUSION

Garlic and derived preparations are often touted as having (or not having) certain physiological effects based on research and clinical studies. Not all studies have employed the same preparations of garlic; rather, products such as powders, oils, macerates and aged garlic extracts have often been used, all containing varying levels of potential physiologically active compounds. Yet, the physiological effects of these garlic products are often grouped together as the effects of garlic and are not separated depending on the type of preparation used. Consequentially, the physiological effects from "garlic" are highly unpredictable and inconsistent. Therefore, ideally, all studies need to be examined in relation to their active components and/or to the specific garlic preparation used and should not be pooled together as the effect of garlic in general.

It is becoming apparent that the physiological/clinical effects (e.g., blood cholesterol lowering) of the same or different garlic preparations can vary considerably across different groups of subjects. These apparent inconsistencies may be due to several factors including dose and duration, dietary components (e.g., fat levels), variability in the levels of the different organosulfur compounds in the garlic supplements, and variability between individuals to generate, *in vivo*, the physiologically active metabolites from the garlic supplements. More research is needed to address these inconsistencies and to demonstrate that the physiological effects are reliable and reproducible for a given garlic preparation. Consumers using garlic preparations for lowering blood cholesterol should be encouraged to have their serum lipids monitored to ensure clinical efficacy of the supplement. Support for the anticancer effects of garlic must await results of further *in vitro* studies and, particularly, human clinical trials. The aforementioned inconsistencies in study results need to be dealt with by developing more reliable and defined garlic products (supplements and functional foods) with dependable and reproducible clinical efficacy.

8. REFERENCES

Adler, A J. and B J. Holub. 1997. "Effect of garlic and fish-oil supplementation on serum lipid and lipoprotein concentrations in hypercholesterolemic men," *Am. J. Clin. Nutr.* 65:445–450.

Agarwal, K C. 1996. "Therapeutic actions of garlic constituents," *Med. Res Rev* 16:111–124.

Ali, M. and M Thomson 1995. "Consumption of a garlic clove a day could be beneficial in preventing thrombosis " *Prostaglandins Leukot. Essent. Fatty Acids.* 53:211–212.

Ali, M., M. Thomson, M. A. Alnaqeeb, J M. al-Hassan, S. H. Khater and S. A. Gomes 1990. "Antithrombotic activity of garlic: Its inhibition of the synthesis of thromboxane-B2 during infusion of arachidonic acid and collagen in rabbits," *Prostaglandins Leukot. Essent Fatty Acids* 41:95–99.

Apitz-Castro, R., J. J. Badimon and L. Badimon 1992. "Effect of ajoene, the major antiplatelet compound from garlic, on platelet thrombus formation," *Thromb Res.* 68:145–155.

Apitz-Castro, R., M. K. Jain, F. Bartoli, E. Ledezma, M. C Ruiz and R. Salas. 1991. "Evidence for direct coupling of primary agonist-receptor interaction to the exposure of functional IIb-IIIa complexes in human blood platelets. Results from studies with the antiplatelet compound ajoene," *Biochim. Biophys. Acta.* 1094:269–280.

Apitz-Castro, R., E. Ledezma, J. Escalante and M K. Jain. 1986. "The molecular basis of the antiplatelet action of ajoene: Direct interaction with the fibrinogen receptor," *Biochem. Biophys. Res Commun.* 141·145–150.

Asero, R , G Mistrello, D Roncarolo, P L. Antoniotti and P. Falagiani. 1998. "A case of garlic allergy," *J. Allergy Clin Immunol* 101·427–428.

Batirel, H. F , S. Aktan, C. Aykut, B C. Yegen and T. Coskun. 1996. "The effect of aqueous garlic extract on the levels of arachidonic acid metabolites (leukotriene C4 and prostaglandin E2) in rat forebrain after ischemia-reperfusion injury," *Prostaglandins Leukotrienes Essent. Fatty Acids.* 54:289–292.

Berthold, H. K., T. Sudhop and K. von Bergmann. 1998. "Effect of a garlic oil preparation on serum lipoproteins and cholesterol metabolism. A randomized controlled trial," *JAMA* 279:1900–1902.

Berthold, H. K and T. Sudhop. 1998. "Garlic preparations for prevention of atherosclerosis," *Curr. Opin. Lipidol.* 9:565–569.

Block, E. 1985 "The chemistry of garlic and onions," *Sci. Am* 252:114–119.

Block, E. 1992. "The organosulfur chemistry of the genus *Allium*—implications for the organic chemistry of sulfur," *Angew Chem Int. Ed. Engl.* 31·1135–1178.

Brady, J. F., H. Ishizaki, J. M. Fukuto, M. C. Lin, A. Fadel, J M. Gapac and C S Yang. 1991. "Inhibition of cytochrome P-450 2E1 by diallyl sulfide and its metabolites," *Chem. Res. Toxicol.* 4.642–647.

Breithaupt-Grogler, K., M. Ling, H. Boudoulas and G. G. Belz. 1997. "Protective effect of chronic garlic intake on elastic properties of aorta in the elderly," *Circulation.* 96 2649–2655.

Cellini, L , E. Di Campli, M. Masulli, S. Di Bartolomeo and N. Allocati. 1996. "Inhibition of *Helicobacter pylori* by garlic extract (*Allium sativum*)," *FEMS Immunol. Med Microbiol.* 13:273–277.

Das, I., N. S. Khan and S. R. Sooranna. 1995a. "Nitric oxide synthase activation is a unique mechanism of garlic action," *Biochem. Soc. Trans.* 23:136S.

Das, I., N. S. Khan and S. R. Sooranna. 1995b. "Potent activation of nitric oxide synthase by garlic: A basis for its therapeutic applications," *Curr. Med. Res. Opin* 13 257–263.

Dorant, E., P. A. van den Brandt and R. A. Goldbohm. 1994. "A prospective cohort study on *Allium* vegetable consumption, garlic supplement use, and the risk of lung carcinoma in The Netherlands," *Cancer Res.* 54:6148–6153.

Dorant, E., P. A van den Brandt and R. A. Goldbohm. 1995. "*Allium* vegetable consumption, garlic supplement intake, and female breast carcinoma incidence," *Breast Cancer Res. Treat.* 33.163–170.

Dorant, E , P. A. van den Brandt and R. A. Goldbohm. 1996. "A prospective cohort study on the relationship between onion and leek consumption, garlic supplement use and the risk of colorectal carcinoma in The Netherlands," *Carcinogenesis.* 17:477–484.

Dorant, E., P. A. van den Brandt, R. A. Goldbohm, R. J. Hermus and F. Sturmans. 1993. "Garlic and its significance for the prevention of cancer in humans: A critical view," *Br. J. Cancer.* 67:424–429.

Ellmore, G S. and R. S. Feldberg. 1994. "Alliin lyase localization in the bundle sheaths of the garlic clove (*Allium sativum*)," *Am. J. Bot.* 81:89–94.

Fehri, B., J. M. Aiache, S. Korbi, M. Monkni, M. Ben Said, A. Memmi, B. Hizaoui and K Boukef. 1991. "Toxic effects induced by the repeat administration of *Allium sativum* L," *J. Pharm. Belg.* 46(6):363–374.

Fenwick, G R. and A B. Hanley 1985a. "Genus *Allium*—Part 1," *CRC Crit. Rev. Food Sci. Nutr.* 22:199–271.

Fenwick, G. R. and A. B. Hanley. 1985b. "Genus *Allium*—Part 2," *CRC Crit. Rev. Food Sci Nutr.* 22.273–377.

Fenwick, G. R. and A. B. Hanley. 1985c. "The genus *Allium*—Part 3," *Crit. Rev. Food Sci. Nutr.* 23:1–73.

Food and Agriculture Organization of the United Nations. 1997. FAO Production Yearbook Vol. 51. FAO Statistics Series No. 142. Rome, Food and Agriculture Organization of the United Nations.

Freeman, F. and Y. Kodera. 1995. "Garlic chemistry: Stability of S-(2-propenyl) 2-propene-1-sulfinothioate (allicin) in blood, solvents, and simulated physiological fluids," *J. Agric. Food Chem.* 43:2332–2338.

Freeman, F. and Y. Kodera. 1997. "Rebuttal on garlic chemistry. Stability of S-(2-propenyl) 2-propene-1-sulfinothioate (allicin) in blood, solvents, and simulated physiological fluids," *J. Agric. Food Chem.* 45:3709–3710.

Fujiwara, M. and T. Natata. 1967. "Induction of tumour immunity with tumour cells treated with extract of garlic (*Allium sativum*)," *Nature.* 216:83–84

Gebhardt, R. and H. Beck. 1996. "Differential inhibitory effects of garlic-derived organosulfur compounds on cholesterol biosynthesis in primary rat hepatocyte cultures," *Lipids.* 31:1269–1276.

Gebhardt, R., H. Beck and K. G. Wagner. 1994. "Inhibition of cholesterol biosynthesis by allicin and ajoene in rat hepatocytes and HepG2 cells," *Biochim. Biophys Acta* 1213:57–62

Hahn, G 1996. "Botanical characterization and cultivation of garlic." In: Garlic: The Science and Therapeutic Application of *Allium sativum* L. and Related Species (Koch, H. P. and L. D. Lawson eds.), pp. 25–36. Baltimore, MD: Williams and Wilkins.

Han, J., L. Lawson, G. Han and P. Han. 1995. "A spectrophotometric method for quantitative determination of allicin and total garlic thiosulfinates," *Anal. Biochem.* 225·157–160.

Heber, D. 1997. "The stinking rose: Organosulfur compounds and cancer [editorial; comment]," *Am. J. Clin. Nutr.* 66:425–426.

Holzgartner, H , U Schmidt and U. Kuhn. 1992. "Comparison of the efficacy and tolerance of a garlic preparation vs. bezafibrate," *Arzneimittelforschung.* 42:1473–1477.

Hopkins, P. N. and R. R. Williams. 1981. "A survey of 246 suggested coronary risk factors," *Atherosclerosis.* 40:1–52.

Horie, T., S Awazu, Y. Itakura and T. Fuwa. 1992. "Identified diallyl polysulfides from an aged garlic extract which protects the membranes from lipid peroxidation," *Planta Med.* 58:468–469.

Hu, X, and S. V. Singh. 1997. "Glutathione S-transferases of female A/J mouse lung and their induction by anticarcinogenic organosulfides from garlic," *Arch Biochem. Biophys.* 340:279–286.

Imai, J., N. Ide, S. Nagae, T. Moriguchi, H. Matsuura and Y. Itakura 1994. "Antioxidant and radical scavenging effects of aged garlic extract and its constituents," *Planta Med.* 60:417–420.

Isaacsohn, J. L., M. Moser, E. A. Stein, K. Dudley, J. A. Davey, E. Liskov and H R. Black. 1998. "Garlic powder and plasma lipids and lipoproteins: A multicenter, randomized, placebo-controlled trial," *Arch. Intern Med.* 158:1189–1194.

Jappe, U., B. Bonnekoh, B. M. Hausen and H. Gollnick. 1999. "Case Report: Garlic-related dermatoses: Case report and review of the literature," *Am. J. Contact. Dermat.* 10:37–39.

Kendler, B. S. 1987. "Garlic (*Allium sativum*) and onion (*Allium* cepa) A review of their relationship to cardiovascular disease," *Prev. Med.* 16.670–685.

Kleijnen, J., P. Knipschild and G. ter Riet. 1989 "Garlic, onions and cardiovascular risk factors A review of the evidence from human experiments with emphasis on commercially available preparations," *Br J. Clin Pharmacol* 28·535–544

Koch, H P. 1996. "Toxicology, side effects and unwanted effects of garlic." In: Garlic. The Science and Therapeutic Application of *Allium sativum* L. and Related Species (Koch, H. P and L D. Lawson, eds), pp 221–228 Baltimore, MD: Williams and Wilkins.

Koch, H. P. and L. D Lawson, eds. 1996. Garlic: The Science and Therapeutic Application of *Allium sativum* L. and Related Species, Second Edition. Baltimore, MD: Williams and Wilkins.

Kosian, A M 1998 "Effect of sulfur nutrition for sulfoxide accumulation in garlic bulbs," *Ukr Biokhim. Zh* 70 105–109.

Lau, B H., P. P Tadi and J. M Tosk 1990. "*Allium sativum* (garlic) and cancer prevention," *Nutr Res* 10·937–948.

Lawson, L. D., Z J. Wang and B. G. Hughes. 1991. "Identification and HPLC quantitation of the sulfides and dialk(en)yl thiosulfinates in commercial garlic products," *Planta Med* 57·363–370.

Lawson, L. D. 1993. "Bioactive organosulfur compounds of garlic and garlic products· Role in reducing blood lipids " In· Human Medicinal Agents from Plants (Kinghorn, A. D. and M. F Balandrin, eds), pp 306–330. Symposium Series 534, Washington, DC· ACS.

Lawson, L D. 1996 "The composition and chemistry of garlic cloves and processed garlic " In: Garlic. The Science and Therapeutic Application of *Allium sativum* L. and Related Species (Koch, H. P. and L. D. Lawson, eds), pp. 37–107. Baltimore, MD: Williams and Wilkins.

Lawson, L. D. 1998a. "Garlic· A review of its medicinal effects and indicated active compounds." In. Phytomedicines of Europe: Their Chemistry and Biological Activity (Lawson, L. D. and R. Bauer, eds.), pp. 176–209. Washington, DC: American Chemical Society.

Lawson, L. D. 1998b. "Garlic powder for hyerlipidemia-analysis of recent negative results," *Quarterly Rev. Nat. Med. Fall.* pp 188–189.

Lawson, L. D and E. Block. 1997. "Comments on garlic chemistry: Stability of S-(2-propenyl) 2-propene-1-sulfinothioate (allicin) in blood, solvents, and simulated physiological fluids," *J. Agric. Food Chem.* 45 542–542

Lawson, L. D., D. K. Ransom and B. G. Hughes. 1992. "Inhibition of whole blood platelet-aggregation by compounds in garlic clove extracts and commercial garlic products," *Thromb. Res* 65.141–156

Lawson, L. D. and Z J. Wang. 1993. "Pre-hepatic fate of the organosulfur compounds derived from garlic (*Allium sativum*)," *Planta Med* 59·A688–A689.

Lee, E. S., M. Steiner and R. Lin. 1994. "Thioallyl compounds: Potent inhibitors of cell proliferation," *Biochim. Biophys. Acta* 1221:73–77.

Liu, J. and J. Milner. 1990 "Influence of dietary garlic powder with and without selenium supplementation on mammary carcinogen adducts," *FASEB J.* 4:A1175.

Makheja, A. N and J. M. Bailey. 1990. "Antiplatelet constituents of garlic and onion," *Agents Actions* 29:360–363

McCrindle, B. W., E. Helden and W. T. Conner. 1998. "Garlic extract therapy in children with hypercholesterolemia," *Arch. Pediatr. Adolesc Med* 152:1089–1094

Mei, X., X. Lin, J Liu, P. Song, J. Hu and X. Liang. 1989. "Gastric inhibition of the formation of N-nitrosaproline in the human body," *Acta. Nutr. Sinica.* 11:141–145.

Mei, X., M. C. Wang, H. X. Xu, X. P. Pan, C. Y. Gao, N. Han and M. Y Fu. 1982. "Garlic and gastric cancer—the effect of garlic on nitrite and nitrite in gastric juice," *Acta. Nutr. Sinica.* 4:53–58

Milner, J. A 1996. "Garlic: Its anticarcinogenic and antitumorigenic properties," *Nutr. Rev.* 54:S82–S86

Mohammad, S. F. and S. C Woodward. 1986. "Characterization of a potent inhibitor of platelet aggregation and release reaction isolated from *Allium sativum* (garlic)," *Thromb. Res.* 44·793–806.

Monmaney, T and S. Roan. 1998. "Hope or Hype? Alternative medicine the 18 billion dollar experiment," LA Times Home Edition, Part A. Aug. 30, 1998. Los Angeles.

Murry, M. T. 1995 The Healing Power of Herbs. Second Edition. pp. 121–131. Rocklin, CA. Prima Publishing.

Neil, H. A., C A Silagy, T. Lancaster, J. Hodgeman, K Vos, J. W. Moore, L. Jones, J Cahill and G. H Fowler. 1996 "Garlic powder in the treatment of moderate hyperlipidaemia· A controlled trial and meta-analysis," *J. R. Coll. Physicians. Lond* 30:329–334.

Ontario Garlic Growers Association. 1998 Personal Communication.

Pedraza-Chaverri, J., E Tapia, O. N. Medina-Campos, A. de los and M. Franco. 1998. "Garlic prevents hypertension induced by chronic inhibition of nitric oxide synthesis," *Life Sci.* 62 PL71-7.

Perez, H A., M. De la Rosa and R. Apitz. 1994 "*In vivo* activity of ajoene against rodent malaria," *Antimicrob. Agents Chemother* 38:337–339.

Phelps, S. and W. S. Harris. 1993. "Garlic supplementation and lipoprotein oxidation susceptibility," *Lipids.* 28:475–477.

Pinto, J. T., C. Qiao, J. Xing, R. S. Rivlin, M. L. Protomastro, M. L. Weissler, Y. Tao, H. Thaler, and W. D. Heston. 1997. "Effects of garlic thioallyl derivatives on growth, glutathione concentration, and polyamine formation of human prostate carcinoma cells in culture," *Am. J. Clin. Nutr.* 66.398–405.

Reuter, H. D., H. P. Koch and L. D. Lawson. 1996. "Therapeutic effects and applications of garlic and its preparations." In: Garlic: The Science and Therapeutic Application of *Allium sativum* L and Related Species (Koch, H. P. and L. D. Lawson, eds.), pp. 135–212. Baltimore, MD. Williams and Wilkins.

Reuter, H. D. and A. Sendl. 1994. "*Allium sativum* and *Allium* ursinum: Chemistry, pharmacology and medicinal applications," *Econ. Med. Plant Res.* 6.56–113.

Riggs, D R., J I. DeHaven and D. L. Lamm. 1997. "*Allium sativum* (garlic) treatment for murine transitional cell carcinoma," *Cancer.* 79:1987–1994

Sakamoto, K , L D. Lawson and J. A. Milner. 1997. "Allyl sulfides from garlic suppress the *in vitro* proliferation of human A549 lung tumor cells," *Nutr. Cancer.* 29:152–156.

Schaffer, E. M., J. Z. Liu, J Green, C. A. Dangler and J. A. Milner. 1996. "Garlic and associated allyl sulfur components inhibit N-methyl-N-nitrosourea induced rat mammary carcinogenesis," *Cancer Lett.* 102:199–204.

Sendl, A., G Elbl, B. Steinke, K. Redl, W. Breu and H. Wagner. 1992. "Comparative pharmacological investigations of *Allium ursinum* and *Allium sativum*," *Planta Med.* 58·1–7.

Sigounas, G., J. Hooker, A. Anagnostou and M. Steiner. 1997a. "S-allylmercaptocysteine inhibits cell proliferation and reduces the viability of erythroleukemia, breast, and prostate cancer cell lines," *Nutr. Cancer.* 27:186–191.

Sigounas, G., J. L Hooker, W. Li, A. Anagnostou and M. Steiner. 1997b. "S-allylmercaptocysteine, a stable thioallyl compound, induces apoptosis in erythroleukemia cell lines," *Nutr. Cancer.* 28·153–159.

Silagy, C. and A. Neil. 1994a. "Garlic as a lipid lowering agent—a meta-analysis," *J.R. Coll Physicians. Lond.* 28:39–45.

Silagy, C A and H A Neil 1994b. "A meta-analysis of the effect of garlic on blood pressure," *J Hypertens* 12 463–468.

Simons, L A , S. Balasubramaniam, M. von Konigsmark, A. Parfitt, J. Simons and W. Peters. 1995. "On the effect of garlic on plasma lipids and lipoproteins in mild hypercholesterolaemia," *Atherosclerosis.* 113:219–225.

Sivam, G P., J. W. Lampe, B. Ulness, S. R. Swanzy and J. D. Potter. 1997 *"Helicobacter pylori—in vitro* susceptibility to garlic (*Allium sativum*) extract," *Nutr. Cancer.* 27:118–121.

Srivastava, K C and O. D. Tyagi. 1993. "Effects of a garlic-derived principle (ajoene) on aggregation and arachidonic acid metabolism in human blood platelets," *Prostaglandins Leukotrienes Essent. Fatty Acids.* 49.587–595

Srivastava, S. K , X Hu, H. Xia, H A. Zaren, M L. Chatterjee, R. Agarwal and S. V. Singh 1997. "Mechanism of differential efficacy of garlic organosulfides in preventing benzo(a)pyrene-induced cancer in mice," *Cancer Lett.* 118.61–67

Steinmetz, K. A., L H Kushi, R. M. Bostick, A. R. Folsom and J. D. Potter. 1994 "Vegetables, fruit, and colon cancer in the Iowa Women's Health Study," *Am J. Epidemiol.* 139:1–15.

Taucher, J , A. Hansel, A. Jordan and W. Lindinger. 1996. "Analysis of compounds in human breath after ingestion of garlic using proton-transfer-reaction mass spectrometry," *J. Agric. Food Chem.* 44:3778–3782.

U.S. Department of Agriculture 1998. USDA Nutrient Database for Standard Reference, Release 12. Food Group 11 Vegetables and Vegetable Products. Nutrient Data Laboratory Home Page, http //www nal usda gov/fnic/foodcomp, Agricultural Research Service.

Venton, D. L , S. O. Kim and G. C Le Breton. 1991. "Antiplatelet activity from plants," *Econ. Med. Plant Res.* 5.323–351.

Warshafsky, S., R S. Kamer and S L Sivak 1993. "Effect of garlic on total serum cholesterol A meta-analysis," *Ann. Intern. Med* 119.599–605.

Wattenberg, L. W , V L. Sparnins and G Barany 1989. "Inhibition of N-nitrosodiethylamine carcinogenesis in mice by naturally occurring organosulfur compounds and monoterpenes," *Cancer Res.* 49.2689–2692.

Yeh, Y. Y and S. M Yeh. 1994. "Garlic reduces plasma lipids by inhibiting hepatic cholesterol and triacylglycerol synthesis," *Lipids* 29.189–193.

CHAPTER 2

Chemistry and Pharmacology of Ginseng and Ginseng Products

D. D. KITTS

1. INTRODUCTION

THE term *ginseng* refers to "the essence of man" and represents perennial herbs that are generally derived from the genus *Panax* and are indigenous to Korea, China (e.g., Asian ginseng or *P. ginseng* C.A. Meyer), Vietnam (e.g., *P. vietnamensis*), Japan, India (e.g., *P. pseudoginseng*) and North America (*P. quinquefolium*) (Table 2.1). *P. ginseng* C.A. Meyer, *P. vietnamensis* and *P. quinquefolium* belong to the *Araliaceae* family (Hu, 1977). Vietnamese ginseng is a wild *Panax* species, whereas North American ginseng is grown commercially in the Canadian provinces of British Columbia and Ontario as well as in the eastern United States. Siberian or Russian ginseng is in fact a different plant (e.g., *Eleutherococcus senticosus*) that has apparent tonic effects similar to ginseng but is commonly referred to as *Eleuthero* ginseng. Traditional oriental folk medicine has claimed various ergogenic and health properties of ginseng when used by man and represents the basis for its use as a natural tonic for restorative properties for thousands of years. In traditional Chinese medicine, a perfect balance between the complementary forces of "yin" and "yang," are required to provide physical and spiritual well-being. Whereas the Asian ginseng is regarded as the yang, providing warmth to offset stressful (e.g., adaptogenic) conditions, the North American ginseng is referred to as the yin, or the converse, providing a cooling effect to counterbalance stress.

The recognized primary bioactive constituents of ginseng fall into the category of triterpene saponins, referred to as ginsenosides that are present in the root, leaf and berry of the plant. The root, being the customary source of ginsenosides, makes it a commodity with significant agricultural, climatic and

TABLE 2.1. Ginseng Source and Country of Origin.

Species of Ginseng	Country of Cultivation
Panax ginseng C A Meyer (Asian ginseng)	Korea, China, Japan, Russia
Panax japonicum, (Japanese ginseng, Bamboo ginseng)	Japan
Panax vietnamensis	Vietnam
Panax notoginsengana F H Chen (Sanchi ginseng)	China (Yunnan and Kwangsi provinces)
Panax guinqefolium (North American ginseng)	Eastern and Western Canada provinces and U.S. states
Panax trifolium C A. Meyer	Japan

growth requirements for a period of four to six years before harvest for optimal yield and quality of product. Asian ginseng (*P. ginseng* C.A. Meyer) is a rich source of ginsenosides, and a standardized concentration of ginsenosides has been prepared from an Asian ginseng extract. The extract, referred to as G115 and marketed as Ginsana, has been tested in several studies for efficacy in combating physical or chemical stress. A similar standardized ginseng extract derived from North American ginseng (*P. quinquefolium*), referred to as CNT2000 from Chai-Na-Ta Corp., Langley, British Columbia is also under evaluation for chemical and physiological activity. It is noteworthy that many of the implied effects of ginseng constituents in combating chronic disease may involve antioxidant properties and activity in quenching free radicals, as shown for *P. ginseng* C.A. Meyer (Kim et al., 1992; Zhang et al., 1996), *P. vietnamensis* (Huong et al., 1998; Thi et al., 1998) and *P. quinquefolium* (Kitts et al., 1999). As a result, some reports have recommended the inclusion of ginseng powder in multivitamin and mineral supplements for possible synergistic effect (Zuin et al., 1987; Marasco et al., 1998).

In lieu of the unique chemical differences between different sources of ginseng and the numerous health claims made by traditional and contemporary Chinese medicine, ginseng is promoted as a potential nutraceutical with tonic, stimulant and aphrodisiac properties. The purpose of this chapter is to describe the chemical and pharmacological properties of ginseng that have been attributed to specific physiological functions.

2. COMPOSITION OF GINSENG

The ginseng plant has a typical seed kernel (e.g., approximately 5–7 mm in length and 4–5 mm in width) that develops into a flower (e.g., 3–7 mm length) and an elongated leaf. Of particular commercial importance is the ginseng root, which is cylindrical in shape and varies in length (e.g., 2–10 cm) and diameter (1–2 cm). Associated with the ginseng root are numerous prong-like fibers that grow outward from the root and are removed prior to marketing. In addition to polysaccharides and mucilaginous substances, protein, amino acids, vitamins and essential oils (Table 2.2), ginseng also contains a complex mixture of glycosides, or steroidal saponins, referred to as ginsenosides. These constituents are attached to various sugar moieties and, along with flavonoids, are generally considered to be the main bioactive components present in ginseng (Zhang et al., 1979a, 1979b). With increased attention given to proposed health claims concerning ginseng and its bioactive components for use as a potential nutraceutical, several reports have shared a concern regarding the great variations in saponin concentration in different ginseng products. Moreover, the apparent absence of standard quality control measures used in preparation of formulations, as evidenced by the common absence of ginsenoside labeling on some commercial ginseng preparations, has raised concerns about the legitimacy of some ginseng products (Phillipson and Anderson, 1984; Baldwin et al., 1986; Cui et al., 1994). The chemical classification of ginseng saponins has been performed on Asian and North American ginsengs (Shibata et al., 1965). More than 30 different ginsenoside saponins have been identified and classified into three groups according to the glyco-chain connection on the aglycone backbone. For example, aglycone of 20-S-protopanaxandiol (e.g., ginsenosides Rb1, Rb2, Rc and Rd) are classified as panaxadiol saponins, whereas the aglycone of 20-S-protopanaxatriol (Re, Rf and Rg1) are in the panaxatriol saponin classification. These two groups of ginsenosides are tetracyclic triterpenoid saponins (Figure 2.1). The third group of saponins is oleanolic acid, a pentocyclic triterpenoid. The chemical compositions of

TABLE 2.2. **Chemical Components of Ginseng.**

Components
Lipid
• triacylglycerides, trilinolein, oligoglycerides
Protein
• peptides, enzymes
Phenolics
• caffeic acid, kaempherol, vanillic acid
• Flavonoids
• β-N-Oxalo-L-α,β,diaminopropionic acid
Vitamins
• ascorbic acid, thiamin, riboflavin, niacin
Carbohydrates
• polysaccharides panaxans
• mucilaginous substances
Others
• steroidal saponins or ginsenosides
• essential oils, β-farnesene, β-gurgenene, β-bisablene
• organic acids

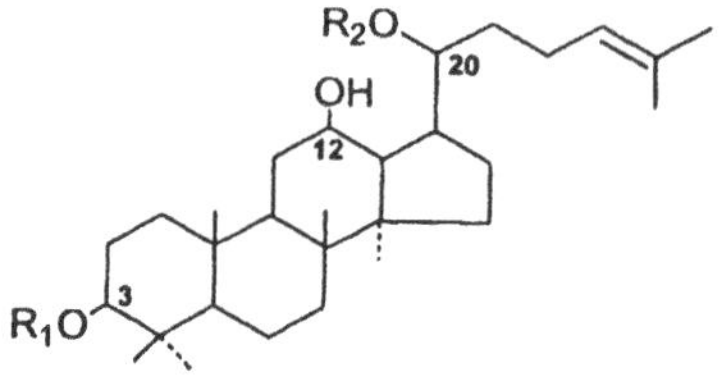

Ra : R1=glucose-6→1-glucose-6→1-glucose
    R2=glucose-3→1-glucose-3→1-glucose

Rb1 : R1=glucose-2→1-glucose
    R2=glucose-6→1-glucose

Rb2 : R1=glucose-2→1-glucose
    R2=glucose-6→1-arabinose(pyr)

Rb3 : R1=glucose-2→1-glucose
    R2=glucose-6→1-xylose

Rc : R1=glucose-2→1-glucose
    R2=glucose-6→1-arabinose(fur)

Rd : R1=glucose-2→1-glucose
    R2=glucose

$R_1=R_2=H$
20-(s)-protopanaxadiol

Re : R1=glucose-2→1-rhamnose
    R2=glucose

Rf : R1=glucose-2→1-glucose
    R2=H

Rg1 : R1=glucose
    R2=glucose

Rg2 : R1=glucose-2→1-glucose
    R2=H

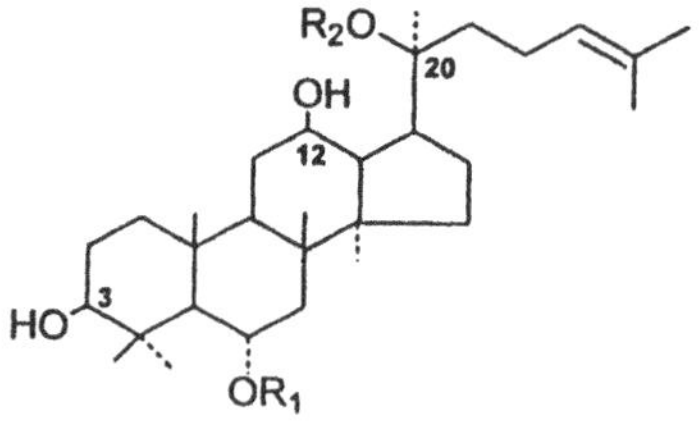

$R_1=R_2=H$
20-(s)-protopanaxatriol

Figure 2.1 Structures of ginsenosides.

individual ginsenosides from Asian and North American ginsengs are very similar except for the different ratio of panaxadiol to panaxatriol of the two species (Hou, 1977). Although more than 30 different ginsenosides have been identified, a group of six ginsenosides, Rb1, Re, Rc, Rd, Rb2 and Rg1, which are named on the basis of individual migration on a thin layer chromatogram, has been chosen for reference standards for ginseng products (Ma et al., 1995). This advance in analytical ginseng chemistry will contribute tremendously to an improved quality control of standardized products.

Recent studies have also reported antioxidant activity associated with a triacylglycerol (e.g., trilinolein) extracted from *P. pseudoginseng* (Chan et al.,

1997). This polyacetylene compound present in ginseng root derived from *Panax* species has also been shown to possess cytotoxic properties (Hirakura et al., 1992; Matsunaga et al., 1989). Metal binding activities of a polypeptide with an amino acid sequence of Glu-Thr-Val-Glu-Ile-Ile-Asp-Ser-Glu-Gly-Gly-Gly-Asp-Ala extracted from *Panax ginseng* have also been reported (Kajiwara et al., 1996).

Ginseng root varies as much as 6.5 times in weight within a homogenous plot of cultivated North American ginseng (*P. quinquefolium*); however, the total ginsenoside content varies less than the harvest weight (e.g., 2.3 times variation) (Smith et al., 1996). Factors influencing total ginsenoside content in root and leaf materials include the genetic diversity within the crop and the geographic location of growth (Li et al., 1996). The ginsenoside content in leaf correlates more closely with soil fertility factors such as soil mineral composition than that in ginseng root (Konsler et al., 1990; Li et al., 1996). High-performance liquid chromatography analyses of ginsenosides have been used for varietal differences of ginseng. Ginseng roots from *P. ginseng* C.A. Meyer, *P. quinquefolium* and *P. vietnamensis* contain ginsenosides Rb1, Rc, Rg1, Rb2, Re and Rd as major saponins (Kitagawa et al., 1989; Ma et al., 1995; Li et al., 1996; Thi et al., 1998). Ginsenoside Rf is present in *Panax ginseng* but is absent in North American ginseng. Ginsenosides Rb1 and Re predominate in North American ginseng and represent about 75% of the total ginsenoside content (Li et al., 1996), particularly in the root periderm, cortex and xylem portions (Smith et al., 1996). Ginseng leaf and stem from *P. quinquefolium* are also good sources of Re ginsenoside, which represents approximately 50% of the total saponin concentration. This corresponds to 1–3% dry weight in a one-month-old leaf and 4–5% in four-month-old leaf material (Li et al., 1996). *P. pseudoginseng* contains Rb1, Rb2 and Rc as principal ginsenosides, reaching concentrations of approximately 3% dry weight. Unique to the *P. vietnamensis* are the ocotillol-type saponins, especially majonoside-R2 (Thi et al., 1998).

3. METABOLIC EFFECTS OF GINSENG

3.1. METABOLISM OF GINSENG

The metabolic clearance of ginseng is relatively unknown in humans and likely difficult to determine due to the varied composition of individual ginsenosides from different ginseng sources. Prolonged use of ginseng results in a few adverse effects referred to as "ginseng abuse syndrome" (Siegel, 1979). Studies conducted with the Asian ginseng standardized extract G115 have indicated that a period of about 10 weeks was required before specific effects of ginseng subsided in healthy competitive athletes (Forgo and Schimert,

1985). Pharmacokinetic studies conducted by Takino et al. (1982) with oral dosing (e.g., 20 mg/animal) of Rg1 ginsenoside to rats, showed a 23% absorption after 2.5 hours, with the majority of dose recovered from the small intestine. Only trace amounts of Rg1 were recovered in heart (<0.1%) and in liver (0.25%) organs. In contrast, 44% of an intravenous dose of Rb1 was recovered unmetabolized in the urine 24 hours after administration, thereby indicating different bioavailabilities for specific ginsenosides. Only a very small amount of ginsenoside (e.g., <1% of dose) was recovered in the fecal matter over this time period. Other workers have identified numerous metabolites of ginsenoside Rg1 and Rg2 in the rat gastrointestinal tract following oral exposure and have also reported very low absorption rates (e.g., 0.1 and 1.9%, respectively, for Rg1 and Rg2) (Karikura et al., 1992). Taken together, these results suggest that absorption efficiencies of ginsenosides are low and metabolic transformation by intestine or first pass metabolism from hepatic tissues may contribute significantly to the overall metabolic clearance characteristics of ginsenoside constituents.

3.2. EFFECTS OF GINSENG ON INTERMEDIARY METABOLISM

Transient regulatory effects of ginseng on carbohydrate, lipid and nucleic acid metabolism have been reported (Popov and Gold, 1973; Yokozawa et al., 1975; Sakakibara et al., 1975; Iijima et al., 1976). These findings support the suggested role of ginseng in increasing nonspecific resistance to different forms of stress or disease (Table 2.3). This activity of ginseng is facilitated by the effectiveness of active constituents to regulate the general metabolism of the organism. Ginseng-induced changes in general metabolism (Table 2.4) are characterized by enhanced carbohydrate utilization or accelerated lipid, protein or nucleic acid synthesis (Yokozawa et al., 1975; Nagasawa et al., 1977; Sotaniemi et al., 1995).

TABLE 2.3. Reported Adaptogenic Effects of Ginseng to Stress.

Source of Stress	Reference
Nonspecific resistance	Brekhamn and Dardymov, 1969
Thermal stress • cold stress • heat stress	Chen et al , 1998 Yuan et al., 1998
Hypobaric conditions	Qu et al , 1988 Cheng et al., 1998
Exercise stress • swimming • bicycling	Grandhi et al., 1994 Saito et al , 1974 Pieralisi et al., 1991
Drug tolerance stress	Takahashi et al , 1992

TABLE 2.4. Metabolic and Physiological Responses to Ginseng.

Parameter	Ginseng (% control)	Placebo (% control)
Cholesterol	0 to −22.5	−1 7
Glucose (serum concentration)	−8 to −12 0	−1 7
Glucose (uptake)	+20 24	N.D.
Lactate (serum concentration[1])	−21 3	−1 2
Lactate (serum concentration[2])	−17 4	−0 9
Lactate production	−65 70	N.D
Pyruvate production	−42 6	N D
Lact/pyruv ratio	−39 2	N D
Fasting insulin (C-peptide)	0	N.D
Oxygen consumption	+5 8	−0 3
Systolic blood pressure	−25.4	−2 1

[1] Before exercise
[2] After exercise, derived from Hassan-Samira et al (1985), Schmidt et al (1978), Pieralisi et al (1991), Sotaniemi et al (1995)

3.2.1. Carbohydrate Metabolism

Standardized ginseng extract G115 derived from Asian ginseng has been shown to increase glucose uptake in rabbit brain tissue, with parallel reductions in lactate and pyruvate concentration ratios (Hassan-Samira et al., 1985). The significance of an *in vitro* dose-dependent increase in glucose uptake in rabbit cerebral cortical tissue has also been shown to correspond directly with a desynchronization of brain tissue electrical activity, characterized by faster and more frequent spindle activity (Hassan-Samira et al., 1985). Other studies have also shown reductions in plasma lactic acid levels in male athletes undergoing rigorous bicycling after an oral dose of 4% ginsenoside (Kirchdorfer, 1985). Moreover, the ginsenoside Rg1 evokes a hypoglycemic effect on alloxan-induced diabetic rats, independent of changes in insulin activity (Gong et al., 1991). In non-insulin-dependent diabetic patients, significant lowering of fasting blood glucose levels, independent of changes in serum insulin, was observed after four weeks of consuming 100 and 200 mg *Panax ginseng* per day (Sotaniemi et al., 1995). These findings collectively suggest that the observed effects of ginseng on carbohydrate metabolism are likely related more to glucose metabolism and a shift of glucose utilization from anaerobic metabolism than to regulation of insulin action.

3.2.2. Lipid Metabolism

The effect of ginseng saponins on lipid metabolism is controversial but may be related to maintaining metabolic homeostatic mechanisms involved in lipid synthesis. Oura and Hiai (1973) reported previously in rats that a saponin-rich preparation of ginseng enhanced the incorporation of ^{14}C-acetate into total hepatic lipids and accumulation of fat in adipose tissue. A similar observation, involving the accelerated rate of disappearance of ^{14}C-cholesterol from the plasma pool, has been attributed to a reduction in the plasma cholesterol of hyperlipidemic rats (Yamamoto, 1973). This effect would demonstrate a potential role for ginseng saponins in stimulating cholesterol clearance by enhancing ^{14}C-cholesterol conversion into bile acids and excreting the radioactive label into the fecal matter. Kang et al. (1995) observed no effect of ginsenoside feeding on plasma lipid levels in normo- and hypercholesterolemic rabbits, a finding confirmed in non-insulin diabetic patients (Sotaniemi et al., 1995). Other workers have demonstrated that the specific ginsenoside, Rb1, can stimulate the rate limiting metabolic step for cholesterol biosynthesis, namely, the conversion of 3-hydroxy-3-methyl glutaryl coenzyme A into mevalonate by the enzyme HMG-CoA reductase (EC 1.1.1.34) in normolipidemic rats (Ikehara et al., 1978). This and other examples suggest that the efficacy of ginseng to modify cholesterol homeostasis is specific to the individual ginsenoside content. For example, Rc produces the greatest reduction of free and esterified cholesterol in liver tissue, much more so than Rb1 and Rg1 (Sakakibara et al., 1975). While Re may stimulate cholesterol esterification, as noted by the marked decrease in free cholesterol, Rd produces the opposite effect by increasing hepatic-free cholesterol content and esterified cholesterol. Individual ginsenosides have also been shown to increase cholesterogenesis. Rb1-treated rats exhibited almost a 10-fold greater incorporation of ^{14}C-acetate into serum cholesterol, compared to controls. The greater effect of Rb1, recorded 30 minutes after administration, was shown to be dose-dependent and much greater than other ginsenosides that were studied (e.g., Rc > Rd > Rg1 > Re). Since Rg1 saponin predominates in ginseng, it may be involved in the noted ginseng-related effects on cholesterol metabolism. Structure-activity explanations underlying the mechanism of Rg1 enhancement of cholesterol biosynthesis may be attributed to the difference in sugar moiety among different ginsenosides. It is noteworthy that Rb1, which contains a glucose molecule, provides greater activity compared to the arabinose-containing Rc ginsenoside. Substituting a glucose in place of arabinose has been shown to enhance cholesterol biosynthesis (Sakakibara et al., 1975). This finding suggests that panaxatriol ginsenosides (e.g., Rg1 and Re) could be less effective than panaxadiol ginsenosides (e.g., Rb1, Rc and Rd) in regulating hepatic cholesterol biosynthesis.

3.2.3. Nucleic Acid and Protein Metabolism

Ginseng saponins have been shown to increase RNA biosynthesis in rat kidney resulting in increased protein synthesis (Nagasawa et al., 1977). Of the different ginsenosides tested, Rd has been shown to induce the largest effect on macromolecular synthesis (Iijima et al., 1976). A general anabolic effect of ginseng intake has been reported for DNA and protein synthesis in such tissues as the testes and adrenal cortex. Gains in body weight of rat and, specifically, significant weight increases in levator muscle occur following a daily dose of ginseng (100 mg/kg) for seven days. This observation is a useful indicator for the anabolic activity of Korean ginseng (Grandhi et al., 1994).

4. PHYSIOLOGICAL ACTIONS OF GINSENG

4.1. ENDOCRINE ACTIVITY

The general assumption that ginseng protects against aging and various diseases has been substantiated to some degree by the capacity of ginseng tonics to regulate physiological adaptation to various stress conditions (Table 2.4). Many of the metabolic effects attributed to ginseng underscore the adaptation of homeostatic mechanisms derived with exposure to stress. A primary target organ complex from this standpoint involves the pituitary-adrenal axis, where ginseng has been shown to induce changes in circulating levels of adrenocorticotrophin (ACTH) (Zhang et al., 1990), to bind to steroid receptors (Pearse et al., 1982) and to alter plasma corticosterone concentrations (Cheng et al., 1998). Hiai et al. (1979) produced evidence to show that the initial effect of ginseng was at the level of the hypothalamus, which in turn triggers release of ACTH from the pituitary gland and initiates subsequent endocrine changes attributed to activity at the pituitary-adrenal axis. Collectively, these results suggest that ginseng-increased adrenal responsiveness to stress is one mechanism for the reported nonspecific resistance afforded by this herb. Indices of altered adrenal function induced by ginseng include changes in adrenal cholesterol and tissue glycogen metabolism (Avakian and Evonuck, 1979), as well as improved recovery of ascorbic acid repletion following exposure to stress (Kim et al., 1970).

The effect of ginseng on reproductive endocrine parameters is confined to one study that reported decreases in luteinizing (LH; -14.3%), follicle-stimulating (FSH; -31.6%) and estradiol (E_2; -12.6%) hormones in women of the 30–39-year age group. Similar aged women given a placebo exhibited increases in LH ($+29\%$), FSH ($+34.9\%$) and E_2 ($+2.2\%$). A similar trend was also seen in older women (40–60 years) administered ginseng or placebo (Forgo et al., 1981). These results indicate that the estrogenic effects observed with ginseng (Punnonen and Lukola, 1980; Hopkins et al., 1988; Duda et al., 1998) may in fact be antiestrogenic.

4.2. NEUROTRANSMITTER ACTIVITY

The suggestion that ginseng constituents modulate stress response by regulating central endocrine and adrenergic metabolism is supported by observations that saponins derived from *Panax ginseng* alter the catecholamine levels in mouse brain (Kim et al., 1985). This finding parallels other studies that show ginseng constituents to be effective inhibitors of dopamine uptake in rat brain synaptosomes (Tsang et al., 1985) and to cause behavioral changes associated with regulation of gamma-aminobutyric (GABA) transmission (Kimura et al., 1994). GABA is an important inhibitory neurotransmitter that mediates the mammalian central nervous system. The ginsenoside, Rb1, specifically increases acetylcholine release from hippocampal slices (Benishine et al., 1991), decreases serotonin metabolism (Zhang et al., 1990) and influences appetite (Oomura et al., 1990). Of the different regions in the brain, ginseng-induced effects on neurotransmitter levels occur specifically at the hypothalamus, with small changes also seen in the striatum, and there are no effects on the frontal cortex, hippocampus or pon regions (Petkov et al., 1992). These workers showed that feeding a ginseng extract, prepared from the stems and leaves, produced a dramatic decline in hypothalamic and striatum noradrenaline concentrations in rats. Paralleling this change was an increase in striatum dopamine but a large decrease in hypothalamic dopamine levels. Serotonin concentrations remained unaltered in all segments of the brain, with the exception of a decline in the hypothalamus region. These biochemical changes corresponded to noticeable indices of increased exploratory behavior, signifying a psychotrophic action or a higher level of neural activity. In relative terms, the ginseng stem-leaf extract was less effective than the standardized G115 extract, derived from *Panax* ginseng root, at modulating learning and memory responses (Petkov et al., 1993).

4.3. GINSENG AND NEUROPHYSIOLOGICAL RESPONSES

Ginseng components have been associated with the regulation of nerve growth (Takemoto et al., 1984) and viability (Himi et al., 1989) as well as the modulation of nerve function (Oomura et al., 1990). Some of the pharmacological actions of ginseng mediated by the central nervous system include inhibition of the conditioned avoidance response and suppression of spontaneous and exploratory movements (Saito et al., 1977; Takagi et al., 1974). Two well-studied neurophysiological responses of ginseng involve its effectiveness in stimulating learning and memory (D'Angelo et al., 1986; Zhang et al., 1990) and enhanced resistance to fatigue and psychophysical performance. The ginsenoside Rb1, inhibits hippocampal long-term potentiation activity (LTP), a form of activity that underlies memory and learning. Unlike Rg1, the ginsenoside Rb1 can facilitate hippocampal LTP in rats (Abe et al., 1994), which in turn is associated with improved learning behavior. This was demonstrated in passive avoidance tests (Zhang et al., 1990). Human subjects given the Ginsana G115 preparation exhibited improved processing (mental arithmetic) and attention skills over counterparts given a placebo, while they showed no difference in pure motor function skills (D'Angelo et al., 1986). These findings corresponded to increased scores for intellectual and cognitive functions from patients receiving the same ginseng supplement equivalent to 500 mg ginseng root over a 56-day period (Rosenfeld, 1989).

Neurophysiological responses of ginseng were not limited to learning enhancement but may also involve blocking the development of tolerance and dependence to narcotics such as morphine and cocaine (Kim et al., 1990b; Tokuyama et al., 1996). This action of ginseng potentially lends itself for the prevention of abuse or adverse effects of narcotics. One explanation for this observation is based on the hypothesis that ginseng is associated with the recovery of dopaminergic dysfunction, since dopaminergic hyperfunction is closely related to behavioral sensitization development with repeated exposure to morphine or cocaine use (Tokuyama et al., 1996). The underlying mechanism for this effect could be related to the effect of ginseng constituents on presynaptic dopamine receptor function, otherwise altered by morphine (Kim et al., 1990; Bhargava and Ramarao, 1991).

4.4. GINSENG EFFECTS ON THE IMMUNE SYSTEM

Possible chemopreventative activities of ginseng have been related to the activity of plant constituents that stimulate the cell immune system. The immunostimulant activity of ginseng is, in general, apparent in treated aged rodents (Kim et al., 1990a; Lui and Zhang, 1995, 1996) and humans (Lui et al., 1995). The mechanism(s) of action of increased immune response to ginseng involve cytokine production and macrophage function (Wang et al., 1980; Jin et al., 1994a, 1994b), selective decreases in specific immunoglobulin titers (Kim et al., 1997), enhancement of natural killer cell activity (Yun et al., 1987; Scaglione et al., 1990) and transformation of T lymphocytes (Scaglione et al., 1990; Xiaoguang et al., 1998). Specifically, the acute administration of Rg1 derived from *Panax ginseng* increases the affinity of recognition of specific antibodies to antigens (Park et al., 1988; Kenarova et al., 1990), whereas the ginseng root polysaccharide stimulates antibody production (Wang et al., 1980). In chronic ginseng feeding studies performed in mice at both 30 mg/kg/day and 150 mg/kg/day dosages, ginseng extract was shown to decrease gamma globulin levels, independent of changes in total protein albumin and α_2 and β-globulin (Kim et al., 1997). Moreover, these workers further demonstrated that only serum IgG_1 isotype was selectively decreased, whereas $IgG_{2\alpha}$ and other Ig isotypes (e.g., IgG3, IgM, IgA) were not affected. The authors speculated, on the basis of the relatively greater cytotoxic activity of

IgG_{2a}, that this selective suppression of serum immunoglobulin (Ig) isotype may be relevant to antibody binding to macrophage Fc receptors associated with increased efficacy of the humoral defense against cancer.

Ginseng extracts obtained from G115 and an alternative source showed different affinities toward stimulating an immune response in humans (Scaglione et al., 1990). Administration of ginseng extract and placebo was required for a four-week period to obtain effective action on blastogenesis, T helper and T suppressor cell subsets. Although not proven, it was suggested that the different ginsenoside composition between the two ginseng extracts was attributed to different actions on the immune system.

4.5. GINSENG AND OXIDATIVE STATUS

4.5.1. Ginseng and Antioxidant Enzyme Interaction

The oxidative status of the subject, as it relates to antioxidant enzyme activities [e.g., superoxide dismutase (SOD), catalase (CAT) and glutathione peroxidase (GSH-Px)] to metabolize oxygen and lipid free radicals is well established in many chronic disease states (Yuan and Kitts, 1996). A link between ginseng and cancer at the molecular level may involve Cu, Zn-superoxide dismutase (SOD), an important endogenous enzyme for the catalysis of superoxide radical dismutation to oxygen and hydrogen peroxide. Kim and coworkers (1996), demonstrated that total saponins and panaxatriol were effective at inducing SOD gene transcription, with no effect on absolute SOD levels. Specifically, they reported that the ginsenoside, Rb2, was particularly effective at activating the SOD gene by inducing AP2, a transcription factor known to be involved in the induction of other antioxidant enzymes, such as heme oxygenase. Complementing this finding is the observation that ginsenosides increase myocardial SOD activity and reduce malonaldehyde, a secondary product of lipid oxidation (Liu and Yjao, 1992), using a heterotopic heart transplantation model for myocardial ischemia and re-perfusion injury. Ginsenosides also protect against cellular toxicity of oxygen and lipid free radicals in dogs with hemorrhagic shock (Li and Zhao, 1989) or in ferrous/cysteine-peroxide induction model systems using hepatic microsomes (Hui-Ling and Jun-tian, 1991). The affinity of Rb1 and Rg1, in particular, to scavenge reactive oxygen species in hepatic and brain microsome preparations exposed to Fenton-type oxidation reactants has been reported to occur at concentrations of 10^{-3}–10^{-4} M (Hui-Ling and Jun-tian, 1991). These same workers observed significant reductions (28%) in malonaldehyde and associated increases in GSH-Px (96.4%) and CAT (47%) after administering 50mg/kg/day Rb1 to rats.

4.5.2. Ginseng and Free Radical Scavenging Activity

Among the oxygen radicals, the hydroxyl radical, formed from the superoxide anion radical and hydrogen peroxide formed through the Haber-Weiss and Fenton reactions, is particularly reactive and initiates damage with adjacent biomolecules (Yuan and Kitts, 1996). The oxidative damage of carbohydrates, proteins, nucleic acids and lipids resulting from contact with hydroxyl radicals is believed to be the source of early detrimental cellular change in the etiology of chronic disease (e.g., cancer, atherosclerosis) and aging. Studies with *Panax ginseng* (Zhang et al., 1996) and North American (*P. quinquefolium*) ginsengs (Kitts et al., 1999) have examined the scavenging effects of ginseng extracts using a number of model systems. First, chronic feeding of rabbits with ginseng has been shown to enhance nitric oxide release, a potent endogenous antioxidant (Kang et al., 1995). The prevention of arachidonic acid loss and formation of secondary lipid oxidation products by ginseng has been associated with hydroxyl radical scavenging activity, as assessed by electron spin trap methods (Zhang et al., 1996). Most notably, antioxidant activity of North American ginseng extract using Fenton reaction site-specific and nonspecific reactions coupled with deoxyribose oxidation and genotoxic endpoint measurements, further demonstrated that a concentration-dependent decrease in the formation of lipid peroxides was in part due to scavenging activity of hydroxyl radicals (Kitts et al., 1999). This mechanism of action (Figure 2.2) depicts the relative resistance of brain tissue and erthyrocyte to hydrogen peroxide-forced peroxidation in mice administered North American ginseng in drinking water for 12 weeks. Tissues from animals fed ginseng exhibited significant resistance to lipid peroxidation induced by the peroxidation agent. In other studies from our laboratory, the scavenging potential of a North American ginseng extract was shown to also correspond to protection against plasmid DNA scissions (Kitts et al., 1999).

4.5.3. Ginseng Effects on the Cardiovascular System

The effects of ginseng constituents on cardiovascular functions may vary with different blood vessels (Chen et al., 1984) but generally include regulation of blood flow and vascular tone. Extracts derived from *Panax ginseng* and North American ginseng produce similar effects on Fourier components of pulses obtained from the right radial artery (Wang et al., 1994). An endothelium-dependent relaxation factor exerts direct actions on vascular smooth muscle as well as regulates vascular tone through modulation of acetylcholine nerve transmission. Studies performed with an artificially digested ginseng extract, prepared from *Panax ginseng* (G115), demonstrated efficacy in producing vasodilation in pulmonary rabbit preparations. It was concluded, there-

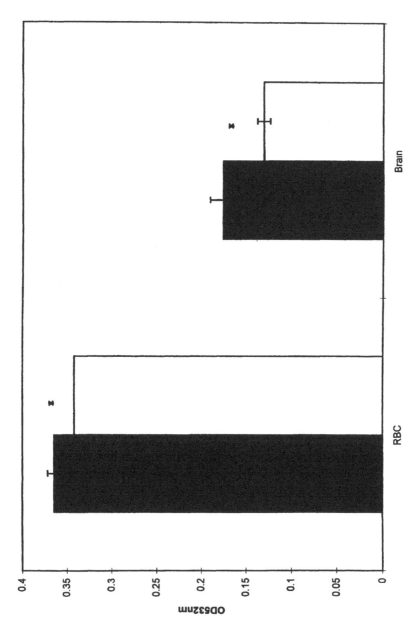

Figure 2.2 TBARS generated from red blood cell and brain homogenate of mice fed North American Ginseng (CNT2000) extract and control treated with 10 mM hydrogen peroxide. * = significant at P < 0.05.

fore, that oral consumption and subsequent digestion of ginseng did not limit bioactivity (Rimar et al., 1996).

The stimulation of nitric oxide triggers guanylate cyclase and elevates tissue cGMP, which in turn, inhibits neurotransmission and relaxation of vascular smooth muscle. This homeostatic mechanism can be impaired by feeding cholesterol to rabbits (Verbeuren et al., 1986). Chronic feeding of ginsenoside to hypercholesterolemic rabbits had no effect on lowering plasma lipids but was effective in inducing endothelium-dependent relaxation (Kang et al., 1995). This result indicates that the protective effect of ginseng on vascular tone is unrelated to hypercholesterolemia. Kim et al. (1995) reported that ginsenoside-induced vasodilating response was associated with enhanced release of nitric oxide and accumulation of cGAMP (e.g., stimulation of the L-arginine-nitric oxide-cyclic GMP pathway). A similar ginseng-induced vasodilation effect attributed to the L-arginine-nitric oxide-cyclic GMP pathway has been observed with rabbit corpus cavernosum (Chen et al., 1993). In addition to relaxing vascular smooth muscle via stimulation of the nitric oxide GMP pathway, ginsenosides have also been shown to inhibit sympathetic neurotransmission (Zhang and Chen, 1987). The ginsenosides derived from the protopanaxatriol group, namely Rg1 and Re, are associated with increased endothelium-dependent relaxation and elevated tissue cyclic-GMP content. Other workers have confirmed that ginsenosides derived from *Panax ginseng* facilitate endogenous nitric oxide release; however, the source of the nitric oxide could be from nonadrenergic, noncholinergic nerves in addition to vascular smooth muscle cells (Peng et al., 1995).

Individual ginsenosides have also been tested for specific affinities to block calcium channels in cultured rat myocardiocytes (Guogan and Yan, 1997). Ginsenosides Rg2 and Rg1 are most effective at inhibiting action potentials (37% and 30%, respectively), with Re (28%), Rh1 (23%) and Rb1 (20%) showing relatively less activity. Ginsenosides Rf, Rd and Ro have no effect on B- and T-type channels.

4.5.4. Ginseng and Cancer

Epidemiological (Yun and Choi, 1990, 1995), and clinical (Lin et al., 1995) studies have associated reduced incidences of human gastric and lung cancers to ginseng consumption, which lends some support to the general conclusion made 25 years earlier that dietary ginseng conveyed a nonspecific resistance against chronic disease (Brekhamn and Dardymov, 1969). Specific reports of the general antimutagenic properties of ginseng in mammalian cell culture (Umnova et al., 1991), effectiveness in reversing the transformation of cultured hepatoma cells (Odsashima et al., 1979) and specific antitumor activity (Matsunaga et al., 1989; Duda et al., 1998) provide mechanisms for the anticarcinogenic potential of ginseng. In addition, other studies have demonstrated a chemopreventive property for ginseng that involves a blocking mechanism toward xenobiotic or precarcinogen-induced tumorigenesis (Yun et al., 1983; Jin et al., 1994a; Xiaoguang et al., 1998). Red ginseng consumption inhibits the incidence and growth of lung adenomas induced by urethane and dimethyl benzathracene as well as the incidence of lung adenoma induced by aflatoxin B1 (Yun et al., 1983). The source and age of ginseng are important variables for anticarcinogenic activity against benzo(a)pyrene-treated mice (Yun and Lee, 1994). For example, six-year-old fresh ginseng derived from extracts of white and red ginseng produces the greatest reduction in benzo(a)pyrene-induced lung adenoma (Yun, 1996). North American ginseng extract components have also been shown to induce the expression of a pS2 gene in breast cancer cell lines, without stimulating cell proliferation, thus providing a potentially protective role against breast cancer (Duda et al., 1998). In addition, the ginsenoside Rh2 provides growth inhibitory effects against human ovarian cancer cells (Tode et al., 1993).

5. CONCLUSIONS

Ginseng has been a staple for folk medicines of Eastern Asian cultures for thousands of years, primarily as a general tonic with pharmacological action or adaptogen-like properties during exposure to physiologically adverse habitats. Ginseng is known for providing anti-stress, immunoenhancement and possible regulation of the central nervous system and aspects of whole body metabolism that respond to stress. Despite the well-regarded pharmacological activities of ginseng, the heightened contemporary interest toward nutraceuticals and functional foods will call for more scientific knowledge to support the perceived benefits of ginseng. Two important examples of this include advances in chemical standardization methods of ginseng to facilitate improved quality control and labeling of a known quantity of bioactive constituents in related products. Moreover, more research concerning the structure-activity relationships between ginseng constituents, acting individually or synergistically in a mixture, is necessary for predicting and ensuring physiological and biochemical efficacy.

6. REFERENCES

Abe, K., S I. Cho, I. Kitagawa, N. Nishiyama and H. Saito 1994. "Differential effects of ginsenoside Rb1 and malonyl-ginsenoside Rb1 on long-term potentiation in the dentate gyrus of rats," *Brain Res.* 649:7–11

Avakian, E. V and E Evonuck. 1979. "Effect of *Panax ginseng* extract on tissue glycogen and adrenal cholesterol depletion during prolonged exercise," *Planta Med.* 36.43–48.

Baldwin, C A , L. A. Anderson and J D Phillipson. 1986 "What pharmacists should know about ginseng," *Pharm. J.* 234:585–586

Benishine, C , R. Lee, L. C H. Wang and H J. Liu 1991. "Effects of ginsenoside Rb1 on central cholinergic metabolism," *Pharmacology* 42 223–229.

Bhargava, J N. and P Ramarao. 1991. "The effect of *Panax ginseng* on the development of tolerance to the pharmacological actions of morphine in the rat," *Gen Pharmacol* 22 521–525

Brekhamn, I I and I. V Dardymov 1969. "New substances of plant origin which increase nonspecific resistance," *Annu. Rev Pharmacol.* 9·419–430

Chan, P , C. Y Hong, B. Tomlinson, N. C Chang, J. P Chen, S. T. Lee and J T. Cheng 1997. "Myocardial protective effect of trilinolein, an antioxidant isolated from the medicinal plant *Panax Pseudoginseng*," *Life Sci.* 61·1999–2006.

Chen, X , F Y Chen and J. J. F. Lee. 1993. "Ginsenosides induce a NO-released relaxation of the corpus cavernosum of the rabbit," *FASEB J.* 7:A770.

Chen, X , C. N. Gillis and R. Moalli. 1984. "Vascular effects of ginsenosides *in vitro*," *Br. J Pharmacol* 82.485–491.

Cheng, X. J , Y L. Liu, Y. S. Deng, G. P Lin and X. T. Luo 1998. "Effects of ginseng root saponins on central transmitters and plasma corticosterone in cold stress mice and rats," *Acta Pharmacol Sin.* 8 486–489.

Cheng, X J., X. R Shi and B. Lin. 1988. "Effects of ginseng root saponins on brain acetylcholine and serum corticosterone in normobaric hypoxia stress mice," 5th Southeast Asian and Western Pacific Regional Meeting of Pharmacologists. Chinese Pharmacological Association. Beijing, p S 31 05

Cui, J., M Garle and P Eneroth. 1994 "What do commercial ginseng preparations contain," (Letter) *Lancet* 334:134.

D'Angelo, L , R Grimaldi, M. Caravaggi, M Marcoli, E Perucca, S Lecchini, G. M. Grigo and A. Crema. 1986. "A double-blind, placebo-controlled clinical study on the effect of a standardized ginseng extract on psychomotor performance in healthy volunteers," *J Ethnopharmacol.* 16 15–22

Duda, R B., Y. Zhong, V Navas, M. Z C. Li, B R Toy, J. Alvarez. 1998. "Synergistic growth inhibitory properties are identified with the concurrent use of American ginseng and breast cancer therapeutic agents in MCF-7 breast cancer cells" (unpublished)

Forgo, I. and G Schimert. 1985. "The duration of the effect of standardized ginseng extract G115 in healthy competitive athletes," *Notabene Med.* 15 636–640.

Forgo, I , J. Kayasseh and J Staub. 1981. "Effect of a standardized ginseng extract on general well-being, reaction capacity, pulmonary function and gonadal hormones," *Medizinische Welt* 19.751–756

Gong, Y. H , J X. Jiang, Z. Li, L. H. Zhu and Z. Z Zhang 1991 "Hypoglycaemic effect of sanchinoside C1 in alloxan diabetic mice." *Acta Pharm. Sin* 26 81–85

Grandhi, A., A M Mujumdar and B Patwardham. 1994. "A comparative pharmacological investigation of Ashwagandha and ginseng," *J. Ethnopharmacol.* 44 131–135.

Guogan, Z and J. Yan. 1997. "Calcium channel blockage and anti-free-radical actions of ginsenosides," *Chin Med. J.* 110:28–29.

Hassan-Samira, M. M., M Attia, M. Allam and O. Elwan 1985 "Effect of the standardized ginseng extract G115 on the metabolism and electrical activity of the rabbit's brain," *J. Int. Med Res* 13 342–348

Hiai, S , S. Sasaki and H. Oura 1979. "Effect of ginseng saponin on rat adrenal cyclic AMP," *Planta Med* 37 15–19

Himi, T , H. Saito and N Nishiyama. 1989. "Effect of ginseng saponins on the survival of cerebral cortex neurons in cell cultures," *Chem Pharm Bull* 37 481–484

Hirakura, K , M. Morita, M. Nakajima, Y Ikeya and H Mitsuhahi. 1992. "Three acetylenic compounds from root of *Panax ginseng*," *Phytochem.* 31:899–903.

Hopkins, M P., H Androff and A. S Benninghoff. 1988. "Ginseng face cream and unexplained vaginal bleeding," *Am. J. Obstet. Gynecol.* 159:1121–1122

Hou, J P. 1977. "The chemical constituents of ginseng plants," *Am. J Chin Med* 5·123–145

Hu, S Y. 1977. "A contribution to our knowledge of Ginseng," *Am J. Chin. Med.* 5·1–23.

Hui-Ling, D. and Z. Jun-tian. 1991. "Anti-lipid peroxidative effect of ginsenoside Rbl and Rgl," *Chin Med. J.* 104:395–398.

Huong, N. T., K. Matsumoto, R. Kasai, K. Yamasaki and H Watanabe. 1998 "*In vitro* antioxidant activity of Vietnamese ginseng saponin and its components," *Biol. Pharm Bull* 21·978–981.

Iijima, M., T. Higashi, S. Sanada and J. Shoji 1976. "Effect of ginseng saponins on nuclear ribonucleic acid metabolism. I. RNA synthesis in rats treated with ginsenosides," *Chem Pharm Bull.* 24:2400–2405.

Ikehara, M., Y Shibata, T Higashi, S Sanada and J. Shoji. 1978 "Effect of ginseng saponins on cholesterol metabolism. III. Effect of ginsenoside-Rbl on cholesterol synthesis in rats fed on high fat diet," *Chem Pharm. Bull.* 26:2844–2849.

Jin, R , L L. Wan, T. Mitsuishi, K Kodama, S. Kruashige. 1994a. "Immuno-modulative effects of Chinese herbs in mice treated with anti-tumour agent cyclophosphamide," *Yakugaku Zasshi-Journal of Pharmaceutical Society of Japan.* 114:533–538.

Jin, R., L. L. Wan, T. Mitsuishi, S. Sato, Y. Akuzawa, K. Kodama and S. Kurashige. 1994b. "Effect of shi-ka-ron and Chinese herbs on cytokine production of macrophage in immunocompromised mice," *Am. J Chin. Med.* 22.255–266

Kajiwara, H., A. M Hemming and H. Hirano. 1996. "Evidence of metal binding activities of pentadecapeptide from *Panax ginseng*," *J. Chromatogr.* 687:443–448.

Kang, S. Y., S H. Kim, V B Schini and N. D Kim. 1995. "Dietary ginsenosides improve endothelium-dependent relaxation in the thoracic aorta of hyper-cholesterolemic rabbit," *Gen. Pharmacol.* 26·483–487

Karikura, M., H. Tanizawa, T. Hirata, T. Miyase and Y. Takino. 1992 "Studies on the absorption, distribution, excretion and metabolism of ginseng saponins VIII Isotope labelling of ginseno-side Rg2," *Chem. Pharm. Bull.* 40.2458–2460.

Kenarova, B , H Neychev, D. Hadjiivanova and V D. Petkov. 1990. "Immuno-modulating activity of ginsenoside Rg₁ from *Panax ginseng*," *Jap. J Pharmacol.* 54:447–454.

Kim, C., C. C. Kim, M. S. Kim, C. Y. Hu and J. S Rhe. 1970. "Influence of ginseng on the stress mechanism," *Lloydia.* 33:43–48

Kim, Y. C., J. H. Lee, M. S. Kim and N. G. Lee. 1985. "Effect of the saponin fraction of *Panax ginseng* on catecholamines in mouse brain," *Arch. Pharm. Res.* 8.45–49.

Kim, J Y., D R. Germolec and N. U. Luster. 1990a. "*Panax* ginseng as a potential immunomodula-tor. studies in mice," *Immunopharmacol. Immunotoxicol* 12:257–276.

Kim, H. Y , C. G. Jang and M K Lee. 1990b. "Antinarcotic effects of the standardized ginseng extract G115 on morphine," *Planta Med* 56:158–163.

Kim, H Y , X Chen and C. N. Gillis. 1992. "Ginsenosides protect pulmonary vascular endothelium against free radical-induced injury." *Biochem Biophys. Res. Comm.* 189.670–676.

Kim, N. D., S. Y. Kang and V. D. Schini 1995 "Ginsenosides evoke endothelium-dependent vascular relaxation in rat aorta," *Gen. Pharmacol.* 25:1071–1077.

Kim, Y H., K. H. Park and H M. Rho. 1996. "Transcriptional activation of the Cu, Zn-superoxide dismutase gene through the AP2 site by ginsenoside Rb₂ extracted from a medicinal plant. *Panax ginseng*," *J. Biol. Chem.* 271.24539–24543.

Kim, Y W., D. K. Song, W. H Kim, M B. Wie, Y H. Kim, S. H. Kee and M. K. Cho 1997. "Long term oral administration of ginseng extract decreases serum gamma-globulin and IgG1 isotype in mice," *J Ethnopharmacol.* 58.55–58

Kimura, T , P. A Saunders, H S. Kim, H. M Rheu, K W Oh and I. K Ho. 1994 "Interactions of ginsenosides with ligand-binding of $GABA_A$ and $GABA_B$ receptors," *Gen Pharmacol.* 25:193–199.

Kirchdorfer, A M. 1985. "Clinical trials with the standardized ginsenoside concentrate G115 " In. *Advances in Chinese Medicinal Materials Research,* Chang, H. M., H W. Yeung, W. W. Tso, A Koo (eds) World Scientific Publ. Co., Singapore, pp. 529–542.

Kitagawa, T , T. Taniyama, M Yoshikawa, Y. Ikenishi and Y. Nagasawa. 1989. "Chemical studies on crude drug processing VI. Chemical structure of malonylginsenoside Rb1, Rb2, Rc and Rd isolated form the root of *Panax ginseng,*" C.A. Meyer (ed.) *Chem. Pharm. Bull* 37.2961–2970

Kitts, D. D , C Hu and A N. Wijewickreme 1999. "Antioxidant activity of a North American Ginseng Extract." In. *Food For Health in the Pacific Rim, 3rd International Conference of Food Science and Technology,* Whitaker, J. R., N. F. Haard, C. F. Shoemaker and P P. Singh (eds.) Food and Nutrition Press, Connecticut, pp. 232–242

Konsler, T R , S W. Zito, J. E. Shelton and E J. Staba. 1990 "Lime and phosphorus effects on American Ginseng. II Root and leaf ginsenoside content and their relationship," *J. Am. Soc Hort. Sci.* 115.575–580.

Li, T. S C , G Mazza, A. C. Cottrell and L. Gao. 1996 "Ginsenosides in roots and leaves of American ginseng," *J. Agric. Food Chem.* 44.717–720.

Li, Y. and X J. Zhao. 1989 "Effects of panaxadiol saponins on the contents of serum enzymes, lipid peroxides and SOD in hemorrhagic shock dogs," *Chin J. Pathophysiol.* 5:539–544.

Lin, S Y , L. M. Liu and L C. Wu. 1995. "Effects of Shenmai injection on immune function in stomach cancer patients after chemotherapy," *Chung-Kuo Chung His I Chieh Ho Tsa Chih.* 15.451–453

Liu, C X and P G. Yjao. 1992. "Recent advances on ginseng research in China," *J. Ethnopharmacol.* 36.27–38

Lui, J , S Wang, H. Liu, L. Yang and G. Nan. 1995. "Stimulatory effect of saponin from *Panax ginseng* on immune function of lymphocytes in the elderly," *Mech. Ageing and Dev.* 83.43–53

Lui, J and J T. Zhang. 1995. "Immunoregulatory effects of ginsenoside Rg1 in aged rats," *Yao Hsueh Hsueh Pao-Acta Pharm. Sin.* 30:818–823.

Lui, J. and J T. Zhang. 1996. "Studies on the mechanisms of immunoregulatory effects of ginsenoside Rg1 in aged rats," *Yao Hsueh Hsueh Pao-Acta Pharm. Sin* 31.95–100.

Ma, Y C , J. Zhu, L Benkrima, M. Luo, L. Sun, S. Sain, K. Kont and Y Y. Plaut-Carcasson 1995 "A comparative evaluation of ginsenosides in commercial ginseng products and tissue culture samples using HPLC." *J. Herbs, Spices and Medicinal Plants.* 3:41–50.

Marasco, C , R. Vargas Ruiz, A. Salas Villagomez and C Begona Infante. 1998. "Double-blind study of a multivitamin complex supplemented with ginseng extract," *Drugs Exp Clin Res.* 22 323–329.

Matsunaga, H., M. Katano, H. Yamamoto, M. Mori, K. Takata. 1989. "Studies on the panaxatriol of *Panax ginseng* C.A. Meyer. Isolation, determination and antitumor activity," *Chem. Pharm Bull.* 37:1281–1291.

Nagasawa, T., H Oura, S. Hiai and K. Nishinaga. 1977. "Effect of ginseng extract on ribonucleic acid and protein synthesis in the rat kidney," *Chem. Pharm. Bull.* (Tokyo) 25.1665–1670.

Odsashima, S , Y. Nakayabu, N. Honjo, H. Abe and S. Arichi. 1979. "Induction of phenotypic reverse transformation by ginsenosides in cultured Morris hepatoma cells," *Euro J Cancer* 15:855–892.

Oomura, Y., K Sasaki, A. Nijima, N. Shimizu, Y. Kai, A. Fukuda and J. Nabekura. 1990. "Effects of ginsenoside functions on alimentation neuron modulators." In· *Recent Advances in Ginseng Studies Proceedings of the International Ginseng Seminar*, Shibata, S , Y. Ohtsuka and H. Saito (eds) Hirokawa, Publishing Company, Tokyo, pp. 63–71.

Oura, H. and S. Hiai. 1973. Cited in Sakakibara, K., Shibata, Y., Higashi, T., Sanada, S and Shoji, J. 1975. "Effect of ginseng saponins on cholesterol metabolism. I. The level and the synthesis of serum and liver cholesterol in rats treated with ginsenosides," *Chem Pharm. Bull.* 23 1017–1024.

Park, H. W., S C. Kim and N. P. Jung. 1988. "The effect of ginseng saponin fractions on humoral immunity of mouse," *Korean J. Ginseng Sci.* 12.63–67.

Pearse, P. T., I Zois, K. N. Wynne and J W. Funder. 1982 "*Panax ginseng* and Eleutherococcus senticosus extracts *in vitro* studies on binding to steroid receptors," *Endocrinol. Jpn* 29:567–573.

Peng, C. F., Y. J. Li, Y J Li and H. W. Deng. 1995. "Effects of ginsenosides on vasodilator nerve actions in the rat perfused mesentery are mediated by nitric oxide," *J. Pharm Pharmacol* 47:614–617.

Petkov, V D , C. Yinglin, I. Todorov, M. Lazarova, D. Getova, S. Stancheva and L. Alova. 1992. "Behavioral effects of stem-leaves extract from *Panax ginseng* C. A. Meyer," *Acta Physiol Pharmacol. Bulg.* 18:41–48.

Petkov, V. D., R Kehayov, S. Belcheva, E Konstantinova, V Petkov, D Getova and V Markovska. 1993 "Memory effects of standardized extracts of *Panax* ginseng (G115), Ginko biloba (GK501) and their combination gincosan (PHL-00701)," *Planta Med* 59 106–114

Phillipson, J D and L A Anderson 1984 "Ginseng-quality, safety and efficacy?" *Pharm. J.* 232 161–165.

Pieralisi, G., P. Ripari and L. Vecchiet. 1991. "Effects of standardized ginseng extract combined with dimethylaminoethanol bitartrate, vitamins, minerals and trace elements on physical performance during exercise," *Clin. Thera.* 13:373–382.

Popov, I M. and W. J. Gold. 1973. "A review of the properties and clinical effects of ginseng," *Am J. Chin Med.* 1 263–270.

Punnonen. R and A Lukola. 1980. "Oestrogen-like effect of ginseng," *Br Med J* 281 1110–1112.

Qu, J. B., Y. N Cao and X. Y Ma 1988. "Effects of ginseng root saponins on animals in acute hypoxia due to negative air pressure," 5th Southest Asian and Western Pacific Regional Meeting of Pharmacologists. Chinese Pharmacological Association. Beijing, p. 614.

Rimar, S , M Lee-Mengel and C. N. Gillis 1996. "Pulmonary protective and vasodilator effects of a standardized *Panax ginseng* preparation following artificial gastric digestion," *Pulm. Pharmacol.* 9·205–209.

Rosenfeld, M. S. 1989. "Evaluation of the efficacy of a standardized ginseng extract in patients with psychophysical asthenia and neurological disorders," *La Semana Medica.* 173:9:148–154

Saito, H., Y. Yoshida and K. Takagi. 1974. "Effect of *Panax ginseng* root on exhaustive exercise in mice," *Jpn. J. Pharmacol.* 24.119–127.

Saito, H., M Tsuchiya, S. Naka and K. Takagi. 1977. "Effect of *Panax ginseng* root on condition avoidance response in rats," *Jpn. J. Pharmacol* 27.509–516.

Sakakibara, K , Y. Shibata, T. Higashi, S Sanada and J. Shoji. 1975. "Effect of ginseng saponins on cholesterol metabolism. I. The level and the synthesis of serum and liver cholesterol in rats treated with ginsenosides," *Chem. Pharm. Bull.* (Tokyo) 23:1009–1016

Scaglione F., F. Ferrara, S. Dugnani, M. Falchi, G. Santorto and F. Fraschini. 1990. "Immunomodulatory effects of two extracts of *Panax ginseng* C.A. Meyer," *Drugs Expt. Clin. Res.* 16·537–542

Schmidt, U. J., I. Kalbe, F. H. Schulz and G. Bruschke. 1978. "Pharmacotherapy and so-called basic therapy in old age," 11th International Congress of Gerontology. Tokyo, Japan, August 20–25, 20 pp.

Shibata, S., O. Tanaka, K Soma, T Ando and H. Nakamura. 1965. "Studies on saponins and sapogenins of ginseng. The structure of panoxachol." *Tetrahedron Lett* 3:207–213.

Siegel, R K 1979 "Ginseng abuse syndrome-problems with the Panacea," *J Am Med Assoc* 241 1614–1615.

Smith, R G , D. Caswell, A Carriere and B. Zielke. 1996. "Variation in the ginsenoside content of American ginseng, *Panax quinquefolium* L., roots," *Can. J Bot* 74 1616–1620.

Sotaniemi, E. A., E. Haapakoski and A. Rautio. 1995 "Ginseng therapy in non-insulin-dependent diabetic patients," *Diabetes Care.* 18:1373–1375.

Takagi, K., H. Saito and M Tsuchiya. 1974. "Effect of *Panax ginseng* root on spontaneous movement and exercise in mice," *Jpn. J. Pharmacol.* 24:41–48.

Takahashi, M., S Tokuyama and H Kaneto. 1992. "Anti-stress effect of ginseng on the inhibition of the development of morphine tolerance in stressed mice," *Jpn J Pharmacol.* 59:399–404

Takemoto, Y , T. Ueyama, H Saito, S. Horio, S Sanada, J Shoji, S. Yahara, O Tanaka and S Shibata 1984. "Potentiation of nerve growth factor-mediated nerve fiber production in organ cultures of chicken embryonic ganglia by ginseng saponins, structure-activity relationship," *Chem. Pharm. Bull.* 32:481–484.

Takino, Y , T. Odani, T. Hisayuki and T Hayashi. 1982. "Studies on the absorption, distribution, excretion and metabolism of ginseng saponins. I. Quantitative analysis of ginsenoside Rg1 in rats," *Chem Pharm Bull* 30·2196–2201

Thi, N , N T. T. Huong, K. Matsumoto, R. Kasai, K. Yamasaki and H. Watanabe. 1998. "*In vitro* antioxidant activity of vietnamese ginseng saponin and its components," *Biol Pharm. Bull.* 21·978–981

Tode, T , Y. Kikuchi, T. Kita, J. Hirata, E Imaizumi and I Nagata 1993. "Inhibitory effects by oral administration of ginsenoside Rh2 on the growth of human ovarian cancer cells in nude mice," *J Cancer Res. Clin. Oncol.* 120:24–26.

Tokuyama, S , M Takahashi and H. Kaneto. 1996. "The effect of ginseng extract on locomotor sensitization and conditioned place preference induced by methamphetamine and cocaine in mice," *Pharmacol Biochem Behav.* 54·671–676.

Tsang, D., H. W. Yeung, W. W Tso and H. Peck. 1985. "Ginseng saponins: Influence on neurotransmitter uptake in rat brain synaptosomes," *Planta Med.* 51:221–224.

Umnova, N V., T. L. Nfichurina, N. L. Sminova, I. V. Aleksandrov and G. G. Poroshenko. 1991. "Study of antimutagenic properties of bio-ginseng in mammalian cells *in vitro* and *in vivo* " *Bull. Exp. Biol. Med.* 111.507–509

Verbeuren J. J., F. H. Jordaens, L. L. Zonnekeyn, D. E. VanHove, M. C Coene, and A G Herman 1986. "Effect of hypercholesterolemia on vascular reactivity in the rabbit," *Circ Res.* 58:552–564.

Wang, B. X , J. C. Cui and A. J. Liu. 1980. "The effect of polysaccharides of root of *Panax ginseng* on the immune function," *Acta Pharm. Sin.* 17 66–68

Wang, W. K., H. L. Chen, T. L. Hsu and Y Y. L Wang. 1994. "Alteration of pulse in human subjects by three chinese herbs." *Am J. Chin. Med.* 22:197–203.

Xiaoguang, C , L. Hongyan, L. Xiaohong, F. Zhaodi, L. Yan, T Lihua and H. Rui 1998 "Cancer chemoprevention and therpeutic activities of red ginseng," *J. Ethnopharmacol* 60:71–78

Yamamoto, M. 1973 Cited in Sakakibara, K., Shibata, Y., Higashi, T., Sanada, S and Shoji, J. 1975. "Effect of ginseng saponins on cholesterol metabolism I. The level and the synthesis of serum and liver cholesterol in rats treated with ginsenosides," *Chem Pharm. Bull.* 23.1017–1024.

Yokozawa, T., H. Seno and H Oura 1975. "Effect of ginseng extract on lipid and sugar metabolism I. Metabolic correlation between liver and adipose tissue," *Chem Pharm Bull* (Tokyo) 23.3095–3100.

Yuan, Y. V. and D. D Kitts 1996. "Endogenous antioxidants. Role of antioxidant enzymes in biological systems." In *Natural Antioxidants: Chemistry, Health Effects and Applications* Shahidi, F. (ed.) AOCS Press, Champaign, Illinois. pp. 258–270.

Yuan, W. X., X. J. Wu and F. X. Yang. 1998 "Effects of ginseng root saponins on brain monoamine and serum corticosterone in heat stressed mice," 5th Southest Asian and Western Pacific Regional Meeting of Pharmacologists. Chinese Pharmacological Association. Beijing, p. S 31.04.

Yun, T. K. 1996. "Experimental and epidemiological evidence of the cancer preventative effects of *Panax ginseng*, C.A Meyer," *Nutr. Rev.* 54 S71–S81.

Yun, T. K. and S. Y. Choi. 1990. "A case-control study of ginseng intake and cancer," *Int. J. Epidemiol.* 19:871–876

Yun, T. K and S. Y. Choi. 1995. "Preventative effect of ginseng intake against various human cancers: A case-control study on 1987 pairs," *Cancer Epidemiology Biomarker and Prevention.* 4:401–408.

Yun, T. K. and Y S Lee. 1994. "Anticarcinogenic effect of ginseng powders depending on the types and ages using Yun's anticarcinogenicity test (I)," *Korean J. Ginseng Sci.* 18:89–94

Yun, T. K., Y. S. Yun and L. W. Han. 1983. "Anticarcinogenic effect of long-term oral administration of red ginseng on newborn mice exposed to various chemical carcinogens," *Cancer Detect Prev* 6:515–525

Yun, Y. S., S K. Jo, H S Moon, Y. J. Kim, Y. R. Oh and T. K. Yun 1987. "Effect of red ginseng on natural killer cell activity in mice and lung adenoma induced by urethane and benzo(a)pyrene," *Cancer Detect. Prev. Suppl.* 1:310–319.

Zhang, F. L. and X. Chen. 1987 "Effects of ginsenosides on sympathetic neuro-transmitter release in pithed rats," *Acta. Pharmacol. Sin.* 8:217–220.

Zhang, G. D, Z. H Zhou, M. Z. Wang and F. Y. Gao 1979a "Analysis of ginseng I," *Acta Pharmacaeut. Sin* 14.309–314.

Zhang, G. D., Z. H Zhou, M. Z. Wang and F Y. Gao. 1979b. "Analysis of ginseng II," *Acta Pharmacaeut. Sin.* 15 175–181.

Zhang, J. T., Z. W. Qu and Y. Liu. 1990. "Preliminary study on anti-amnestic mechanism of ginsenoside Rg1 and Rb1," *Chin. Med J.* 103:932–938.

Zhang, D., T. Yasuda, Y. Yu, P. Zheng, T. Kawabata, Y. Ma and S. Okada. 1996. "Ginseng scavenges hydroxyl radical and protects unsaturated fatty acids from decomposition caused by iron-mediated lipid peroxidation," *Free Radical Biol. & Med.* 30:145–150.

Zuin, M., Z. M. Battezzati, M. Camisasca, D. Riebenfield and M. Podda. 1987. "Effects of a preparation containing a standardized ginseng extract combined with trace elements and multi-vitamins against hepatotoxin-induced chronic liver disease in the elderly," *J. Int. Med. Res* 15:276–281.

CHAPTER 3

Chemistry, Pharmacology and Clinical Applications of *Echinacea* Products

R. BAUER

1. INTRODUCTION

ECHINACEA products represent the most popular herbal immunostimulants in North America and Europe (Grünwald and Büttel, 1996). In recent years, *Echinacea* products have been the best-selling herbal products in natural food stores in the U.S. with 11.9% (1996: 9.6%) of herbal supplement sales in 1997 (Richman and Witkowski, 1997; Brevoort, 1998). In North America, mainly encapsulated powders from roots and aerial parts and also tinctures and extracts from the roots and aerial parts are used. Most products are sold as dietary supplements (U.S.) or as natural health products (Canada). In some European countries, the situation is different. In Germany, most *Echinacea* products are registered as drugs and are sold in pharmacies. Most of them contain the expressed juice of *Echinacea purpurea* aerial parts, or hydroalcoholic extracts of *E. pallida* or *E. purpurea* roots (Bauer and Wagner, 1990). Clinical studies have demonstrated the effectiveness of the expressed juice from aerial parts of *E. purpurea* as adjuvant in the therapy of relapsing infections of the respiratory and urinary tracts, as well as the hydroalcoholic extracts of *E. pallida* and *E. purpurea* roots in the therapy of common cold and flu (Bauer and Liersch, 1993; Melchart et al., 1994; Bauer, 1998a, 1998b; Melchart and Linde, 1998).

2. BOTANICAL ASPECTS AND TRADITIONAL USE

The genus *Echinacea* Moench (Compositae) is endemic to North America,

45

where it occurs in the Great Plains between the Appalachian Mountains in the east and the Rocky Mountains in the west. Taxonomically, the genus *Echinacea* is assigned to the *Heliantheae* that is the largest tribe within the Compositae (Asteraceae) family. According to Stuessy (1977), *Echinacea* is part of the "Verbesininae line" in the subtribe *Helianthinae*, where it forms a small subgroup with the genera *Rudbeckia* L., *Ratibida* RAF. and *Dracopis* CASS. However, Robinson (1981) revised the taxonomy of the *Heliantheae* and divided the tribe into 35 subtribes based on chemical and micromorphological characters. This resulted in a fundamental reinterpretation of the phylogeny of the genus *Echinacea*, and its taxonomic relationship to the other genera. Thus, *Echinacea* was transferred from the *Helianthinae* to the *Ecliptinae*, while *Rudbeckia*, *Ratibida* and *Dracopis* were placed in the new subtribe, *Rudbeckiinae*.

The current taxonomy of the genus *Echinacea*, also used in the present National List of Scientific Plant Names, is based on a comparative morphological and anatomical study (McGregor, 1968), according to which the genus comprises nine species and two varieties (Table 3.1). But, only *E. purpurea* (L.) Moench *E. angustifolia* DC. and *E. pallida* (Nutt.) are widely used medicinally.

The medical application of *Echinacea* can be traced to the American Indians, who regarded *Echinacea* as one of the most favorable remedies for treating wounds, snakebites, headache and the common cold (Moerman, 1998). The territories of the tribes that most frequently used *Echinacea* show a close correspondence with the distribution range of *E. angustifolia*, *E. pallida* and *E. purpurea* (Figure 3.1). But, possibly, other *Echinacea* species have also been used.

In the second half of the last century, European settlers took over the use of *Echinacea*. H. G. F. Meyer, a German quack doctor, distributed a tincture, named "Meyer's Blood Purifier," which he praised for rheumatism, neuralgia, headache, erysipelas, dyspepsia, tumors and boils, open wounds, vertigo, scrofula and bad eyes, as well as "poisoning by herbs" and rattlesnake bites (Lloyd, 1904). The plant used was identified as *E. angustifolia* by Lloyd (1893).

When the Lloyd Brothers Inc. started to produce *Echinacea* tinctures in a large scale at the beginning of this century, a second plant became important: *E. pallida*. This species is more abundant and much taller with bigger roots than *E. angustifolia*. In the monograph of the National Formulary of the U.S. in 1916, the roots of both species were officially listed, with the result that differentiation between these two species was subsequently neglected (Schindler, 1940). Even recently, most *E. angustifolia* available on the market and in botanical gardens in Europe was in fact *E. pallida* (Bauer and Wagner, 1991).

E. purpurea was introduced as a medicinal plant in Europe in the middle of this century by Madaus (1939). A juice preparation expressed from the

TABLE 3.1. Taxonomic Formation of the Genus *Echinacea*, According to McGregor (1968).

Echinacea angustifolia DC var *angustifolia*
    Synonyms
    Brauneria angustifolia HELLER
    Echinacea pallida var *angustifolia* (DC) CRONQ
Echinacea angustifolia DC. var *strigosa* McGREGOR
Echinacea atrorubens NUTT
    Synonym.
    Rudbeckia atrorubens NUTT
Echinacea laevigata (BOYNTON & BEADLE) BLAKE
    Synonyms
    Brauneria laevigata BOYNTON & BEADLE
    Echinacea purpurea (L) MOENCH var *laevigata* CRONQ
Echinacea pallida (NUTT) NUTT.
    Synonyms
    Echinacea angustifolia HOOKER
    Rudbeckia pallida NUTT
    Brauneria pallida BRITTON
    Echinacea pallida (NUTT.) NUTT f *albida* STEYERM
Echinacea paradoxa (NORTON) BRITTON var. *paradoxa*
    Synonyms.
    Brauneria paradoxa NORTON
    Echinacea atrorubens NUTT var *paradoxa* (NORTON) CRONQ
Echinacea paradoxa (NORTON) BRITTON var *neglecta* McGREGOR
Echinacea purpurea (L) MOENCH
    Synonyms
    Rudbeckia purpurea L
    Rudbeckia hispida HOFFMGG
    Rudbeckia serotina SWEET
    Echinacea purpurea (L.) MOENCH var *arkansana* STEYERM.
    Echinacea purpurea (L) MOENCH f *ligettii* STEYERM
    Echinacea speciosa PAXTON
    Echinacea intermedia LINDLEY
    Brauneria purpurea (L) BRITTON
Echinacea simulata McGREGOR
    Synonym.
    Echinacea speciosa McGREGOR
Echinacea sanguinea NUTT.
Echinacea tennesseensis (BEADLE) SMALL
    Synonyms.
    Brauneria tennesseensis BEADLE
    Echinacea augustifolia DC. var. *tennesseensis* (BEADLE) BLAKE

fresh aerial parts proved to be very active, and this species subsequently achieved medicinally equal status with *E. angustifolia* and *E. pallida* (review by Hahn and Mayer, 1984). The roots had already been used in North America in the treatment of saddle sores on horses and of syphilis (King and Newton, 1852).

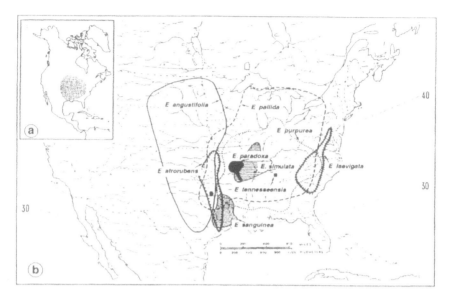

Figure 3.1 Distribution map of the genus *Echinacea* according to McGregor (1968) and McGregor and Barkley (1977)

During the last 50 years, adulterations and substitutions of *E. purpurea* roots with *Parthenium integrifolium* roots have occurred. The sesquiterpene esters, echinadiol-, epoxyechinadiol-, echinaxanthol- and dihydroxy-nardol-cinnamate, described as constituents of *E. purpurea* roots (Bauer et al., 1985), are in fact derived from *P. integrifolium,* which was mistakenly processed at that time. Since both have different constituents, HPLC and TLC methods have been developed to distinguish the roots chemically (Bauer et al., 1987b). *P. integrifolium* is characterized by sesquiterpene esters, which is absent in *E. purpurea.* Nevertheless, adulterations of *E. angustifolia* and *E. purpurea* roots with *P. integrifolium* still occur.

3. BIOLOGICALLY ACTIVE CONSTITUENTS

The constituents of *Echinacea* species cover a wide range of polarity, from the polar polysaccharides and glycoproteins, via the medium polar caffeic acid derivatives and flavonoids to the rather lipophilic polyacetylenes and alkamides.

3.1. POLYSACCHARIDES AND GLYCOPROTEINS

Investigations of the aqueous extract of the aerial parts of *E. purpurea* led to the isolation of two polysaccharides (PS I and PS II) with immunostimulatory properties (Wagner and Proksch, 1981; Stimpel et al., 1984). They stimulate phagocytosis *in vitro* and *in vivo* and enhance the production of oxygen radicals by macrophages in a dose-dependent way (Stimpel et al., 1984; Lohmann-Matthes and Wagner, 1989). Structural analysis revealed PS I to be a 4-O-methyl-glucuronoarabinoxylan with an average MW of 35 kDa, while PS II was identified as an acidic arabinorhamnogalactan of MW 45 kDa (Proksch and Wagner, 1987). A xyloglucan, MW 79.5 KDa, was isolated from the leaves and stems of *E. purpurea,* and a pectin-like polysaccharide was isolated from the expressed juice (Stuppner, 1985). Raw polysaccharide fractions have also been obtained from *E. purpurea* and *E. angustifolia (E. pallida)* roots (Wagner et al., 1985). Recently, arabinogalactan-like polysaccharides have also been detected in the expressed juice of *E. purpurea* (Blaschek et al., 1998). Giger et al. (1989) investigated the fructan content in the roots of *E. angustifolia* and *E. purpurea* and found that the total fructose was lowest in May, increasing during the summer and autumn. The formation of highly polymerized fructans occurred earlier in *E. purpurea* than in *E. angustifolia, E. purpurea* was characterized by the accumulation of fructosans during the winter, mainly of polymerization grade 4.

Three homogeneous polysaccharides have been isolated from the growth medium of *E. purpurea* tissue cultures, i.e., two neutral fucogalactoxyloglucans with MW of 10 and 25 kDa, and an acidic arabinogalactan, MW 75 kDa (Wagner et al., 1987, 1988). These *E. purpurea* polysaccharides have also been produced on an industrial scale (75,000 L fermenter) from cell cultures (Wagner et al., 1989). The isolated polysaccharides stimulated macrophages (secretion of TNF and IFNβ_2) and provided protection against *Candida* and *Listeria* infections in mice (Luettig et al., 1989; Roesler et al., 1991a, 1991b; Steinmüller et al., 1993) and were able to activate the nuclear transcription factor NFκB, which controls the expression of genes related to the immunological response (Emmendörffer et al., 1998). Polysaccharides from *E. angustifolia* also exhibited anti-inflammatory activity (Tubaro et al., 1987; Tragni et al., 1988). The pharmacological performance of *Echinacea* polysaccharides have recently been reviewed by Emmendörffer et al. (1998), and their efficacy in humans is currently undergoing clinical trials (Melchart et al., 1993, 1998).

Routine analytical methods for the specific analysis of the pharmacologically active polysaccharides in phytopreparations are not yet available. An exact characterization can only be achieved by gas chromatographic determination of the sugar sequences and linkages after isolation and purification (Proksch and Wagner, 1987). Identification and quantification of monosaccharides after hydrolysis, recently suggested as a standardization method (Wagner, 1997), does not possess sufficient specificity for the active polysaccharides. It is more promising to develop fluorescence or radioactivity labeled polysaccharides or antibodies for a specific assay (Kraus et al., 1996).

Three glycoproteins, with MW of 17, 21 and 30 kDa, respectively, containing ca. 3% protein, have been isolated from *E. angustifolia* and *E. purpurea* roots. The main sugars were arabinose (64–84%), galactose (1.9–5.3%) and glucosamines (6%). The protein moiety contained aspartate, glycin, glutamate and alanin (Beuscher et al., 1987). An ELISA method developed for the detection and determination of these glycoproteins in preparations (Egert and Beuscher, 1992) showed that *E. angustifolia* and *E. purpurea* roots contain similar amounts of glycoproteins, while *E. pallida* contains less (Beuscher et al., 1995).

Purified root extracts containing a glycoprotein-polysaccharide complex exhibited B-cell stimulating activity and induced the release of interleukin 1, TNF and IFN in macrophages. The results could also be reproduced in mice (Beuscher et al., 1987, 1995; Bodinet and Beuscher, 1991).

3.2. CAFFEIC ACID DERIVATIVES

Caffeic acid derivatives are also a major group of constituents in *Echinacea* species (Figure 3.2). Stoll et al. (1950) isolated echinacoside from the roots of *E. angustifolia*, and Becker et al. (1982) elucidated the chemical structure of this caffeic acid derivative. At a concentration of 0.3–1.7% echinacoside is the major polar constituent of the roots of *E. angustifolia* (Schenk and Franke, 1996; Bauer et al., 1988c; Bauer and Remiger, 1989a). In *E. pallida*, it occurs at a similar concentration and is therefore not suitable for the discrimination of these two species. However, they can be distinguished by the occurrence of 1,3- and 1,5-*O*-dicaffeoyl-quinic acids that are present only in the roots of *E. angustifolia* (Bauer et al., 1988c). In *E. pallida* roots, 6-O-caffeoyl-echinacoside has also been detected in low quantities (Cheminat et al., 1988).

The roots of *E. purpurea* lack echinacoside but contain cichoric acid (2*R*,3*R*-*O*-dicaffeoyl-tartaric acid), a compound also present in the aerial parts of *E. purpurea*, *E. pallida* and *E. angustifolia* (Becker and Hsieh, 1985; Bauer et al., 1988a). Cichoric acid can also be synthesized via a facile route (Zhao and Burke, 1998).

In *Echinacea*, cichoric acid occurs especially in high concentrations in the flowerheads (ligules) of the three medicinally used species and in the roots of *E. purpurea* (1.2–3.1% and 0.6–2.1%, respectively). Leaves and stems contain lower amounts of cichoric acid (Figure 3.3). *E. angustifolia* contains the lowest amount of cichoric acid (Bauer et al., 1988a). The content of cichoric acid depends on the season and the stage of development of the plant (Alhorn, 1992; Bauer and Vom Hagen-Plettenberg, 1999) and is highest at the beginning of the vegetation period and decreases during plant growth (Figure 3.4).

Cichoric acid regularly undergoes rapid decomposition during the preparation of tinctures or expressed juices, and it is believed that enzymatic degrada-

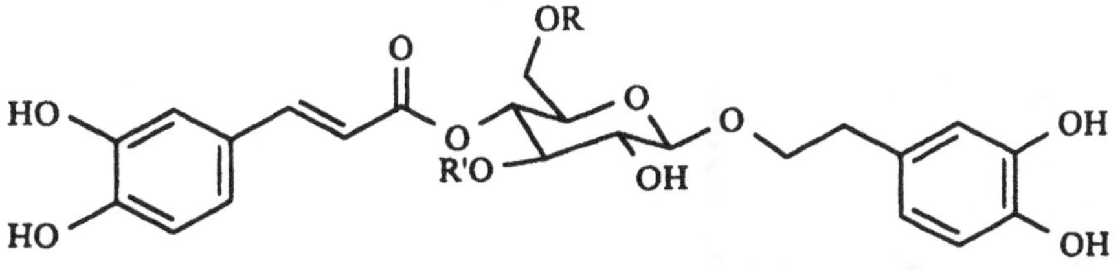

Echinacoside: R = Glucose (1,6-); R' = Rhamnose (1,3-)

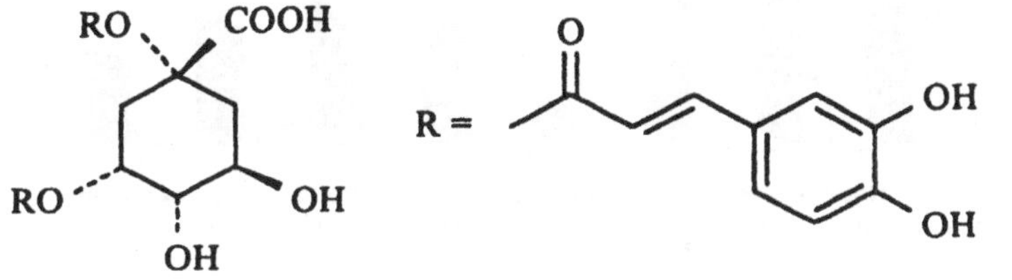

1,3-Dicaffeoyl-quinic acid (Cynarine)

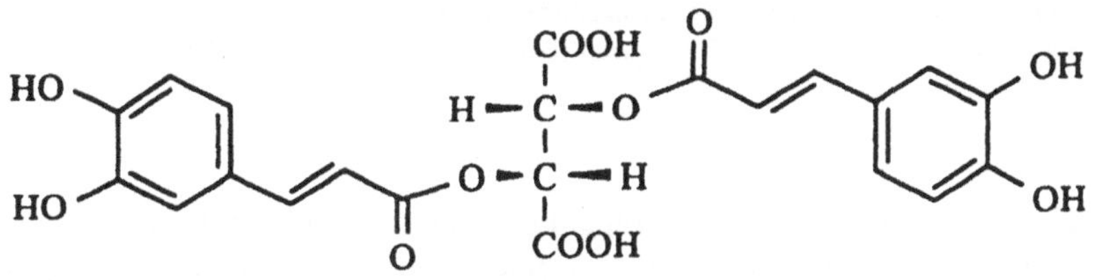

1,2-Dicaffeoyl-tartaric acid (Cichoric acid)

Figure 3.2 Caffeic acid derivatives found in *Echinacea* species.

tion occurs during the extraction process (Remiger, 1989). Similar degradation has been observed during the regular preparation of *E. purpurea* expressed juice (Bauer, 1997). It has now been found that a phenoloxidase is mainly responsible for this degradation process (Nüβlein et al., 1999).

For quality control regarding caffeic acid derivatives, extracts can be analyzed by either TLC (silica gel; ethylacetate/formic acid/acetic acid/water 100:11:11:27; detection: natural product reagent/UV 360 nm) or RP-HPLC (C18 column; solvent, 5–25% acetonitrile/phosphoric acid; flow: 1.0 mL/min; detection; 280 nm) (Bauer et al., 1988a; Bauer, 1997). HPLC has also been used for the analysis of phenolic acids in *Echinacea* preparations (Glowniak et al., 1996). Determination of total polyphenols, however, is rather unspecific and is only a very rough parameter. Echinacoside can be used as an analytical marker to test batch-to-batch consistency of preparations from *E. angustifolia*

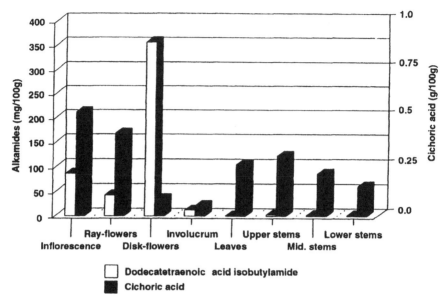

Figure 3.3 Content of cichoric acid and alkamides (dodeca-2E,4E,8Z,10E,Z-tetraenoic acid isobutylamides) in different parts of *E. purpurea* (Bauer and Vom Hagen-Plettenberg, 1999).

or *E. pallida* roots. However, preparations should not be optimized regarding echinacoside, since it is not a very pharmacologically relevant compound. Capillary electrophoresis (micellar electrokinetic chromatography MEKC) has also been successfully applied for the analysis of caffeic acid derivatives in *Echinacea* extracts (Pietta et al., 1998). This method provides an excellent resolution, a very high sensitivity and enables the discrimination of the species (Figure 3.5).

Recently automated multiple development (AMD) thin layer chromatography has been used for the analysis of caffeic acid derivatives in *E. angustifolia* root extracts (Trypsteen et al., 1989; Gocan et al., 1996), and a fast-atom-bombardment tandem mass spectrometric method has been developed for screening of echinacoside in crude plant extracts (Facino et al., 1991, 1993).

Cichoric acid, unlike echinacoside, verbascoside and 2-caffeoyl-cichoric acid has been shown to possess phagocytosis stimulatory activity *in vitro* and *in vivo* (Bauer et al., 1989b). Cichoric acid also inhibits hyaluronidase (Facino et al., 1993) and protects type III collagen from free-radical-induced degradation (Facino et al., 1995). It could, therefore, contribute to the immunomodulatory activity of *Echinacea* extracts. Recently, cichoric acid was found to selectively inhibit human immunodeficiency virus type 1 (HIV-1) integrase (Robinson et al., 1996; Neamati et al., 1997; McDougall et al., 1998; Robinson,

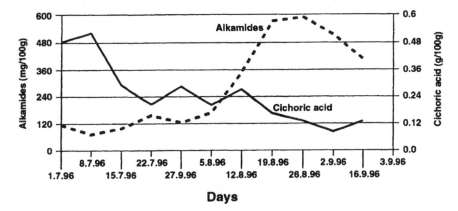

Figure 3.4 Content of cichoric acid and dodeca-2E,4E,8Z,10E,Z-tetraenoic acid-isobutylamide in *E. purpurea* aerial parts during the vegetation period, July to September 1996 (Bauer and Vom Hagen-Plettenberg, 1999).

1998; King and Robinson, 1998). However, cichoric acid is a rather labile compound, and its bioavailability therefore needs to be checked.

Echinacoside has low antibacterial and antiviral activity but does not possess immunostimulatory effects (Stoll et al., 1950; Cheminat et al., 1988). It also protects the free-radical-induced degradation of type III collagen (dose dependently) by a reactive oxygen scavenging effect (Facino et al., 1995) that led these authors to conclude a protective activity of *Echinacea* polyphenols against photodamage of the skin. Echinacoside and other phenylpropanoid glycosides from *Pedicularis* are able to protect against oxidative hemolysis *in vitro* (Li et al., 1993).

3.3. LIPOPHILIC CONSTITUENTS

The lipophilic constituents of *Echinacea* consist mainly of essential oil compounds, alkamides and/or ketoalkenyns (Figures 3.6 and 3.7). There are striking differences in the constituents between the roots of *E. angustifolia* and *purpurea* (alkamides) and *E. pallida* (ketoalkenyns). The lipophilic constituents of *E. pallida* roots have been identified mainly as ketoalkenes and ketoalkyns with a carbonyl group in the 2-position, namely, tetradeca-8Z-ene-11,13-diyn-2-one, pentadeca-8Z-ene-11,13-diyn-2-one, pentadeca-8Z,13Z-diene-11-yn-2-one, pentadeca-8Z,11Z,13E-triene-2-one, pentadeca-8Z,11E, 13Z-triene-2-one and pentadeca-8Z,11Z-diene-2-one (Schulte et al., 1967; Khan, 1987; Bauer et al., 1988c). They are not abundant in *E. angustifolia* and *E. purpurea* roots and so are suitable as markers for the identification of *E. pallida* roots.

Autoxidation of these compounds has been observed when roots are stored

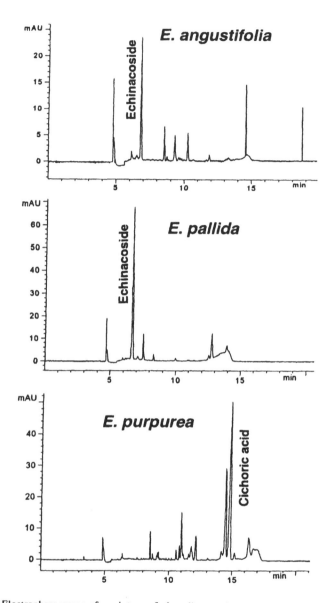

Figure 3.5 Electropherograms of a mixture of phenolic constituents in *Echinacea* root extracts separated by MEKC analysis (Pietta et al., 1998).

Separation parameters: 3DCE system with diode array detector (Hewlett-Packard, Waldbronn, Germany), uncoated fused-silica capillary 58 cm (50 cm to the detector) × 50 μm i d ; 3D extended light path (bubble cell) from Hewlett-Packard, running buffer: 25 mM tetraborate, pH 8 6, containing 30 mM SDS; injections by positive pressure (50–200 mbar × sec corresponding to about 1–4 nL); voltage: + 20 kV; temperature. 30°C; detection 320 nm.

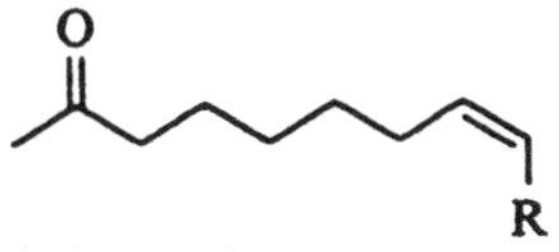

Tetradeca-8Z-ene-11,13-diyn-2-one   R =

Pentadeca-8Z-ene-11,13-diyn-2-one   R =

Pentadeca-8Z,13Z-diene-11-yn-2-one   R =

Pentadeca-8Z,11Z,13E-triene-2-one   R =

Pentadeca-8Z,11E,13Z-triene-2-one   R =

Pentadeca-8Z,11Z-diene-2-one   R =

Heptadeca-8Z,11Z-diene-2-one   R =

Pentadeca-8Z-ene-2-one   R =

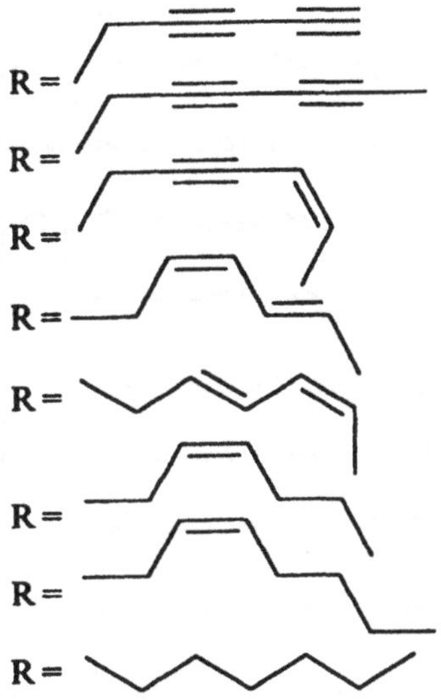

Artifacts:

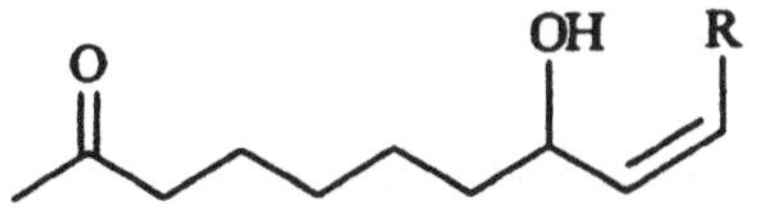

8-Hydroxytetradeca-9E-ene-11,13-diyn-2-one   R =

8-Hydroxypentadeca-9E-ene-11,13-diyn-2-one   R =

8-Hydroxypentadeca-9E,13Z-diene-11-yn-2-one   R =

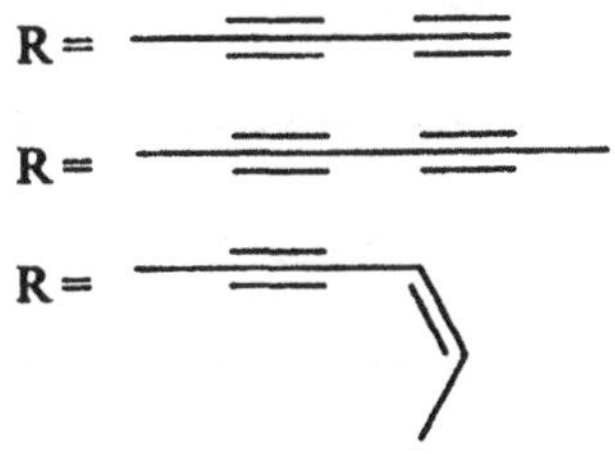

Figure 3.6 Ketoalkenes and ketoalkynes in *E. pallida* roots.

in powdered form. In powdered bulk material, the hydroxylated artifacts, 8-hydroxy-tetradeca-9*E*-ene-11,13-diyn-2-one, 8-hydroxy-pentadeca-9*E*-ene-11,13-diyn-2-one and 8-hydroxypentadeca-9*E*,13*Z*-diene-11-yn-2-one are dominating, often with only small residual quantities of the native compounds (Bauer et al., 1987a). Therefore, the roots are best stored whole.

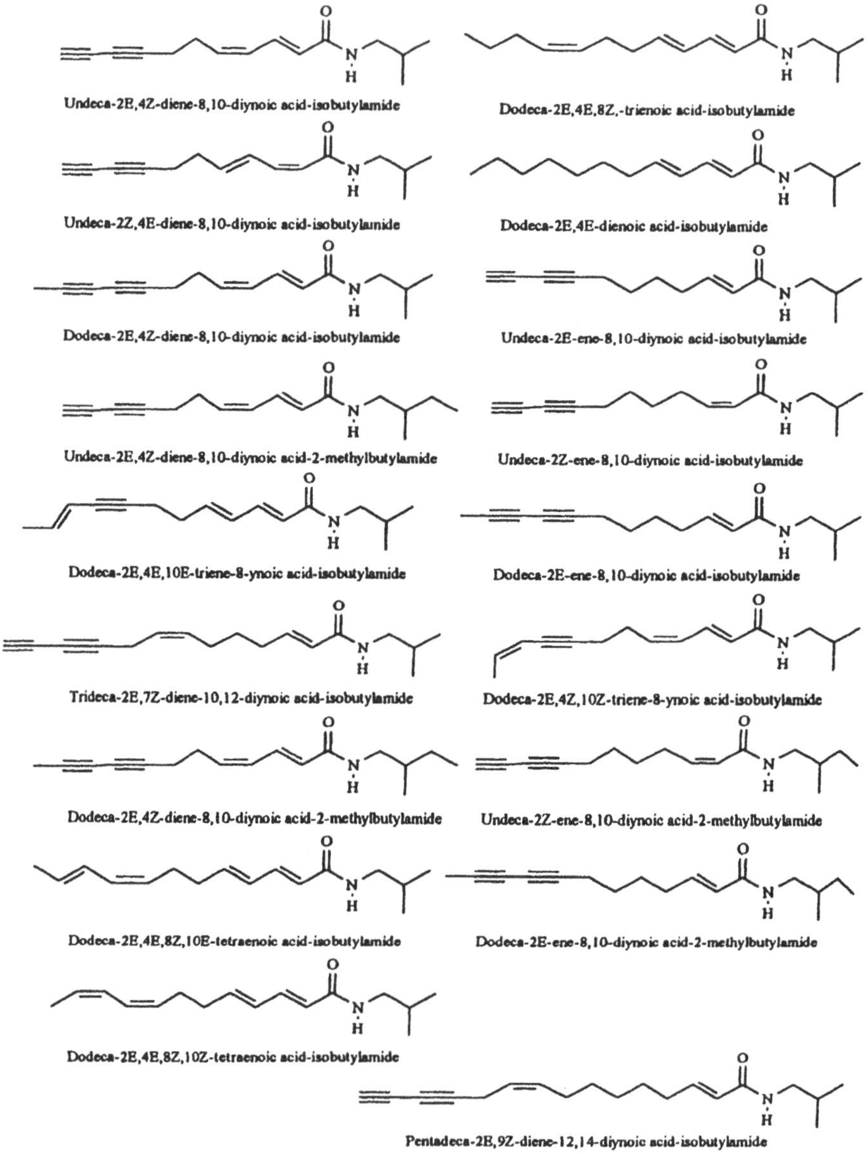

Undeca-2E,4Z-diene-8,10-diynoic acid-isobutylamide

Undeca-2Z,4E-diene-8,10-diynoic acid-isobutylamide

Dodeca-2E,4Z-diene-8,10-diynoic acid-isobutylamide

Undeca-2E,4Z-diene-8,10-diynoic acid-2-methylbutylamide

Dodeca-2E,4E,10E-triene-8-ynoic acid-isobutylamide

Trideca-2E,7Z-diene-10,12-diynoic acid-isobutylamide

Dodeca-2E,4Z-diene-8,10-diynoic acid-2-methylbutylamide

Dodeca-2E,4E,8Z,10E-tetraenoic acid-isobutylamide

Dodeca-2E,4E,8Z,10Z-tetraenoic acid-isobutylamide

Dodeca-2E,4E,8Z,-trienoic acid-isobutylamide

Dodeca-2E,4E-dienoic acid-isobutylamide

Undeca-2E-ene-8,10-diynoic acid-isobutylamide

Undeca-2Z-ene-8,10-diynoic acid-isobutylamide

Dodeca-2E-ene-8,10-diynoic acid-isobutylamide

Dodeca-2E,4Z,10Z-triene-8-ynoic acid-isobutylamide

Undeca-2Z-ene-8,10-diynoic acid-2-methylbutylamide

Dodeca-2E-ene-8,10-diynoic acid-2-methylbutylamide

Pentadeca-2E,9Z-diene-12,14-diynoic acid-isobutylamide

Figure 3.7 Alkamides in *Echinacea* species

56

About 15 alkamides have been identified as major lipophilic constituents of *E. angustifolia* roots (Bohlmann and Grenz, 1966; Jacobson, 1967; Bauer et al., 1989a). They are derived mainly from undeca- and dodecanoic acid and differ in the degree of unsaturation and the configuration of the double bonds. The main structural type is a 2-monoene-8,10-dienoic acid isobutylamide, but some 2′-methyl-butylamides have also been found. The major compounds are the isomeric mixture of dodeca-2*E*,4*E*,8*Z*,10*E*,*Z*-tetraenoic acid isobutylamides.

In *E. purpurea* roots, ca. 11 alkamides have been identified (Bohlmann and Grenz, 1966; Bauer et al., 1988a). In contrast to those of *E. angustifolia,* most of these alkamides possess a 2,4-diene moiety. Therefore, *E. purpurea* and *E. angustifolia* can also be clearly distinguished by their lipophilic constituents (Bauer et al., 1988b; Bauer and Remiger, 1989b).

The aerial parts of all three *Echinacea* species contain alkamides of the type found in *E. purpurea* roots (Bohlmann and Hoffmann, 1983; Bauer et al. 1988a) and differ only in the concentration of the constituents (*E. purpurea* > *E. pallida* > *E. angustifolia*).

HPLC analysis with photodiode array detection is an excellent method for the analysis of alkamides, because the different types can be identified from their UV spectra (Bauer and Remiger, 1989b; Tobler et al., 1994; Perry et al., 1997). LC-MS methods for alkamide analysis have also been recently published (He et al., 1998).

Small amounts of alkamides can also be found in the expressed juice of fresh *E. purpurea* aerial parts (Bauer, 1997). Considerable variation in alkamides exists among different products in the market and even between batches of the same product (Bauer, 1997). Seasonal variation of the alkamide content, low at the beginning of the growing season and high in mid August (Figure 3.4) (Bauer and Vom Hagen-Plettenberg, 1999), and different levels of alkamides in the various parts of the plant may account for this variation. Alkamides accumulate especially in the flowerheads and in the tubulous flowers and achenes (Figure 3.3) (Bauer et al., 1988a; Giger, 1990; Bauer and Vom Hagen-Plettenberg, 1999). Therefore, the date and mode of harvest play an important role in the quality of *Echinacea* preparations.

Purified alkamide fractions from *E. purpurea* and *E. angustifolia* roots enhance phagocytosis in the Carbon-Clearance-Test by a factor of 1.5 to 1.7 (Bauer et al., 1989b) and, hence, contribute to the immunostimulatory activity of *Echinacea* tinctures. Since the main constituent, dodecatetraenoic acid-isobutylamide, exhibits only weak activity, the most effective constituents remain to be found. Alkamides also inhibit 5-lipoxygenase and cyclooxygenase, and therefore, may also be an anti-inflammatory principle of *Echinacea* extracts (Müller-Jakic et al., 1994).

Flowering aerial parts of *E. purpurea* contain less than 0.1% essential oil (Kuhn, 1939; Neugebauer, 1949; Schindler, 1953; Heinzer et al., 1988) that consists of borneol, bornyl acetate, pentadeca-8-en-2-one, germacrene D, caryophyllene, caryophyllene epoxide and palmitic acid (Bos et al., 1988). *E. angustifolia* and *E. pallida* contain identical constituents, and discrimination via the essential oil is therefore difficult.

The germacrene alcohol, isolated by Bauer et al. (1988e) from the aerial parts of *E. purpurea*, is also a component of the essential oil. It has not been detected in the dried drug, and HPLC analysis showed that it is a characteristic indicator for fresh plant extracts and is regularly present as a major component in homeopathic tinctures (Remiger, 1989).

Several constituents, like the alkamides, cichoric acid, glycoproteins and polysaccharides, may contribute to the immunostimulatory activity of *Echinacea* extracts. Therefore, the application of extracts appears reasonable. However, their standardization is necessary in order to generate reproducible products with reliable activity.

4. PHARMACOLOGICAL EFFECTS OF ECHINACEA EXTRACTS

The influence of alcoholic extracts of *E. purpurea* roots was studied by Vömel (1985) on the activity of Kupffer cells isolated from perfused rat liver. It was shown that the phagocytosis of erythrocytes was significantly improved and influenced phagocytosis-dependent metabolism.

When the alcoholic extracts obtained from aerial parts and roots were tested for phagocytosis-stimulating activity, the lipophilic fractions showed the highest activity, indicating that the nonpolar constituents may also represent an active principle (Bauer et al., 1988d, 1989b). Ethanolic extracts of aerial parts of *E. angustifolia* and *E. purpurea* showed immunomodulatory activity on the phagocytic, metabolic and bactericidal activities of peritoneal macrophages in mice (Bukovsky et al., 1993a, 1993b, 1995).

Extracts of *E. purpurea* and *Panax ginseng* were evaluated for stimulatory effects on cellular immune function by peripheral blood mononuclear cells (PBMC) from normal individuals and patients with either the chronic fatigue or the acquired immunodeficiency syndrome. PBMC isolated on a Ficoll-hypaque density gradient were tested in the presence or absence of varying concentrations of each extract for natural killer (NK) cell activity versus K562 cells and antibody-dependent cellular cytotoxicity (ADCC) against human herpes virus 6 infected H9 cells. Both *Echinacea* and *Ginseng*, at concentrations ≥ 0.1 or 10 μg/kg, respectively, significantly enhanced NK-function of all groups. Similarly, the addition of either herb significantly increased ADCC of PBMC from all subject groups (See et al., 1997).

In a double-blind study with 24 healthy humans, an ethanolic extract (1:5) of *E. purpurea* roots was tested for phagocytosis stimulatory effects *ex vivo*. The extract or a placebo was administered orally at a dose of 3×30 drops daily for five days, representing about 1 mg cichoric acid and alkamides per day. Granulocyte phagocytosis was measured by the modified Brandt test. Maximal stimulation with 120% of the starting value was found at day five. After discontinuation of the extract, phagocytosis activity decreased to a normal level within three days. The other immune parameter (EG) used to monitor the course of immunoactivity, as well as leukocytes and the BKS values, remained in normal biological ranges. Tolerance was good in all cases. Only two probands showed a slight temperature increase of about 0.5°C (Jurcic et al., 1989).

An ethanolic extract of *E. purpurea* roots displayed a dose-dependant inhibition of the collagen gel contraction in collagen lattices populated with C3H10T1/2 fibroblasts. As time elapsed between preparation of gel and addition of extract increased, inhibition of elongation of fibroblasts and of processes leading to collagen linking decreased. Addition of the extract one hour after gel preparation showed no effect (Zoutewelle and Van Wijk, 1990).

An alcoholic extract (30% ethanol) from the roots of *E. pallida*, standardized on glycoproteins/polysaccharides by an ELISA method, was shown to enhance production of Ig M, IL1, IL6, TNFα and IFNα,β in NMRI- and C3H/HeJ mouse spleen cell lines and increased IL1 concentration in the serum of mice *in vivo* (Beuscher et al., 1995).

Mice treated for five days with the ethanolic extract of the aerial parts of *E. purpurea* showed immunostimulatory effects on the phagocytic, metabolic and bactericidal activities of peritoneal macrophages and increases in total weight of the spleens when activity was tested (Bukovsky et al., 1993, 1995).

An extract prepared with 90% ethanol from fresh plants (final ethanol concentration 65%) significantly inhibited the contraction of collagen seeded with C3H10T1/2 fibroblasts. A corresponding amount of ethanol showed no effect. The elongation of fibroblasts and the processes leading to the linking of collagen were inhibited, depending on the time of addition of the extract. No effect was observed when the extract was added one hour after starting the collagen linking that led the authors to conclude an influence on wound healing (Zoutewelle and Van Wijk, 1990).

Most pharmacological investigations with the aerial parts of *E. purpurea* have been performed with a preparation containing the expressed juice of the fresh plant material (Echinacin®), which has been on the market in Germany for over 50 years. This preparation of *E. purpurea* on phagocytosis in granulocytes was measured by a chemiluminescence method, and the reaction of the granulocytes was found to depend on the doses and methods of application. This led Gaisbauer et al. (1990) to conclude that standardized methods and investigations of various immunoparameters are necessary to prove the immunostimulatory effects of such preparations.

Dilutions of expressed juice (1:5 and 1:10) (Myo-Echinacin® 5%) improved granulocyte phagocytosis of yeast cells *in vitro* at the same intensity as identical dilutions of Intraglobin F (Tympner, 1981; Fanselow, 1981). A lyophylizate of the expressed juice of *E. purpurea* at the concentration of 5 mg/mL increased the number of phagocyting human granulocytes from 79% to 95% (p ≤ 0.001) and significantly stimulated phagocytosis of yeast cells by more than 50% (p ≤ 0.01). At the highest tested dose of 12.5 mg/mL, the number of phagocyting granulocytes and the phagocytosis index decreased (Stotzem et al., 1992).

A preparation of *E. purpurea* expressed juice increased the *in vitro* phagocytosis of *Candida albicans* by granulocytes and monocytes from healthy donors by 30–45%. The chemotactic migration of granulocytes in the Boyden Chamber increased by 45%. However, the preparation had no effect on intracellular killing of bacteria or yeasts or *in vitro* transformation of lymphocytes (Wildfeuer and Mayerhofer, 1994).

Commercial preparations of *Echinacea* fresh pressed juice and dried juice at concentrations ranging from 0.012 to 10 μg/mL were compared to endotoxin stimulated and unstimulated controls. Cytokine production measured by ELISA after 18 h of incubation for IL-1 and 36 and 72 h for TNF-α, IL-6 and IL-10 showed that macrophages cultured in *Echinacea* pressed juice at low concentrations (0.012 μg/mL) produced significantly higher levels of IL-1, TNF-α, IL-6 and IL-10 than unstimulated cells. The high levels of IL-1, TNF-α and IL-10 induced by very low levels of *Echinacea* pressed juice are consistent with an immune activated antiviral effect. *Echinacea* pressed juice induced lower levels of IL-6 in comparison to the other measured cytokines (Burger et al., 1997).

Recently, a topical microbiocide, consisting of a blend of benzalkonium chloride and an extract from *E. purpurea* was shown to have good antiviral activity against resistant and susceptible strains of herpes simplex virus, HSV-1 and HSV-2 (Thompson, 1998).

5. CLINICAL STUDIES

Although many case reports on the clinical effectiveness of *Echinacea* preparations have been published (see reviews by Lloyd, 1917; Madaus, 1939; Hahn and Mayer, 1984; Bauer and Wagner, 1990; Bauer, 1994), only few controlled clinical studies have been performed to demonstrate the therapeutic value of *Echinacea* preparations. The latter have recently been reviewed by Melchart and Linde (1998).

A placebo-controlled double-blind study was performed to demonstrate the therapeutic effectiveness of an alcoholic extract of *E. purpurea* roots on 180 patients with the common cold (Bräunig et al., 1992). Symptoms such as irritated nose, frontal headache, lymph node swelling, coated tongue and rale were scored at the beginning of the study and after three to four and eight to

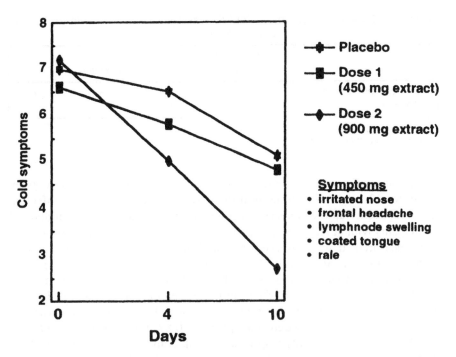

Figure 3.8 Placebo-controlled double-blind study with an alcoholic extract of *E. purpurea* roots on 180 patients with the common cold (Bräunig et al., 1992).

ten days. The group receiving a dose equivalent to 900 mg root/day (3 × 60 drops) had significantly reduced symptoms compared to those receiving the placebo and to a group treated with 450 mg/day (3 × 30 drops) (Figure 3.8).

A similar placebo-controlled double-blind study with 160 patients has been performed with an aqueous alcoholic tincture (1:5) of *E. pallida* roots (Bräunig and Knick, 1993; Dorn et al., 1997). The group of patients treated with a dose of 90 drops *E. pallida* tincture (equivalent to 900 mg roots) recovered significantly faster from infections of the upper respiratory tract than the placebo group (Table 3.2).

The efficacy of an anti-cold remedy including vitamin C and *E. purpurea* root extract in the treatment of the common cold was investigated in a random-ized, single-blind and placebo-controlled study with 32 subjects (17 male and 15 female, aged between 18 and 71 years). The length of the common cold was 3.37 days for treated subjects and 4.37 days for those given the placebo (p < 0.01) (Scaglione and Lund, 1995).

A double-blind, placebo-controlled randomized trial with 302 healthy volun-teers has recently been performed with *Echinacea* root extracts for the preven-tion of upper respiratory tract infections (Melchart et al., 1998). Ethanolic

TABLE 3.2.   Duration of Common Cold (Days) After Treatment with an Extract of *E. pallida* Roots.

	Bacterial Infection	n	Viral Infection	n
Extr *E pallidae* radix	9 8	10	9 1	70
Placebo	13 0	36	12.9	44

* n = Duration of illness in days
Source Bräunig and Knick (1993)

extracts from *E. purpurea* roots, *E. angustifolia* roots or placebo, were given orally for 12 weeks. The time until the first upper respiratory tract infection (time to event) and secondary outcome measures such as the number of participants with at least one infection, global assessment and adverse effects were measured. It was observed that the time of the first upper respiratory tract infection was 66 days (95% confidence interval [CI], 61–72 days) in the *E. angustifolia* group, 69 days (95% CI, 64–74 days) in the *E. purpurea* group and 65 days (95% CI, 59–70 days) in the placebo group (P = 0.49) that had a 36.7% infection rate. In the treatment groups, 32% in the *E. angustifolia* group (relative risk compared with placebo, 0.87; 95% CI, 0.59–1.30) and 29.3% in the *E. purpurea* group (relative risk compared with placebo, 0.80; 95% CI, 0.53–1.31) had an infection. It was concluded that the investigated *Echinacea* extracts were not effective as a prophylactic and that a larger study would be needed to prove this effect (Melchart et al., 1998). From a practical standpoint, neither the ethanolic extract from *Echinacea purpurea* roots, nor that from *E. angustifolia* roots represent a commercial extract or product on the market.

Clinical investigations on the immunomodulatory effects also exist with a combination of *Echinacea, Baptisia* and *Thuja* extracts (Esberitox®). In a placebo-controlled double-blind study, Vorberg (1984) investigated the improvement of symptoms of febrile respiratory tract infections of 100 patients. Half the number of patients were treated with tablets containing a hydroalcoholic extract from *E. purpurea* and *E. pallida* roots, *Baptisia tinctoria* roots and *Thuja occidentalis* twig tips, the other half was treated with placebo. In the treated group, the main symptoms like weariness, catarrh and sore throat, improved significantly after three days (Vorberg, 1984). The immunomodulatory effects were also confirmed in another, more recent study (Henneicke-von Zeppelin et al., 1997).

The effect of adjuvant application of *E. purpurea* expressed juice on the recurrence of vaginal *Candida* infections within six months was tested in a comparative study (Table 3.3). Econazol nitrate was used as a local therapeutic agent over six days. The recurrence frequency in the group treated with econazol nitrate was only 60.5%. This decreased to 5–16% (depending on the

TABLE 3.3. Adjuvant Treatment of Recurrent *Candida* Infections with *E. purpurea* Expressed Juice Preparation (Echinacin®). Rate of Recurrence Within Six Months.

Therapeutic Scheme	Number of Patients	Recurrences	Frequency of Recurrence (%)
Antimycotic, topically*	43	26	60 5
Antimycotic + Echinacin-amp s.c.	20	3	15
Antimycotic + Echinacin-amp i m.	60	3	5
Antimycotic + Echinacin-amp i.v	20	3	15
Antimycotic + Echinacin liquidum	60	10	16.7

* Econazol nitrate
Source Coeugniet and Kühnast (1986)

mode of application) after adjuvant application of *E. purpurea* expressed juice (Coeugniet and Kühnast, 1986).

In a retrospective study with 1,280 children with acute bronchitis, treatment with *E. purpurea* expressed juice prompted a faster healing process than treatment with antibiotics (Table 3.4) (Baetgen, 1988). This fast process may be explained by the fact that acute bronchitis is mostly due to a viral infection against which antibiotics have no effect, while *Echinacea* extracts may be active via stimulation of the unspecific immune system.

Recently, a randomized, double-blind, placebo-controlled, single center clinical trial was carried out to investigate the therapeutic efficacy of an expressed juice preparation of *Echinacea purpurea* aerial parts in 120 patients with initial symptoms of acute, uncomplicated upper airways infection (Hoheisel et al., 1997). An intention-to-treat analysis revealed that 24/60 patients (40%) in the treated group, but 36/60 (60%) in the placebo group, experienced a "real" cold, i.e., fully expressed disease ($p = 0.04$). In the subgroup of patients with a "real" cold, the median time for improvement was four days (treatment, $n = 24$) and eight days (placebo, $n = 36$), respectively (Figure 3.9).

In another recent double-blind study, 42 triathletes received 8 mL of an expressed juice preparation from *E. purpurea* aerial parts. After four weeks, the number of immunocompetent cells slightly increased in the treated group. After a competition, which led to a temporary immunosuppression, a significant increase of interleukin 6 was observed in the urine of only the treated group. The concentration of soluble IL-2R receptors was significantly reduced. Two months after the competition, common colds were registered only in the placebo group (Berg et al., 1998).

Studies on the effectiveness of *Echinacea* preparations in the adjuvant

TABLE 3.4. Retrospective Study of Children with Acute Bronchitis Treated with *E. purpurea* Expressed Juice (Echinacin®) and/or Antibiotics.

Group of Treatment	Improvement (%) Within			No Improvement	Deterioration	Recurrence
	5 Days	10 Days	>10 Days			
Echinacin exclusively (n = 468)	45.7	87.2	93.4	5.6	0.4	0.6
Echinacin + antibiotic (n = 330)	25.5	82.4	97.9	1.2	0.9	—
Antibiotic exclusively (n = 482)	16.0	72.0	96.0	2.3	—	1.7

Source: Baetgen (1988).

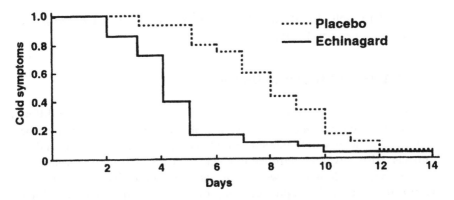

Figure 3.9 Randomized, double-blind, placebo-controlled, single-center clinical trial on the therapeutic efficacy of an expressed juice preparation of *E. purpurea* aerial parts in 120 patients with initial symptoms of acute, uncomplicated upper airways infection. Time course of improvement in the subgroup of patients with a "real" cold Source: O. Hoheisel, M Sandberg, S. Bertram, M. Bulitta and M. Schäfer (1997).

treatment of cancer (Lersch et al., 1990, 1992, 1994; Nagel, 1995), have been carried out, but results have not been significant.

Melchart et al. (1994), reviewing the clinical studies performed with *Echinacea* preparations up to 1994, concluded that *Echinacea* preparations can be effective for the enhancement of the body's defense mechanisms, but further investigations are necessary to find the best dosage and application form.

6. SIDE EFFECTS AND TOXICOLOGICAL CONSIDERATIONS

Only few data exist on side effects and toxicological risks of *Echinacea*. All clinical reports provide indications of good tolerance. However, allergy to plants of the Compositae family can be a contraindication. Few cases of intolerance and anaphylactic reactions to *Echinacea* have been reported (Becker et al., 1988; Mullins, 1998; Myers and Wohlmuth, 1998). This problem has to be seriously watched.

Echinacea preparations, as any other immunomodulators, should not be used in cases of progredient systemic diseases like tuberculosis, leukoses, collagenoses, multiple sclerosis, AIDS, HIV infection and other autoimmune diseases (Shohan, 1985). In the case of diabetes, the metabolic status may worsen.

Mutagenicity studies have indicated no tumor initiating properties (Mengs et al., 1991; Schimmer et al., 1989). The acute toxicity of *E. purpurea* root extract and aerial parts expressed juice is extremely low (Mengs et al., 1991;

Bauer and Liersch, 1993), and the expressed juice of *E. purpurea* is well-tolerated in long-term treatment (Parnham, 1996).

7. SUMMARY AND CONCLUSION

In many studies, several constituents have been identified in the different medicinally used *Echinacea* species, and pharmacological assays have been performed to demonstrate immunostimulatory properties of extracts and constituents. Also, several controlled clinical studies have been performed. As a consequence, in Germany, two *Echinacea* preparations have been approved as effective drugs: the expressed juice of *E. purpurea* aerial parts for the internal adjuvant therapy of relapsing infections of the respiratory and derivative urinary tract and for external therapy of poorly healing superficial wounds; and tinctures of *E. pallida* roots for adjuvant therapy of common cold-like infections. Extracts of *E. purpurea* roots have not yet been accepted but could also be effective in the treatment of the common cold (Kommission E, 1989, 1992; Blumenthal, 1998). It is still necessary to confirm the efficacy of *Echinacea* preparations by further clinical studies, to perform investigations for the optimum dose and to elucidate the molecular mode of action.

8. REFERENCES

Alhorn, R. 1992. Phytochemische und vegetationsperiodische Untersuchungen von *Echina-cea purpurea* (L.) MOENCH unter Berücksichtigung der Kaffeesäurederivate. PhD thesis, Universität Marburg/Lahn

Baetgen, D. 1988 Behandlung der akuten Bronchitis im Kindesalter *TW Paediatrie*. 1.66–70.

Bauer, R. 1994. *Echinacea*—Eine Arzneidroge auf dem Weg zum rationalen Phyto-therapeutikon. *Dtsch. Apoth. Ztg*. 134·94–103.

Bauer, R. 1997. Standardisierung von *Echinacea purpurea*-Preßsaft auf Cichoriensäure und Alkamide. *Z. Phytother*. 18:270–276.

Bauer, R. 1998a. Chemistry, analysis and immunological investigations of *Echinacea* phytopharmaceuticals. In. *Immunomodulatory Agents from Plants* (Ed. H. Wagner), pp. 41–88. Birkhäuser Verlag, Basel

Bauer, R. 1998b. The *Echinacea* story—The scientific development of an herbal immunostimulant. In. *Plants for Food and Medicine—Modern Treatments and Traditional Remedies* (Eds. H. D. V. Prendergast, N L. Etkin, D. R. Harris, P. J. Houghton), p. 317–332, The Royal Botanic Gardens, Kew.

Bauer, R. and Liersch, R. 1993. *Echinacea*. In. *Hagers Handbuch der Pharmazeutischen Praxis, Vol. 5, Drogen E—O* (Eds. R. Hänsel, K. Keller, H Rimpler, G. Schneider), pp. 1–34, Springer Verlag, Berlin, Heidelberg, New York.

Bauer, R and Remiger, P. 1989a Der Einsatz der HPLC bei der Standardisierung von *Echinacea*-Drogen. *Arch. Pharm*. 322, 324

Bauer, R. and Remiger, P. 1989b. TLC and HPLC analysis of alkamides in *Echinacea* drugs. *Planta Med* 55:367–371.

Bauer, R and Vom Hagen-Plettenberg, F 1999. Der Einfluß von Erntezeitpunkt und Blütenanteil auf die Qualität von Echinacea purpurea-Frischpflanzen-Preßsäften In *Fachtagung Arznei- und Gewürzpflanzen* (Eds R Marguard and E. Schubert), pp 93–100, Fachverlag Köhler, Giessen, Germany

Bauer, R. and Wagner, H. 1990 *Echinacea*—Ein Handbuch für Ärzte, Apotheker und andere Naturwissenschaftler. Wissenschaftliche Verlagsgesellschaft, Stuttgart

Bauer, R and Wagner, H. 1991. *Echinacea* species as potential immunostimulatory drugs. In *Economic and Medicinal Plant Research, Vol V*, (Eds. H. Wagner, N R Farnsworth), pp 253–321, Academic Press, London

Bauer, R , Khan, I. A., Lotter, H , Wagner, H and Wray, V. 1985. Structure and stereochemistry of new sesquiterpene esters from *Echinacea purpurea* (L) MOENCH, *Helv. Chim. Acta.* 68:2355–2358

Bauer, R , Khan, I A., Wray, V and Wagner, H 1987a Two acetylenic compounds from *Echinacea pallida* roots *Phytochemistry.* 26.1198–1200

Bauer, R., Khan, I. A and Wagner, H. 1987b *Echinacea*—Nachweis einer Verfälschung von *Echinacea purpurea* (L) Moench mit *Parthenium integrifolium* L. *Dtsch. Apoth Ztg* 127·1325–1330

Bauer, R., Remiger, P. and Wagner, H. 1988a. *Echinacea*—Vergleichende DC- und HPLC-Analyse der Herba-Drogen von *Echinacea purpurea*, E. pallida und *E. angustifolia. Dtsch. Apoth Ztg.* 128 174–180.

Bauer, R., Remiger, P and Wagner, H 1988b Alkamides from the roots of *Echinacea purpurea.* Phytochemistry. 27 2339–2342.

Bauer, R , Khan, I A. and Wagner, H. 1988c. TLC and HPLC analysis of *Echinacea pallida* and *E. angustifolia* roots *Planta Med.* 54:426–430

Bauer, R., Jurcic, K , Puhlmann, J. and Wagner, H. 1988d. Immunologische *In-vivo-* und *In-vitro*-Untersuchungen mit *Echinacea*-Extrakten *Arzneim-Forsch.* 38·276–281.

Bauer, R., Remiger, P , Wray, V. and Wagner, H. 1988e. A germacrene alcohol from fresh aerial parts of *Echinacea purpurea.* Planta Med 54.478–479

Bauer, R , Remiger, P and Wagner, H. 1989a Alkamides from the roots of *Echinacea angustifolia.* Phytochemistry 28:505–508.

Bauer, R., Remiger, P , Jurcic, K. and Wagner, H. 1989b. Beeinflussung der Phagozytose-Aktivität durch *Echinacea*-Extrakte. *Z. Phytother* 10·43–48.

Becker, H. and Hsieh, W.-C 1985 Chicoree-Säure und deren Derivate aus *Echinacea*-Arten. *Z. Naturforsch* 40c 585–587

Becker, H., Hsieh, W.-C , Wylde, R., Laffite, C., and Andary, C. 1982. Struktur von Echinacosid. *Z. Naturforsch.* 37:351–353

Becker, K. P., Ditter, B., Nimsky, C., Urbaschek, R and Urbaschek, B. 1988. Endotoxin contents of phytopharmaceuticals. Correlation with clinically observed side effects. *Dtsch. Med. Wschr.* 3:83–87.

Berg, A , Northoff, H., König, D., Weinstock, C., Grathwohl, D , Parnham, M J., Stuhlfauth, I. and Keul, J. 1998 Influence of Echinacin (EC31) treatment on the exercise-induced immune response in athletes. *J. Clin Res.* 1 367–380.

Beuscher, N., Kopanski, L. and Ernwein, C. 1987. Modulation der Immunantwort durch polymere Substanzen aus *Baptisia tinctoria* und *Echinacea angustifolia.* Adv. Biosci. 68:329–336.

Beuscher, N., Bodinet, C., Willigmann, I., and Egert, D. 1995. Immunmodulierende Eigenschaften von Wurzelextrakten verschiedener *Echinacea*-Arten. *Z. Phytother.* 16:157–166

Blaschek, W , Döll, M., and Franz, G. 1998. *Echinacea* Polysaccharide—Analytische Untersuchungen an Preßsaft und am Fertigarzneimittel Echinacin. *Z. Phytother.* 19.255–262.

Blumenthal, M. 1998. *The Complete German Commission E Monographs.* American Botanical Council, Austin, Texas.

Bodinet, C. and Beuscher, N. 1991. Antiviral and immunological activity of glycopoteins from *Echinacea purpurea* radix. *Planta Med* 57 Suppl. 2:A33–A34.

Bohlmann, F. and Grenz, M. 1966. Über die Inhaltsstoffe aus *Echinacea*-Arten. *Chem Ber.* 99.3197–3200

Bohlmann, F. and Hoffmann, H. 1983. Further amides from *Echinacea purpurea.* Phytochemistry. 22:1173–1175.

Bos, R., Heinzer, F. and Bauer, R. 1988. Volatile Constituents of the Leaves of *Echinacea purpurea, E. pallida* and *E. angustifolia.* Poster at the 19th International Symposium on Essential Oils and Other Natural Substrates, 7.10.9.1988 in Zürich.

Bräunig, B. and Knick, E. 1993. Therapeutische Erfahrungen mit *Echinacea pallida* bei grippalen Infekten. *Naturheilpraxis,* 1:72–75.

Bräunig, B., Dorn, M. and Knick, E. 1992 *Echinacea*e purpureae radix: Zur Stärkung der körpereigenen Abwehr bei grippalen Infekten. *Z. Phytother.* 13:7–13.

Brevoort, P. 1998. The booming U.S. botanical market. *Herbalgram.* 44:33–46.

Bukovsky, M., Vaverkova, S., Kostalova, D. and Magnusova, R 1993a. Immuno-modulating activity of ethanol-water extracts of the roots of *Echinacea* gloriosa L., *Echinacea angustifolia* DC. and *Rudbeckia* speciosa Wenderoth tested on the immune system in C57BL6 inbred mice (in slovak. lang) *Cesk. Farm.* 42:184–187.

Bukovsky, M., Kostalova, D., Magnusova, R , and Vaverkova, S. 1993b. Testing for immunomodulating effects of ethanol-water extracts of the above-ground parts of the plants *Echinacea* (Moench) and *Rudbeckia* L. *Cesk-Farm* 42:228–231.

Bukovsky, M., Vaverkova, S. and Kostalova, D 1995. Immunomodulating activity of *Echinacea* gloriosa L., *Echinacea angustifolia* DC. and *Rudbeckia* speciosa Wenderoth ethanol-water extracts. *Pol. J. Pharmacol.* 47:175–177.

Burger, R. A., Torres, A. R., Warren, R. P., Caldwell, V. D. and Hughes, B. G. 1997. *Echinacea*-induced cytokine production by human macrophages. *Int. J Immunopharmacol.* 19:371–379.

Cheminat, A., Zawatzky, R., Becker, H. and Brouillard, R. 1988. Caffeoyl conjugates from *Echinacea* species: Structures and biological activity. *Phytochemistry.* 27:2787–2794.

Coeugniet, E. G. and Kühnast, R. 1986. Adjuvante Immuntherapie mit verschiedenen Echinacin®-Darreichungsformen. *Therapiewoche,* 36:3352–3358.

Dorn, M., Knick, E and Lewith, G. 1997. Placebo-controlled, double-blind study of *Echinacea*e *pallida* radix in upper respiratory tract infections. *Compl. Ther. Med.* 3:40–42.

Egert, D. and Beuscher, N. 1992. Studies on antigen specificity of immunoreactive arabinogalactan proteins extracted from *Baptisia tinctoria* and *Echinacea purpurea.* Planta Med. 58:163–165.

Emmendörffer, A. C., Wagner, H. and Lohmann-Matthes, M.-L. 1998. Immunologically active polysaccharides from *Echinacea purpurea* plant and cell cultures. In: *Immunomodulatory Agents from Plants* (Ed. H. Wagner), pp. 89–104, Birkhäuser Verlag, Basel.

Facino, R. M., Sparatore, A., Carini, M., Gioia, B., Arlandini, E. and Franzoi, L. 1991. Field desorption mass spectrometry, fast atom bombardment mass spectrometry and fast atom bombardment tandem mass spectrometry of echinacoside, the main caffeoyl-glycoside from *Echinacea angustifolia* roots (Asteraceae). *Org. Mass Spectrom* 26:951–955

Facino, R. M., Carini, M., Aldini, C., Marinello, C., Arlandini, E., Franzoi, L., Colombo, M., Pietta, P. and Mauri, P. 1993. Direct characterization of caffeoyl esters with antihyaluronidase activity in crude extracts from *Echinacea angustifolia* roots by fast atom bombardment tandem mass spectrometry. *Farmaco* 48:1447–1461.

Facino, R M., Carini, M., Aldini, G., Saibene, L., Pietta, P. and Mauri, P. 1995. Echinacoside and caffeoyl conjugates protect collagen from free radical-induced degradation: A potential use of *Echinacea* extracts in the prevention of skin photodamage. *Planta Med.* 61.510–514.

Fanselow. G. 1981. Der Einfluß von Pflanzenextrakten (*Echinacea purpurea*, Aristolochia clematitis) und homöopathischen Medikamenten (Acidum formicicum, Sulfur) auf die Phagozytoseleistung humaner Granulozyten *in vitro*. MD thesis, University, Munich.

Gaisbauer, M , Schleich, T., Stickl, H A. and Wilczek, I. 1990 The effect of *Echinacea purpurea* Moench on phagocytosis in granulocytes measured by chemiluminescence. *Arzneim.-Forsch.* 40:594–598

Giger, E. 1990 *Echinacea purpurea* und *Echinacea angustifolia*—Biomasse, Alkamide und Fructane in Abhängigkeit von Jahreszeit, Alter und Nachbarschaftssituation. PhD thesis, Universität Zürich.

Giger, E , Keller, F. and Baumann, T. W. 1989 Fructans in *Echinacea* and in its phytotherapeutic preparations. *Planta Med.* 55:638–639

Glowniak, K., Zgorka, G. and Kozyra, M. 1996. Solid-phase extraction and reversed-phase high-performance liquid chromatography of free phenolic acids in some *Echinacea* species. *J. Chromatogr.* A 730.25–29.

Gocan, S., Cimpan, G. and Muresan, L. 1996. Automated multiple development thin layer chromatography of some plant extracts. *J. Pharm. Biomed. Anal.* 14:1221–1227.

Grünwald, J. and Büttel, K. 1996. Der europäische Markt für Phytotherapeutika—Zahlen, Trends, analysen *Pharm. Ind* 58 209–214.

Hahn, G. and Mayer, A. 1984 *Echinacea*—Igelkopf oder Sonnenhut. *Österr. Apoth. Ztg.* 38:1040–1046.

He, X.-G., Lin, L.-Z., Bernart, M. W. and Lian, L -Z 1998 Analysis of alkamides in roots and achenes of *Echinacea purpurea* by liquid chromatography-electrospray mass spectrometry. *J. Chromatogr.* A. 815:205–211.

Heinzer, F., Chavanne, M., Meusy, J.-P., Maitre, H.-P., Giger, E. and Baumann, T. W. 1988. Ein Beitrag zur Klassifizierung der therapeutisch verwendeten Arten der Gattung *Echinacea Pharm. Acta Helv.* 63:132–136.

Henneicke-von Zeppelin, H. H., Hentschel, C., Kohnen, R., Köhler, R., and Wüstenberg, P. 1997 Placebokontrollierte, randomisierte, doppelblinde, multizentrische klinische Prüfung zur therapeutischen Wirksamkeit und Sicherheit von Esberitox® N Tabletten bei Patienten mit akutem viralem Atemwegsinfekt (Erkältung). Abstract, 8. Kongreß der Gesellschaft für Phytotherapie in Würzburg.

Hoheisel, O , Sandberg, M., Bertram, S. Bulitta, M. and Schäfer, M. 1997. Echinagard treatment shortens the course of the common cold: a double-blind, placebo-controlled clinical trial. *Eur. J. Clin. Res.* 9·261–269.

Jacobson, M 1967. The structure of echinacein, the insecticidal component of American coneflower roots. *J. Org. Chem.* 32:1646–1647.

Jurcic, K., Melchart, D., Holzmann, M., Martin, P., Bauer, R., Doenicke, A. and Wagner, H. 1989. Zwei Probandenstudien zur Stimulierung der Granulozytenphagozytose durch *Echinacea*-Extrakt-haltige Präparate. *Z. Phytother.* 10:67–70.

Khan, I. A. 1987. *Neue Sesquiterpenester aus Parthenium integrifolium L. und Polyacetylene aus Echinacea pallida NUTT.* PhD thesis, Universität München.

King, J. and Newton, R. S. 1852. The Eclectic Dispensatory of the United States, p. 351, H. W. Derby & Co., Cincinnati, Ohio.

King, P. J. and Robinson, W. E., Jr. 1998. Resistance to the anti-human immunodeficiency virus type 1 compound L-chicoric acid results from a single mutation at amino acid 140 of integrase. *J. Virol.* 72·8420–8424.

Kommission, E. 1989 *Echinaceae* purpureae herba. *BAnz.* 43, p 1070, 02 03.1989

Kommission, E. 1992. *Echinaceae* pallidae radix. *BAnz.* 162, p 7360, 29 08.1992

Kraus, S., Wagner, H and Liptak, A. 1996 Labelling of bioactive polysaccharides for resorption studies. Abstract of the 2nd International Congress on Phytomedicine, Munich

Kuhn, A 1939 Chemistry of *Echinacea*. In· *Echinacea purpurea* Moench (Ed. G. Madaus), Med. Biol. Schriftenreihe, Vol. 13, Verlag Rohrmoser, Radebeul/Dresden

Lersch, C., Zeuner, M , Bauer, A., Siebenrock, K., Hart., R , Wagner, F , Fink, U , Dancygier, H. and Classen, M. 1990. Stimulation of the immune response in outpatients with hepatocellular carcinomas by low doses of cyclophosphamide (LDCY), *Echinacea purpurea* extracts (Echinacin) and thymostimulin. *Arch Geschwulstforsch* 60:379–383

Lersch, C , Zeuner, M , Bauer, A., Siemens, M , Hart, R., Drescher, M , Fink, U., Dancygier, H and Classen, M 1992. Nonspecific immunostimulation with low doses of cyclophosphamide (LDCY), thymostimulin, and *Echinacea purpurea* extracts (Echinacin) in patients with far advanced colorectal cancers. Preliminary results. *Cancer-Invest* 10 343–348

Lersch, C., Gain, T., Lorenz, R. and Classen, M. 1994. Chemoimmunotherapy of malignancies of the gastrointestinal tract. *Immun Infekt* 22:58–59.

Li, J., Wang, P. F , Zheng, R., Liu, Z. M. and Jia, Z 1993. Protection of phenylpropanoid glycosides from *Pedicularis* against oxidative hemolysis *in vitro* Planta Med. 59·315–317

Lloyd, C. G 1893. Should discoveries made by physicians be protected *Ann Eclect. Med. Surg.* 4·332–333.

Lloyd, J. U. 1904. History of *Echinacea angustifolia*. Pharm Review 22:9–14

Lloyd, J U 1917 A Treatise on *Echinacea angustifolia*, Drug Treatise No XXX, Lloyd Brothers Inc., Cincinnati, Ohio.

Lohmann-Matthes, M.-L. and Wagner, H. 1989. Aktivierung von Makrophagen durch Polysaccharide aus Gewebekulturen von *Echinacea purpurea*. Z Phytother 10:52–59

Luettig, B., Steinmüller, C., Gifford, G. E., Wagner, H and Lohmann-Matthes, M -L 1989 Macrophage activation by the polysaccharide arabinogalactan isolated from plant cell cultures of *Echinacea purpurea* J Natl Cancer Inst. 81:669–675.

Madaus, G 1939 *Echinacea purpurea* Moench. Med. Biol Schriftenreihe, Issue 13. Verlag Rohrmoser, Radebeul/Dresden.

McDougall, B., King, P. J., Wu, B. W , Hostomsky, Z , Reinecke, M G. and Robinson, W. E. Jr. 1998. Dicaffeoylquinic and dicaffeoyltartaric acids are selective inhibitors of human immunodeficiency virus type 1 integrase. *Antimicrob. Agents Chemother.* 42.140–146.

McGregor, R L 1968 The taxonomy of the genus *Echinacea* (Compositae). *The University of Kansas Sci. Bull.* 48:113–142.

McGregor, R. L. and Barkley, T M. 1977. Atlas of the Flora of the Great Plains The Iowa State University Press, Ames, Iowa.

Melchart, D. and Linde, K. 1998. Clinical investigations of *Echinacea* phytopharmaceuticals In. *Immunomodulatory agents from plants* (Ed. H. Wagner), pp. 105–118, Birkhäuser Verlag, Basel.

Melchart, D., Worku, F , Linde, K., Flesche, C., Eife, R , and Wagner, H. 1993 Erste Phase-I-Untersuchung von *Echinacea*-Polysaccharid (EPO VIIa/EPS) bei i.v. Application. *Erfahrungskeilkunde*, 316–323

Melchart, D., Linde, K., Worku, F., Bauer, R. and Wagner H. 1994 Immunomodulation with *Echinacea*—a systematic review of controlled clinical trials. *Phytomedicine.* 1.245–254.

Melchart, D., Walther, E., Linde, K , Brandmaier, R. and Lersch. C. 1998. *Echinacea* root extracts for the prevention of upper respiratory tract infections: A double-blind, placebo-controlled randomized trial *Arch Fam Med* 7:541–545.

Mengs, U , Clare, C. B and Poiley, J. A. 1991 Toxicity of *Echinacea purpurea*. Acute, subacute and genotoxicity studies *Arzneim-Forsch* 41:1076–1081.

Moerman, D. E 1998. *Native American Ethnobotany.* pp 205–206, Timber Press, Portland, Oregon

Müller-Jakic, B., Breu, W., Pröbstle, A., Redl, K., Greger, H. and Bauer, R 1994. *In vitro* inhibition of cyclooxygenase and 5-lipoxygenase by alkamides from *Echinacea* and *Achillea* species. *Planta Med* 60.37–40

Mullins, R. J. 1998. *Echinacea*-associated anaphylaxis. *Med. J. Aust* 168 170–171

Myers, S. P and Wohlmuth, H 1998. *Echinacea*-associated anaphylaxis [letter] *Med J Aust* 168.583–584

Nagel, G. A 1995. The cancer patient between traditional medicine and alternative medicine: Therapeutic concepts *Arch. Gynecol. Obstet.* 257.283–294

Neamati, N , Hong, H., Sunder, S., Milne, G. W. and Pommier, Y. 1997. Potent inhibitors of human immunodeficiency virus type 1 integrase. Identification of a novel four-point pharmacophore and tetracyclines as novel inhibitors. *Mol. Pharmacol* 52.1041–1055

Neugebauer H 1949. Zur Kenntnis der Inhaltsstoffe von *Echinacea*. *Pharmazie,* 4 137–140.

Nüßlein, B , Kurzmann, M , Bauer, R. and Kreis, W. 1999. Enzymatic Degradation of Caffeic Acid Derivatives in *Echinacea purpurea* Herba. In preparation.

Parnham, M. J. 1996 Benefit-risk assessment of the squeezed sap of the purple coneflower (*Echinacea purpurea*) for long-term oral immunostimulation *Phytomedicine* 3.95–102.

Perry, N B , Van Klink, J. W , Burgess, E. J and Parmenter, G. A. 1997. Alkamide levels in *Echinacea purpurea* A rapid analytical method revealing differences among roots, rhizomes, stems, leaves, and flowers *Planta Med.* 63.58–62.

Pietta, P , Mauri, P. and Bauer, R. 1998. MEKC Analysis of different *Echinacea* species *Planta Med* 64 649–652.

Proksch, A. and Wagner, H. 1987. Structural analysis of a 4-O-methylglucuronoarabinoxylan with immunostimulating activity from *Echinacea purpurea*. Phytochemistry. 26.1989–1993

Remiger, P 1989. *Zur Chemie und Immunologie neuer Alkylamide und anderer Inhaltsstoffe aus Echinacea purpurea, Echinacea angustifolia und Echinacea pallida.* PhD thesis, Universität München.

Richman, A. and Witkowski, J. P. 1997 Herbs . . . by the numbers. *Whole Foods.* October 1997:20–28.

Robinson, H 1981. A revision of the tribal and subtribal limits of the *Heliantheae* (Asteraceae). *Botany* 51.1–102

Robinson, W. E. 1998. L-chicoric acid, an inhibitor of human immunodeficiency virus type 1 (HIV-1) integrase, improves on the *in vitro* anti-HIV-1 effect of Zidovudine plus a protease inhibitor (AG1350). *Antiviral Res* 39.101–111

Robinson, W. E Jr., Reinecke, M G., Abdel-Malek, S., Jia, Q and Chow, S. A 1996 Inhibitors of HIV-1 replication [corrected, erratum to be published] that inhibit HIV integrase. *Proc. Natl. Acad Sci. USA.* 93.6326–6331.

Roesler, J , Steinmüller, C , Kiderlen, A., Emmendörfer, A., Wagner, H. and Lohmann-Matthes, M. L 1991a. Application of purified polysaccharides from cell cultures of the plant *Echinacea purpurea* to mice mediates protection against systemic infections with *Listeria* monocytogenes and *Candida albicans Int. J. Immunopharmacol.* 13:27–37.

Roesler, J., Emmendörffer, A., Steinmüller, C , Luettig, B , Wagner, H and Lohmann-Matthes, M. L. 1991b. Application of purified polysaccharides from cell cultures of the plant *Echinacea purpurea* to test subjects mediates activation of the phagocyte system *Int. J Immunopharmacol.* 13 931–941.

Scaglione, F. and Lund, B. 1995. Efficacy in the treatment of the common cold of a preparation containing an *Echinacea* extract *Int. J. Immunother.* 11.163–166.

Schenk, R. and Franke, R 1996. Content of echinacoside in *Echinacea* roots of different origin. *Beitr Züchtungsforsch.* 2:64–67.

Schimmer, O., Abel, G. and Behninger, C. 1989. Untersuchungen zur genotoxischen Potenz eines neutralen Polysaccharids aus *Echinacea*-Gewebekulturen in menschlichen Lymphozytenkulturen. *Z Phytother* 10:39–42.

Schindler, H. 1940. Geschichte, Systematik und Verbreitung der therapeutisch wichtigen *Echinacea* Arten: *E. angustifolia* DC, *E. pallida* Nutt und *E. purpurea* Moench. *Pharm Zentralhalle*, 81:579–583.

Schindler, H. 1953. Die Inhaltsstoffe von Heilpflanzen und Prüfungsmethoden für pflanzliche Tinkturen—58. *Echinacea. Arzneim-Forsch.* 3:485–488.

Schulte, K. E., Rücker, G., and Perlick, J. 1967. Das Vorkommen von Polyacetylen-Verbindungen in *Echinacea purpurea* MOENCH und *Echinacea angustifolia* DC. *Arzneim-Forsch* 17.825–829.

See, D. M., Broumand, N , Sahl, L. and Tilles, J. G. 1997. *In vitro* effects of *Echinacea* and *Ginseng* on natural killer and antibody-dependent cell cytotoxicity in healthy subjects and chronic fatigue syndrome or acquired immunodeficiency syndrome patients. *Immunopharmacol.* 35:229–235

Shohan, J. 1985. Specific safety problems of inappropriate immune responses to immunostimulating agents. *TIPS.* 6:178–182.

Steinmüller, C., Roesler, J., Gröttrup, E., Franke, G., Wagner, H and Lohmann-Matthes, M.-L. 1993. Polysaccharides isolated from plant cell cultures of *Echinacea purpurea* enhance the resistance of immunosuppressed mice against systemic infections with *Candida albicans* and *Listeria* monocytogenes *Int. J. Immunopharmacol* 15.605–614

Stimpel, M., Proksch, A., Wagner, H. and Lohmann-Matthes, M -L 1984. Macrophage activation and induction of macrophage cytotoxicity by purified polysaccharide fractions from the plant *Echinacea purpurea. Infect Immun.* 46 845–849.

Stoll, A., Renz, J. and Brack, A. 1950. Isolierung und Konstitution des Echinacosids, eines Glykosids aus den Wurzeln von *Echinacea angustifolia* D.C. *Helv Chim. Acta*, 33.1877–1893.

Stotzem, C. D., Hungerland, U. and Mengs, U. 1992. Influence of *Echinacea purpurea* on the phagocytosis of human granulocytes. *Med. Sci. Res.* 20:719–720.

Stuessy, T. F. 1977 *Heliantheae*—Systematic Review. *The Biology and Chemistry of the Compositae* (Eds. V. H. Harborne, J. B. Harborne, and B. L. Turner), Vol. II, pp. 622–671, Academic Press, London.

Stuppner, H. 1985 *Chemische und immunologische Untersuchungen von Polysacchariden aus der Gewebekultur von Echinacea purpurea (L.)MOENCH.* PhD thesis, Universität München.

Thompson, K. D. 1998. Antiviral activity of Viracea against acyclovir susceptible and acyclovir resistant strains of herpes simplex virus. *Antiviral Res.* 39:55–61.

Tobler, M., Krienbühl, H., Egger, M., Maurer, C and Bühler, U. 1994. Charakteristik von Frischpflanzengesamtextrakten. Teill: Ergebnisse analytischer Untersuchungen. *Schweiz Zschr. Ganzheitsmedizin*, 257–266.

Tragni, E., Galli, C. L , Tubaro, A., Del Negro, P. and Della Logia, R. 1988. Anti-inflammatory activity of *Echinacea angustifolia* fractions separated on the basis of molecular weight. *Pharm Res. Comm.* 20, Suppl. V. 87–90.

Trypsteen, M. F M., Van Severen, R. G. E. and De Spiegeleer, B. M. J. 1989. Planar chromatography of *Echinacea* species extracts with automated multiple development *Analyst.* 114:1021–1024.

Tubaro, A , Tragni, E , Del Negro, P., Galli, C L and Della Loggia, R 1987. Anti-inflammatory activity of a polysaccharide fraction of *Echinacea angustifolia*. *J. Pharm. Pharmacol* 39 567–569.

Tympner, K. D. 1981 Der immunbiologische Wirkungsnachweis von Pflanzenextrakten *Z angew Phytother.* 2·181–184.

Vömel, Th. 1985. Der Einfluß eines pflanzlichen Immunstimulans auf die Phagozytose von Erythrozyten durch das retikulohistozytäre System der isoliert perfundierten Rattenleber. *Arzneim-Forsch* 35.1437–1439.

Vorberg, G. 1984. Bei Erkältung unspezifische Immunabwehr stimulieren. *Ärztliche Praxis,* 36·97–98.

Wagner, H. 1997. *Echinacea*-Polysaccharide und immunologische Wirkung. In: Bauer R *Echinacea*—Pharmazeutische Qualität und therapeutischer Wert. *Z. Phytother* 18:207–214.

Wagner, H. and Proksch, A. 1981. Über ein immunstimulierendes Wirkprinzip aus *Echinacea purpurea* MOENCH. *Z Angew. Phytother.* 2:166–171.

Wagner, H., Proksch, A , Riess-Maurer, I., Vollmar, A., Odenthal, S., Stuppner, H., Jurcic, K., LeTurdu, M. and Fang, J. N. 1985. Immunstimulierend wirkende Polysaccharide (Heteroglykane) aus höheren Pflanzen *Arzneim.-Forsch.* 35.1069–1075.

Wagner, H., Stuppner, H., Puhlmann, J., Jurcic, K., Zenk, M. A. and Lohmann-Matthes, M.-L. 1987. Immunstimulierend wirkende Polysaccharide aus Zellkulturen von *Echinacea purpurea* (L) Moench. *Z Phytother.* 8.125–126.

Wagner, H., Stuppner, H., Schäfer, W. and Zenk, M. A. 1988. Immunologically active polysaccharides of *Echinacea purpurea* cell cultures. *Phytochemistry* 27:119–126.

Wagner, H., Stuppner, H , Puhlmann, J , Brümmer, B , Deppe K. and Zenk, M. A. 1989. Gewinnung von immunologisch aktiven Polysacchariden aus *Echinacea*-Drogen und -Gewebekulturen. *Z. Phytother.* 10:35–38.

Wildfeuer, A. and Mayerhofer, D. 1994 The effects of plant preparations on cellular functions in body defense. *Arzneim -Forsch.* 44:361–366.

Zhao, H. and Burke, T. R. 1998. Facile syntheses of (2R,3R)-(−)- and (2S,3S)-(+)-chicoric acids *Synthetic Commun* 28:737–740.

Zoutewelle, G. and Van Wijk, R. 1990. Effects of *Echinacea purpurea* extracts on fibroblast populated collagen lattice contraction *Phytotherapy Res.* 4:77–84.

CHAPTER 4

Ginger for Drug and Spice Purposes

H. KIKUZAKI

1. INTRODUCTION

THIS chapter focuses mainly on the chemistry and biological activity of common ginger (*Zingiber officinale* Roscoe), an important spice and folk medicine. It also discusses some tropical gingers that are used for drug and spice purposes mainly in Southeast Asia.

2. COMMON GINGER AS A SPICE

"Common ginger" is the rhizome of *Zingiber officinale*, and it is one of the most popular spices all over the world. In Arabic and western cooking, dried ginger is mainly used for breads, cookies, cakes, teas, beers and wines, while in most Asian countries (e.g., China, India, Thailand, Indonesia and Japan), fresh ginger is mainly used as an ingredient for cooking to add a distinctive flavor to a dish. It is indispensable for curry powders and some other spice blends. In Japan, pickled rhizomes and young ginger shoots are widely used as a garnish.

3. GINGER FAMILY

Common ginger (*Zingiber officinale*) is a perennial plant belonging to the family *Zingiberaceae* that consists of 47 genera including *Zingiber*, *Curcuma*, *Alpinia*, *Amomum*, *Elettaria*, *Kaempferia* and *Hedychium*, and of about 1,400

75

species (Hegnauer, 1963). Ginger is native to Southeast Asia and has been cultivated in countries such as India and China for over 3,000 years. Ginger seems to have been traded to ancient Egypt, Greece and Rome by the Phoenicians and to have spread throughout Europe by the ninth century. In the sixteenth century, it was introduced into the West Indies by the Spaniards (Norman, 1990). Today, it is widely cultivated in the tropical and temperate zone. The leading producers are Indonesia, India, China and Jamaica, while the leading importers are Saudi Arabia, Yemen, United States, Japan and Europe (Schulick, 1993).

Turmeric (*Curcuma longa, C.* domestica) belongs to the same family as common ginger, is also a well-known spice and is usually available as yellow powder that is prepared through boiling, peeling, drying and then grinding. Turmeric is essential in preparing curry powder and is used as flavoring for many Southeast Asian dishes. In the Okinawa area of Japan, turmeric has been traditionally added to drinks. Another member of the family is galangal (*Alpinia galanga*). The rhizome of this plant, with its pleasant aroma, is also used in food preparations, especially in Thailand and Malaysia, instead of common ginger. It is an essential ingredient in "Tom Yam Goong," a famous spicy prawn soup of Thailand. Cardamom (*Elettaria cardamomum*), which is one of the most expensive spices in the world and is widely used to flavor sweet and savory dishes, cakes, pastries or coffee, also belongs to the ginger family. The small, dark greenish-brown seeds are used as a spice having a strong and sweetish lemon-like odor. In Southeast Asia, the rhizomes or seeds of many other gingers are available as spice. Furthermore, the buds of some gingers such as myoga (*Zingiber mioga*) in Japan and torch ginger (*Phaeomeria elatior*) in Malaysia and Thailand, and the leaves and young stems of some other gingers are also edible. The leaves of *Alpinia speciosa* and *A. formosana* have a characteristic pleasant aroma and are sometimes used as wrapping material to flavor rice cakes. Some gingers (e.g., *Curcuma aeruginosa, C. amada, C. longa, C. xhanthorrhiza* and *C. zedoaria*) are good sources of starch because of their high carbohydrate content. Table 4.1 shows the uses of some gingers in Asia.

4. GINGER FOR DRUG PURPOSES

Most gingers are used not only as spice but also for drug purposes as shown in Table 4.1. Many gingers have appeared in prescriptions of Chinese medicine, Ayurveda in India and Jamu in Indonesia and have a long history as traditional medicine. In Chinese traditional medicine, the fresh rhizome of *Zingiber officinale* is called "shoukyo" while the dried rhizome is called "kankyo," and their application as drugs are quite different from each other. "Shoukyo" is prescribed as an antiemetic, as a cough and cold remedy, as an antitoxic and as a digestive stimulant. On the other hand, "kankyo" is regarded as a good remedy for stomach-ache. *Z. cassumunar, Z. spetabile* and *Z. zerumbet* belong to the same genus as *Z. officinale* and are also applied as drugs. *Curcuma longa* has very important pharmacological properties including anti-inflammatory, carminative, anti-allergy and antihepatotoxic properties. Other species of the genus *Curcuma* (*C. aeruginosa, C. aromatica, C. xanthorrhiza* and *C. zedoaria*) are also traditional medicines in tropical Asia. The seeds of many plants of the genera *Amomum* and *Alpinia* as well as cardamom are aromatic digestive stimulants.

The recent remarkable progress of science has brought out developments in studies on the chemistry, pharmacology and biochemistry of ginger. Experimental evidences of the bioactivities of ginger that people know from experience and have utilized traditionally, have been increasing. The succeeding discussions deal with some of the chemical components of the plants of the family *Zingiberaceae* and the physiological effects of these constituents.

5. CHEMISTRY OF BIOACTIVE COMPONENTS OF *ZINGIBER OFFICINALE*

5.1. GINGEROLS

[6]-gingerol (1) is known as the main pungent component of the rhizome of *Zingiber officinale* (Lapworth et al., 1917). As shown in Figure 4.1, [6]-gingerol is a noncyclic decane having a 4-hydroxy-3-methoxyphenyl group at C-1, a carbonyl group at C-3 and a hydroxyl group at C-5. Furthermore, the stereochemistry at C-5 was determined to be *S* configuration (Connell and Sutherland, 1969). The name [6]-gingerol was derived from the fact that alkaline hydrolysis of gingerol afforded n-hexanal, a six-carbon aldehyde (Nomura and Iwamoto, 1928). The amount of [6]-gingerol in fresh rhizomes of ginger cultivated in China and Japan was approximately 0.3~0.5% and those of [8]-gingerol (2) and [10]-gingerol (3), having two and four more methylenes than [6]-gingerol, respectively, were each 5~20% of the [6]-gingerol content (Yoshikawa et al., 1993). Other gingerols including [3]-, [4]-, [5]-, [7]-, [12]- and [14]-gingerol were found in ginger by gas chromatography-mass spectrometry (GC-MS) or high pressure liquid chromatography (HPLC) (Masada et al., 1973; Harvey, 1981; Chen et al., 1986; Farthing and O'Neill, 1990). Furthermore, [6]-(4), [8]-, [10]- and [12]-methylgingerol, having a methoxyl group instead of a hydroxyl group at C-4, and [6]-demethoxygingerol (5), lacking a methoxyl group at C-3 of the benzene ring, were also found in ginger rhizome (Harvey, 1981; Kikuzaki et al., 1994).

TABLE 4.1. Uses of Some Spices and Herbs Belonging to the Family Zingiberaceae.

Species	Origin	Parts Used	Usage	
			For Food	For Drug
Genus *Alpinia* (*Languas*)				
A. *galanga*	Tropical Asia	rhizome	spice	stomach-ache, anorexia eczema, anticonvulsant
A. *mutica*	Malaya	flower rhizome bud	food spice food	stomach-ache
A. *speciosa*	India Burma	leaves fruit	flavoring	stomach-ache
Genus *Amomum*				
A. *kepulaga* (*A. cardamomum*)	Indonesia	fruit rhizome	spice	cough, tonsilitis antiemetic, gastritis tonic
A. *tsao-ko*	China Indochina	fruit	spice	stomach-ache
Genus *Curcuma*				
C. *aeruginosa*	Burma Indochina	rhizome	spice source of starch	anthelmintic, scabies after childbirth
C. *aromatica*	India	rhizome	spice	
C. *coloraia*	Malaya	rhizome	spice	
C. *longa* (*C. domestica*)	Tropical Asia	rhizome	spice, dye source of starch	eczema, diarrhea constipation puerperal fever rheumatism jaundice, anemia
		leaves	food	

TABLE 4.1 (continued)

Species	Origin	Parts Used	Usage		
			For Food	For Drug	
C. *mangga*	Malaysia	rhizome	spice	abdominalgia	
C. *xanthorrhiza*	Indonesia	rhizome	spice, dye	anticonvulsant	
			source of starch	malaria, anorexia	
				eczema, galactagogue	
C. *zedoaria*	India	rhizome	spice	after childbirth	
	Indonesia		source of starch		
Genus *Elettaria*					
E. *cardamomum*	India	fruit	spice	stomach-ache	
Genus *Kaempferia*					
K. *galanga*	India	rhizome	spice	swelling, cough, cold	
				colic, antiemetic	
				gastritis, tetanus	
Genus *Zingiber*					
Z. *cassumunar*	India	rhizome	food	obesity, headache	
	Malaya			constipation, jaundice	
				anticonvulsant	
Z. *officinale*	Southeast Asia	rhizome	spice	anorexia, stomach-ache	
			food	antiemetic, cough, cold	
				neuropathy, swelling	
Z. *zerumbet*	Tropical Asia	rhizome	spice	abdominalgia, anorexia	
			food	renal calculus, asthma	
				anticonvulsant	

Source: Corner and Watanabe (1978), Kitano (1984), Aburada (1987), Takahashi (1988).

5.2. SHOGAOLS

Shogaol, a monodehydrated gingerol, was reported by Nomura (1917) to be a pungent component of ginger. So far, [4]-, [6]-(6), [8]- and [10]-shogaol have been found in ginger (Masada et al., 1973; Harvey, 1981; Kikuzaki et al., 1994). Usually, the fresh rhizome contains none or very small amounts of shogaols, while the dried rhizome is rich in shogaols. This suggests that shogaols are formed through dehydration during processing or storage of the fresh rhizome (Govindarajan, 1982). Kano et al. (1986) reported that the ratio of [6]-gingerol to [6]-shogaol in fresh rhizome was about 7:1, and this ratio did not change after drying at 40°C. When steamed for 10 hours followed by drying, the ratio was about 1:1 that implied the dependence of the conversion of [6]-gingerol to [6]-shogaol with time. The ratio of [6]-gingerol to [6]-shogaol of prepared ginger available commercially in Japanese markets was as follows: about 10:1 in fresh rhizome, from 10:1 to 5:1 in dried rhizome, and about 1:1 in steamed-dried rhizome. It seems that the amounts of [6]-gingerol and [6]-shogaol are affected by processing of the rhizome. [6]-(7) and [10]-methylshogaol, having a methoxyl group instead of a hydroxyl group, and [6]-demethoxyshogaol (8) lacking a methoxyl group at C-3 of the benzene ring were also found in ginger rhizome (Harvey, 1981; Kikuzaki et al., 1994).

5.3. OTHER GINGEROL-RELATED COMPOUNDS

Figure 4.1 includes other gingerol-related compounds found in ginger rhizome. [6]-gingerdiol (9) was isolated first by Murata et al. (1972). The stereochemistry at C-3 and C-5 was determined to be 3R, 5S because the spectral data of natural [6]-gingerdiol were determined to be the same with those of the synthetic 3R, 5S-gingerdiol that was obtained by reduction of [6]-gingerol (5S configuration) with NaBH$_4$ (Kikuzaki et al., 1992). [4]-, [8]- and [10]-gingerdiol and [6]-methylgingerdiol were detected in ginger by GC-MS (Masada et al., 1974). Other [6]-gingerdiol analogues, mono- or diacetyl derivatives (10) and [6]-methylgingerdiacetate were isolated (Masada et al., 1974; Kikuzaki et al., 1992). Other gingerol-related compounds are [6]-paradol (11) (Masada et al., 1974), [6]-(12), [10]-dehydrogingerdione, [6]-(13), [10]-gingerdione (Kiuchi et al., 1982), 3,6-epoxy-1-(4-hydroxy-3-methoxyphenyl)-3,5-decadiene homologues (14), [6]-hydroxyshogaol (15) (Nakatani and Kikuzaki, 1992), [6]-gingesulfonic acid (16) (Yoshikawa et al., 1994) and [6]-(17), [8]- and [10]-dehydroshogaol (Wu et al., 1998). The amounts of these gingerol-related compounds were estimated to be 1/100 to 1/10 compared with that of gingerols.

The biosynthesis of gingerol was proposed by Denniff et al. (1980) based on experiments incorporating several radioactive precursors. Figure 4.2 shows the pathway in which condensation of ferulic acid derived from phenylalanine with malonyl-CoA and n-hexanoic acid yielded [6]-dehydrogingerdione fol-

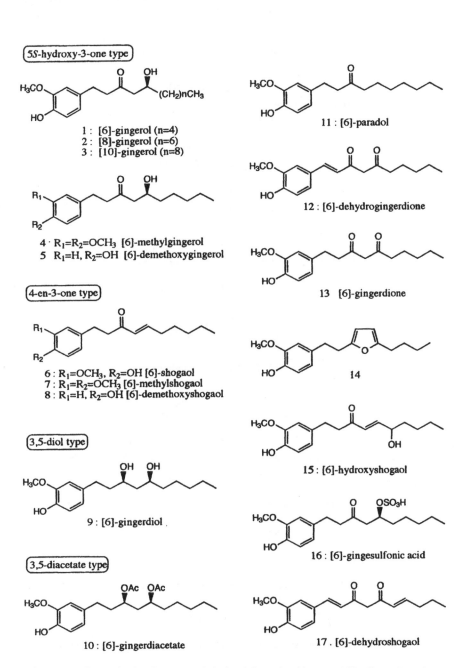

Figure 4.1 Gingerol-related compounds isolated from the rhizome of *Zingiber officinale*.

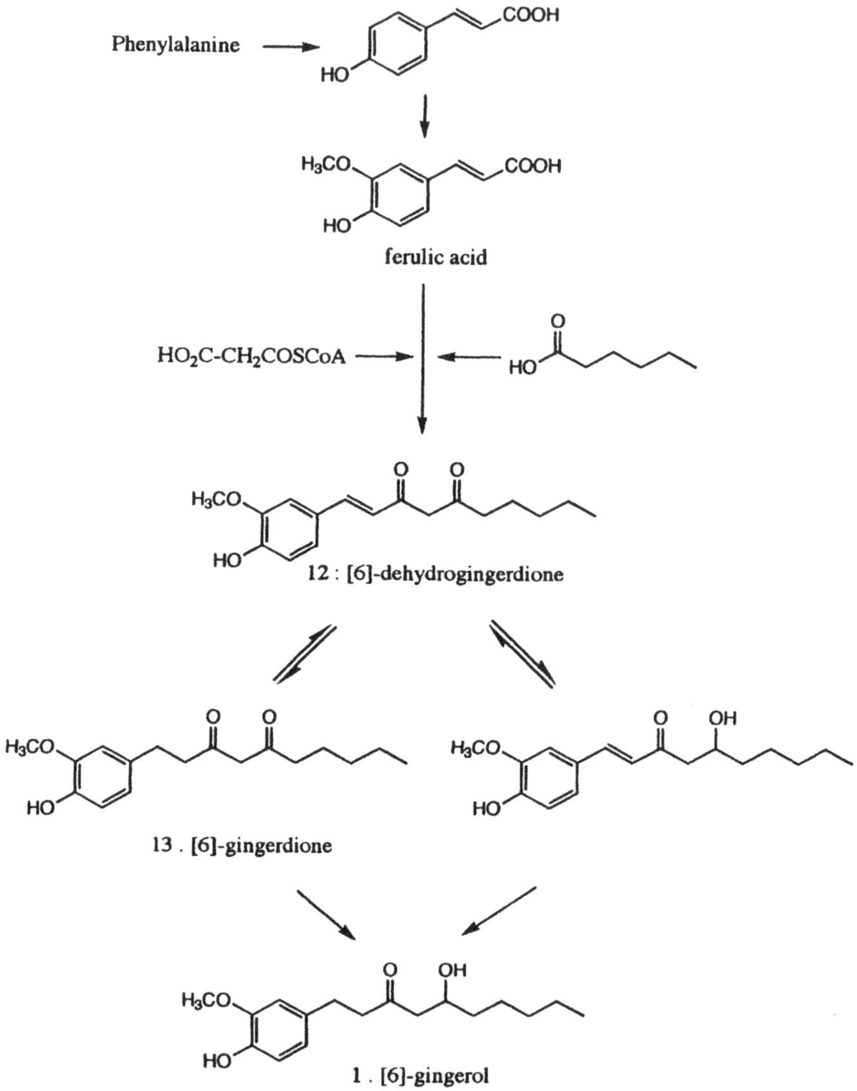

Figure 4.2 Proposed biosynthesis pathway of gingerol. (Adapted from Denniff et al., 1980.)

lowed by a two-step reduction to give [6]-gingerol. This proposal was supported by the finding of natural [6]-dehydrogingerdione and [6]-gingerdione in ginger rhizome.

Ginger contains high amounts of gingerols that make them its principal pungent components. Denniff et al. (1981) reported that [8]-gingerol was the most pungent among the members of the gingerol series, which indicated that the chain length plays an important role in pungency. Recently, Yamahara et al. (1992) examined the pungency of gingerol-related compounds, and they reported that pungency decreased in the order of [6]-shogaol > [6]-paradol = [6]-gingerdione > [6]-gingerol > [6]-gingerdiol, and the threshold of [6]-gingerol was estimated to be about 1/100 to [6]-shogaol.

5.4. DIARYLHEPTANOIDS

Ginger rhizome contains many diarylheptanoids, heptanes with phenyl groups at C-1 and C-7, although the amounts are less than 0.005 percent each. Among the diarylheptanoids, hexahydrocurcumin (18) is the main component, and it has a carbonyl group at C-3 and a hydroxyl group at C-5 (5S configuration) corresponding to gingerol (Murata et al., 1972; Itokawa et al., 1985b). Figure 4.3 shows the structures of some diarylheptanoids (19–32) isolated from ginger rhizome to date (Endo et al., 1990; Kikuzaki et al., 1991a, 1991b; Kikuzaki and Nakatani, 1996; Yamahara et al., 1992). Like gingerol-related compounds, these diarylheptanoids could be structurally classified according to the substitution pattern of heptane into four types: 5S-hydroxy-3-one, 4-en-3-one, 3,5-diol and 3,5-diacetate. In the case of diol and diacetate types, diarylheptanoids having both 3R, 5S and 3S, 5S configurations were found in ginger, while only 3R, 5S types of gingerdiol and gingerdiacetate have been found so far. In addition, there were five unique cyclic diarylheptanoids like 31 and 32 that were oxygenated at C-1, C-3 and C-5 of the heptane chain and cyclized between C-1 and C-5 (Kikuzaki and Nakatani, 1996). Other minor gingerol analogues having 3,4-dimethoxyphenyl or 4-hydroxyphenyl groups were found, and various diarylheptanoids with 4-hydroxyphenyl, 3,4-dihydroxyphenyl, 3,4-dihydroxy-5-methoxyphenyl or 3,5-dimethoxy-4-hydroxyphenyl groups were isolated. The gingerol-related compounds and diarylheptanoids having the 4-hydroxy-3-methoxyphenyl group are, however, the main components of ginger.

5.5. TERPENOIDS

Ginger rhizome possesses many terpenoids, including the monoterpenoids geraniol, geranylacetate, geranial and 1,8-cineole, and the sesquiterpenoids α-zingiberene (33), β-sesquiphellandrene (34), β-bisabolene (35) and ar-curcumene (36) (Nigam et al., 1964; Sakamura and Hayashi, 1978) (Figure 4.4). The amount and composition of ginger terpenoids vary considerably according to the production area, harvest season and preparation. The Japanese ginger generally called "kintoki" contains some labdane-type diterpenoids (37–40) (Kano et al., 1990; Kawakishi et al., 1994). These diterpenoids exist only in "kintoki" species found in Japan, but are not detected in gingers from China, Taiwan and Vietnam (Yoshikawa et al., 1993).

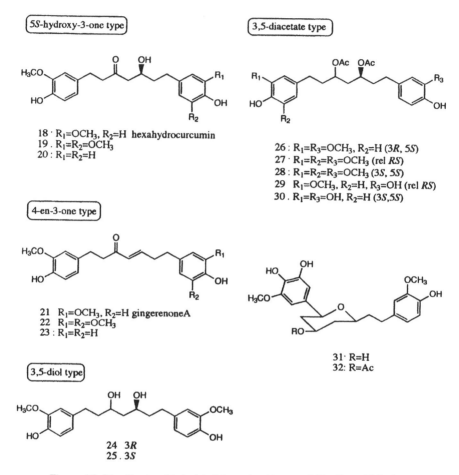

18 · R$_1$=OCH$_3$, R$_2$=H hexahydrocurcumin
19 . R$_1$=R$_2$=OCH$_3$
20 : R$_1$=R$_2$=H

26 : R$_1$=R$_3$=OCH$_3$, R$_2$=H (3R, 5S)
27 · R$_1$=R$_2$=R$_3$=OCH$_3$ (rel RS)
28 : R$_1$=R$_2$=R$_3$=OCH$_3$ (3S, 5S)
29   R$_1$=OCH$_3$, R$_2$=H, R$_3$=OH (rel RS)
30 . R$_1$=R$_3$=OH, R$_2$=H (3S,5S)

21   R$_1$=OCH$_3$, R$_2$=H gingerenoneA
22   R$_1$=R$_2$=OCH$_3$
23 : R$_1$=R$_2$=H

31· R=H
32: R=Ac

24   3R
25 . 3S

Figure 4.3 Diarylheptanoids isolated from the rhizome of *Zingiber officinale*.

5.6. OTHER COMPOUNDS

Three monoacyldigalactosylglycerols and (+)-angelicoidenol-2-*O*-β-D-glu-copyranoside were isolated from ginger cultivated in Taiwan (Yoshikawa et al., 1994).

6. BIOACTIVITY OF GINGER CONSTITUENTS

6.1. ANTIOXIDANT ACTIVITY

Spices and herbs possess potential antioxidant activity (Chipault et al.,

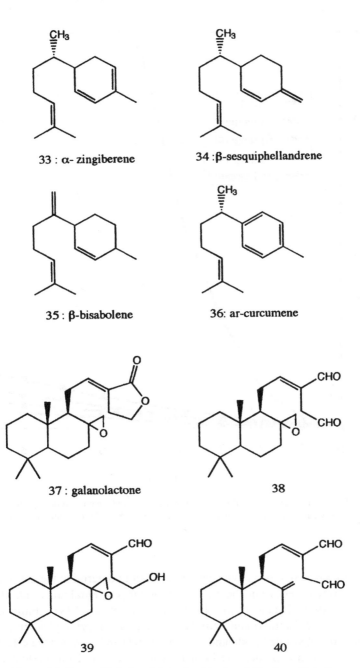

33 : α- zingiberene

34 : β-sesquiphellandrene

35 : β-bisabolene

36: ar-curcumene

37 : galanolactone

38

39

40

Figure 4.4 Terpenoids isolated from the rhizome of *Zingiber officinale*.

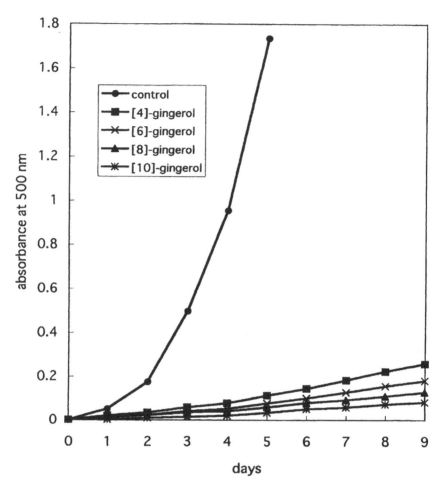

Figure 4.5 Effect of gingerol homologues with different alkyl chain length on inhibition of linoleic acid autoxidation (ferric thiocyanate method; conc. 100 μM). (Adapted from Kikuzaki et al , 1994)

1952). Many investigations have been carried out to determine the antioxidant effect of various spices and herbs including ginger (Hirahara et al., 1974; Saito, et al., 1976; Huang et al., 1981; Al-Jalay et al., 1987). Fujio et al. (1969) reported the antioxidant activity of the pungent principles, zingerone and shogaol in dehydrated pork, and Lee and Ahn (1985) showed the effectiveness of gingerol in a β-carotene-linoleic acid-water emulsion system as an antioxidant. We also examined the antioxidant effect of some gingerol-related compounds and diarylheptanoids using linoleic acid as the substrate in ethanol-phosphate buffer solution. Figure 4.5 shows the antioxidant activity of gingerol homologues with different alkyl chain lengths. These gingerols exhibited effectiveness as an antioxidant, with higher activity being correlated with longer alkyl chains. Shogaol homologues had similar tendencies as gingerols (Kikuzaki et al., 1994). Gingerol-related compounds with the same alkyl chain length and with a 4-hydroxy-3-methoxyphenyl group, efficiency increased in the order [6]-dehydrogingerdione < [6]-gingerol < [6]-shogaol ≤ [6]-ginerdiol ≤ [6]-gingerdiacetate (Figure 4.6). Similar results were obtained for diarylheptanoids. When [6]-gingerol was compared with hexahydrocurcumin, the activity of the latter was stronger than the former. Furthermore, comparison of the activity of the diacetate-type of diarylheptanoids (26, 27 and 30) indicated that a compound with a 4-hydroxy-3-methoxyphenyl group (26) was more active than that with a 3,5-dimethoxy-4-hydroxyphenyl (27) or 3,4-dihydroxyphenyl group (30), which showed a tendency to promote the initial stages

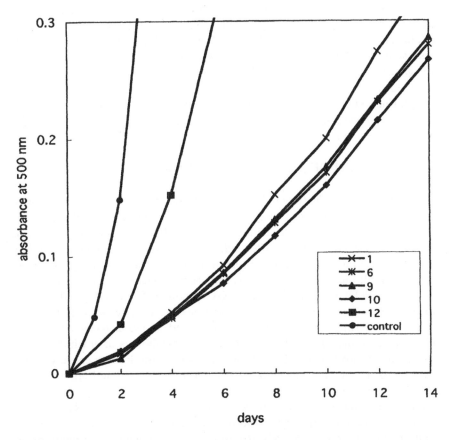

Figure 4.6 Effect of the gingerol-related compounds with the same alkyl chain length and with a 4-hydroxy-3-methoxyphenol group on the inhibition of linoleic acid autoxidation (ferric thiocyanate method; conc. 100 μM). (Adapted from Kikuzaki and Nakatani, 1993.)

of linoleic acid autoxidation (Kikuzaki and Nakatani, 1993). These results suggest that the alkyl chain length, the substitution pattern of alkyl chain and benzene ring of gingerol-related compounds and diarylheptanoids are important in order to be active against autoxidation of linoleic acid.

6.2. ANTIMICROBIAL ACTIVITY

Spices and herbs have been used throughout history not only for flavoring but also for preservative purposes. Previous studies have shown that the essential oils of most spices and herbs generally have strong antimicrobial effect. However, the essential oil of ginger did not exhibit very strong activity. On the other hand, there are several reports concerning the antimicrobial activity of the nonvolatile components of ginger. Gingerols showed antibacterial effects on *Bacillus subtilis*, a gram positive bacteria, and *Escherichia coli*, a gram negative bacteria. Chain lengths of 12 and 14 carbons such as [8]- and [10]-gingerol probably play an important role in the antimicrobial activity (Yamada et al., 1992). [6]-shogaol and [6]-paradol inhibited growth of the disease germs of the genus *Mycobacterium* (*M. chelonei*, *M. intracellular*, *M. smegmatix* and *M. senopi*) (Galal, 1996). Recently, [10]-gingerol (3) was reported as an inhibitor of *M. avium* and *M. tuberculosis* (Hiserodt et al., 1998). Endo et al. (1990) reported that gingerenone A (21), a diarylheptanoid, showed anticoccidium activity *in vitro* and exhibited strong antifungal effect with *Pyricularia oryzae*.

6.3. INHIBITION OF THE EICOSANOID CASCADE

Eicosanoid cascade (arachidonic cascade) is divided into two pathways depending on the catalyzing enzymes (Figure 4.7). One is the cyclooxygenase pathway where prostaglandin endoperoxide (PGH_2) is produced from arachidonic acid by cyclooxygenase (PG endoperoxide synthase), and subsequently, several kinds of prostaglandins are formed by PG synthetases. They are formed throughout all the organs for various important bioactivities. For example, in the blood platelet, the thromboxane (TXA_2) produced from PGH_2 by thromboxane synthetase has very strong platelet aggregation activity, while in the blood vessel endodermis, PGI_2 is biosynthesized and shows activity opposite of TXA_2. Loss of the balance between these two metabolites of arachidonic acid results in arteriosclerosis. The second pathway is the lipoxygenase pathway where leukotrienes are biosynthesized by 5-lipoxygenase and are associated with inflammation and allergy.

Ginger has been known to be an important modulator or balance of the eicosanoid cascade (Srivastava, 1986). Many gingerol-related compounds inhibit prostaglandin biosynthesis. [6]-gingerol (1), shogaol (6), gingerdiol (9), gingerdiacetate (10), dehydrogingerdione (12) and gingerdione (13) showed

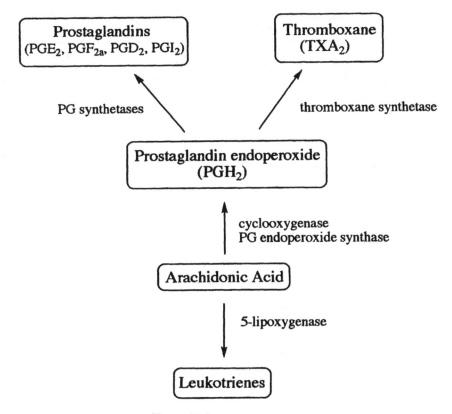

Figure 4.7 Eicosanoid cascade.

strong inhibition of PG-synthetase activity (Kiuchi et al., 1982; Sankawa, 1987). The gingerol series ([2]-gingerol~[16]-gingerol), including natural and synthetic ones, were examined for their inhibitory effect on PG-synthetase, in which [10]-gingerol exhibited the minimum concentration required for inhibition while the higher or lower homologues were less active. In addition, the inhibitory effect of the gingerol series on 5-lipoxygenase increased as the length of the alkyl chain increased and reached a plateau in the homologues higher than [10]-gingerol (Table 4.2) (Kiuchi et al., 1992).

Another unique inhibitory effect on the eicosanoid cascade was reported by Kawakishi et al. (1994) who found that two labdane-type diterpene dialdehydes (38, 40) obtained from ginger strongly inhibited platelet aggregation without suppressive action against TXA_2. These compounds had inhibitory activity only on 5-lipoxygenase in leukocytes, which suggested a different mechanism for inhibition of platelet aggregation from TXA_2 biosynthesis suppression. Among many kinds of inducers on platelet aggregation, these

TABLE 4.2. Inhibition of PG-Synthetase and 5-Lipoxygenase by the Gingerol-Related Compounds.

Compounds	PG-Synthetase IC_{50} (μM)[a]	5-lipoxygenase IC_{50} (μM)[a]
[6]-gingerol (1)	5 5	—
[6]-shogaol (6)	1 1	—
[6]-gingerdiol (9)	6 6	—
[6]-gingerdiacetate (10)	4 3	—
[6]-dehydrogingerdione (12)	1 0	—
[6]-gingerdione (13)	1 6	—
[2]-gingerol	7% (100 μM)[b]	34% (10 μM)[b]
[4]-gingerol	35% (100 μM)[b]	39% (10 μM)[b]
[6]-gingerol (1)	4 6	3 0
[8]-gingerol (2)	5 0	0 36
[10]-gingerol (3)	2 5	0 053
[12]-gingerol	4 1	0 046
[14]-gingerol	5 7	0 042
[16]-gingerol	8 6	0 055

[a] 50% inhibitory concentration
[b] Percent inhibition at the concentration indicated
Source Sankawa (1987), Kiuchi et al (1992)

diterpenoids inhibited platelet aggregation induced by ADP. Therefore, it was proposed that the mechanism for inhibition of platelet aggregation of labdane-type diterpenes may include their masking actions of the ADP binding site on platelets.

6.4. ANTIHEPATOTOXIC ACTION

To investigate the antihepatotoxic action of ginger constituents, a model hepatitis *in vitro* assay using carbon tetrachloride- and galactosamine-induced cytotoxicity in primary cultured rat hepatocytes is often used. Hikino et al. (1985) assessed the antihepatotoxic effect of some gingerol-related compounds, natural or synthetic ones, based on glutamic pyruvic transaminase (GPT) activity. Gingerol-related compounds were more effective on tetrachloride-induced cytotoxicity than galactosamine-induced cytotoxicity. When the activity of gingerols and other gingerol-related compounds with the same length of alkyl chain were compared, it was observed that gingerols showed significantly stronger activity than the corresponding shogaols, dehydrogingergiones and gingerdiones. On the other hand, when gingerols having alkyl side chains of 5 to 20 carbons were compared, it was observed that the activity reached a maximum at the 12 carbon side chain ([8]-gingerol) and decreased thereafter, suggesting that the activity is dependent on the length of the alkyl chain. Hexahydrocurcumin also exerted antihepatotoxic action.

6.5. ANTI-ULCER ACTIVITY

Ginger has beneficial effects on the digestive system and is used as an important traditional medicine for the treatment of stomach ache, vomiting, indigestion and stomach ulcers. Some ginger components were investigated for their anti-ulcer activity in rats having HCl-ethanol-induced gastric ulcer. The active components included [6]-gingerol and [6]-shogaol, and the sesquiterpenoids, α-zingiberene (33), β-sesquiphellandrene (34), β-bisabolene (35) and ar-curcumene (36) (Table 4.3). In addition, [6]-gingesulfonic acid (16), a water soluble component of ginger, exhibited a dramatically stronger effect than [6]-gingerol (Yamahara et al., 1992; Yoshikawa et al., 1994).

6.6. ANTITUMOR PROMOTING ACTIVITY

Recently, [6]-gingerol was reported to have antitumor promoting activity in a two-stage mouse skin carcinogenesis model (Park et al., 1998). [6]-gingerol significantly inhibited 12-*O*-tetradecanoylphorbol-13-acetate (TPA)-induced inflammation, epidermal ornitine decarboxylase activity and skin tumor promotion in mice.

6.7. OTHER BIOACTIVE EFFECTS OF GINGER COMPONENTS

The components of ginger have also been found to enhance gastrointestinal motility (Yamahara et al., 1990) and inhibit serotonin-induced hypothermia and diarrhea (Huang et al., 1990). [6]-gingerol and [6]-shogaol also showed suppression of the central nervous system and had antipyretic, analgesic and antitussive effects. [6]-shogaol showed significantly stronger efficacy in these actions than [6]-gingerol (Aburada, 1987). Gingerols showed a positive inotro-

TABLE 4.3. Effect of the Gingerol-Related Compounds and Sesquiterpenoids Isolated from Ginger on HCl/Ethanol-Induced Gastric Ulcer in Rat.

Compounds	Dose (mg/kg)	Inhibition (%)
[6]-gingerol (1)	150	57 5
[6]-shogaol (6)	150	70 2
[6]-gingersulfonic acid (16)	150	92 7
α-zingiberene (33)	100	53 6
β-sesquiphellandrene (34)	100	61 1
β-bisabolene (35)	100	57.3
ar-curcumene (36)	100	44.5

Source Yamahara et al (1992), Yoshikawa et al (1994)

pic effect on the guinea pig isolated atria, and cardiotonic activity appears to be in the decreasing order, [8]- > [10]- > [6]-gingerol (Shoji et al., 1982). In addition, [6]-gingerol showed molluscicidal and antischistosomal activities against *Schistosoma mansoni* miracidia and cercariae in *Biomphalaria glabrata* and mice (Adewunmi et al., 1990).

(*E*)-8β,17-epoxylabd-12 ene-15,16-dial (38), one of the diterpenoids characterizing Japanese ginger, showed cholesterol biosynthesis inhibitory effect in *in vitro* and *in vivo* experiments. Further, oral administration of this diterpenoid resulted in dose-dependent decrease of cholesterol biosynthesis in rat liver (Tanabe et al., 1993).

Table 4.4 summarizes several bioactivities of ginger and their associated components.

7. CHEMISTRY AND BIOACTIVITY OF CURCUMINOIDS OF THE GENUS *CURCUMA*

7.1. CHEMISTRY OF CURCUMIN-RELATED COMPOUNDS

Turmeric (*Curcuma longa*) is the most popular ginger of the genus *Curcuma*. The value of turmeric as a spice is in its vivid orange-yellow color in addition to its aroma. Turmeric has been used as a dye from early times, and today it is used mainly as food coloring. The characteristic yellow color is derived primarily from curcumin (41), which is a diarylheptanoid. It can also be derived from monodemethoxycurcumin (42) and bisdemethoxycurcumin (43) (Srinivasan, 1953). Fresh rhizome of turmeric contains a total of about 0.8% of these three curcuminoids. In addition, some other curcumin-related compounds (44–46) (Masuda et al., 1993; Nakayama et al., 1993), cyclocurcuminoids (47) (Kiuchi et al., 1993) and diarylpentanoids (48, 49) (Masuda et al., 1993) were isolated from the rhizome (Figure 4.8). It has been mentioned earlier that in Southeast Asia many tropical gingers of the genus *Curcuma* besides *Curcuma longa* have been used as spices or traditional medicines. Some of these gingers, such as *C. aromatica, C. colorata, C. manga* and *C. xanthorrhiza*, have yellow color. Jitoe et al. (1992) reported that the fresh rhizome of *C. xanthorrhiza* contains approximately 0.2% of the three curcuminoids (41–43) as determined by HPLC. Still, some other gingers, such as *C. aeruginosa*, do not contain curcuminoids and thus are not yellow although they belong to the same genus as *C. longa*. As shown in Figure 4.8, other curcuminoids (45, 50–53) have been isolated from the rhizome of *C. xanthorrhiza*, in addition to the three curcuminoids (41–43) (Uehara et al., 1987; Masuda et al., 1992; Claeson et al., 1993).

TABLE 4.4. Principal Bioactivities of Ginger and Their Associated Components.

Bioactivity	Bioactive Component	Reference
Antioxidant	[4]-, [6]-, [8]-, [10]-gingerol	Lee and Ahn, 1985
	[4]-, [6]-, [8]-, [10]-shogaol	Kikuzaki et al., 1993, 1994
	[6]-gingerdiol, [6]-gingerdiacetate	
	hexahydrocurcumin, gingerenone A	
	24, 26, 27	
Antimicrobial	[6]-, [8]-, [10]-gingerol	Yamada et al., 1992
	[6]-shogaol, [6]-paradol	Galal, 1996
	gingerenone A, [12]-	Endo et al., 1990
Inhibition of PG-synthetase	[6]-, [8]-, [10]-gingerol	Kiuchi et al., 1992
	[6]-shogaol, [6]-gingerdiol	
	[6]-gingerdiacetate	
	[6]-dehydrogingerdione	
	[6]-gingerdione	
Inhibition of 5-lipoxygenase	[6]-, [8]-, [10]-, [12]-gingerol	Kiuchi et al., 1992
	38, 40	Kawakishi et al., 1994
Inhibition of platelet aggregation	38, 40	Kawakishi et al., 1994
Antihepatotoxic	[1]-, [16]-gingerol	Hikino et al., 1985
	[6]-gingerol, [6]-shogaol	
Anti-ulcer	[6]-gingesufonic acid	Yamahara et al., 1992
	α-zingiberene, β-bisabolene	Yoshikawa et al., 1994
	β-sesquiphellandrene, ar-curcumene	
Antitumor	[6]-gingerol	Park et al., 1998
Gastrointestinal motility enhancing	[6]-gingerol	Yamahara et al., 1990
Inhibition of serotonin-induced hypothermia, diarrhea	[6]-gingerol	Huang et al., 1990
Suppression of central nervous system	[6]-gingerol, [6]-shogaol	Aburada, 1987
Cardiotonic	[6]-, [8]-, [10]-gingerol	Shoji et al., 1982
Inhibition of cholesterol biosynthesis	38	Tanabe et al., 1993

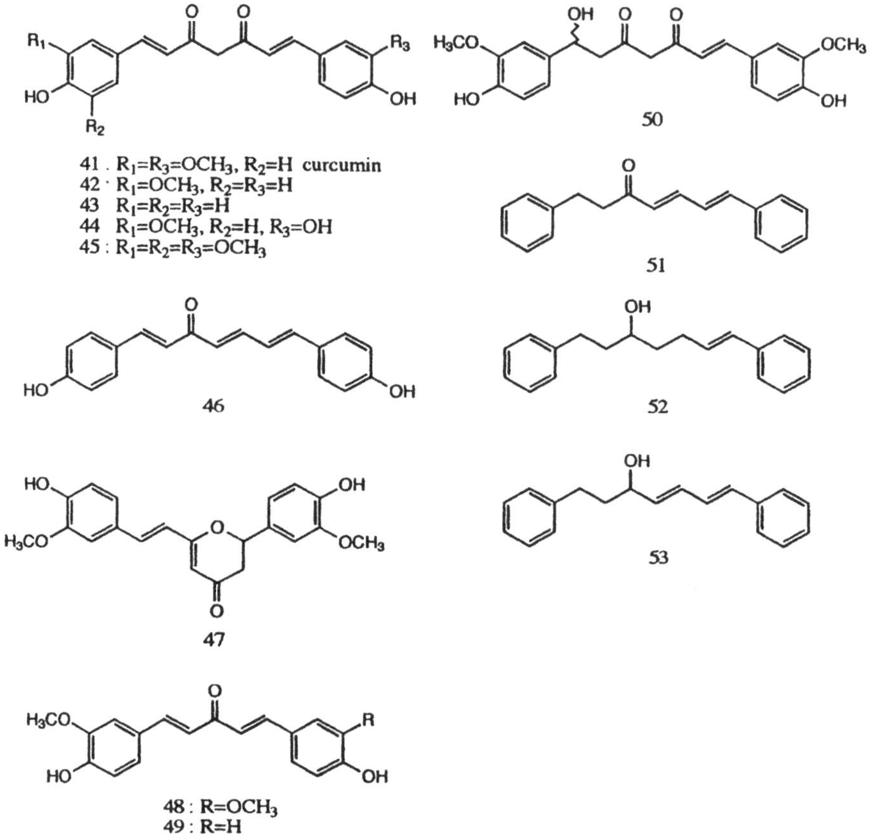
41 . R_1=R_3=OCH_3, R_2=H curcumin
42 · R_1=OCH_3, R_2=R_3=H
43 R_1=R_2=R_3=H
44 R_1=OCH_3, R_2=H, R_3=OH
45 : R_1=R_2=R_3=OCH_3
46
47
48 : R=OCH_3
49 : R=H
50
51
52
53

Figure 4.8 Curcuminoids isolated from the plants of the genus *Curcuma*.

8. BIOACTIVITY OF CURCUMINOIDS OF THE GENUS *CURCUMA*

8.1. ANTIOXIDANT ACTIVITY

It is well-known that turmeric has strong antioxidant activity (Chipault et al., 1952), and curcumin is the main compound responsible for the activity. Toda et al. (1985) examined the antioxidant effect of three curcuminoids (41–43) on linoleic acid and indicated that it increased in the order bisdemethoxy-curcumin < demethoxycurcumin < curcumin. It was speculated that the antioxidant activity was associated with the stabilization of the radical that resulted from the delocalization of the unpaired electron to ketone function, because curcumin, formed by condensation of two ferulic acids, showed stronger activity than ferulic acid (Cuvelier et al., 1992). Furthermore, curcumin was biologically a beneficial antioxidant for hemolysis and lipid peroxidation of mouse erythrocytes induced by hydrogen peroxide (Toda et al., 1988), oxidation of phosphatidylcholine liposomal membranes and rat liver homogenate induced by free radicals (Noguchi et al., 1994) and oxidation of rat brain homogenate (Sharma, 1976). Curcumin analogues (45, 48–50) also showed antioxidant activity (Masuda et al., 1992, 1993).

8.2. ANTI-INFLAMMATORY AND ANTITUMOR PROMOTING EFFECTS

Environmental chemical substances play an important part in the initiation and promotion of cancer. In the initial stage includes DNA base damage by carcinogenic substances. In the second stage, called promotion, some chemical substances promote the formation of the tumor (Bertram et al., 1987). This two-stage carcinogenesis can be applied to all internal organs. Experimental carcinogenesis is systematically established in skin. In mouse skin, tumor formation is carried out by only one application of benzo[a]pyrene or 7,12-dimethylbenze[a]anthracene (DMBA) in the initial stage and then repeated topical application of 12-O-tetradecanoylphorbol-13-acetate (TPA) at regular intervals in the promotion stage. The latter stage is accompanied by many biological, cytological and morphological changes in mouse epidermis, including skin inflammation.

The most important biological activities of curcumin are its anti-inflammatory and antitumoral promoting effects. In the late 1980s, curcumin was found to inhibit tumor initiation and promotion (Huang et al., 1988; Nishino et al., 1987). Huang et al. (1992a) reported that topical application of curcumin inhibited benzo[a]pyrene-mediated DNA adduct formation in the epidermis of mouse skin in the initial stage (Table 4.5). Furthermore, the antitumor promoting activity of curcumin was established by its inhibition of the TPA-induced increase in epidermal ornithine decarboxylase activity (Table 4.6) (Huang et al., 1992b), TPA-induced epidermal thickness and leukocyte infiltration (Huang et al., 1992c). Since the effects of curcumin were found, attention has suddenly been focused on it all over the world. During the past five years, more than 100 papers have been published about the anti-inflammatory and

TABLE 4.5. Inhibition of the Covalent Binding of Benzo[a]pyrene (B[a]P) to Epidermal DNA by Topical Application of Curcumin (Dose of [³H]B[a]P:20 nmol).

Pretreatment	[³H]B[a]P-DNA Adducts (pmol/mg DNA)	% Inhibition
Acetone	0.738 ± 0.133	—
Curcumin (3 μmol)	0.450 ± 0 005	39
Curcumin (10 μmol)	0 285 ± 0.035	61

Adapted from Huang et al (1992a)

TABLE 4.6. Inhibitory Effects of Curcumin on TPA-Induced Mouse Ornithine Decarboxylase Activity (Dose of TPA: 5 nmol).

Treatment	Ornithine Decarboxylase (pmol CO_2/mg protein/hr)	% Inhibition
Acetone	31 ± 12	—
TPA	3,831 ± 682	—
TPA ± curcumin (0 5 μM)	2,659 ± 894	31
TPA + curcumin (1 0 μM)	2,082 ± 544	46
TPA + curcumin (3 0 μM)	647 ± 186 ($p < 0 05$)	84
TPA + curcumin (10 μM)	123 ± 54 ($p < 0 05$)	98

Reprinted from Huang et al (1992b), with permission from CRC Press, Boca Raton, Florida

antitumor promoting effects of curcumin with the objective of elucidating the molecular mechanism of their effects. Table 4.7 shows some properties of curcumin concerning anti-inflammatory and antitumor promoting effects.

The curcumin analogues (51–53) of *C. xanthorrhiza* also showed anti-inflammatory activity in the assay for carrageenin-induced hind paw edema in rats (Claeson et al., 1993).

9. BIOACTIVE SESQUITERPENOIDS OF THE GENUS *CURCUMA*

Generally, the rhizomes of the plants of the genus *Curcuma* contain many sesquiterpenoids. The rhizome of *C. longa* contains more than 40 volatile components (Sharma et al., 1997) including some sesquiterpenoids: α-curcumene, zingiberene, β-bisabolene, β-sesquiphellandrene, α-turmerone, ar-turmerone, β-turmerone and germacrone, which are the major components. The amount and composition of sesquiterpenoids depend on the production area (Uehara et al., 1992). Ar-turmerone displayed mosquitocidal activity on *Aedes aegypttlarvae* (Roth et al., 1998).

The bisabolane sesquiterpenoids, curcumene, ar-turmerone, β-atlantone and xanthorrhizol isolated from *C. xanthorrhiza* showed antitumor activity against sarcoma 180 in the mouse (Itokawa et al., 1985a). Xanthorrhizol also showed pronounced toxicity against neonate larvae of *Slittoralis* in contact residue bioassay (Pandji et al., 1993). The sesquiterpenoids germacrone, curzerenone and germacrone epoxide showed central nervous system depressant properties (Shin et al., 1989).

10. ANTIOXIDANT AND ANTI-INFLAMMATORY CURCUMINOIDS OF *ZINGIBER CASSUMUNAR*

Although *Zingiber cassumunar* botanically belongs to the genus *Zingiber*, its rhizome is yellow, like turmeric. The rhizome contains about 0.5% of

TABLE 4.7. Bioactive Properties of Curcumin Concerning Its Anti-inflammatory and Antitumor Promoting Effects.

Scavenging effect of superoxide anion free radicals	Kunchandy and Rao, 1990
Scavenging effect of nitric oxide	Sreejayan and Rao, 1997
Inhibition of nitric oxide synthase in RAW 264 7 macrophages activated with lipopolysaccharide and interferon-γ	Brouet and Ohshima, 1995
Reduction of inducible nitric oxide synthase—RNA expression in the livers of lipopolysaccharide-injected mice by oral treatment of curcumin	Chan et al , 1998
Inhibition of activation of the necrosis factor-kappa B (NF-κB)	Lienhard et al , 1998
Inhibition of the covalent binding of benzo [a]pyrene (B[a]P) to epidermal DNA by topical application of curcumin	Huang et al., 1992a
Suppression of skin tumor formation induced by B[a]P and DMBA in mice	Huang et al., 1992a
Inhibitory effect of dietary curcuminoid on skin carcinogenesis in mice induced by DMBA and TPA	Limtrakul et al., 1997
Inhibition of platelet aggregation induced by arachidonate, adrenaline and collagen	Srivastava et al., 1995
Inhibition of *in vitro* lipoxygenase and cyclooxygenase activities in mouse epidermis	Huang et al., 1991
Suppression of TPA-induced skin inflammation of edema of mouse ears	Huang et al , 1988
Inhibition of TPA-induced ornithine decarboxylase activity	Huang et al , 1988
Inhibition of TPA-induced protein kinase C activity	Lin et al , 1998
Suppression of TPA-induced expression of *c-fos, c-jun* and *c-myc* proto-oncogenes messenger RNAs in mouse skin	Kakar and Roy, 1994
Inhibition of the proliferation and cell cycle progression of human umbilical vein endothelial cell	Singh et al , 1996
Inhibitory effect of dietary curcumin on azosymethane-induced colon tumorigenesis in mice or rats	Huang et al., 1994, Pereira et al., 1996

curcumin and six additional curcumin-related compounds (54–59), complexes of curcumin and phenylbutenoid, have also been isolated. All of these compounds were more active than curcumin in inhibiting autoxidation of linoleic acid as well as in suppressing TPA-induced anti-inflammation (Masuda and Jitoe, 1994; Masuda et al., 1995) (Figure 4.9). Moreover, (*E*)-4-(3′,4′-dimethoxyphenyl)but-3-en-2-ol, a phenylbutenoid, showed a significant inhibition

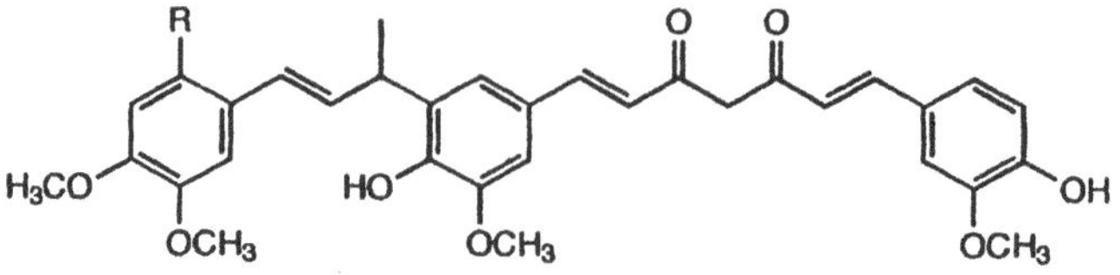

54 : R=H cassumunin A (83%)
55 · R=OCH₃ cassumunin B(76%)

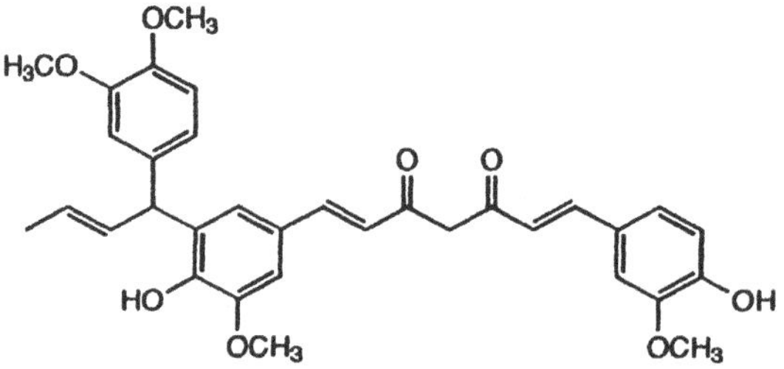

56 cassujunin C (75%)

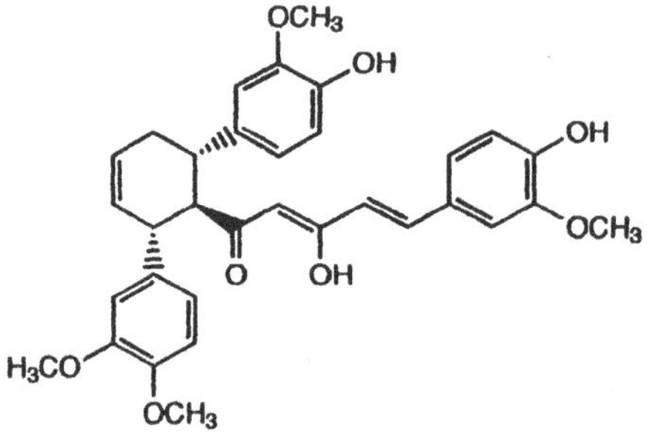

57 cassumunarin A(75%)

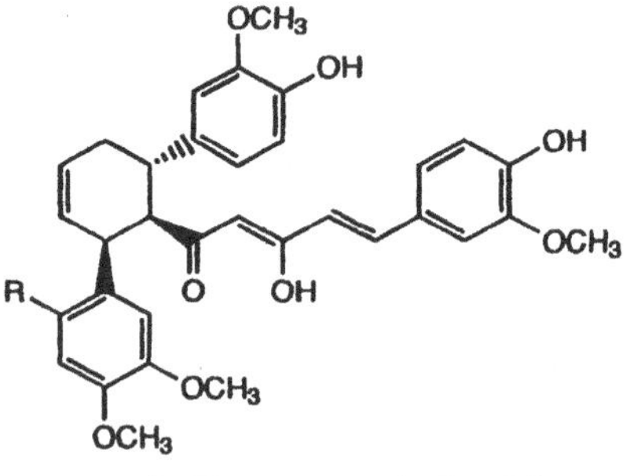

58 : R=H cassumunarin B (62%)
59 : R=OCH₃ cassumunarin C (62%)

Figure 4.9 Curcuminoids isolated from the rhizome of *Zingiber cassumunar* and their TPA-induced anti-inflammatory effects [() means % inhibition when that of curcumin is 51%] (Masuda and Jitoe, 1994; Masuda et al., 1995).

98

on carrageenin-induced inflammation including rat paw edema, leukocyte accumulation and prostaglandin biosynthesis (Panthong et al., 1997).

11. ANTITUMOR PROMOTING EFFECT OF THE COMPONENT OF *ALPINIA GALANGA*

(1′S)-Acetoxychavicol acetate (ACA) is known as a characteristic flavor compound of *Alpinia galanga*. Several groups of workers have examined the antitumor promoting effect of ACA as well as its mechanism. ACA efficiently inhibited DMBA-TPA-induced two-stage carcinogenesis in mouse skin (Murakami et al., 1996) and oral and colon carcinogenesis (Ohnishi et al., 1996; Tanaka et al., 1997) at the initial and promotion stages. The antitumor promoting has been in effect of ACA is believed to occur by suppression of the generation of active oxygen derived from neutrophil leukocyte and generation of nitric oxide (NO) associated with carcinogenesis because ACA inhibits xanthine oxidase, TPA-induced superoxide generation in differentiated human leukemic (HL-60) cells and the induction of nitric oxide syntase in RAW 264.7 macrophages activated with lipopolysaccharide (Murakami et al., 1996; Noro et al., 1988; Ohata et al., 1998).

12. CONCLUSIONS

Many plants of the family *Zingiber*aceae traditionally used as spices and drugs possess a variety of important biological properties. Previous studies on these plants have demonstrated the presence of many potential bioactive components, especially in ginger (*Z. officinale*) and turmeric (*C. longa*), and their mechanisms of action at the molecular level have been proposed. However, studies involving supplementation of human diet have been carried out only with curcumin, which has been found to possess antitumor activity in the skin and the colon. This observation indicates that dietary curcumin plays a positive role in the reduction of cancer risk. Further investigations on the prevention of disease and improvement of human health by these plants and/ or their bioactive components are required.

13. REFERENCES

Aburada, M. 1987. "Pharmacological Effects of *Zingiberis* Rhizoma," *J. Traditional Sino-Japanese Medicine,* 8(1)·45–50.

Adewunmi, C. O., B. O. Oguntimein and P. Furu. 1990 "Molluscicidal and Antischistosomal Activities of *Zingiber* Offcinale," *Planta Medica,* 56(4) 374–376.

Al-Jalay, B., G. Blank, B. McConnell and M. Al-Khayat. 1987. "Antioxidant Activity of Selected Spices Used in Fermented Meat Sausage," *J. Food Protection,* 50(1).25–27.

Bertram, J S., L. N. Kolonel and F L. Meyskens. 1987. "Rational and Strategies for Chemoprevention of Cancer in Humans," *J Cancer Res.,* 47(11):3012–3031.

Brouet, I. and H. Ohshima. 1995. "Curcumin, an Anti-Tumor Promoter and Anti-Inflammatory Agent, Inhibits Induction of Nitric Oxide Synthase in Activated Macrophages," *Biochem. Biophys. Res. Commun.,* 206(2):533–540.

Chan, M. M -Y., H. -I. Huang, M. R. Fenton and D. Fong. 1998. "*In Vivo* Inhibition of Nitric Oxide Synthase Gene Expression by Curcumin, a Cancer Preventive Natural Product with Anti-Inflammatory Properties," *Biochem. Pharmacol.,* 55(12):1955–1962.

Chen, C. -C., R. T. Rosen and C. -T. Ho. 1986. "Chromatographic Analyses of Gingerol Compounds in Ginger (*Zingiber officinale* Roscoe) Extracted by Liquid Carbon Dioxide," *J. Chromatography,* 360:163–173.

Chipault, J. R., G. R. Mizuno, J. M. Hawkins and W. O. Lundberg. 1952. "The Antioxidant Properties of Natural Spices," *Food Res.,* 17:46–54.

Claeson, P, A. Panthong, P. Tuchinda, V. Reutrakul, D. Kanjanapothi, W. C. Taylor and T. Santisuk. 1993. "Three Non-Phenolic Diarylheptanoids with Anti-Inflammatory Activity from *Curcuma xanthorrhiza,*" *Planta Medica,* 59(5):451–454.

Connell, D. W and M. D. Sutherland. 1969. "A Re-Examination of Gingerol, Shogaol, and Zingerone, the Pungent Principles of Ginger (*Zingiber officinale* Roscoe)," *Aust. J. Chem. Soc.,* 22:1033–1043.

Corner, E. J. H. and K. Watanabe. 1978. *Illustrated Guide to Tropical Plants* Tokyo: Hirokawa Publishing Co., pp. 1069–1086.

Cuvelier, M. E., H. Richard and C. Berset. 1992. "Comparison of the Antioxidative Activity of Some Acid-phenols: Structure-Activity Relationship," *Biosci Biotech. Biochem.,* 56(2):324–325.

Denniff, P., I. Macleod and D. A. Whiting. 1980. "Studies in the Biosynthesis of [6]-Gingerol, Pungent Principle of Ginger (*Zingiber officinale*)," *J. C. S. Perkin I,* 2637–2644.

Denniff, P., I. Macleod and D. A. Whiting. 1981. "Syntheses of the (±)-[*n*]-Gingerols (Pungent Principles of Ginger) and Related Compounds through Regioselective Aldol Condensations: Relative Pungency Assays," *J. C. S. Perkin I,* 82–87.

Endo, K., E. Kanno and Y. Oshima. 1990. "Structures of Antifungal Diarylheptenones, Gingerenones A, B, C and Isogingerenone B, Isolated from the Rhizomes of *Zingiber officinale,*" *Phytochemistry,* 29(3):797–799.

Farthing, J. E. and M. J. O'Neill. 1990. "Isolation of Gingerols from Powdered Root Ginger by Countercurrent Chromatography," *J. Liquid Chromatography,* 13(5):941–950.

Fujio, H., A. Hiyoshi, T Asari and K. Suminoe. 1969. "Studies on the Preventive Method of Lipid Oxidation in Freeze-Dried Foods Part III. Antioxidative Effects of Spices and Vegetables," *Nippon Shokuhin Kogyo Gakkaishi,* 16(6):241–246.

Galal, A. M. 1996. "Antimicrobial Activity of 6-paradol and Related Compounds," *Int. J. Pharmacogn.,* 34(1):64–69.

Govindarajan, V. S. 1982. "Ginger-Chemistry, Technology and Quality Evaluation," *CRC Critical Reviews of Food Sciences and Nutrition,* 17:1–258.

Harvey, D. J. 1981. "Gas Chromatographic and Mass Spectrometric Studies of Ginger Constituents," *J. Chromatography,* 212:75–84.

Hegnauer, R. 1963. "*Zingiberaceae,*" In: Chemotaxonomie der Pflauzen Band II. Basel und Stuttgart: Birkhaser Verlag, pp. 451–471.

Hikino, H., Y. Kiso, N Kato, Y Hamada, T. Shioiri, R. Aiyama, H Itokawa, F. Kiuchi and U. Sankawa 1985. "Antihepatotoxic Actions of Gingerols and Diarylheptanoids," *J. Ethnopharmacology*, 14 31–39

Hirahara, F, Y. Takai and H Iwao. 1974 "Antioxidative Activity of Various Spices on Oils and Fats (I)," *Eiyogakuzasshi*, 32(1):1–8.

Hiserodt, R. D., S G Franzblau and R. T. Rosen. 1998. "Isolation of 6-, 8-, 10-Gingerol from Ginger Rhizome by HPLC and Preliminary Evaluation of Inhibition of *Mycobacterium* avium and *Mycobacterium* tuberculosis," *J. Agric Food. Chem.*, 46(7).2504–2508.

Huang, J. K, C. S. Wang and W. H. Chang. 1981. "Studies on the Antioxidative Activities of Spices Grown in Taiwan (I)," *J. Chinese Agric. Chem. Soc.*, 19(3–4):200–207.

Huang, M. T., R. C. Smart, C. Q. Wong and A. H. Conney. 1988. "Inhibitory Effects of Curcumin, Chlorogenic Acid, Caffeic Acid, and Ferulic Acid on Tumor Promotion in Mouse Skin by 12-*O*-tetradecanoylphorbol-13-acetate," *Cancer Res.*, 48(21):5941–5946.

Huang, Q, H. Matuda, K. Sakai, J. Yamahara and Y. Tamai. 1990. "The Effect of Ginger on Serotonin Induced Hypothermia and Diarrhea," *Yakugaku Zasshi*, 110(12).936–942.

Huang, M. T, T. Lysz, T. Ferraro, T. F. Abidi, J. D. Laskin and A. H. Conney. 1991. "Inhibitory Effects of Curcumin on *In Vitro* Lipoxygenase and Cyclooxygenase Activities in Mouse Epidermis," *Cancer Res.*, 51(3):813–819.

Huang, M. T., Z. Y. Wang, C. A. Georgiadis, J. D. Laskin and A H. Conney. 1992a. "Inhibitory Effects of Curcumin on Tumor Initiation by Benzo[*a*]pyrene and 7,12-Dimethylbenz[*a*]anthracene," *Carcinogensis*, 13(11):2183–2186.

Huang, M T., T Lysz, T. Ferraro and A. H. Conney. 1992b. "Inhibitory Effects of Curcumin on Tumor Promotion and Arachidonic Acid Metabolism in Mouse Epidermis," In. *Cancer Chemoprevention*. L. Wattenberg, M. Lipkin, C. W. Boone and G. J. Kelloff (eds), Boca Raton· CRC Press, pp. 375–391.

Huang, M. T., F. M. Robertson, T. Lysz, T. Ferraro, Z. Y. Wang, C. A. Georgiadis, J. D. Laskin and A. H. Conney. 1992c. "Inhibitory Effects of Curcumin on Carcinogenesis in Mouse Epidermis, in Phenolic Compounds in Food and Their Effects on Health II. Antioxidants & Cancer Prevention," ACS Symposium Series 507. M. T. Huang, C. T. Ho and C. Y Lee (eds.), Washington, D.C.: American Chemical Society, pp. 338–349.

Huang, M. T, Y. R Lou, W. Ma, H. L. Newmark, K. R. Reuhl and A. H. Conney. 1994. "Inhibitory Effects of Dietary Curcumin on Forestomach, Duodenal and Colon Carcinogenesis in Mice," *Cancer Res*, 54:5841–5847.

Itokawa, H., F. Hirayama, K. Funakoshi and K. Takeya. 1985a. "Studies on the Antitumor Bisabolane Sesquiterpenoids Isolated from *Curcuma xanthorrhiza*," *Chem. Pharm. Bull*, 33(8):3488–3492.

Itokawa, H., H. Morita, I. Midorikawa, R. Aiyama and M. Morita. 1985b. "Diarylheptanoids from the Rhizome of *Alpinia officinarum* Hance," *Chem. Pharm. Bull.*, 33(11):4889–4893.

Jitoe, A, T Masuda, I. G. P. Tengah, D. N. Suprapta, I. W. Gara and N. Nakatani. 1992. "Antioxidant Activity of Tropical Ginger Extracts and Analysis of the Contained Curcuminoids," *J. Agric Food Chem.*, 40(8):1337–1340.

Kakar, S. S. and D. Roy. 1994. "Curcumin Inhibits TPA Induced Expression of *C-fos*, *C-jun* and *C-myc* Proto-Oncogenes Messenger RNAs in Mouse Skin," *Cancer Letters*, 87:85–89.

Kano, Y., K Saito, T. Sakurai, S. Kanemaki, M. Tanabe and M. Yasuda. 1986 "On the Evaluation of the Preparation of Chinese Medicinal Prescriptions (I) 6-gingerol in *Zingiberis* Rhizoma," *Shoyakugaku Zasshi*, 40(3):333–339.

Kano, Y, M Tanabe and M. Yasuda. 1990. "On the Evaluation of the Preparation of Chinese Medicinal Prescriptions (V) Diterpenes from Japanese Ginger Kintoki," *Shoyakugaku Zasshi*, 44(1)·55–57.

Kawakishi, S., Y. Morimitsu and T. Osawa 1994. "Chemistry of Ginger Components and Inhibitory Factors of the Arachidonic Acid Cascade, in Food Phytochemicals for Cancer Prevention II, Teas, Spices, and Herbs," ACS Symposium Series 547. C. T. Ho, T. Osawa, M. T. Huang and R. T. Rosen (eds.), Washington, D.C.: American Chemical Society, pp 244–250.

Kikuzaki, H. and N. Nakatani. 1993. "Antioxidant Effects of Some Ginger Constituents," *J. Food Sci.*, 58(6):1407–1410.

Kikuzaki, H and N. Nakatani. 1996. "Cyclic Diarylheptanoids from Rhizomes of *Zingiber officinale*," *Phytochemistry*, 43(1).273–277.

Kikuzaki, H., J. Usuguchi and N. Nakatani. 1991a. "Constituents of *Zingiber*aceae. I Diarylheptanoids from the Rhizomes of Ginger (*Zingiber officinale* Roscoe)," *Chem. Pharm. Bull.*, 39(1)·120–122.

Kikuzaki, H., M Kobayashi and N. Nakatani. 1991b. "Diarylheptanoids from Rhizoomes of *Zingiber officinale*," *Phytochemistry*, 30(11)·3647–3651.

Kikuzaki, H., S. M. Tsai and N. Nakatani. 1992. "Gingerdiol Related Compounds from the Rhizomes of *Zingiber officinale*," *Phytochemistry*, 31(5):1783–1786.

Kikuzaki, H., Y. Kawasaki and N Nakatani. 1994 "Structure of Antioxidative Compounds in Ginger, in Food Phytochemicals for Cancer Prevention II, Teas, Spices, and Herbs, ACS Symposium Series 547 C. T. Ho, T. Osawa, M. T. Huang, and R. T. Rosen (eds.), Washington, D.C.: American Chemical Society, pp. 237–243.

Kitano. S. (ed.) 1984. *Handbook of Tropical Plants and Trees.* Tokyo. Youkendo, pp. 544–551.

Kiuchi, F., M. Shibuya and U. Sankawa. 1982. "Inhibitors of Prostaglandin Biosynthesis from Ginger," *Chem. Pharm. Bull.*, 30(2)·754–757.

Kiuchi, F., S. Iwakami, M. Shibuya, F. Hanaoka and U. Sankawa. 1992. "Inhibition of Prostaglandin and Leukotriene Biosynthesis by Gingerols and Diarylheptanoids," *Chem. Pharm. Bull.*, 40(2):387–391

Kiuchi, F., Y. Goto, N. Sugimoto, N. Akao, K. Kondo and Y. Tsuda. 1993. "Nematocidal Activity of Turmeric: Synergistic Action of Curcuminoids," *Chem. Pharm. Bull.*, 41(9):1640–1643

Kunchandy, E. and M. N. A. Rao. 1990. "Oxygen Radical Scavenging Activity of Curcumin," *Int. J. Pharm.*, 38:237–240

Lapworth, A., L. K. Pearson and F. A Royle. 1917. "Pungent Principle of Ginger. I Chemical Characters and Decomposition Products of Thresh's ginger," *J. Chem. Soc. III*, 777–790

Lee, I K. and S. Y. Ahn. 1985. "The Antioxidant Activity of Gingerol," *Korean J Food Sci. Technol.*, 17(2):55–59.

Lienhard, S. M., S. P. Hehner, S. Bacher, W. Droege and M. Heinrich. 1998. "Transcription Factor NF-kappaB," *Dtsch. Apoth. Ztg.*, 138(50):4881–4886.

Limtrakul, P., S Lipigorngoson, O. Namwong, A Apisariyakul and F. W. Dunn 1997. "Inhibitory Effect of Dietary Curcumin on Skin Carcinogenesis in Mice," *Cancer Lett.*, 116(2) 197–203

Lin, J. K., Y. C. Chen, Y. T. Huang and S. Y. Lin-Shiau. 1998. "Suppression of Protein Kinase C and Nuclear Oncogene Expression as Possible Molecular Mechanisms of Cancer Chemoprevention by Apigenin and Curcumin," *J. Cell. Biochem.*, Suppl. 1997 28–29:39–48.

Masada, Y., T. Inoue, K. Hashimoto, M. Fujioka and K. Shiraki. 1973. "Studies on the Pungent Principles of Ginger (*Zingiber officinale* Roscoe) by GC-MS," *Yakugaku Zasshi*, 93(3):318–321.

Masada, Y., T. Inoue, K. Hashimoto, M. Fujioka and C. Uchino. 1974. "Studies on the Constituents of Ginger (*Zingiber officinale* Roscoe) by GC-MS," *Yakugaku Zasshi*, 94(6):735–738

Masuda, T., J Isobe, A. Jitoe and N. Nakatani. 1992. "Antioxidative Curcuminoids from Rhizomes of *Curcuma xanthorrhiza*," *Phytochemistry*, 31(10):3645–3647.

Masuda, T., A. Jitoe, J. Isobe, N Nakatani and S Yonemori 1993 "Anti-Oxidative and Anti-Inflammatory Curcumin-Related Phenolics from Rhizomes of *Curcuma domestica*," *Phytochemistry*, 32(6) 1557–1560.

Masuda, T and A. Jitoe 1994. "Antioxidative and Anti-inflammatory Compounds from Tropical Gingers: Isolation, Structure Determination, and Activities of Cassumunins A, B, and C, New Complex Curcuminoids from *Zingiber cassumunar*," *J. Agric. Food Chem*, 42(9) 1850–1856.

Masuda, T., A Jitoe and T J Mabry 1995. "Isolation and Structure Determination of Cassumunarins A, B, and C. New Anti-Inflammatory Antioxidants from a Tropical Ginger, *Zingiber cassumunar*," *JAOCS*, 72(9).1053–1057.

Murakami, A., S Ohura, Y. Nakamura, K. Koshimizu and H Ohigashi 1996. "1'-Acetoxychavicol Acetate, a Superoxide Anion Generation Inhibitor, Potently Inhibits Tumor Promotion by 12-O-tetradecanoylphorbol-13-acetate in ICR Mouse Skin," *Oncology*, 53:386–391.

Murata, T, M. Shinohara and M. Miyamoto 1972. "Isolation of Hexahydrocurcumin, Dihydrogingerol and Two Additional Pungent Principles from Ginger," *Chem. Pharm. Bull.*, 20(10):2291–2292

Nakatani, N. and H Kikuzaki 1992. "New Phenolic Compounds of Ginger (*Zingiber officinale* Roscoe)," *Chemistry Express*, 7(3):221–224.

Nakayama, R, Y. Tamura, H Yamanaka, H. Kikuzaki and N Nakatani. 1993. "Two Curcuminoid Pigments from *Curcuma domestica*," *Phytochemistry*, 33(2).501–502.

Nigam, M C., I. C Nigam, L. Levi and K. L. Handa. 1964. "Essential Oils and Their Constituents XXII. Detection of New Trace Components in Oil of Ginger," *Can J. Chem.*, 42(11).2610–2615

Nishino, H, A. Nishino, J Takayasu and T. Hasegawa. 1987. "Antitumor-Promoting Activity of Curcumin, a Major Constituent of the Food Additive Turmeric Yellow," *Kyoto-furitsu Ika Daigaku Zasshi*, 96(8) 725–728.

Noguchi, N., E. Komuro, E. Niki and R. L. Willson. 1994. "Action of Curcumin as an Antioxidant against Lipid Peroxidation," *J. Jpn Oil Chem. Soc*, 43(12).1–7.

Nomura, H 1917. "Pungent Principles of Ginger. I A New Ketone, Zingerone (4-hydroxy-3-methoxyphenylethyl MethylKetone) Occurring in Ginger " *J. Chem. Soc III*, 769–776.

Nomura, H. and K Iwamoto 1928. "Pungent Principles of Ginger V. Distillation of Methylgingerol," *Sci Rep. Tohoku Imp. Univ*, 17:973–984

Norman, J. 1990. *The Complete Book of Spices* New York: Viking Penguin, pp. 62–63.

Noro, T., T Sekiya, M Katoh, Y. Oda, T. Miyase, M Kuroyanagi, A. Ueno and S Fukushima 1988. "Inhibitors of Xanthine Oxidase from *Alpinia galanga*," *Chem. Pharm. Bull*, 36(1): 244–248.

Ohata, T, K Fukuda, A Murakami, H. Ohigashi, T. Sugimura and K. Wakabayashi. 1998. "Inhibition by 1'-Acetoxychavicol Acetate of Lipopolysaccharide- and Interferon-γ-Induced Nitric Oxide Production through Suppression of Inducible Mitric Oxide Synthase Gene Expression in Raw 264 Cells," *Carcinogenesis*, 19(6):1007–1012.

Ohnishi, M., T. Tanaka, H. Makita, T. Kawamori, H. Mori, K Satoh, A Hara, A. Murakami, H. Ohigashi and K. Koshimizu. 1996 "Chemopreventive effect of a Xanthine Oxidase Inhibitor 1'-Acetoxychavicol Acetate on Rat Oral Carcinogenesis," *Jpn J Cancer Res.*, 87(4).349–356.

Pandji, C., C. Grimm, V. Wray, L. Witte and P. Proksch 1993. "Insecticidal Constituents from Four Species of the *Zingiber*aceae," *Phytochemistry*, 34(2).415–419.

Panthong, A., D. Kanjanapothi, W Niwatananant, P Tuntiwachwuttikul and V. Reutrakul. 1997. "Anti-Inflammatory Activity of Compound D{(E)-4-(3,4-dimethoxyphenyl)but-3-en-2-ol} isolated from *Zingiber cassumunar*," *Phytomedicine*, 4(3).207–212

Park, K. K., K. S Chun, J. M. Lee, S. S Lee and Y. J Surh 1998. "Inhibitory Effects of [6]-Gingerol, a Major Pungent Principle of Ginger, on Phorbor Ester-Induced Inflammation,

Epidermal Ornithine Decarboxylase Activity and Skin Tumor Promotion in ICR Mice," *Cancer Letter,* 129(2):139–144.

Pereira, M. A , C. J. Grubbs, L. H. Barners, H Li, G. R Olson, I. Eto, M. Juliana, L. M. Whitake and J Gary. 1996. "Effects of the Phytochemicals, Curcumin and Quercetin, upon Azosymethane-Induced Colon Cancer and 7,12-Dimethylbenz[a]anthracene-Induced Mammary Cancer in Rats," *Carcinogenesis,* 17(6)·1305–1311.

Roth, G. N., A. Chandra and M. G. Nair. 1998. "Novel Bioactivities of *Curcuma longa* Constituents," *J. Natural Products,* 61(4)·542–545.

Saito, Y., Y. Kimura and T Sakamoto 1976. "Studies on the Antioxidative Properties of Spices. III. The Antioxidative Effects of Petroleum Ether Soluble and Insoluble Fractions from Spices," *Eiyo to Shokuryo,* 29(9).505–510.

Sakamura, F. and S. Hayashi 1978 "Constituents of Essential Oil from Rhizomes of *Zingiber officinale* Roscoe," *Nippon Nogeikagaku Kaishi,* 52(5):207–211

Sankawa, U. 1987 "Biochemistry of *Zingiber*is Rhizoma," *The J. Traditional Sino-Japanese Medicine,* 8(1).57–61.

Schulick, P. 1993. *Ginger Common Spice & Wonder Drug.* Brattleboro, VT: Herbal Free Press, Ltd., p. 7.

Sharma, O. P. 1976. "Antioxidant Activity of Curcumin and Related Compounds," *Biochem. Pharmacol.,* 25(15):1811–1812.

Sharma, R. K , B. P Misra, T. C. Sarma, A. K Bordoloi, M G. Pathak and P. A. Leclereq. 1997. "Essential Oils of *Curcuma longa* L. from Bhutan," *J. Essent. Oil Res.,* 9(5).589–592.

Shin, K. H., O N Kim and S Won. 1989. "Isolation of Hepatic Drug Metabolism Inhibitors from the Rhizomes of *Curcuma zedoaria,*" *Arch. Pharmacal Res.,* 12(3):196–200.

Shoji, N , A. Iwasa, T. Takemoto, Y. Ishida and Y. Ohizumi. 1982. "Cardiotonic Principles of Ginger (*Zingiber officinale* Roscoe)," *J. Pharm. Sci.,* 71(10)·1174–1175.

Singh, A. K., G. S. Sidhu, T Deepa and R. K. Maheshwari. 1996. "Curcumin Inhibits the Proliferation and Cell Cycle Progression of Human Umbilical Vein Endothelial Cell," *Cancer Lett ,* 107(1):109–115

Sreejayan and M. N. A Rao 1997. "Nitric Oxide Scavenging by Curcuminoids," *J. Pharm. Pharmacol.,* 49(1).105–107.

Srinivasan, K. R. 1953. "Chromatographic Study of the Curcuminoids in *Curcuma longa,*" *J. Pharm. Pharmacol ,* 5:448–457.

Srivastava, K. C. 1986. "Isolation and Effects of Some Ginger Components on Platelet Aggregation and Eicosanoid Biosynthesis," *Prostaglandins, Leukotrienens Med.,* 25(2–3):187–198.

Srivastava, K. C., A. Bordia and S. K. Verma. 1995. "Curcumin, a Major Component of Food Spice Turmeric (*Curcuma longa*) Inhibits Aggregation and Alters Eicosanoid Metabolism in Human Blood Platelets," *Prostaglandins, Leukotrienes Essent. Fatty Acids,* 52(4).223–227.

Takahashi, S 1988 *Jamu.* Tokyo: Hirakawa Shuppansha, pp 141–180.

Tanabe, M., Y D. Chen, K. Saito and Y Kano. 1993. "Cholesterol Biosynthesis Inhibitory Component from *Zingiber officinale* Roscoe," *Chem. Pharm. Bull.,* 41(4):710–713.

Tanaka, T., K. Kawabata, M Kakumoto, H Makita, K. Matsunaga, H. Mori, K. Satoh, A Hara, A. Murakami, K Koshimizu and H. Ohigashi 1997. "Chemoprevention of Azosymethane-Induced Rat Colon Carcinogenesis by a Xanthine Oxidase Inhibitor, 1'-Acetoxychavicol Acetate," *Jpn J. Cancer Res.,* 88(9):821–830.

Toda, S , T. Miyase, H. Arichi, H. Tanizawa and Y Takino 1985. "Natural Antioxidants. III Antioxidative Components Isolated from Rhizome of *Curcuma longa* L.," *Chem. Pharm. Bull.,* 33(4):1725–1728.

Toda, S , M Ohnishi, M Kimura and K. Nakashima. 1988. "Action of Curcuminoids on the Hemolysis and Lipid Peroxidation of Mouse Erythrocytes Induced by Hydrogen Peroxide," *J. Ethnopharmacology,* 23 105–108

Uehara, S., I Yasuda, K Akiyama, H Morita, K. Takeya and H Itokawa 1987. "Diarylheptanoids from the Rhizomes of *Curcuma xanthorrhiza* and *Alpinia officinarum,*" *Chem. Pharm. Bull ,* 35(8):3298–3304.

Uehara, S , I. Yasuda, K. Takeya and H. Itokawa. 1992. "Comparison of the Commercial Tumeric and Its Cultivated Plant by Their Constituents," *Shoyakugaku Zasshi,* 46(1):55–61.

Wu, T. S., Y. C. Wu, P. L. Wu, C. Y Chem, Y. L. Leu and Y Y. Chan. 1998 "Structure and Synthesis of [n]-Dehydroshogaols from *Zingiber officinale,*" *Phytochemistry,* 48(5):889–891

Yamada, Y., H. Kikuzaki and N. Nakatani 1992. "Identification of Antimicrobial Gingerols from Ginger (*Zingiber officinale* Roscoe)," *J. Antibact. Antifung. Agents,* 20(6).309–311.

Yamahara, J., Q. Huang, Y. Li, L. Xu and H. Fujimura. 1990 "Gastrointestinal Motility Enhancing Effect of Ginger and Its Active Constituents," *Chem. Pharm Bull.,* 38(2):430–431.

Yamahara, J., S. Hatakeyama, K. Taniguchi, M. Kawamura and M. Yoshikawa. 1992. "Stomach-ache Principles in Ginger. II. Pungent and Anti-Ulcer Effects of Low Polar Constituents Isolated from Ginger, the Dried Rhizoma of *Zingiber officinale* Roscoe Cultivated in Taiwan. The Absolute Stereostructure of a New Diarylheptanoid," *Yakugaku Zasshi,* 112(9)·645–655.

Yoshikawa, M., S. Hatakeyama, N Chatani, Y. Nishino and J. Yamahara. 1993. "Qualitative and Quantitative Analysis of Bioactive Principles in *Zingiber*is Rhizoma by Means of High Performance Liquid Chromatography and Gas Liquid Chromatography. On the Evaluation of *Zingiber*is Rhizoma and Chemical Change of Constituents during *Zingiber*is Rhizoma Processing," *Yakugaku Zasshi,* 113(4):307–315.

Yoshikawa, M , S. Yamaguchi, K Kunimi, H. Matsuda, Y. Okuno, J Yamahara and N. Murakami. 1994. "Stomach ache Principles in Ginger. III. An Anti-Ulcer Principle, 6-Gingesulfonic Acid, and Three Monoacyldigalactosylglycerols, Gingerglycolipids A, B, and C, from *Zingiber*is Rhizmoma Originating in Taiwan," *Chem. Pharm. Bull.,* 42(6):1226–1230.

CHAPTER 5

Chemistry and Pharmacology of Fenugreek

Y. SAUVAIRE
P. PETIT
Y. BAISSAC
G. RIBES

1. INTRODUCTION

IT is generally acknowledged that plants have a highly advanced chemical potential, especially with respect to the synthesis of different, often complex, molecules. Indeed, a plant could be considered as a "model laboratory" that traps energy from the sun via photosynthesis and provides mankind with a wide range of useful substances. Some of these substances are already being utilized, while others are being developed for medicinal and dietary applications.

Fenugreek (*Trigonella foenum-graecum* L.), which has been the focus of our research for several years at the University of Montpellier, has turned out to be a source of phytochemicals with unique chemical structures and innovative biological and pharmacological properties. We will discuss these features in the present review while concentrating on fenugreek seeds which are the most investigated and utilized part of this plant.

1.1. THE PLANT

Fenugreek[1] is an herbaceous annual plant that grows up to 40–60 cm high with alternate trifoliate leaves and pale yellow triangular flowers (Figure 5.1).

[1]Fenugreek names in other languages Amharic = Abish, Ancient Greek = aillis; Arabic = helba or helbeh, Armenian = hambala, Chinese = hou lou pa, French = fenugrec; German = griechisches heu or bockschornklee Hindu = methi, Italian = fienogreco, Persian = schemlit, Russian = pazhitnik, Spanish = fenogreco or alholva

Figure 5.1 Fenugreek, *Trigonella foenum-graecum* L

Its pods are curved and range from 10–15 cm in length and 0.5 cm in width, with narrow ends of 2–3 cm in length.

Fenugreek is a dicotyledon belonging to the Leguminosae family, the Trifoliae tribe, the Trigonellinae subtribe and the *Trigonella* genus. The pod contains 10 to 12 brown or yellowish seeds that are hard and have an unusual smell. The lozenge-shaped seeds are 4–5 mm long and 2–3 mm wide. The 1,000-seed weight is 9–23 g depending on the cultivar. When cut lengthwise, the seed coat, endosperm and embryo can be seen (Figure 5.2). The root system has nitrogen fixing nodules resulting from symbiosis with *Rhizobium meliloti*.

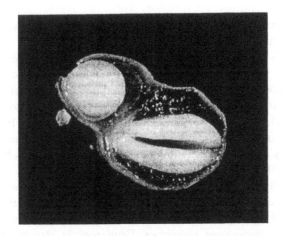

Figure 5.2 A section of a fenugreek seed showing seed coat, axis, endosperm and cotyledons. Magnification × 10.

Fenugreek is thought to originate from India or the Middle East where it grows naturally in some areas. It was first mentioned at the time of the Pharaons, when it was used to make a type of incense or was mixed with other plants or resins to embalm mummies. Around 200 BC, Cato the Elder mentioned the name fenugreek in the oldest known book on agriculture. He was the first to call it *foenum-graecum* (Greek hay) and recommended sowing it as a forage crop for cattle.

Today, fenugreek is grown primarily in the Mediterranean countries and India, although there is limited production in parts of Canada, especially Saskatchewan.

In trials carried out in Montpellier, fall-sown fenugreek produced much higher seed yields than spring-sown crops, with yields exceeding 3,000 kg/ ha for some cultivars. The different biochemical analyses showed that seed composition was relatively constant (Sauvaire, 1984). This crop is easy to mechanize, and the trials demonstrated that equipment used for cereals is perfectly suitable. As fenugreek is characterized by early emergence and a high nitrogen fixing capacity, two complementary crops can be grown on the same land in one year. For example, it is a good crop to grow before maize or sunflower in a rotation. In its vegetative state, fenugreek can be grown for green manure as an alternate crop and used as fodder for cattle and sheep.

1.2. TRADITIONAL USES

Fenugreek is, above all, known for its seeds and their aromatic and seasoning properties. In fact, fenugreek is found in most curry recipes and in certain culinary dishes such as *pastrama* or *helbé*. Fenugreek is also used as a base for a maple syrup substitute. In addition, some perfumes are known to contain compounds found in fenugreek.

Fenugreek's nutritional and tonic properties have also been tapped since ancient times. North African women still eat fenugreek seed to get plump, especially before marriage! Fenugreek-based preparations stimulate the appetite and promote weight gain in people with anemia, anorexia, tuberculosis or asthenia. These properties are also useful for livestock. Fenugreek seeds have been advocated for appetite stimulation, particularly in cattle and horses. However, they should not be fed fenugreek within a few days before slaughter or the meat will have a strong smell.

Fenugreek has numerous medicinal properties, most of which have been known for a very long time. Over and above their tonic qualities, fenugreek seeds are used for digestive disorders as they are reputed to have laxative, antiseptic and carminative properties. Fenugreek is also known to increase milk production in animals. When used externally, fenugreek may have emollient and resolvent properties.

Other interesting pharmacological applications for diabetes or hyperlipidemia will be discussed in the pharmacology section.

2. CHEMICAL COMPOSITION OF FENUGREEK SEEDS

2.1. FREE AMINO ACIDS AND PROTEINS

Fenugreek has a very unusual amino acid composition (Table 5.1) characterized by the predominance of 4-hydroxyisoleucine. This unusual substance

4-Hydroxyisoleucine

represents 0.6% of the seed weight and more than 85% of the free amino acids in the seeds; it is not found in the storage proteins (Sauvaire et al., 1984b). Two isomers coexist: the major ($\approx$ 90%), with the configuration (2S, 3R, 4S)-4 hydroxyisoleucine and the minor ($\approx$ 10%) with the configuration (2R, 3R, 4S) (Alcock et al., 1989; Sauvaire et al., 1998).

Compared to other pulses, very little research has been performed on fenugreek proteins. The high protein content of 25–32% in the seed led us to investigate the chemistry of fenugreek. Experiments carried out in Montpellier

TABLE 5.1. Amino Acid Composition of Fenugreek Seed Proteins Before and After Extraction of Free Amino Acids (FAA).

Amino Acid	Seed (g/16 g N)	Seed FAA (g/16 g N)	Free Amino Acid (g/100 g)
Aspartic acid	10 5	10 7	0 9
Threonine	3.3	3 3	0 3
Serine	4 7	4.7	0 3
Asparagine	—	—	1 2
Glutamic acid	16 9	16 9	, 1 4
Glutamine	—	—	tr
Proline	3.7	3 7	1 2
Glycine	4 5	4 5	tr
Alanine	3 0	3 9	tr
Valine	3 5	3 5	0 3
Cystine	1.4	1 7	tr
Methionine	1 0	0 8	tr
Isoleucine	3 9	4 4	0 2
Leucine	6 2	6 8	0.2
Tyrosine	2 8	2 4	0 9
Phenylalanine	3 8	4.0	1.3
Aminobutyric acid	—	—	0 4
Histidine	2 1	1 9	tr
Lysine	6 6	6 0	0 1
Arginine	8 8	8 9	1 7
Tryptophan	0 7	1 0	tr
Hydroxyisoleucine	3 6	0 0	0 63

tr traces
Source: Adapted from Sauvaire (1984)

at sites with different environmental and cropping conditions revealed that the protein content of harvested seed was relatively constant over a five-year period and ranged from 28.2 to 32.1%. By comparison, the protein content of other leguminous seeds were soybean, 38%; pea, 22.5%; and lupin, 31.2%.

Ninety percent of fenugreek proteins are located in cotyledons and embryos (C+E) that are separated from the seed coat and the endosperm on the basis of their differences in density. Most of these proteins are water soluble and belong to the albumin class (Sauvaire et al., 1984a).

Of the essential amino acids, lysine, for example, fluctuates between 5.8 and 6.6 grams per 16 g nitrogen with the mean value (6.3 ± 0.4g N) very similar to that of soybean. Compared to other grains used as livestock feed (Table 5.2), fenugreek ranks between horse bean and lupin in terms of its protein content and nutritional value (Sauvaire et al., 1976). Its amino acid composition is comparable to that of soybean with a high lysine content, twice that of wheat proteins. However, as with all legumes, fenugreek is deficient in methionine and cystine, relative to cereal proteins.

TABLE 5.2. Protein Content and Essential Amino Acid Composition of Fenugreek and Other Legumes Used in Animal Nutrition.

	Fenugreek	Soybean	Pea	Fababean	Lupin
Proteins (%)	29.8	38 0	22 5	23.4	31 2
Amino acids					
(g. 16gN¹)					
Lysine	6 3	6 4	7 5	6 5	5.3
Threonine	3 4	3 9	4 1	3 4	3 6
Valine	3 6	4 8	4 7	4 4	4 0
Cystine	1 5	1 3	1 1	0 8	1 4
Methionine	1 0	1 3	0 9	0 7	0 8
Isoleucine	4 3	4 5	4.3	4 0	4 4
Leucine	6 6	7 8	6 8	7 1	7 2
Tyrosine	2 9	3 1	2 7	3 2	3 5
Phenylalanine	4 0	4 9	4 6	4 3	3 7
Tryptophane	0 8	1 3	0 9	0 9	1 0

Source Adapted from Sauvaire (1984)

In addition, fenugreek contains trypsin inhibitors (13.6 unit/mg sample) at concentrations similar to those found in peas and much lower than those in soybean (70–100 unit/mg sample) (Sauvaire, 1984).

2.2. FLAVOR

Fenugreek seed has a very characteristic smell which is used for its pleasant appetizing aroma. Toasted and ground fenugreek seed is an essential ingredient in curry powders. It is used as a seasoning in food products such as pickles, chutneys, vanilla extracts, artificial maple syrup, coffee extract and as a flavoring in tobacco. Over 50 volatile constituents have been detected, 39 of which have been identified as sesquiterpenes, γ and δ-lactones, n-alkanes and unsaturated aliphatic hydrocarbons (Girardon et al., 1985).

Their contribution to the aroma is variable, but none of them are responsible for the seed's characteristic persistent quality. 3-hydroxy-4,5-dimethyl-2(5H)-furanone (sotolone) was established as an important volatile constituent

Sotolone

of fenugreek due to the characteristic impact of its flavor (Girardon et al.,

1986; Blank et al., 1996). Sotolones occur predominantly in the (5S) enantiomeric form, and about 2–25 mg/kg were found in fenugreek of different origins using the isotope dilution technique (Blank et al., 1997). It is now well established that sotolone is generated from 4-hydroxyisoleucine, the most abundant free amino acid in fenugreek seeds, as a result of an oxidative deamination process (Sauvaire et al., 1984b; Sauvaire et al., 1993; Blank et al., 1997).

2.3. TRIGONELLINE

Trigonelline or N-methylnicotinic acid was first isolated from *Trigonella*

Trigonelline

foenum-graecum in 1895 and was subsequently isolated from other biological sources such as coffee.

Fenugreek seeds contain approximately 0.1–0.15% trigonelline. In plants, there is sufficient evidence to indicate that this compound is a phytohormone in the classical sense and is responsible for predominant cell arrest in G2 in *Pisum sativum* roots (Evans and Tramontano, 1981).

2.4. LIPIDS

Fenugreek seeds contain 5.5–7.5% total lipids (extracted by hexane). The total lipids consist of 84.1% neutral lipids, 5.4% glycolipids and 10.5% phospholipids. The neutral lipids consist mainly of triacylglycerols (86%), diacylglycerols (6.3%) and small quantities of monoacylglycerols, free fatty acids and sterols (Hemavathy and Prabhakar, 1989). Linoleic acid is the predominant fatty acid (~40%), followed by linolenic acid (~25%), oleic acid (~14%) and palmitic acid (~11%) (Baccou et al., 1978).

2.5. STEROLS, SAPOGENINS AND STEROID SAPONINS

The steroid composition of the fenugreek seed is unique. Here we will examine the steroid substances successively as a function of their structural difference.

As in other plants, there are four classes of sterols. Free sterols and steryl esters are predominant proportionally (32% and 43%, respectively) compared to steryl glycosides (14%) and acylsteryl glycosides (11%) (Brenac, 1993). Sitosterol and 24-methyl-cholesterol are the main sterols. The total amount of sterols represents approximately 0.2% in the fenugreek seed (Artaud et al., 1988; Brenac and Sauvaire, 1996a). The sterolic composition (Table 5.3) is also characterized by a very low level of stigmasterol and the presence of Δ^7 sterols and pollinastanol (14 α-methyl-9B,19-cyclo-5α-cholestan-3β-ol), as determined by GC-MS and 1H NMR spectroscopy. This uncommon sterol, which has never been found in plant tissues other than pollens, could be considered as a chemotaxonomic marker.

A review of the literature on the subject revealed that the presence of special steroid substances in fenugreek seeds was first reported in 1919 by Wunschendorff. This discovery was confirmed in various follow-up studies, notably by Marker et al. (1947) who detected the presence of diosgenin after hydrolysis of the plant material.

Steroidal sapogenins are mainly found as glycosides, and although they have been isolated almost exclusively from plants, very few plants accumulate these compounds in their seeds. Fenugreek seeds have been recognized as a potential source of diosgenin, a basic compound in the hemisynthesis of steroid drugs such as cortisone and sex hormones. This substance is mainly extracted from Dioscoreaceae tubers.

Using more sophisticated analytic techniques, we detected and identified 10 different sapogenins including diosgenin (Brenac and Sauvaire, 1996a, 1996b; Taylor et al., 1997).

TABLE 5.3.   Sterols in Fenugreek Seeds.

Sterol	Content (% Total Sterols)
Cholesterol	4.7
Δ^7-Cholesterol	2 0
Pollinastanol	2 6
24-Methyl-cholesterol	11.9
24-Methylene-cholesterol	0.4
Stigmasterol	0.7
Δ^7-Campestenol	0 3
Silosterol	71.7
Fucosterol	0.7
Δ^5-Avenasterol	1 6
Δ^7-Stigmasterol	0 7
3 unknowns	2 7

Source Adapted from Brenac and Sauvaire (1996a)

TABLE 5.4. Composition of Steroidal Sapogenins of Fenugreek Seed.

Steroidal Sapogenin	Content (% Total Sapogenins)
Spirosta-3,5-diens	0 7
Smilagenin	0 6
Sarsasapogenin	1 2
Diosgenin	39.0
Tigogenin + yamogenin	30 2
Neotigogenin	7.0
Yuccagenin + lilagenine	5 2
Gitogenine	10 8
Neogitogenine	5.3

Source Adapted from Brenac and Sauvaire (1996a)

These steroidal sapogenins are structurally related (Table 5.4). The 10 compounds identified by GC-MS differ in terms of the presence or absence of a double bond at C-5 and position 5α or 5β of the hydrogen atom, resulting from a reduction in this double bond; the presence or absence of a second hydroxyl group at C-2 (2α); and the conformational position R or S of the methyl group at C-25.

Fenugreek seed contains 1.5% of steroidal sapogenin. Diosgenin [(25R)-spirost-5-en-3β-ol] is the major compound and, along with 25S epimer and yamogenin (not separated from tigogenin [(25R)-5a-spirostan-3β-ol], accounted for approximately 70% of total sapogenins in our analysis (Brenac and Sauvaire, 1996b). Fenugreek is also characterized by the presence of dihydroxylated sapogenins: yuccagenin [(25R)-spirost-5-en-2α, 3β-diol] and its 25S epimer lilagenin and gitogenin [(25R)-5a-spirostan-2α, 3β-diol] and its 25S epimer neogitogenin. In addition, the presence of a very low level of spirosta-3,5-diens confirmed the efficacy of our conditions for hydrolysis. These products are formed under acid hydrolysis when sapogenins with a double bond between C-5 and C-6 lose the hydroxyl group at C-3.

The first study on saponins in fenugreek was performed by Heintz in 1959. Different saponins were identified and submitted for structural analyses by four research teams: Bogacheva et al. (1976, 1977) (Russia), Hardman et al. (1980) (UK), Gupta et al. (1984, 1985a, 1985b, 1986) (India) and Yoshikawa et al. (1997, 1998) (Japan).

Saponins isolated using column chromatography or droplet countercurrent chromatography and 1N NMR and FAB-MS have shed light on the structures. These bidesmosidic saponins have two sugar chains: one bonded at C-3 and one attached by an ether linkage at C-26 with a D-glucose (furostanol saponins) (see p. 116). The structures of the steroidal saponins of fenugreek seed are shown in Table 5.5.

R : Sugars   Steroid saponins
(furostanol glycosides)

R : Sugars   Steroid saponins
(spirostanol glycosides)

R : H   Diosgenin

Steroid saponins were obtained from fenugreek seed by extraction, with several purification stages, including dialysis (Sauvaire et al., 1994; 1996). The product obtained, total furostanol saponin, contained only steroid saponin. Saponins occur in furostanic form, and fenugreek seeds contain approximately 4–6% saponins (Leconte, 1996).

2.6. GALACTOMANNAN

Trigonella foenum-graecum seeds have a seed coat that is separated from the embryo by a well-developed endosperm with an aleurone layer. The aleurone cells, which form the outermost cell layer of the endosperm, are small and thick-walled. The rest of the endosperm is composed of large cells with thin primary walls that appear to be completely filled with galactomannan, a non-starch storage polysaccharide (Reid, 1971).

This polymer can be isolated by treating the endosperm or the whole seed with water or dilute acid. Ethanol is then added to the filtrate (up to 50%) and a white hair-like precipitate is formed, dried and weighed. The mature fenugreek seed contains approximately 15% of galactomannan (Ribes et al., 1986).

Structurally, this polymer is based on a linear (1 → 4) β linked D-mannan

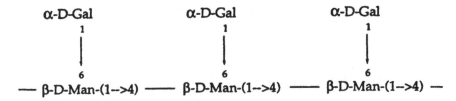

Galactomannan

"backbone" to which D galactosyl residues are attached as single-unit side chains by (1 → 6) α-linkages. The mannose:galactose ratio in fenugreek is approximately 1.1 or 1.2 (1.6 in guar, 3.4 in locust bean) (Buckeridge and Reid, 1996).

The composition of guar and fenugreek, with their high proportion of galactosyl residues, makes galactomannan readily soluble in cold water. However, hot water is needed to dissolve the galactomannan from the carob bean. Given the strong similarity between the structures and properties of fenugreek and guar, it is not surprising that some effort is being made to commercialize this new source of galactomannan (Garti et al., 1997). Products containing

TABLE 5.5. Structures of Steroid Saponins of Fenugreek Seed.

Steroid Saponins	Structure	References
Trigonelloside C	Rha⁴ Glc³ [(25S)-furost-5-en-3β,22,26-triol]26 Glc 2	Bogacheva et al., 1976, 1977
Furostanol glycoside	Glc³ Glc³ [(25S)-5α-furostan-3β,22,26-triol]²⁶ Glc 2	Hardman et al., 1980
Trigofoenoside A	Glc³ [(25S)-furost-5-en-3β,22,26-triol]²⁶ Glc 2	Gupta et al., 1985a
Trigofoenoside B	Rha⁴ Glc³ [(25S)-5α-furostan-2α,3β,22,26-tetraol]²⁶ Glc	Gupta et al., 1986
Trigofoenoside C	Rha⁴ Glc³ [(25R)-5α-furostan-2α,3β,22,26-tetraol]²⁶ Glc 2	Gupta et al., 1986
Trigofoenoside D	Glc³ Glc³ [(25S)-furost-5-en-3β,22,26-triol]²⁶ Glc 2	Gupta et al., 1985a
Trigofoenoside E	Xyl⁴ Glc³ [(25R)-5α-furostan-3β,22,26-triol]²⁶ Glc 2	Gupta et al., 1985b
Trigofoenoside F	Glc⁶ Glc³ [(25R)-furost-5-en-3β,22,26-triol]²⁶ Glc 2	Gupta et al., 1984
Trigofoenoside G	Xyl⁴ Glc⁶ Glc³ [(25R)-furost-5-en-3β,22,26-triol]²⁶ Glc 2	Gupta et al., 1984
Trigoneoside Ia	Xyl⁶ Glc³ [25S]-5α-furostan-2α,3β,22,26-tetraol]²⁶ Glc	Yoshikawa et al., 1997
Trigoneoside Ib	Xyl⁶ Glc³ [(25R)-5α-furostan-2α,3β,22,26-tetraol]²⁶ Glc	Yoshikawa et al., 1997
Trigoneoside IIa	Xyl⁶ Glc³ [(25S)-5b-furostan-3β,22,26-triol]²⁶ Glc	Yoshikawa et al., 1997
Trigoneoside IIb	Xyl⁶ Glc³ [(25R)-5β-furostan-3β,22,26-triol]²⁶ Glc	Yoshikawa et al., 1997
Trigoneoside IIIa	Glc³ [(25S)-5α-furostan-3β,22,26-triol]²⁶ Glc 2-Rha	Yoshikawa et al., 1997
Trigoneoside IIIb	Glc³ [(25R)-5α-furostan-3β,22,26-triol]²⁶ Glc 2 Rha	Yoshikawa et al., 1997
Trigoneoside IVa	Glc⁴ Glc³ [(25S)-furost-5-ene-3β; 22,26-triol]²⁶ Glc 2 Rha	Yoshikawa et al., 1998
Trigoneoside Va	Glc⁶ Glc³ Glc⁴ Glc³ [(25S)-furost-5-ene-3β,22,26-triol]²⁶ Glc 4 2-Xyl—Rha	Yoshikawa et al., 1998
Trigoneoside Vb	Glc⁶ Glc³ Glc⁴ Glc³ [(25R)-furost-5-ene-3β,22,26-triol]26, Glc 4 2 Xyl Rha	Yoshikawa et al., 1998
Trigoneoside VI	Xyl⁶ Glc³ Glc⁴ Glc³ [furost-5,25(27)-diene-3β,22,26-triol]²⁶ Glc 4 2 Xyl Rha	Yoshikawa et al., 1998
Trigoneoside VIIb	Xyl⁶ Glc³ Glc⁴ Glc³ [(25R)-furost-5-ene-3β,22,26-triol]²⁶ Glc 4 2 Xyl Rha	Yoshikawa et al., 1998
Trigoneoside VIIIb	Glc⁶ Glc³ Glc⁴ Glc³ [(25R)-5a-furostane-3β,22,26-triol] Glc-4 2 Xyl Rha	Yoshikawa et al., 1998

galactomannan are used as thickeners, texture improvers or emulsifiers in the food-processing industry, especially in low-fat products.

In terms of functions for the plant, it has been shown that galactomannan is hydrolyzed during germination, and the free simple sugars can then be metabolized and even temporarily transformed into starch in the cotyledons. In addition to its water retention properties, galactomannan can act as a water reservoir, buffering the germinating embryo against desiccation during subsequent temporary periods of drought (Reid and Bewley, 1979).

2.7. OTHER CONSTITUENTS

In a recent study on isoflavonoid and lignan content of fenugreek, Mazur et al. (1998) found that the levels of phytoestrogen were very low and ranged from 8.6 μg/100 g for secoisolariciresinol to 9.8 mg/100 for genistein and 10.2 mg/100 g for diadzein, respectively.

Fenugreek has also been suggested as a potential source of natural antioxidants (Hettiarachchy et al., 1996).

3. PHARMACOLOGY OF FENUGREEK

3.1. EFFECTS ON GLUCOSE HOMEOSTASIS

Fenugreek seeds are traditionally known to have antidiabetic properties (Moissides, 1939). The glucose lowering effect was demonstrated *in vivo* and attributed to the defatted fraction of the seed, which induced hyperglycemia in normal animals and decreased hyperglycemia and glycosuria in alloxan-induced diabetic dogs (Ribes et al., 1984). These properties were further investigated in insulin-dependent diabetic dogs, and the testa and endosperm fraction of the seed, which has high viscosity and is particularly rich in fibers, was found to be the active subfraction (Ribes et al., 1986). A dose-related hypoglycemic effect was also noted in normal and alloxan-induced diabetic rats (Khosla et al., 1995) as well as in normal and alloxan-diabetic mice (Ajabnoor and Tilmisany, 1988).

Clinical studies with fenugreek seed revealed an improvement of glucose tolerance in healthy volunteers (Sharma, 1986) as well as in type 2 (non-insulin dependent) diabetic patients (Madar et al., 1988; Sharma and Raghuram, 1990). In 60 type 2 diabetic patients, a diet containing fenugreek seed powder administered for 24 weeks lowered fasting blood glucose levels, glycosylated hemoglobin and urinary sugar excretion and improved glucose tolerance at lower plasma insulin levels (Sharma et al., 1996b). In a recent placebo-controlled study, fenugreek did not affect fasting or postprandial blood sugar in healthy individuals; it significantly reduced glycemia in mild but not in severe cases of type 2 diabetes (Bordia et al., 1997). Moreover, in type 1 (insulin-dependent) diabetic patients, a fenugreek diet reduced fasting blood sugar and urinary glucose excretion and improved glucose tolerance (Sharma et al., 1990).

The antidiabetic effect of fenugreek has been attributed mainly to the high fiber content of the seeds, with slower gastric emptying and subsequent reduction of glucose intestinal absorption (Madar and Thorne, 1987). The soluble dietary fiber fraction in fenugreek, with galactomannan as a major constituent, was shown to reduce glycemia after glucose ingestion (Ali et al., 1995). Hence, by slowing gastric emptying and forming a nonabsorbable viscous gum when mixed with water (viscosity effect), the fiber may reduce or delay intestinal absorption of glucose, which subsequently may improve glycemic control.

The major alkaloid trigonelline was previously reported to have a hypoglycemic effect (Mishkinsky et al., 1967; Shani et al., 1974). Fenugreek was also reported to increase insulin binding sites of erythrocytes, and it may improve peripheral glucose utilization (Raghuram et al., 1994).

In most studies with animals or human subjects, a significant reduction in plasma glucose concentrations following fenugreek administration has been observed with no significant elevation or even with a reduction in plasma insulin concentrations. However, after subchronic administration of a fenugreek seed extract in normal rats, an increase in morning plasma insulin level was observed in overnight fed animals, suggesting the presence of an insulin stimulating compound in the extract (Petit et al., 1993). Indeed, the amino acid 4-hydroxyisoleucine was extracted and purified by sequential chromatography from defatted fenugreek seeds. This compound is not found in mammalian tissues and is only present in plants, particularly in *Trigonella* species. It induced concentration-dependent stimulation of insulin secretion *in vitro* from rat incubated Langerhans islets in the micromolar range of concentrations (200–1,000 μmol/L). At 200 μmol/L concentration, 4-hydroxyisoleucine induced a biphasic insulin response in rat isolated perfused pancreas in the presence of a slightly stimulating glucose concentration (8.3 mmol/L) (Sauvaire and Ribes, 1992–1994). It was ineffective in the presence of 5.0 mmol/ L glucose and induced a glucose-dependent response in the presence of intermediate to high glucose concentrations (Sauvaire et al., 1998) (Figure 5.3). The amino acid was effective *in vivo* in conscious fasted dogs in improving oral glucose tolerance after oral administration (Hillaire-Buys et al., 1993) and in type 2 diabetic rats (Broca et al., 1998). It also increased insulin secretion in human islets at micromolar concentrations, similar to results obtained in animal models (Fernandez-Alvarez et al., 1996; Sauvaire et al., 1998). The coupling mechanism of the secretagogue action of 4-hydroxyisoleucine remains to be clearly established. The drug only partially affects the diazoxide-induced increase in potassium permeability of the B cell plasma membrane in the presence of 3.0 mmol/L glucose, without any significant

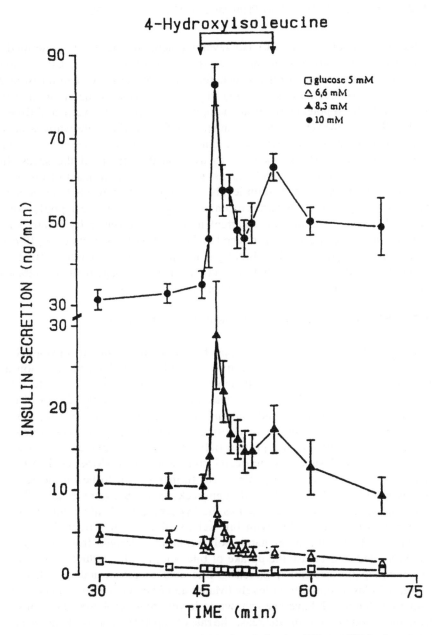

Figure 5.3 Kinetics of insulin secretion of 4-hydroxyisoleucine (200 mmol/L) in the presence of different glucose concentrations (5, 6.6, 8.3 and 10 mmol/L) Adapted from Sauvaire et al. (1998).

effect per se (Petit et al., 1995b). However, it potentiates the effect of intermediary glucose concentrations on calcium signaling (Petit et al., 1997). Concerning structure-activity relationships, it was reported that the major natural isomer of 4-hydroxyisoleucine (2S, 3R, 4S), representing 97% of the compound in seeds, is the most potent structural analogue tested so far (Sauvaire and Ribes, 1996–1997). However, different synthetic monomethylated derivatives have been found to be more potent than leucine and isoleucine (Ribes et al., 1996).

Most studies on fenugreek have focused on the effects of the seeds. In subjects fed a diet of fenugreek leaves, there were no significant differences in either blood glucose or insulin levels as compared to subjects fed a control diet (Sharma, 1986). In contrast, a recent investigation reported a hypoglycemic effect of an aqueous extract of fenugreek leaves when given orally or intraperitoneally in normal and alloxan-induced diabetic rats (Abdel-Barry et al., 1997).

Taken together, these data highlight the potentially beneficial effect of fenugreek seed as a nutritional supplement in the management of diabetes and of some of its components for pharmaceutical uses.

3.1.1. Effects on Plasma Cholesterol Concentration

Fenugreek was shown to have hypocholesterolemic properties (Singhal et al., 1982) and was shown to prevent a diet-induced cholesterol elevation in rats (Sharma, 1984). In normal and diabetic dogs, a lipid extract was ineffective, and the hypocholesterolemic effect was attributed to the defatted part of the seeds, which is rich in fibers and contains steroid saponins (Valette et al., 1984). The fiber-rich fraction (testa and endosperm) induced an hypocholesterolemic effect, and the saponin- and protein-rich fraction (cotyledon and axes) was shown to reduce plasma cholesterol and triglyceride levels (Ribes et al., 1987). Saponins, alone or with diosgenin, were further shown to be implicated in the hypocholesterolemic effect of fenugreek seeds (Sauvaire et al., 1991). Saponins have also been identified as the hypocholesterolemic component of fenugreek seeds, interacting with bile salts in the digestive tract (Stark and Madar, 1993). On the other hand, polysaccharides derived from fenugreek were reported to decrease the uptake of bile acid, i.e., the reduced efficiency of enterohepatic circulation increases bile acid excretion and may lead to decreased plasma cholesterol levels (Madar and Shomer, 1990). Galactomannans from fenugreek have been shown to lower cholesterol concentrations in liver and blood plasma, and decrease the rate of hepatic synthesis of cholesterol (Evans et al., 1992).

In a clinical study in hyperlipidemic nondiabetic subjects, incorporation of defatted fenugreek in the diet resulted in a significant reduction of serum total cholesterol, LDL- and VLDL-cholesterol as well as triglyceride levels. HDL-cholesterol levels were not altered, but the ratio with total cholesterol and LDL and VLDL cholesterol were significantly increased, suggesting a beneficial effect in the lipid profile (Sharma et al., 1991). Fenugreek seeds have been found to exert hypocholesterolemic activity in diabetic patients (Sharma et al., 1990). In type 2 diabetic patients, administration of fenugreek seed powder resulted in a sustained and long-lasting reduction of total cholesterol, LDL- and VLDL-cholesterol and triglyceride levels, while a slight but insignificant rise in HDL-cholesterol was observed (Sharma et al., 1996c).

These data suggest that some fenugreek components may have a beneficial effect on the lipid profile of diabetic subjects. The ability of fenugreek to reduce the LDL and VLDL fractions of total cholesterol may be beneficial in preventing atherosclerosis.

3.2. OTHER PHARMACOLOGICAL EFFECTS

Fenugreek seeds are traditionally assumed to have restorative and nutritive properties. When used as a dietary supplement, Fenugreek seeds did not alter the food intake of animals (Udayasekhara Rao et al., 1996). In a long-term study in type 2 diabetic patients, food consumption and mean energy intake during control and experimental (fenugreek seed administration) periods were reported to be almost similar and constant with no significant change in body weight (Sharma et al., 1996a). However, continued administration of a fenugreek seed extract in rats was shown to increase the animal's appetite and food consumption (Petit et al., 1993). A regular treatment with purified steroid saponins from fenugreek seeds was shown to increase food intake and appetite in normal rats, while modifying the circadian rhythm of feeding behavior. This treatment was able to stabilize food consumption in diabetic animals, resulting in progressive weight gain in these animals, as compared to untreated diabetic controls (Sauvaire et al., 1994; Petit et al., 1995a).

An aqueous extract of fenugreek leaves was recently shown to produce antinociceptive effects in a dose-dependent manner in rats, and both central and peripheral mechanisms were suggested (Javan et al., 1997).

3.3. ADVERSE EFFECTS AND TOXICITY

Among the various antidiabetic plants that have been proposed, fenugreek is generally considered to be nontoxic (Marles and Farnsworth, 1995). A safety evaluation of fenugreek seeds was performed in rats whose diet was supplemented with fenugreek seed flour (up to 20%) for 90 days; fenugreek appeared to be essentially nontoxic (Udayasekhara Rao et al., 1996). In a 24-week clinical investigation in type 2 diabetic patients, a diet containing 25 g fenugreek seed powder appeared to be suitable, i.e., some patients complained initially of minor gastrointestinal symptoms, such as diarrhea and excess flatulence, which subsided after a few days, and no other adverse effect was observed (Sharma et al., 1996a, 1996b, 1996c).

TABLE 5.6. Principal Pharmacological Effects of Fenugreek.

Reported Activity	Species	Reference
Glucose homeostasis (antidiabetic and/or insulinotropic properties)	Dogs	Ribes et al , 1984 Ribes et al , 1986
	Rats	Shani et al , 1974 Khosla et al., 1995 Abdel-Barry et al , 1997 Sauvaire et al , 1998 Broca et al., 1998
	Mice	Ajabnoor and Tilmisany, 1988
	Humans	Sharma, 1986 Madar et al , 1988 Sharma and Raghuram, 1990 Fernandez-Alvarez et al , 1996 Sharma et al , 1996b Bordia et al , 1997
Lipidemia (hypocholesterolemic and hypotriglyceridemic properties)	Rats	Singhal et al , 1982 Sharma, 1984 Madar and Shomer, 1990 Evans et al , 1992 Stark and Madar, 1993
	Dogs	Valette et al , 1984 Ribes et al., 1987 Sauvaire et al , 1991
	Humans	Sharma and Raghuram, 1990 Sharma et al., 1990 Sharma et al , 1991 Sharma et al., 1996c
Feeding behavior (food intake and motivation to eat)	Rats	Petit et al , 1993 Petit et al , 1995a
Nociception (antinociceptive properties)	Rats	Javan et al , 1997

Allergic reactions after consumption of spices are well known. Two cases of allergy to fenugreek after inhalation of the seed powder or after skin application of a paste were recently documented (Patil et al., 1997).

4. CONCLUSIONS

Fenugreek seeds contain several compounds such as 4-hydroxyisoleucine, trigonelline as well as certain aromatic compounds and steroidal substances that have not been found in other plants. All of these components, alone or in combination, provide this plant with a number of pharmacological and therapeutic properties (Table 5.6). In addition, fenugreek is one of the few plants of the Leguminosae family that contains diosgenin, a key component used for producing steroidal drugs through hemisynthesis. It is, thus, quite likely that the components of fenugreek will find applications for the treatment and prevention of a wide range of diseases.

5. REFERENCES

Abdel-Barry, J. A., Abdel-Hassan, I A and Al-Hakiem, M. H. H. 1997. "Hypoglycaemic and antihyperglycaemic effects of *Trigonella foenum-graecum* leaf in normal and alloxan induced diabetic rats," *J. Ethnopharmacol* 58·149–155

Ajabnoor, M A. and Tilmisany, A K. 1988. "Effect of *Trigonella foenum-graecum* on blood glucose levels in normal and alloxan-diabetic mice," *J Ethnopharmacol.* 22:45–49.

Alcock, N W , Crout, D. H G., Gregorio, M. V M , Lee, E , Pike, G. and Samuel, C. J 1989 "Stereochemistry of the 4-hydroxyisoleucine from *Trigonella foenum-graecum*," *Phytochemistry* 28·1835–1841

Ali, L., Azad Khan, A K., Hassan, Z., Mosihuzzaman, M., Nahar, N., Nasreen, T , Nur-e-Alam, M. and Rokeya, B. 1995. "Characterization of the hypoglycemic effects of *Trigonella foenum-graecum* seed," *Planta Med* 61 358–360

Artaud, J., Iatrides, M C , Baccou, J. C and Sauvaire, Y 1988. "Particularités de la composition stérolique des huiles de deux *Trigonella*," *Rev. Fse Corps Gras.* 35:435–440.

Baccou, J. C , Sauvaire, Y., Olle, M., Petit, J 1978 "L'huile de fenugrec. composition, propriétés, possibilités d'utilisation dans l'industrie des peintures et vernis," *Rev. Fse Corps Gras.* 25:353–359.

Blank, I , Lin, J., Devaud, S , Fumeaux, R. and Fay, L B. 1997. "The principal flavor components of fenugreek (*Trigonella foenum-graecum* L.)." In: *Spices, Flavor Chemistry and Antioxidant Properties,* S. J. Risch and C T. Ho (eds.), ACS symposium series 660, pp. 12–28.

Blank, I , Lin, J , Fumeaux, R., Welti, D H and Fay, L. B. 1996. "Formation of 3-hydroxy-4,5-demethyl-2(5H)- furanone (sotolone) from 4-hydroxy-L-isoleucine and 3-amino-4,5-dimethyl-3,4-dihydro-2(5H)-furanone," *J. Agric. Food Chem.* 44·1851–1856.

Bogacheva, N. G., Kiselev, V. P and Kogan, L. M 1976. "Isolation of 3,26 bisglycoside of yamogenin from *Trigonella foenum-graecum*," *Khim. Prir. Soed.* 2:268–269 (Chem. Abstr., 1976, 85, 106634)

Bogacheva, N. G., Sheichenko, V. L and Kogan, L. M. 1977 "Structure of yamogenin tetroside from *Trigonella foenum-graecum* seeds," *Khim. Farm Zh.* 11:65–69 (Chem. Abstr , 1977, 11, 180685)

Bordia, A., Verma, S. K. and Srivastava, K C. 1997. "Effect of ginger (*Zingiber officinale* Rosc) and fenugreek (*Trigonella foenum-graecum* L.) on blood lipids, blood sugar and platelet aggregation in patients with coronary artery disease," *Prostaglandins Leukot Essent Fatty Acids.* 56·379–384.

Brenac, P. 1993. "Sterols et sapogenines stéroidiques du fenugrec (*Trigonella foenum-graecum* L.). Dynamique de l'accumulation de ces métabolites dans les graines". Ph. D. Thesis, University of Montpellier II, France

Brenac, P. and Sauvaire, Y. 1996a. "Chemotaxonomic value of sterols and steroidal sapogenins in the genus *Trigonella*," *Biochem. Syst. Ecol.* 24.157–164.

Brenac, P. and Sauvaire, Y 1996b. "Accumulation of sterols and steroidal sapogenins in developing fenugreek pods: Possible biosynthesis in situ," *Phytochemistry.* 41·415–422

Broca, C., Gross, R , Petit, P., Sauvaire, Y., Manteghetti, M , Masiello, P., Gomis, R. and Ribes, G. 1998. "4-Hydroxyisoleucine improves glucose tolerance in normal and NIDDM animals," *Diabetologia.* 41 (suppl. 1)·A239.

Buckeridge, M. S. and Reid, J. S G. 1996. "Major cell wall storage polysaccharides in legume seeds. Structure, catabolism and biological functions," *Cienc. Cult. (Sao Paulo).* 48:153–162.

Evans, A. J., Hood, R. L., Oakenfull, D G. and Sidhu, G. S 1992. "Relationship between structure and function of dietary fibre: A comparative study of the effects of three galactomannans on cholesterol metabolism in the rat," *Br J. Nutr.* 68·217–229.

Evans, L. S. and Tramontano, W A 1981. "Is trigonelline a plant hormone?" *Amer. J Bot* 68:1282–1289.

Fernandez-Alvarez, J., Sauvaire, Y., Petit, P., Casamitjana, R., Ribes, G. and Gomis, R. 1996. "Could 4-hydroxyisoleucine be used as a hypoglycaemic agent in the treatment of type 2 diabetes mellitus?" *Diabetologia.* 39 (Suppl 1). A 234.

Garti, N., Madar, Z., Aserin, A. and Sternheim, B. 1997 "Fenugreek galactomannans as food emulsifiers," *Lebensm. - Wiss. u - Technol.* 30:305–311.

Girardon, P., Baccou, J. C , Sauvaire, Y. and Bessière, J. M. 1985. "Volatile constituents of fenugreek seeds," *Planta Med.* 6.533–547

Girardon, P., Sauvaire, Y., Baccou, J C. and Bessière, J. M 1986. "Identification de la 3-hydroxy-4,5-diméthyl-2(5H)-furanone dans l'arôme des graines de fenugrec (*Trigonella foenum-graecum* L.)", *Lebensm - Wiss. u - Technol.,* 19.44–45.

Gupta, R. K , Jain, D. C. and Thakur, R. S. 1984. "Furostanol glycosides from *Trigonella foenum-graecum* seeds," *Phytochemistry.* 23:2605–2607

Gupta, R. K., Jain, D. C. and Thakur, R. S 1985a. "Furostanol glycosides from *Trigonella foenum-graecum* seeds," *Phytochemistry.* 24:2399–2401.

Gupta, R. K., Jain, D C. and Thakur, R. S. 1985b "Trigofoenoside E-1, a new furostanol saponin from *Trigonella foenum-graecum*," *Indian J. Chem.* 24B:1215–1217.

Gupta, R. K., Jain, D. C. and Thakur, R S. 1986. "Two furostanol saponins from *Trigonella foenum-graecum*," *Phytochemistry.* 25·2205–2207.

Hardman, R., Kosogi, J and Parfitt, R. J. 1980. "Isolation and characterization of a furostanol glycoside from fenugreek," *Phytochemistry* 19 698–700.

Heintz, S. 1959. "Les saponosides des graines de fenugrec", *C.R. Acad. Sci.* 248:283–286.

Hemavathy, J and Prabhakar, J. V. 1989 "Lipid composition of fenugreek (*Trigonella foenum-graecum* L.) seeds," *Food Chem.* 31:1–7

Hettiarachchy, N. S., Glenn, K. C., Gnanasambandam, R. and Johnson, M. G. 1996 "Natural antioxidant extract from fenugreek (*Trigonella foenum-graecum*) for ground beef patties," *J. Food Sci.,* 61:516–519.

Hillaire-Buys, D., Petit, P., Manteghetti, M., Baissac, Y., Sauvaire, Y and Ribes, G. 1993 "A recently identified substance extracted from fenugreek seeds stimulates insulin secretion in rat," *Diabetologia.* 36 (Suppl 1) A 119

Javan, M., Ahmadiani, A., Semnanian, S., and Kamalinejad, M. 1997. "Antinociceptive effects of *Trigonella foenum-graecum* leaves extract," *J Ethnopharmacol* 58·125–129.

Khosla, P., Gupta, D. D and Nagpal, R K 1995. "Effect of *Trigonella foenum-graecum* (Fenugreek) on blood glucose in normal and diabetic rats," *Indian J. Physiol Pharmacol.* 39:173–174

Leconte, O 1996. "Etude des saponines stéroïdiques du fenugrec (*Trigonella foenum-graecum* L.). Activité antifongique et approches allélopathiques *in vitro*," Ph D Thesis, University of Montpellier II France.

Madar, Z., Abel, R., Samish, S. and Arad, J 1988. "Glucose-lowering effect of fenugreek in noninsulin dependent diabetics," *Eur. J. Clin. Nutr* 42 51–54.

Madar, Z. and Shomer, I 1990. "Polysaccharide composition of a gel fraction derived from fenugreek and its effect on starch digestion and bile acid absorption in rats." *J Agric. Food Chem* 38.1535–1539.

Madar, Z and Thorne, R. 1987 "Dietary fiber," *Prog Food Nutr. Sci* 11 153–174.

Marker, R. E., Wagner, R. B., Ulshaffer, P R., Wittbecker, E L., Goldsmith, D. P. J and Ruof, C H. 1947. "New sources for sapogenins," *J. Am Chem Soc.* 69:2242.

Marles, R. J. and Farnsworth, N R 1995 "Antidiabetic plants and their active constituents," *Phytomedicine.* 2.137–189

Mazur, W. M., Duke, J. A., Wähälä, K., Rasku, S and Adlercreutz, H. 1998 "Isoflavonoids and lignans in legumes. Nutritional and health aspects in humans," *J Nutr Biochem* 9.193–200

Mishkinsky, J., Joseph, B. and Sulman, F G 1967. "Hypoglycaemic effect of trigonelline," *Lancet* 16.1311–1312

Moissides, M. 1939 "Le fenugrec autrefois et aujourd'hui," *Janus,* 43.123–130.

Patil, S P., Niphadkar, P. V. and Bapat, M M. 1997 "Allergy to fenugreek (*Trigonella foenum-graecum*)," *Ann. Allergy Asthma Immunol.* 78·297–300.

Petit, P., Liu, Y J., Broca, C., Sauvaire, Y., Ribes, G and Gylfe, E. 1997. "Calcium signalling is involved in the insulin-releasing effect of 4-hydroxyisoleucine," *Diabetologia.* 40 (Suppl 1).A 112.

Petit, P., Sauvaire, Y., Hillaire-Buys, D., Leconte, O M., Baissac, Y., Ponsin, G and Ribes, G 1995a "Steroid saponins from fenugreek seeds. Extraction, purification, and pharmacological investigation on feeding behavior and plasma cholesterol," *Steroids* 60 674–680.

Petit, P., Sauvaire, Y., Hillaire-Buys, D., Manteghetti, M., Baissac, Y., Gross, R. and Ribes, G 1995b "Insulin stimulating effect of an original amino acid, 4-hydroxyisoleucine, purified from fenugreek seeds," *Diabetologia.* 38 (Suppl 1)·A 101.

Petit, P., Sauvaire, Y., Ponsin, G., Manteghetti, M., Fave, A and Ribes, G. 1993 "Effects of a fenugreek seed extract on feeding behaviour in the rat: Metabolic-endocrine correlates," *Pharmacol. Biochem. Behav.* 45.369–374

Raghuram, T. C., Sharma, R. D., Sivakumar, B. and Sahay, B. K. 1994. "Effect of fenugreek seeds on intravenous glucose disposition in non-insulin dependent diabetic patients," *Phytotherapy Res.* 8.83–86.

Reid, J. S. G. 1971 "Reserve carbohydrate metabolism in germinating seeds of *Trigonella foenum-graecum* L (Leguminosae)," *Planta* 100.131–142.

Reid, J. S. G. and Bewley, J. D D G 1979 "A dual role for the endosperm and its galactomannan reserves in the germinative physiology of fenugreek (*Trigonella foenum-graecum* L.) an endospermic leguminous seed," *Planta* 147:145–150.

Ribes, G., Broca, C., Petit, P., Jacob, M., Baissac, Y., Manteghetti, M , Roye, M and Sauvaire, Y 1996. "Structure activity analysis of different analogues of the new insulinotropic agent 4-hydroxyisoleucine," *Diabetologia* 39 (Suppl 1): A 234.

Ribes, G., Da Costa, C , Loubatières-Mariani, M. M , Sauvaire, Y and Baccou, J. C 1987 "Hypocholesterolaemic and hypotriglyceridaemic effects of subfractions from fenugreek seeds in alloxan diabetic dogs," *Phytotherapy Res* 1:38–43.

Ribes, G , Sauvaire, Y., Baccou, J. C., Valette, G., Chenon, D., Trimble, E. R. and Loubatières-Mariani, M. M 1984. "Effects of fenugreek seeds on endocrine pancreatic secretions in dogs," *Ann. Nutr. Metab.* 28·37–43.

Ribes, G., Sauvaire, Y., Da Costa, C., Baccou, J. C. and Loubatières-Mariani, M. M 1986. "Antidiabetic effects of subfractions from fenugreek seeds in diabetic dogs," *Proc. Soc. Exp Biol Med* 182.159–166.

Sauvaire, Y. 1984. "Le fenugrec : son intérèt comme source de sapogénines stéroidiques, de protéines, d'huile . . Essais de valorisation," Ph D. Thesis, University of Montpellier II. France

Sauvaire, Y., Baccou, J. C. and Besançon, P 1976 "Nutritional value of the proteins of a leguminous seed: Fenugreek (*Trigonella foenum-graecum* L.)," *Nutr. Rep. Int.* 14.527–537.

Sauvaire, Y , Baccou, J C. and Kobrehel, K. 1984a. "Solubilization and characterization of fenugreek seed proteins," *J. Agri. Food Chem* 32 41–47.

Sauvaire, Y , Baissac, Y., Leconte, O., Petit, P. and Ribes, G 1996. "Steroid saponins from fenugreek and some of their biological properties," *Adv. Exp. Med Biol.* 405·37–46.

Sauvaire, Y., Brenac, P., Guichard, E and Fournier, N. 1993 "Relation entre la composition en acides aminés libres des graines de fenugrec et la qualité aromatique," *Aspects fondamentaux et appliqués de la biologie des semences*. Eds. D. Come and F Corbineau, in ASFIS, Paris, pp 201–206

Sauvaire, Y , Girardon, P , Baccou, J C and Risterucci, A. 1984b. "Changes in growth, proteins and free amino acids of developing seed and pod of fenugreek," *Phytochemistry* 23 479–486

Sauvaire, Y , Petit, P., Broca, C., Manteghetti, M , Baissac, Y , Fernandez-Alvarez, J , Gross, R , Roye, M , Leconte, A , Gomis, R. and Ribes, G 1998. "4-Hydroxyisoleucine, a novel amino acid potentiator of insulin secretion," *Diabetes.* 47·206–210.

Sauvaire Y and Ribes G. 1992–1994. "Composition capable of stimulating insulin secretion intended for the treatment of noninsulin-dependent diabetes," French patent 2,695,317; Eur. pat appl. EPO, 587,476; US patent 5,470,879; Japanese patent 217,588/93; Canadian patent 2,105,502; Indian patent 244/DEL/94.

Sauvaire Y. and Ribes G. 1996–1997. "Antidiabetic composition containing (2S, 3R, 4S)4-Hydroxyisoleucine," French patent 96,02,955; CT Int Appl. WO 97 32,577.

Sauvaire, Y., Ribes, G , Baccou, J. C. and Loubatières-Mariani, M M 1991. "Implication of steroid saponins and sapogenins in the hypocholesterolemic effect of fenugreek," *Lipids.* 26 191–197

Sauvaire Y., Ribes, G., Baissac, Y and Petit, P. 1994. "Composition de saponines et/ou de leurs formes aglycones et leurs applications comme médicaments," French patent 94,09,056.

Shani (Mishkinsky), J , Goldschmied, A., Joseph, B., Ahronson, Z. and Sulman, F G. 1974. "Hypoglycaemic effect of *Trigonella foenum-graecum* and *Lupinus termis* (Leguminosae) seeds and their major alkaloids in alloxan-diabetic and normal rats," *Arch. Int. Pharmacodyn.* 210.27–37.

Sharma, R. D. 1984 "Hypocholesterolaemic activity of fenugreek (*Trigonella foenum-grae-cum*).An experimental study in rats," *Nutr Rep. Int.* 30.221–231.

Sharma, R. D 1986. "Effect of fenugreek seeds and leaves on blood glucose and serum insulin responses in human subjects," *Nutr. Res.* 6.1353–1364.

Sharma, R. D. and Raghuram, T. C 1990. "Hypoglycaemic effect of fenugreek seeds in non-insulin dependent diabetic subjects," *Nutr. Res* 10:731–739

Sharma, R. D , Raghuram, T C and Dayasagar Rao, V. 1991 "Hypolipidaemic effect of fenugreek seeds. A clinical study," *Phytotherapy Res.* 5·145–147

Sharma, R. D., Raghuram, T C. and Sudhakar Rao, N 1990. "Effect of fenugreek seeds on blood glucose and serum lipids in type I diabetes," *Eur J Clin Nutr.* 44·301–306.

Sharma, R D , Sarkar, A , Hazra, D. K., Misra, B., Singh, J B. and Maheshwari, B B 1996a "Toxicological evaluation of fenugreek seeds A long term feeding experiment in diabetic patients," *Phytotherapy Res.* 10·519–520

Sharma, R. D., Sarkar, A , Hazra, D K., Mishra, B., Singh, J. B , Sharma, S K , Maheshwari, B. B and Maheshwari, P. K 1996b "Use of fenugreek seed powder in the management of non-insulin dependent diabetes mellitus," *Nutr. Res.* 16.1331–1339

Sharma, R D , Sarkar, A , Hazra, D K , Misra, B., Singh, J. B., Maheshwari, B. B and Sharma, S K. 1996c. "Hypolipidaemic effect of fenugreek seeds: A chronic study in non-insulin dependent diabetic patients," *Phytotherapy Res.* 10:332–334.

Singhal, P C , Gupta, R. K. and Joshi, L. D. 1982. Hypocholesterolaemic effect of *Trigonella foenum-graecum* (Methi)," *Curr Sci* 51.136–137.

Stark, A and Madar, Z 1993 "The effect of an ethanol extract derived from fenugreek (*Trigonella foenum-graecum*) on bile acid absorption and cholesterol levels in rats," *Br. J. Nutr* 69·277–287.

Taylor, W G., Zaman, M S., Mir, Z., Mir, P., Acharya, S. N , Mears, G J. and Elder, J L. 1997. "Analysis of steroidal sapogenins from amber fenugreek (*Trigonella foenum-graecum*) by capillary gas chromatography and combined gas chomatography/mass spectrometry," *J Agric. Food Chem.* 45 753–759.

Udayasekhara, Rao, P , Sesikeran, B., Srinivasa Rao, P , Nadamuni Naidu, A , Vikas Rao, V. and Ramachandran, E. P. 1996 "Short term nutritional and safety evaluation of fenugreek," *Nutr. Res* 16·1495–1505.

Valette, G., Sauvaire, Y , Baccou, J. C and Ribes, G. 1984. "Hypocholesterolaemic effect of fenugreek seeds in dogs," *Atherosclerosis.* 50:105–111.

Wunschendorff, M. H. E 1919 "La saponine des graines de fenugrec," *J Pharm. Chim.* 20.183–185.

Yoshikawa, M., Murakami, T , Komatsu, H., Murakami, N , Yamahara, J. and Matsuda, H. 1997. "Medicinal foodstuffs IV Fenugreek seeds (1) structures of trigoneosides Ia, Ib, IIa, IIb, IIIa and IIIb, new furostanol saponins from the seeds of Indian *Trigonella foenum-graecum* L.," *Chem. Pharm. Bull* 45:81–87.

Yoshikawa, M , Murakami, T , Komatsu, M., Yamahara, J and Matsuda, M 1998. "Medicinal foodstuffs. VIII. Fenugreek seeds (2): structures of six new furostanol saponins, trigoneosides IVa, Va, Vb, VI, VIIb, and VIIIb, from the seeds of Indian *Trigonella foenum-graecum* L ," *Heterocycles.* 47·397–405.

CHAPTER 6

Chemistry, Pharmacology and Clinical Applications of St. John's Wort and Ginkgo Biloba

G. MAZZA

B. D. OOMAH

1. ST. JOHN'S WORT

1.1. INTRODUCTION

ST. John's wort, *Hypericum perforatum* L., is a shrubby, aromatic perennial herb belonging to the family of Hypericaceae. It is native to Europe, western Asia, North Africa, Madeira and the Azores and is naturalized in many parts of the world, particularly North America and Australia (Hobbs, 1989). The medicinal parts of the plant include the fresh buds and flowers, the aerial parts collected during the flowering season and the whole fresh flowering plant (Anonymous, 1998a).

Historically, *Hypericum* has been used for its sedative, anti-inflammatory, anxiolitic and astringent qualities. Other folkloric uses include treatment of burns, insomnia, shocks, concussions, hysteria, gastritis, hemorrhoids, kidney disorders, scabies and wounds (Hobbs, 1989; Chavez and Chavez, 1997; Upton, 1997). Currently, St. John's wort is best known for the treatment of mild to moderate depression. In Germany, where it is licenced for the treatment of depression, anxiety and insomnia, sales outnumber all other antidepressants combined, and it outsells Prozac by more than seven to one (Chavez and Chavez, 1997). In the United States, 7.3 million people took St. John's wort extract in 1997 (Giese, 1999).

Clinical studies conducted over the past few years have indicated that whole extract of St. John's wort has antidepressant effects, and treatment of patients with mild and moderate depression appears promising (Linde et al., 1996; Hippius, 1998). In the 14 double-blind placebo-controlled trials that have been performed to date, 55.1% of patients receiving *Hypericum* were classified as responders (defined as those patients showing a 50% reduction in the severity of depression from baseline), compared with 22.3% of patients receiving placebo (Wheatley, 1998).

In addition, extracts of flowering tops of *Hypericum* have shown other effects including anti-inflammatory activity, wound- and burn-healing action, antibacterial and antifungal activity, antiviral effects, immunological effects, antioxidant activity and antispasmodic effects (Hobbs, 1989; Wichtl, 1994; Chavez and Chavez, 1997; Anonymous, 1998a).

1.2. CHEMICAL COMPOSITION AND CHEMISTRY

St. John's wort is a chemically complex material with a large number of bioactive components. In Table 6.1 are listed the major classes of compounds that have been identified in *H. perforatum* together with reported pharmacological properties. According to Hobbs (1989), Duke (1992), Wichtl (1994) and Anonymous (1998a), the bioactive agents in St. John's wort of greatest interest are the phloroglucinols, anthrocene derivatives, flavonoids and essential oils.

1.2.1. Phloroglucinols

This class of compounds consists predominantly of the prenylated derivatives of the phloroglucinols (1,3,5-benzenetriols) hyperforin and adhyperforin (Brondz et al., 1983; Hobbs, 1989). Hyperforin (Figure 6.1), one of the main components (2–4%) of the dried herb *H. perforatum,* has recently been shown to inhibit the reuptake of serotonin, dopamine and norepinephrine, and it has been identified as the leading candidate for a single antidepressant compound in *Hypericum* (Muller et al., 1998; Chatterjee et al., 1998a, 1998b; Bhattacharya et al., 1998). In addition, hyperforin is well known for its antibiotic properties and is believed to be the agent that inhibits *S. aureus* (Negrash and Pochinok, 1972; Brondz et al., 1983). Hyperforin was first isolated from *Hypericum* in the early 1970s (Gurevich et al., 1971), and the chemical structure was established in 1983 by Brondz et al. (1983). A recent paper by Orth et al. (1999) describes the latest methodology for isolation, analysis and stability of hyperforin from *H. perforatum.* According to these authors, isolation is achieved by extraction of frozen blossoms with n-hexane followed by separation of lipophilic substances on a silica gel column, purification of the relevant fraction by preparative HPLC, evaporation of the mobile phase under reduced pressure, removal of the remaining water by freeze-drying, and storage of hyperforin at −20°C under nitrogen. Gradient HPLC can also be used to detect polar analogues of hyperforin in St. John's wort oils in which hyperforin has decomposed (Maisenbacher and Kovar, 1992). Because of its high sensitivity to oxidation, hyperforin should be stored under nitrogen at −20°C or *lower*

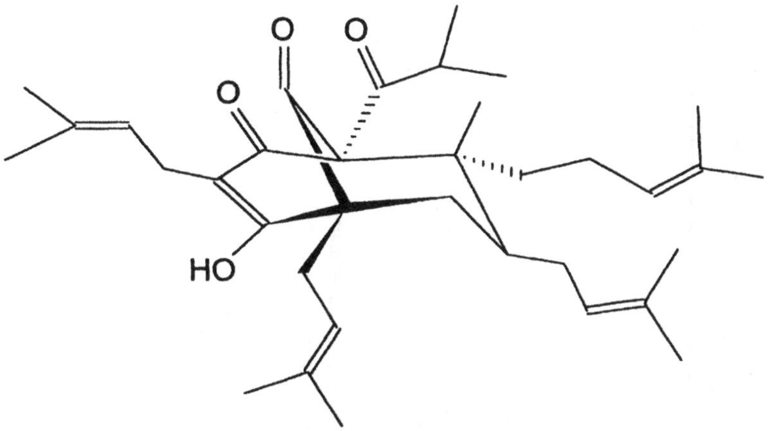

Figure 6.1 Structure of hyperforin.

temperatures. A highly sensitive LC/MS/MS method suitable for determining hyperforin in plasma after administration of alcoholic *H. perforatum* extracts containing hyperforin has also been reported (Biber et al., 1998).

1.2.2. Naphthodianthrones

The naphthodianthrones include hypericin, pseudohypericin, isohypericin, and their chemical precursor *proto-hypericin* and hypericodehydrodianthone (Hobbs, 1989; Upton, 1997). Of these compounds, hypericin and pseudohypericin are the most studied. Hypericin (Figure 6.2), pseudohypericin and related compounds constitute 0.05 to 3.0% of the total plant weight. Hypericin and pseudohypericin were formerly believed to function as inhibitors of monoamine oxidase (MAO), the enzyme(s) associated with the breakdown of noradrenalin and serotonin. Today, however, most researchers believe that the flavonoids are the MAO inhibitors (Bladt and Wagner, 1994). Both hypericin and pseudohypericin are photosensitizing agents, with pseudohypericin being less photosensitizing than hypericin (Meruelo et al., 1988).

1.2.3. Flavonoids

Flavonoids constitute about 11.7% of the leaves and 7.4% of the stalks (Upton, 1997). These include the flavonols: kaempferol, luteolin, myricetin,

TABLE 6.1. Constituents and Biological Activities of *Hypericum perforatum*.

Constituent	Biological Activity	References
Naphthodianthrones		
hypericin, pseudohypericin frangula-emodin anthranol (and a mixture of the precursors, proto-hypericin and hypericodehydrodianthrone)	photodynamic, antidepressive (MAO inhibitor), antiviral	*
Flavanols		
(+)-catechin, procyanidins, (−)-epicatechin (total tannin content is 6.5–15%)	astringent, anti-inflammatory, antiviral	**
Flavonols		
hyperoside (hyperin), quercetin, isoquercetin, rutin, methylhesperidin, iso-quercitrin, quercitrin, I-3/II-8-biapidenin, kaempferol	capillary-strengthening, anti-inflammatory, diuretic, cholagogic, dilate coronary arteries, sedative, tumor inhibition, antitumoral, antidiarrheal	***
Xanthones		
xanthonolignoid compound (roots)	generally, xanthones exhibit antidepressant, antitubercular, choleretic, diuretic, antimicrobial, antiviral and cardiotonic activity	****
Coumarins		
umbelliferone, scopoletin	—	+
Phenolic carboxylic acids		
caffeic acid, chlorogenic acid, gentisic acid, ferulic acid	—	++
Phloroglucinol derivatives		
hyperforin, adhyperforin	antidepressant, antibacterial	+++
Essential oil components		
Monoterpenes α-pinene, β-pinene, myrcene, limonene	antimicrobial, anti-inflammatory, and antitumoral	++++
Sesquiterpenes caryophyllene, humulene		

TABLE 6.1. (continued)

Constituent	Biological Activity	References
n-Alkanes		
methyl-2-octane, n-nonae, methyl-2-decane, n-undecane, all in the series C16–C29 (especially nonacosane)	—	$
n-Alkanols		
1-tetracosanol, 1-hexacosanol, 1-octacosanol, 1-triacontanol	metabolic stimulants	$$
Phytosterols		
β-sitosterol	hypocholesterolemic	$$$

Adapted from Hobbs (1989).

*   Mathis and Ourisson (1963), Dorossev (1985), Okpanyi and Weischer (1987), Meruelo et al. (1988).
: :   Derbentseva (1972), Kitanov (1983).
: : :   Hoelzl and Ostrowski (1987), Berghoefer and Hoelzl (1987).
: : : : Neilsen and Arends (1978), Suzuki et al. (1984), Rocha (1994).
+   Karryev and Komissarenko (1980).
+ +   Ollivier et al. (1985).
+ + +   Brondz et al. (1983), Gurevich et al. (1971), Negrash and Pochinok (1972), Chatterjee et al. (1998a, 1998b), Crowell (1999).
+ + + + Sticher (1977), Chalva et al. (1981), Mathis and Ourisson (1964).
$ Brondz et al. (1983), Mathis and Ourisson (1964).
$$ Brondz and Greibrokk (1983), Snider (1984), Mori (1982), Gonsette (1982).
$$$ Mathis and Ourisson (1964), Lees et al. (1977).

135

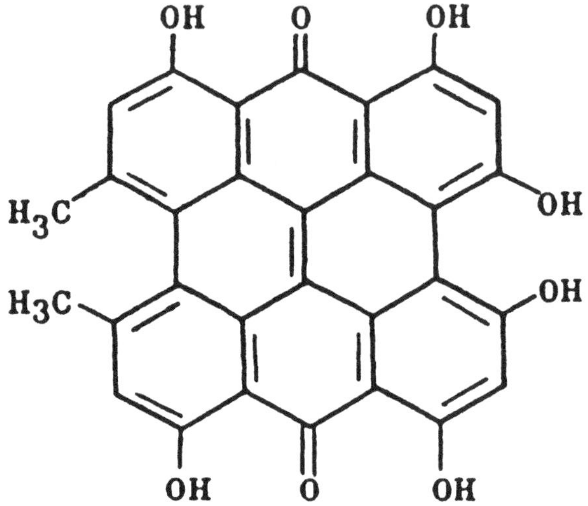

Figure 6.2 Structure of hypericin.

quercetin (2%); the flavonol glycosides hyperoside, also called hyperin (0.7–1.1%); quercitrin (0.3–0.5%); isoquercitrin (0.3%); amentoflavone, also known as I3′, II8-biapigenin (0.01–0.05% in flowers); I3-II8-biapigenin (0.1–0.5%); and luteolin and rutin (0.3%) (Razinskaite, 1971; Hobbs, 1989; Nahrstedt and Butterweck, 1997; Cracchiolo, 1998). Comparative HPLC analyses of amentoflavone and hypericin in extracts of *H. perforatum, H. hirsutum, H. patulum* and *H. olympicum* have been reported by Baureithel et al. (1997). Flavonoids are potent antioxidants, free radical scavengers and metal chelators, and they inhibit lipid peroxidation. Certain biflavonoids including amentoflavone (Figure 6.3) were previously reported to have inhibitory effects on the group II phospholipase A2 activity. Amentoflavone has also been found to inhibit cyclooxygenase from guinea pig epidermis without affecting lipoxygenase, and it showed potent anti-inflammatory activity in the rat carrageenan paw edema model (Kim et al., 1998). Epidemiological studies show an inverse correlation between flavonoids and mortality from coronary heart diseases (Cook and Samman, 1996).

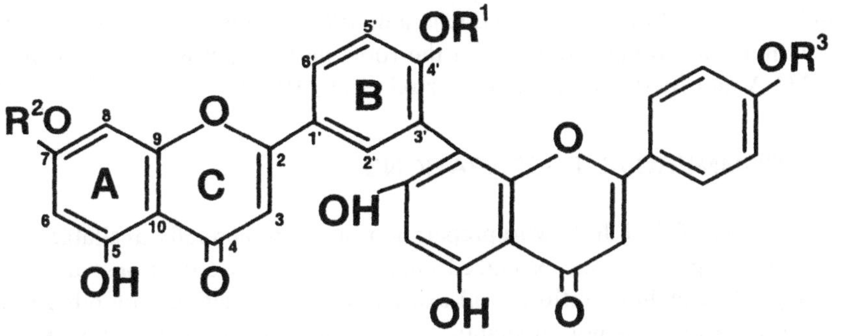

1	Amentoflavone	$R^1 = R^2 = R^3 = H$
2	Bilobetin	$R^1 = Me, R^2 = R^3 = H$
3	Ginkgetin	$R^1 = R^2 = Me, R^3 = H$
4	Isoginkgétin	$R^1 = R^3 = Me, R^2 = H$
5	Sciadopitysin	$R^1 = R^2 = R^3 = Me$

Figure 6.3 Structure of amentoflavone, bilobetin, ginkgetin, isoginkgetin and sciadopitysin.

1.2.4. Essential Oils

Essential oils constitute about 0.06–0.35% of the plant. The major components are 2-methyloctane (16.4%), α-pinene (10.6%), β-pinene, limonene, myrcene, caryophyllene and humulene. Trace components include 2-methyldecane, 2-methybutenol, undecane and various n-alkanes and n-alkanols (Mathis and Ourisson, 1964; Sticher, 1977; Chialva et al. 1981; Upton, 1997). A number of these monoterpenes have antitumoral activity. For example, d-limonene has been shown to have chemopreventive activity against rodent mammary, skin, liver, lung and forestomach cancer (Crowell, 1999).

1.2.5. Other Constituents

Proanthocyanidins or condensed tannins are found in large concentrations in the aerial portions of the plant and consist primarily of dimers, trimers, tetramers and high polymers of catechin and epicatechin. These chemicals are known to posses antioxidant, antimicrobial, antiviral and cardioprotective effects (Cook and Samman, 1996; Hollman et al., 1996). Other bioactives known to occur in St. John's wort are phytosterols, the coumarins umbellifone and scopoletin, xanthones, carotenoids and several phenolic acids including caffeic, chlorogenic, p-coumaric, ferulic, isoferulic and gentisic acid (Mathis and Ourisson, 1964; Costes and Chantal, 1967; Hobbs, 1989; Upton, 1997). Xanthones are found primarily in the roots and have been shown to inhibit MAO-A and MAO-B enzymes (Suzuki et al., 1984).

1.3. COMMERCIAL PREPARATIONS

A variety of St. John's wort preparations are commercially available (Table 6.2). These products are produced under a variety of conditions such as air drying, olive and/or sunflower oil extraction and extraction with other media including water, ethanol, methanol, glycerol and supercritical carbon dioxide. Tea bags containing two grams of the raw herb are also available. A St. John's wort tea is normally made by pouring about one cup of boiling water over two teaspoons (2–4 grams) of chopped raw herb, steeping for 5–10 minutes, then straining.

Little is known about the impact of processing and storage stability on the quality of St. John's wort products. The reason is the uncertainty of the various components of St. John's wort that may be responsible for the desired biological effects, especially antidepressant effects. Until recently, it was believed that the antidepressant activity was related to the level of hypericin (Suzuki et al., 1984). Now, it appears that the antidepressant constituent of *Hypericum* is hyperforin (Chatterjee et al., 1998a, 1998b; Dimpfel et al., 1998; Laakmann et al., 1998; Muller et al., 1998). Nonetheless, most of the commercially available St. John's wort products are still standardized to a certain content of hypericin (Table 6.2), and the few studies on the influence of processing and storage stability on the quality of St. John's wort products have used hypericin as the marker compound. Thus, Adamski and Styp-Rekowska (1971) reported that the hypericin content of juice of *H. perforatum* and powdered extract dropped by 14% during one year, and the dry extract remained stable, when stored at 20°C. When stored at 60°C, the hypericin level decreased by 33%, 33% and 47% from powder extract, tablets and liquid juice, respectively. Similarly, Araya and Ford (1981) reported that drying of *H. perforatum* plants in sunlight destroyed 80% of hypericin.

Similarly, very little is known about the influence of different cultivation methods on the biochemical activities of *Hypericum* extracts. According to Denke et al. (1999), nitrogen fertilization and cultivar influence yield and quality of the plant material. In this experiment, the most active extract was from plants of a broad-leaf cultivar that were non-fertilized with nitrogen and that were extracted with methanol. In another recently published study, Büter et al. (1998) reports that genetic factors strongly affect plant yield and concentration of secondary metabolites of *Hypericum*. Thus, HPLC analysis of amentoflavone, biapigenin, hyperforin, hypericin, hyperoside, pseudohypericin, quercetin and rutin contents were significantly lower in the first year of

TABLE 6.2. Commercially Available St. John's Wort Products.

Product	Processing	Content of St. John's Wort and Notes
Fresh herb	None	—[na]
Dried herb	Drying	—
Juice	Extraction	—
Oil-extracted	Grinding/extraction	—
Water-extracted	Grinding/extraction	—
Ethanol-extracted	Grinding/extraction	—
Ethanol-water-extracted	Grinding/extraction	—
Methanol-extracted	Grinding/extraction	—
Carbon dioxide-extracted	Grinding/extraction	—
Tablets	Drying/grinding	—
Capsules (NW)	Grinding/extraction	375 mg/*/cap; 2–3 caps/days; $0.16/cap; recommended for modulation of mood
Capsules (NS)	Grinding/extraction	400 mg/cap; 3 caps/day; $0.25/cap
Capsules (CSN)	Grinding/extraction	300 mg/cap; 3 caps/day; $0.15/cap
Liquid concentrate	Grinding/extraction/concentration	Ethanol-water-extract: $21.50/2 oz. bottle

[na] Not available.

*375 mg St. John's wort leaf and stem extract standardized to 0.3% hypericin delivering 0.9 mg hypericin. According to several published reports, the standard dose is 2 to 4 grams of raw herb or 0.2 to 1.0 mg of extracted hypericin/day; prices are in US $ for selected products sold on the Internet in August 1999.

139

cultivation ranging from 12% (hyperin) to 83% (hyperforin) of the contents measured in the two-year-old crop.

Almost all clinical studies have been conducted using a methanol extract manufactured by Willmar Schwabe GmbH & Co. of Karlsruhe, Germany, and marketed under the trade name of Jarsin (TM). This is also called LI 160 and research grade *Hypericum*. It should be noted that the alcohol used in the extraction process is removed from the end product that is essentially methanol-free. At the present time, it is not known if other products, such as chopped raw St. John's wort or a tea, are effective in treating depression or other illness.

1.4. PHARMACOLOGICAL EFFECTS

1.4.1. Antidepressant Action

The recent increase in popularity of St. John's wort in North America is due primarily to its antidepressant action. This increase in popular attention came as a result of a June 1997 positive feature on the herb by ABC television news magazine program 20/20. Nevertheless, St. John's wort does indeed appear to be effective in the treatment of mild to moderate cases of depression (Linde et al., 1996; Hippius, 1998).

Some of the most convincing evidence for the efficacy and safety of St. John's wort was published in the British Medical Journal by Linde et al. (1996). The aim of this study was to investigate if extracts of *H. perforatum* are more effective than placebo in the treatment of depression, are as effective as standard antidepressive treatment and have fewer side effects than standard antidepressant drugs. In the study, the authors conducted a systematic review and meta-analysis of 23 randomized trials including a total of 1,757 outpatients with mainly mild or moderately severe depressive disorders: 15 (14 testing single preparations and one a combination with other plant extracts) were placebo controlled and eight (six testing single preparations and two combinations) compared *Hypericum* with another drug treatment. The results revealed that *Hypericum* extracts were significantly superior to placebo (95% confidence interval) and similarly effective as standard antidepressants. There were two (0.8%) dropouts for side effects with *Hypericum* and seven (3.0%) with standard antidepressant drugs. Side effects occurred in 50 (19.8%) patients on *Hypericum* and 84 (52.8%) patients on standard antidepressants. It was concluded that extracts of *Hypericum* are more effective than placebo for the treatment of mild to moderately severe depressive disorders.

The herbal preparations used in the studies reviewed by Linde et al. (1996) were standardized on the basis of hypericin content, and most used the methanol extract of *Hypericum* called LI 160 or Jarsin. However, in the reviewed studies, the dose of the whole herb varied considerably (from 300 mg to 1,000 mg/day) as did the dose of hypericin (0.4 to 2.7 mg/day), and all of the studies were of short duration. Most were four to eight weeks in length. Also, most antidepressants become effective after a few weeks; but it may take longer than eight weeks for antidepressants to build up to maximum effectiveness in certain individuals. Similarly, the doses of antidepressants used in the control groups were relatively low, and it is not clear as to how effective St. John's wort is when compared to a high dose of the synthetic drugs. In addition, even though the products tested were standardized for hypericin content, it now appears that hypericin is not among the phytochemicals responsible for St. John's wort's antidepressant effects.

In a separate commentary accompanying the Linde et al. (1996) article, De Smet and Nolen (1996) called for further studies aimed at finding the most effective treatment dose of *Hypericum* and for longer studies to evaluate the risk of relapse and late-emerging side effects. They also called for trials in severely depressed patients.

In 1997, the National Institutes of Health (NIH) launched the first U.S. clinical trial of St. John's wort. The three-year study, sponsored by NIH's Office of Alternative Medicine (OAM), the National Institute of Mental Health (NIMH) and the Office of Dietary Supplements (ODS), will include 336 patients with major depression who will be randomly assigned to one of three treatment arms for an eight-week trial. One-third of the patients will receive a uniform dose of St. John's wort, another third will be given placebo and the final third will take sertraline (Zoloft), a selective serotonin reuptake inhibitor (SSRI). This study will permit the use of relatively high doses of both St. John's wort and sertraline and will include severely depressed patients (NIH, 1997). This study will use a standardized preparation containing a 900 mg daily dose of the herb. In addition, study participants who respond positively will be followed for another 18 weeks. The goal of the follow-up is to determine if patients given St. John's wort have fewer relapses than patients given placebo.

Very recently, the findings of a randomized double-blind multicenter comparative study were reported by Lenoir et al. (1999). The aim of this study was to investigate the efficacy and tolerability of a new standardized fresh-plant extract obtained from the shoot tips of St. John's wort in the treatment of mild to moderate depression. Outpatients (348 total: 259 female, 89 male) with mild to moderate depression were given one tablet of a *Hypericum* preparation standardized to either 0.17 mg (114 patients), 0.33 mg (115 patients) or 1 mg (119 patients) total hypericin per day for six weeks, three times a day. The main outcome measure was the Hamilton Psychiatric Rating Scale for Depression, which at the end of treatment decreased from 16–17 to 8–9. That is, a relative reduction of about 50% was observed in all groups (280 patients, for protocol analysis). Overall, the intergroup comparison revealed no significant differences, indicating that this *Hypericum* preparation was effective at all three doses. Tolerability was found to be excellent, with mild adverse reactions occurring in only seven of the 348 patients (2%).

The finding by Lenoir et al. (1999) that their *Hypericum* preparation was equally effective at the 0.17, 0.33 and 1.0 mg total hypericin dose per day clearly suggests that antidepressant effects do not appear to be directly related to the intake of hypericin. Hyperforin, on the other hand, has recently been shown to be a dose-related marker for antidepressant efficacy in humans (Dimpfel et al., 1998; Shellenberg et al., 1998; Laakmann et al., 1998).

Shellenberg et al. (1998) conducted a randomized, double-blind and placebo-controlled parallel-group study on three groups of 18 volunteers each to determine the effects on EEG of two extracts with the same hypericin content and differing amounts of hyperforin. They found no effects on EEG for hypericin, but they found the effects on EEG to be proportional to hyperforin content. They suggest that *Hypericum* products containing high amounts of hyperforin have a "shielding" effect on the central nervous system. The same team observed a similar effect in rats (Dimpfel et al., 1998).

Results from a randomized, double-blind and placebo-controlled multicenter study (on the clinical efficacy of extracts of St. John's wort in 147 outpatients suffering from mild or moderate depression) reported by Laakmann et al. (1998) also showed that patients receiving a 5% hyperforin solution exhibited the largest reduction in depression followed by the group receiving 0.5% hyperforin and then by the placebo group.

Thus, the likely constituent of *Hypericum* responsible for the antidepressant effects is hyperforin, a unique phloroglucinol derivative found in the plant (Chatterjee et al., 1998a, 1998b). The mechanism by which St. John's wort exerts antidepressant effects is unclear. However, according to recent reports (Muller et al., 1998; Chatterjee et al., 1998a), the action is probably via inhibition of the reuptake of serotonin, norepinephrine and dopamine. In other words, by preventing the brain from reabsorbing the neurotransmitters in question, thus, keeping serotonin, norepinephrine and dopamine levels in the brain at a higher level. It also appears to inhibit reuptake of GABA and L-glutamine, which are the primary inhibitory neurotransmitters in the brain (Chatterjee et al., 1998a). Very recently, Singer et al. (1999) attempted to characterize the mechanism of serotonin reuptake inhibition using kinetic analyses in synaptosomes of mouse brain. Their findings show that hyperforin inhibits serotonin uptake by elevating free intracellular sodium.

Another component of St. John's wort believed to have antidepressant effects is the flavonoid amentoflavone (Nahrstedt and Butterweck, 1997; Baureithel et al., 1997). Recently, Baureithel et al. (1997) suggested that amentoflavone exerts its antidepressant action by binding benzodiazepine receptors in the brain.

1.4.2. Antiviral Activity

In vitro, St. John's wort and/or hypericin have been shown to possess antiviral activity against several viruses including HIV, influenza, herpes and equine infectious anemia virus (Meruelo et al., 1988; Lopez-Bazzocchi et al., 1991; Lavie et al., 1994, 1995; Upton, 1997; Cracchiolo, 1998; Gulick et al., 1999). Recent research has focused on the use of hypericin for the treatment of AIDS. There have been at least two AIDS clinical trial group studies that used synthetic hypericin as a treatment for AIDS, and both trials appear to have had difficulties with photosensitivity reactions (Chavez and Chavez, 1997). A Phase I/II clinical trial in Thailand found that hypericin administered once a day for 28 days to 12 HIV-infected patients at a dose of 0.05 mg/kg, resulted in 10 of the 12 patients exhibiting reduced viral load (Chavez and Chavez, 1997).

However, a recent Phase I study with 30 HIV-infected simplex 1 and 2, Sindhis virus, murine cytomegalovirus, para-influenza 3 virus and vesicular stomatitis virus patients with CD4 counts less than 350 cells/mm^3 showed that intervention with intravenous hypericin, 0.25 or 0.5 mg/kg of body weight twice weekly or 0.25 mg/kg three times weekly or oral hypericin, 0.5 mg/kg daily, caused significant phototoxicity and had no antiretroviral activity in the patients studied (Gulick et al., 1999). Severe coetaneous phototoxicity was observed in 48% of the patients.

1.4.3. Antibacterial and Antifungal Activity

Several studies with St. John's wort extracts have demonstrated their antibacterial and antifungal activities. For instance, two Russian preparations of *Hypericum* were found to be more effective than sulfonilamide against *Staphylococcus aureus* infection *in vivo* and *in vitro* (Derbentseva and Rabinovich, 1968; Aizenman, 1969; Gurevich et al., 1971; Hobbs, 1989). The phloroglucinol derivative hyperforin is known to be an important antibiotic constituent of *Hypericum* and is believed to be the agent that inhibits *S. aureus* (Negrash and Pochinok, 1972; Brondz et al., 1983). Recently, Rocha et al. (1995) reported on antibacterial phloroglucinols and flavonoids from *H. brasiliense*. A resin fraction of the methanol extract of St. John's wort LI 160 has been shown to have antifungal activity and significant action against gram positive bacteria (Upton, 1997). Tannins and flavonoids in St. John's wort have been reported to inactivate *Escherichia coli* at dilutions of 1:400 and 1:200 (Upton, 1997). Essential oil from *H. perforatum* also displays antimicrobial activity (Hobbs, 1989).

1.4.4. Anti-Inflammatory Activity

Preparations of *Hypericum* have been found to suppress inflammation *in vivo* (Hobbs, 1989; Anonymous, 1998a). The likely phytochemicals responsible for this action are the flavonoids, which constitute 11.7% of the flowers and 7.4% of the stems and leaves (Lietti et al., 1976; Middleton and Kandaswami, 1992;

Upton, 1997). The mechanisms for the anti-inflammatory activity of flavonoids include inhibition of arachidonic acid metabolism (Ferrandiz and Alcaraz, 1991) and inhibition of the prostaglandin synthase cyclooxygenase activity (Hoult et al., 1994). The property of flavonoids to decrease the fragility and permeability of blood capillaries was originally identified by Albert Szent-Gyorgyi, who coined the term Vitamin P for those compounds that reduce capillary permeability.

Several prescription and nonprescription pharmaceutical products containing flavonoids as the active principle are used to control capillary permeability and fragility (Lietti et al., 1976). Clinical applications include treatment of visual disorders related to nighttime visual acuity, to aid in adapting to low light conditions, and to decrease recovery time after exposure to glare.

1.4.5. Antioxidant Activity

Antioxidant activity is one of the most important mechanisms for preventing or delaying the onset of major degenerative diseases of aging, including cancer, heart disease, cataracts and cognitive dysfunction. The antioxidants are believed to exert their effects by blocking oxidative processes and free radicals that contribute to the causation of these chronic diseases (Ames, 1983; Block, 1992). Several phenolics, known to be present in *Hypericum*, especially catechins, flavonols and tannins have been shown to perform these functions (Block, 1992; Mackerras, 1995; Bors et al., 1996; Mazza, 1997). It has recently been reported that St. John's wort extracts exhibit antioxidant activity (Anonymous, 1997).

1.4.6. Wound-Healing Activity

In a number of studies, St. John's wort extracts have demonstrated wound-healing activity. For instance, a 1975 patent from Germany claims that an ointment containing an extract of St. John's wort flowers shortened healing time of burns and showed antiseptic activity (Saljic, 1975). According to the report, first degree burns healed in 48 hours when treated with the ointment, while second and third degree burns healed three times faster than burns treated by conventional methods. St. John's wort has also been compared with *Calendula*, another herb commonly used to heal wounds, and according to Rao et al. (1991), St. John's wort applied topically was found to be more effective than *Calendula*. The effects of St. John's wort on wounds and burns are probably due to the anti-inflammatory and the antibacterial and antifungal effects mentioned above. It has also been speculated that the wound-healing properties attributed to *Hypericum* may be due to the high content of tannins in the plant/extracts that act as an astringent and have the ability to complex and precipitate proteins (Chavez and Chavez, 1997).

1.4.7. Other Effects

Other reported effects of St. John's wort include enhancement of coronary flow, inhibition of receptor tyrosine kinase activity, inhibition of release of arachidonic acid and leukotriene B4 and increase in the production of nocturnal melatonin (Upton, 1997). One double-blind, placebo-controlled study, conducted with 12 older, healthy volunteers, has found that *Hypericum* extract LI 160 (300 mg three times daily) increased deep sleep and slightly decreased REM sleep (Schulz and Jobert, 1994).

1.5. SAFETY

St. John's wort is well known for its ability to cause photosensitivity to grazing animals, particularly cattle, sheep, horses and goats, and also rabbits and rats (Hobbs, 1989; Wichtl, 1994; Chavez and Chavez, 1997; Upton, 1997). Photosensitivity in livestock is referred to as hypericism or "light sickness" (Bombardelli and Morazzoni, 1995). Reported reactions are mainly dermatological, such as severe erythema and edema of skin, conjunctiva and bucal mucous membranes, which can lead to restlessness, psychomotor excitement blindness and refusal to eat by the animals (Araya and Ford, 1981). Because of refusal to eat and loss of appetite, the threat of *Hypericum* intoxication to livestock is reduced, which makes the absorption of the photodynamic pigment, hypericin, self-limiting (Araya and Ford, 1981).

Phototoxicity in humans appears to be rare; most authors express no concerns (Hobbs, 1989; Bombardelli and Morazzoni, 1995). This viewpoint is supported by the results of a recent placebo-controlled randomized clinical trial in which 13 volunteers received 900, 1,800 or 3,600 mg of a standardized *Hypericum* extract (LI 160) containing zero, 2.81, 5.62 and 11.25 mg of total hypericin (total hypericin is the sum of hypericin and pseudohypericin) (Brockmoller et al., 1997). Nonetheless, most authors recommend caution when using large quantities of St. John's wort extract for therapeutic uses, particularly for people with fair skin, who should not expose themselves to strong sunlight during *Hypericum* therapy (Hobbs, 1989; Bombardelli and Morazzoni, 1995).

Recently, a systematic review on adverse effects of *H. perforatum* by Rand et al. (1998) concluded that St. John's wort is well tolerated. It has an encouraging safety profile with an incidence of adverse reactions similar to that of placebo. The most common adverse effects are gastrointestinal symptoms, dizziness/confusion and tiredness/sedation. Photosensitivity appears to occur extremely rarely. Similarly, in a recent comparative trial, 63% of patients receiving St. John's wort reported that they experienced no adverse effects during treatment. In contrast, only 36% of amitriptyline-treated patients reported no adverse effects. *Hypericum* has also been associated with significantly less dry mouth and drowsiness than amitriptyline (Wheatley, 1998).

Similar results were reported by Stevinson and Edzard (1999) in a comparison of St. John's wort with the conventional antidepressants dothiepin, fluoxetine, moclobemide and mirtazapine. In this study, the authors also found that *Hypericum* was associated with fewer and milder adverse drug reactions in clinical trials than any of the other drugs. Data on the safety of *Hypericum* in overdose and on interactions with other drugs are as yet scarce.

1.6. CONCLUSIONS

Extracts of the plant *H. perforatum* have been used in herbal medicine for centuries. Traditional uses include treatment of nervous disorders, insomnia, burns, shocks, hysteria, gastritis, hemorrhoids, urinary disorders, scabies and wounds. Currently, St. John's wort is best known for the treatment of mild to moderate depression. The constituent of *Hypericum* extracts responsible for the antidepressant action is, however, unknown, and does not appear to be hypericin, the constituent by which the extracts are currently standardized. Based upon limited studies, St. John's wort appears to be an acceptable alternative to traditional antidepressant therapy. In the 14 double-blind placebo-controlled trials that have been performed to date, 55% of patients receiving *Hypericum* were classified as responders (defined as those patients showing a 50% reduction in the severity of depression from baseline), compared with 22% of patients receiving placebo. In similar comparative trials, *Hypericum* has been shown to be as effective as standard tricyclic antidepressants (imipramine, amitriptyline and maprotiline). Data indicate that *Hypericum* is a well-tolerated alternative to synthetic drugs for the treatment of mild-to-moderate depression, particularly in patients who are intolerant of standard antidepressants. Trials comparing the effect of St. John's wort with selective serotonin (5-hydroxytryptamine; 5-HT) reuptake inhibitors and other newer antidepressants and assessing the effect of higher dosages in patients with severe depression are required to fully determine the place of *Hypericum* in the treatment of depressive illness.

In vitro investigations of *Hypericum* show antiviral activity, although there is evidence that these promising results might not occur *in vivo*. Photosensitivity has been reported in animals that have eaten large quantities of *Hypericum*, however, no cases of photosensitivity have been reported in humans.

2. GINKGO BILOBA

2.1. INTRODUCTION

Ginkgo (*Ginkgo biloba* L.), also known as maidenhair-tree, of the *Ginkgoacea* family, is a dioecious tree reaching up to 30 m tall, with a bole circumference of up to 9 m, and is believed to be native to China. The name of the genus *Ginkgo* is derived from a mistranslation of the Japanese name, Yin-Kwo, meaning "silver fruit." The leaves are deciduous, alternate or in clusters of three to five on short twigs, petiole, fan, thickened at the margins, 5–10 cm across and bilobed, hence, the species *biloba* appellation. The seed is yellow, round, about 2.5 cm long, with bad-smelling pulp surrounding the thin-shelled white nut that contains an edible sweet kernel.

The leaf extract of *Ginkgo biloba* is one of the oldest natural therapeutic agents still used today. The *Ginkgo biloba* tree has long been part of the traditional Chinese pharmacopeia, first cited as a medicinal agent about 5,000 years ago (Michel, 1986). It was first cited in Chinese herbals around the fourteenth century AD for its "fruit" that was consumed raw or cooked. In traditional Chinese medicine, *Ginkgo* seeds are prescribed as a remedy against asthma, cough, bladder inflammation, blenorrhagia and alcohol abuse (Chang and But, 1987). Anticarcinogenic and vermifugal properties have also been claimed for the raw nuts. The medicinal uses of *Ginkgo* leaves in cardiovascular disorders and asthma dates back to 1550 according to traditional Chinese medicine. At present, extracts of *Ginkgo* leaves are used extensively for treatment of memory disorders associated with aging, including Alzheimer's disease and vascular dementia. In the United States, nearly 11 million people took *Ginkgo* extract in 1997 (Giese, 1999). The physiological benefits of *Ginkgo* are generally attributed to the antioxidant, vasoregulating and neuro-protective properties of some of its constituents.

2.2. CHEMICAL COMPOSITION AND CHEMISTRY

Ginkgo contains numerous compounds with documented biological activity. Constituents that have been most studied include the ginkgolides, flavonoids and ginkgolic acids.

2.2.1. Ginkgolides

2.2.1.1. Structural Characteristics

*Ginkgo*lides are molecules that only occur naturally in the leaves and roots of *Ginkgo biloba*. They were first isolated in 1932 from the bitter principles of *Ginkgo biloba*, but their structure was only resolved in 1967 (Furukawa, 1932; Maruyama et al., 1967a). In a series of studies, Maruyama et al. (1967b, 1967c, 1967d) characterized the ginkgolides as 20 carbon cage molecules, incorporating a t-butyl group and six, five-membered rings A to F including a spironane, a tetrahydrofuran cycle and three lactone rings. The ginkgolides possess three lactone rings and differ only in the number and positions of substitutes. Four structures were identified differing only by the number and

Compound n°	Ginkgolide	IHB nomenclature	R_1	R_2	R_3
2	A	BN 52020	OH	H	H
3	B	BN 52021	OH	OH	H
4	C	BN 52022	OH	OH	OH
5	J	BN 52024	OH	H	OH
6	M	BN 52023	H	OH	OH
7	synthetic	BN 50580	OH	OMe	H
8	synthetic	BN 50585	OH	OEt	H

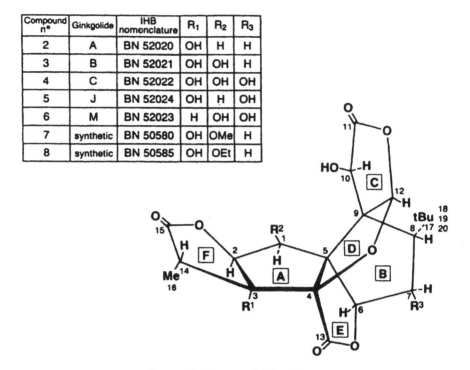

Figure 6.4 Structure of ginkgolides.

position of hydroxyl groups on the C1, C3 or C7 of the spironane framework. These compounds were named ginkgolides A, B, C, M and J (Figure 6.4). Termed BN 52020, BN 52021, BN 52022, BN 52023 and BN 52024, respectively, in Institut Henri Beaufour (IHB) internal nomenclature, these represent the four platelet activating factor (PAF) antagonists initially characterized in *Ginkgo biloba* leaf extract EGb 761.

Ginkgolide B (BN 52021), the most active PAF antagonist found in this class, is a specific and potent competitive antagonist of the binding of PAF to its membrane receptor in human platelet and lung preparations. Ginkgolides A (BN 52020) and C (BN 52022) are about one order of magnitude less potent than BN 52021, the less active compound being BN 52022. The PAF antagonist activity of ginkgolide B is significantly reinforced by combining it with ginkgolides A and C. BN 52063, a standardized mixture of ginkgolides A, B and C (in the ratio 2:2:1) is the first drug that has been shown to be a potent PAF antagonist in humans (Guinot et al., 1988).

Ginkgolides are derived from one of the most complex natural molecular frameworks in which mevalonate and methionine are incorporated. The ginkgolide structure bears a striking resemblance to that of bilobalide, another constituent of *Ginkgo biloba*. This incorporates a tertbutyl group and five fused cyclopentanoid rings, among which are again three γ-lactone (but no tetrahydrofuran) and only one cyclopentane and, hence, no spiro units.

2.2.1.2. Occurrence and Pathophysiology

Ginkgolides have been extracted from various organs of the *Ginkgo* tree. They are found in the leaves, roots and root bark. The biosynthesis of ginkgolides occurs at a very early stage in the development of the tree. Because leaves are a renewable resource that can be harvested every year, it is the organ of choice for large-scale extraction of ginkgolides.

The best source of the ginkgolides is the root bark, yielding 0.01% ginkgolide A (GA), 0.01% ginkgolide B (GB), 0.02% ginkgolide C (GC) and 0.0002% GM after extraction, chromatography and multiple (10–15 steps) fractional recrystallization steps (Nakanishi, 1988).

Seasonal variations in ginkgolide content of leaves have been observed. It reaches a maximum in late summer and falls to a minimum at the end of autumn. There are also variations in the ratios of individual ginkgolides with each other in leaves from different geographical origins that may be attributed to variation in climate, conditions of culture and differences within species.

According to Braquet (1988), ginkgolides have therapeutic effects in different pathophysiological processes and models of disease, such as inflammation, airway hyperactivity, the cardiovascular system, endotoxemia and other models of shock, gastrointestinal ulceration, the renal system, CNS functions, immune processes, ocular diseases and skin diseases (Table 6.3).

Since a potential role for PAF has been ascribed in various pathologies such as inflammation, allergy, sepsis or thrombosis, the ginkgolides might be of therapeutic interest, not only by inhibiting the direct effects of PAF such as lung tissue contraction, vascular tone increase or aggregation, but also by blocking the formation of the other mediators, thromboxane A2 or leukotrienes. In addition, the role of PAF in late asthmatic responses could trigger the recruitment of inflammatory cells such as neutrophils, macrophages/monocytes and eosinophils seen in bronchial tissue from asthmatic patients. Ginkgolides may thus provide new insight in the treatment of bronchial asthma.

Topical or systematic administration of the ginkgolide BN 52021, a specific inhibitor of PAF-acether, not only inhibits thrombus formation but equally prevents recruitment of leukocytes and recurrence of the thrombic phenomenon attributed to the generation of PAF-like activity in the vessel wall. Since BN 52021 potently inhibits the PAF-acether-induced eosinophil chemotaxis, it (BN 52021) may be useful in preventing inflammation and in treating bronchial asthma. BN 52021 is able to block [Ca^{++}] transient in the presence and absence of extracellular calcium for human platelets. Thus, extract of *Ginkgo biloba* leaves (EGb) and its constituent ginkgolide B prevent neuronal damage through

TABLE 6.3. Therapeutic Effects of Ginkgolides.

Disease	Physiological Response
Airway hyperactivity	reduces cell inflammation
	inhibits weal and flare in human skin
	inhibits cutaneous response
	modifies allergic skin responses and allergic asthma
	inhibits bronchial allergen challenge in humans
	antagonizes inflammation responses in humans
Cardiovascular system	inhibits coronary vasoconstriction
	inhibits hypotension
	prevents or reverses pulmonary derangements
	prevents anaphylactic shock of the heart
	inhibits vasodilation of mesenteric microvessels
CNS function	mediates noradregenic system
	improves cerebral metabolism and reverses cerebral impairment
	alleviates post-stroke syndrome by reducing cerebral edema
	inhibits deterioration of regional cerebral blood flow caused by injury
	reduces blood-brain barrier permeability
	inhibits endotoxic shock
Gastrointestinal ulceration	reduces endotoxin-induced gastrointestinal leakage
	reduces endotoxin-induced impairment in stomach and small intestine
	ameliorates mucosal damage associated with the small intestine
Immune process	regulates immune reactivity
	modulates immune response
	prevents suppression of deleterious immunological reactions
	suppresses rejection of bone marrow transplants
	prevents or inhibits cellular rejection
	prolongs graft survival

TABLE 6.3. (continued)

Disease	Physiological Response
Inflammation	reduces edema
	blocks inflammatory action
Leukocytes and endothelial cells	inhibits degranulation, superoxide generation and chemotaxis of neutrophils
	protects viability of endothelium
	modulates physico-chemical state of cytoskeleton via calcium
	increases vascular permeability
Ocular disease	protects retinal tissue against argon laser damage
	improves healing of corneal wounds and decreases corneal swelling
	modulates treatment of retinal functional impairments
	inhibits inflammatory response of the anterior segments of the eye
Platelets and thrombosis	antagonizes platelet binding
	interferes with fibrinogen binding and aggregation
	prevents the activation of transmembrane events
	blocks the formation of several mediators (thromboxane A_2 and leukotrienes)
Renal system	protects tissue from post-ischemia oxidative damage
	blocks glomerular filtration and urinary sodium excretion
	reverses glomeruli and mesangial cell extraction
	affords protection against renal damage
Shock	interferes with changes in blood pressure
	restores the capability of lung parenchymal strips
	prevents increase of vascular permeability
	prevents arterial hypotension
	inhibits intestinal lesions induced by endotoxins
	induces progressive increase in glycemia
	inhibits the release of postaglandins
Skin disease	inhibits candida killing activity of human keratinocytes
	suppresses allergic dermatitis
	reduces swelling of cutaneous tissues during contact dermatitis

Compiled from Braquet (1988).

reduction of the rise in $[Ca^{2+}]I$ (Zhu et al., 1997). BN 52021 (20 mg/kg) given orally one hour before the application of a skin test had a small but significant effect on inhibiting the late phase obstruction airway response. The PAF antagonist BN 52021 has no effect on the pulmonary or systemic circulation per se. However, it completely blocks the increase in pulmonary arterial pressure and vascular resistance as well as the decrease in cardiac output produced by PAF. It seems that BN 52021 is especially potent in blocking the renal receptor of PAF. At low doses (< 25 μg/kg), BN 52021 efficiently blocks the effect of PAF on glomerular filtration in dogs. In the dog, ginkgolide B inhibits the decrease of the glomerular filtration rate and urinary sodium excretion induced by PAF injection.

BN 52063, an extract of ginkgolides containing a mixture of BN 52020, BN 52021 and BN 52022, has been shown to be a selective PAF-antagonist in humans. It inhibits PAF-induced weal and flare in human skin while it has no effect on histamine-induced weal and flare responses (Page and Robertson, 1988). BN 52063 dose-dependently inhibits the coronary vasoconstriction caused by PAF-acether. It also elicits a decrease in heart rate with a maximum effect after 2 min. In humans, BN 52063 has been shown to inhibit PAF-induced skin plasma exudation and platelet aggregation.

Ginkgolides significantly inhibit PAF-acether effects and various immune disorders both *in vivo* and *in vitro* in humans and in animals. BN 52021, the most potent of ginkgolide, inhibits PAF-binding to rabbit- and human-washed platelets, PAF-induced aggregation and degranulation of human isolated neutrophils, PAF-induced contraction of guinea pig lung parenchymal strips or IgG-induced hypotension, hemo concentration and extravasation in the rat. BN 52021 also antagonizes edema, cell filtration, eicosanoid release and lipid peroxidation of intestine and heart ischemia (Spinnewyn et al., 1988).

2.2.2. Bilobalide

Bilobalide is a sesquiterpene lactone believed to be a degraded ginkgolide (loss of carbon 2, 3, 14, 15 and 16) (Nakanishi, 1988). Bilobalide is unstable under basic conditions and may play a significant role in protecting *Ginkgo biloba* against herbivorous insects or mammals.

Chatterjee et al. (1986) indicated that bilobalide-containing agents might provide therapy for neurological disorders that are caused by, or associated with, pathological changes in the myelin sheaths of nerve fibers. In pharmacological models, bilobalide is thought to posses anti-edematous, astrocyte-stimulating and myelin-protecting effects (Chatterjee et al., 1986). Bilobalide has also been found to be suitable for the treatment of certain nervous diseases such as neuropathies, encephalopathies and myelopathies in animal models (Chatterjee et al., 1990).

Bilobalide, as well as ginkgolide B, are known to protect cultured neurons from hypoxia- and glutamate-induced damage. Recently, Ahlemeyer et al.

(1999) reported that among the constituents of *Ginkgo biloba* extract, bilobalide has the most potent anti-apoptotic capacity.

2.2.3. Flavonoids

Ginkgo-flavone glycosides are mono-, di- or tri-glycosides whose aglycon is a flavonol (quercetin, kaempferol or isorhamnetin) and whose glycosidic constituents are glucose and rhamnose. Different classes of flavonoids, including dimeric flavonoids, flavonols, flavonol glycosides and coumaric esters of flavonol glycosides have been isolated from *Ginkgo biloba* leaves (Van Beek et al., 1998). The biflavones, amentoflavone, bilobetin, sequojaflavone, ginkgetin, isoginkgetin and sciadopitysin are present in small amounts but are of major therapeutic interest. The biflavone ginkgetin strongly reduces arthritic inflammation in the rat and may, therefore, be a potential antiarthritic agent with analgesic activities (Kim et al., 1999). The structure and chemistry of these biflavones together with that of a I-5′-methoxybilobetin have been elucidated (Joly et al., 1980) (Figure 6.3).

Flavonoids isolated from commercial extract (EGb 761) taken at a dose of 80 mg, three times a day in a placebo-controlled double-blind study with quadruple crossover design in 12 participants showed cerebral electrical activity in the frontal region suggesting that the flavonoids were bioavailable to the brain (DeFeudis, 1991). The antithrombotic and vasoregulatory activities of commercial *Ginkgo* extracts have been attributed to the antioxidant and oxygen scavenging activities of the flavonoid constituent of these extracts.

Two flavonols, quercetin and rutin, present in *Ginkgo biloba* leaves are potent antithrombotic agents *in vivo*, and hence, it follows that these flavonoids and their glycosides could be relevant constituents responsible for the thrombolytic and vasoprotective actions of commercial extracts observed in clinical settings (DeFeudis, 1991). In addition, quercetin and kaempferol have the same hydrogen-donating antioxidant mechanism as *Ginkgo biloba* extract (Shi and Niki, 1998). These flavonols consumed in the form of *Ginkgo biloba* tablets are eliminated in the human urine as glucuronides (Watson and Oliveira, 1999). A mixture of flavonols (quercetin, kaempferol, isorhamnetin) extracted from *Ginkgo biloba* leaves has been shown to inhibit contraction of isolated guinea pig intestine and, therefore, may exert inhibitory effects on spasms of the intestine (Peter et al., 1966).

Flavonoids are also potent inhibitors of cyclic-AMP-phosphodiesterase, thereby enhancing the effects of prostaglandin I2 or prostacyclin, mediators of vasoprotection and thrombosis. The degree of cyclic-AMP-phosphodiesterase inhibition by *Ginkgo biloba* bioflavones in rat adipose tissue was found to be amentoflavone > bilobetin > sequoiaflavone > ginkgetin = isoginkgetin, and sciadopitysin was almost inactive (Saponara and Bosisio, 1998). Amentaflavone is also known to exhibit five to 10 times greater potency than papaverine in inhibiting the degradation of cyclic-GMP, which in turn leads to endothelium-dependent relaxation of vascular smooth muscle (DeFeudis, 1991).

Quercetin and kaempferol coumaroyl glycosides are found only in *Ginkgo biloba*, and the latter is considered particularly essential for the efficacy of commercial extracts. Some of the flavonol glycosides are unique to *Ginkgo biloba* and have, therefore, been used for the standardization of commercial extracts. Several HPLC methods developed for the qualitative and quantitative determination of flavonoids of *Ginkgo biloba* have recently been summarized (Van Beek et al., 1998). The flavonoid glycosides are generally acid hydrolyzed, cleaned by solid-phase extraction cartridges or diluted prior to quantitatively determining the corresponding aglycones (quercetin, kaempferol and isorhamnetin) (Stricher, 1993). Fingerprinting of all *Ginkgo biloba* flavonoids using RP-HPLC allows the separation and identification of 33 flavonoids (22 flavonoid glycosides, six flavonoid aglycones and five biflavones) by elution order and diode-array UV spectra (Van Beek et al., 1998). The bioflavone constituents of *Ginkgo biloba* leaves have been identified and confirmed using HPLC-MS with a thermospray interface (Gobbato et al., 1996).

Rutin and quercetin as well as mixtures of flavonoids from *Ginkgo biloba* extracts protect cerebellar granular cells from oxidative damage and apoptosis due to the scavenging activity of their hydroxyl radicals (Chen et al., 1999). The biflavone ginkgetin isolated from *Ginkgo biloba* leaves may be a potential antiarthritic agent with analgesic activity. It strongly reduces arthritic inflammation and inhibits wreathing in the animal model (Kim et al., 1999). Flavonoids in *Ginkgo biloba* extract have been credited with the protection of endothelial cells against hyperoxia and hypoxia-reoxygenation due to their antioxidant activity. These protective effects of the flavonoids account for the wide use of *Ginkgo biloba* extracts in venous diseases (Remacle et al., 1990).

2.2.4. Ginkgolic Acids

The chemistry and biology of ginkgolic acid and related alkyphenols from *Ginkgo biloba* have recently been reviewed (Jaggy and Koch, 1997). *Ginkgo*lic acids are 6-alkyl salicylic acids and are the major components of the lipid fraction in the nutshells of *Ginkgo biloba*. This class of substances is also present in *Ginkgo* leaves. The 3-alkyl phenol is called ginkgol and the 5-alkyl resorcins, derived via decarboxylation of 6-alkyl resorcylic acids, are represented by bilobol and hydrobilobol (Figure 6.5). The simple unsaturated C15-ginkgolic acid is the main component of the nutshells and leaves (3.1 and 1.2%, respectively), followed by C17:1-acid (0.22 and 0.44%, respectively). A ginkgolic acid content of 1.73% was determined in dried *Ginkgo biloba* leaves by HPLC.

Contact with *Ginkgo biloba* fruit is frequently reported to induce allergic skin reactions generally attributed to the presence of ginkgolic acids. The

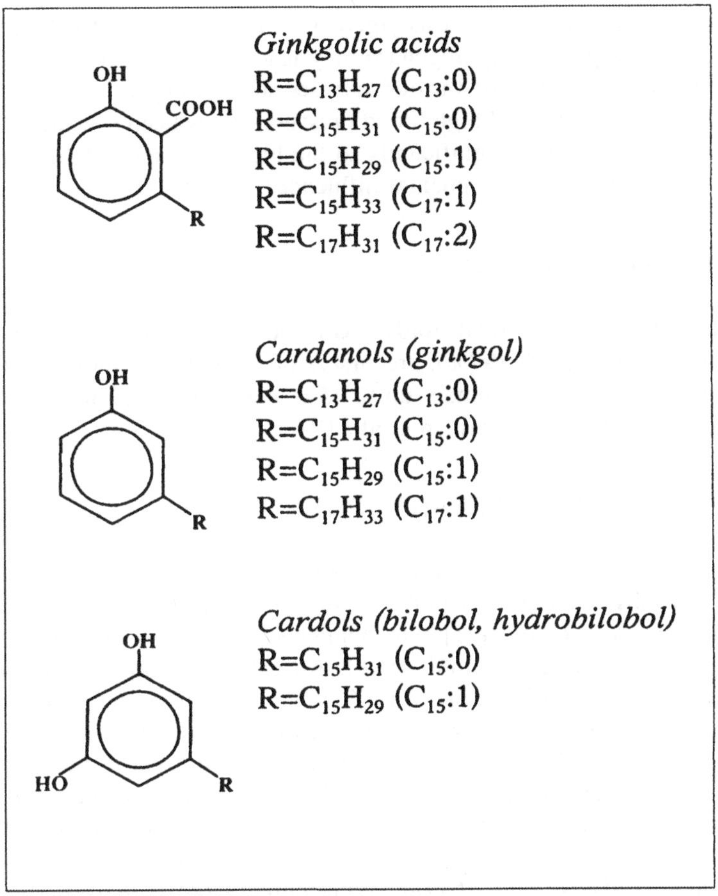

Ginkgolic acids
$R=C_{13}H_{27}$ (C_{13}:0)
$R=C_{15}H_{31}$ (C_{15}:0)
$R=C_{15}H_{29}$ (C_{15}:1)
$R=C_{15}H_{33}$ (C_{17}:1)
$R=C_{17}H_{31}$ (C_{17}:2)

Cardanols (ginkgol)
$R=C_{13}H_{27}$ (C_{13}:0)
$R=C_{15}H_{31}$ (C_{15}:0)
$R=C_{15}H_{29}$ (C_{15}:1)
$R=C_{17}H_{33}$ (C_{17}:1)

Cardols (bilobol, hydrobilobol)
$R=C_{15}H_{31}$ (C_{15}:0)
$R=C_{15}H_{29}$ (C_{15}:1)

Figure 6.5 Structure of ginkgolic acids, cardanols and cardols.

alkylphenolic acids are also known to cause gastrointestinal disturbances following the consumption of *Ginkgo biloba* fruit (Becker and Skipworth, 1975). The allergenic properties of alkylphenols are presumably based on their chemical structure, which makes them capable of being incorporated in cell membranes. Ginkgolic acid has immunotoxic properties and causes swelling of the lymph nodes in mice (Jaggy and Koch, 1997). It has very high sensitizing potential even at very low concentrations (125 μg ginkgolic acids). Ginkgolic acids have several biological activities, including dehydrogenase inhibitory activity, antimicrobial activity against *Mycobacterium smegmatis*, *Bacillus subtilis* and *Staphylococcus aureus*, molluscicidal activity, antitumoral activity and an inhibitory effect on seed germination (Jaggy and Koch, 1997). Ginkgolic acids are effective inhibitors of glycerol-3-phosphate dehydrogenase, a key enzyme in the synthesis of triacylglycerol (Irie et al., 1996). Inhibition of glycerol-3-phosphate dehydrogenase may lead to low accumulation of lipids resulting in reduced risk of various diseases such as hyperglycemia, diabetes, myocardial infarction and high blood pressure. It has been speculated that the alkylphenols contribute significantly to the high resistance of *Ginkgo biloba* to different adverse environmental influences.

2.2.5. Other Constituents

A lipid fraction (5% total lipids) extracted from *Ginkgo* seed contains 4-hydroxyanacardic acids (97%), precursors of 5-alkylresorcinol. The lipids of immature *Ginkgo* seeds contain up to 75% anacardic acids that have proven antibiotic properties (Gellerman et al., 1976).

2.3. COMMERCIAL PREPARATIONS

Extracts from *Ginkgo biloba* are prepared from dried leaves using acetone/water and subsequent purification and concentration steps aimed at increasing the concentration of bioactive phytochemicals (Anonymous, 1998b; DeFeudis, 1991; Kleijnen and Knipschild, 1990). An extract of *Ginkgo biloba* leaves was first registered in Germany in 1965. It was introduced in the market under the Tebonin® trade name as drops, tablets and later as ampoules. Development of a highly purified extract at single doses of 40 mg led to the registration in 1974 of the oral preparation under the trade name Tanakan® in France. Preparations of this extract were introduced in Germany as Rökan® in 1978 and as Tebonin® forte in 1982. The extract obtained the code name EGb 761 (Extractum *Ginkgo biloba*e 761) and is currently marketed in over 30 countries under different trade names (DeFeudis, 1991). EGb 761 is the most studied and commonly used *Ginkgo* preparation in clinical trials.

The pharmacological activities and clinical applications of *Ginkgo biloba* extract EGb 761 have been extensively reviewed by DeFeudis (1991) and more recently by Clostre (1999). The extract is standardized on the amount of ginkgo-flavone glycosides (24%) and terpenoids (ginkgolides A, B, C, J and bilobalide) (6%). Ginkgolides A, B and C, collectively account for about 3.1%, ginkgolide J ≤0.5% and bilobalide about 2.9% of EGb 761 extract. Organic acids account for about 5–10% of EGb 761 and give the extract an acidic character, thereby increasing its water solubility (DeFeudis, 1991). Another *Ginkgo* preparation often encountered in clinical trial is Kaveri (also called LI 1370) that is standardized on the same ingredients in doses similar to EGb 761 (25% ginkgo-flavone glycosides, and recently, also 6% terpenoids) (Kleijnen and Knipschild, 1990). The composition of *Ginkgo* preparations may vary depending on the manufacturing process used, and hence, different effects can result. More than 24 different brands of *Ginkgo biloba* extract are sold in the United States.

2.4. PHARMACOLOGICAL EFFECTS

Currently, *Ginkgo* is widely used in Europe for the treatment of memory disorders associated with aging, including Alzheimer's disease (AD) and vascular dementia. Over 5 million prescriptions were issued mostly by physicians for *Ginkgo* in 1988 in Germany (Oken et al., 1998). In 1994, a standardized dry extract of *Ginkgo biloba* leaves (SeGb) was approved by German health authorities for the treatment of primary degenerative dementia and vascular dementia. *Ginkgo biloba* extract is already widely used in the United States as an alternative therapy for AD. The wide use of *Ginkgo biloba* extracts is based on the pharmacological action resulting from the combined activities of several active principles (primarily, the flavonoids and terpenoids). This "polyvalent" (terminology used by DeFeudis, 1991) action explains the vaso- and tissue-protective and the cognition-enhancing benefits of the *Ginkgo biloba* extract (Van Beek et al., 1998).

2.4.1. Clinical Trials

The principal clinical trials concerning the therapeutic applications of *Ginkgo biloba* extract have been extensively reviewed (Braquet, 1988; De-Feudis, 1991; Van Beek et al., 1998; Kleijnen and Knipschild, 1990, 1992). According to these reviews, the therapeutic effect of *Ginkgo biloba* can be generally attributed to its antioxidant, anti-ischemic, cardioprotective and vasoregulating activities and neuroprotective properties. The pharmacological studies on the *Ginkgo biloba* extract are still ongoing. Several new clinically controlled studies confirming the efficacy of *Ginkgo biloba* extract in important pathologies have appeared since the publications of the above cited reviews and are summarized in Table 6.4 and briefly discussed here.

Chung et al. (1999) recently evaluated a possible therapeutic effect of *Ginkgo biloba* extract (GBE) on glaucoma patients that may benefit from improvements in ocular blood flow. A Phase I crossover trial of GBE with placebo control in 11 healthy volunteers (eight women, three men: age, 34 $\pm$ 3 years, mean $\pm$ SE) was performed. Patients were treated with either GBE 40 mg or placebo three times daily, orally, for two days. Color Doppler imaging (Siemens Quantum 2000) was used to measure ocular blood flow before and after treatment. There was a two-week washout period between GBE and placebo treatment. *Ginkgo biloba* extract significantly increased end diastolic velocity (EDV) in the ophthalmic artery (OA) (baseline vs. GBE treatment; 6.5 $\pm$ 0.5 vs. 7.7 $\pm$ 0.5 cm/s, 23% change, $p = 0.023$), with no change seen in placebo (baseline vs. GBE treatment: 7.2 $\pm$ 0.6 vs.

TABLE 6.4. Recent Clinical Studies with *Ginkgo biloba* Extract.

Authors	Type of Study	No. Patients	Diagnosis	Dosage and Duration	Improvement	Overall Assessment
Chung et al., 1999	Phase I crossover trial	11 (8 w/3 m)	Glaucoma	40 mg, 3X daily for two days	Increased end diastolic velocity in ophthalmic artery 23/3% change	Increase in ocular blood flow
Janssens et al., 1999	Randomized double-blind	—	Patients with primary chronic venous insufficiency	4 wk treatment with Ginkor Fort	% decrease in circulating endothelial cell count 14.5/8.4	Beneficial action of GBE on venous wall
Lingaerde et al., 1999	Randomized double-blind	27	SAD (seasonal affective disorder)	PN246 Tablet 10 wks	No significant improvement in symptoms of winter depression	Development of the symptoms of winter depression not preventable by GB
Rigney et al., 1999	Randomized double-blind	31 (ages 30–59)	Asymptomatic volunteers	120–300 mg (50–100 mg t.d.s.) for two days	GBE 120 mg pronounced improvement in aspects of cognition	Cognitive enhancing effects of GBE likely more apparent in 50–59 yr individuals
Castelli et al., 1998	Double-blind vs. placebo	22 (women of 22–55 yrs)	Allergic contact dermatitis	GBE and sodium carboxy methyl-beta-1,3-glucan applied to intact skin 2X a day for 2 wks prior to contact allergen	68% reduction in skin reactivity of treated vs. placebo	GBE formulation mitigates allergic contact dermatitis

158

TABLE 6.4. (continued)

Authors	Type of Study	No. Patients	Diagnosis	Dosage and Duration	Improvement	Overall Assessment
Cesarani et al., 1998	Open, controlled vs. placebo	44	Vertiginous syndrome	80 mg, 2X daily, three months	Improved viscovestibular ocular reflex	Beneficial action of GBE in treatment of equilibrium disorders
Ivaniv, 1998	—	208	Discirculatory encephalopathy (DE)	—	Positive changes in cognitive behavior	Positive in rehabilitation of patients affected with DE
Li et al., 1998	Controlled trial	24 (12/12, diabetic/nondiabetic)	Peripheral arterial occlusive disease (PAOD)	240 mg · d⁻¹¹, PO, 48 wks	Pain-free walking distance 3.3–3.8-fold increase	Therapeutic effect in PAOD patients
Li et al., 1997	Randomized controlled trial	—	Asthma	—	Reduced airway hyperreactivity	Improved symptoms of pulmonary functions

GBE = *Ginkgo biloba* extract.

159

7.1 ± 0.5 cm/s, 3% change, $p = 0.892$). No side effects related to GBE were found. *Ginkgo biloba* extract did not alter arterial blood pressure, heart rate or IOP. Since *Ginkgo biloba* extract significantly increased EDV in the OA, Chung et al. (1999) suggest that GBE may possibly find application in the treatment of glaucomatous optic neuropathy and other ischemic ocular diseases.

One possible mechanism that accounts for the alterations observed in varicose veins is the activation of endothelial cells by ischemia occurring in the leg veins during blood stasis and the cascade of reactions that follows. Because *in vitro* data suggest that endothelium alteration is a key event in the development of the pathology, Janssens et al. (1999) conducted a clinical trial to confirm this hypothesis. The authors used the number of circulating endothelial cells (CECs) detached from the vascular wall as a criterion of the endothelium injury in patients with chronic venous insufficiency (CVI) and determined the protective effect of *Ginkgo biloba* extract (Ginkor Fort), by a randomized, double-blind, placebo-controlled clinical trial. Their results showed that in the active-treatment group, the mean values of the CEC count decreased by 14.5% after a four-week treatment, whereas in the placebo group, the decrease was less (8.4%).

Ginkgo biloba extract PN246, in tablet form (brand name Bio-Biloba) has also been used to test the hypothesis that it may prevent the symptoms of winter depression (WD) in patients with seasonal affective disorder (SAD) (Lingaerde et al., 1999). In this study, a total of 27 SAD patients were randomized to receive double-blind placebo or Bio-Biloba for 10 weeks or until they developed symptoms of WD, starting in a symptom-free phase about one month before expected WD symptoms. An extended Montgomery-Asberg Depression Rating Scale was completed before and immediately after termination of medication. The patients also self-rated some key symptoms on a visual analogue scale every two weeks during the trial. Since differences between the treatment groups in the number of patients who developed treatment-requiring WD, or in the development of single key symptoms during the trial were not significant, it was concluded that *Ginkgo biloba* was unable to prevent the development of the symptoms of winter depression.

A study on the effects of acute doses of a *Ginkgo biloba* extract (GBE) on memory and psychomotor performance in a randomized, double-blind and placebo-controlled five-way crossover design was carried out by Rigney et al. (1999). In this study, 31 volunteers aged 30–59 years received GBE 150 mg (50 mg t.d.s.), GBE 300 mg (100 mg t.d.s.), GBE 120 mg mane and GBE 240 mg mane and placebo for two days. Following baseline measures, the medication was administered at 0900 h for the single doses and at 0900, 1500 and 2100 h for the multiple doses. The psychometric test battery was administered pre-dose (0830 h) and then at frequent intervals until 11 h post-dose. The results demonstrate pronounced effects of GBE extract on aspects of cognition in asymptomatic volunteers for memory, particularly working memory. They also show that these effects may be dose dependent though not in a linear dose-related manner, and that GBE 120 mg produces the most evident effects of the doses examined. Additionally, the results suggest that the cognitive enhancing effects of GBE are more likely to be apparent in individuals aged 50–59 years (Rigney et al., 1999).

The clinical efficiency of mitigating contact dermatitis with a *Ginkgo biloba* extract and carboxymethyl-beta-1,3-glucan formulation was investigated in a double-blind versus placebo study using 22 subjects (Caucasian women aged 22–55 years) with allergic contact dermatitis from various substances in the European standard series (Castelli et al., 1998). The formulation was applied to intact skin two times a day for two weeks ("in use" application) prior to a single application of a selected contact allergen under a Finn Chamber for 24 h. Readings were carried out in a blind study by a dermatologist two and three days after patch removal. Representative photographs were taken of treated, placebo and untreated test areas. Significantly reduced skin reactivity was shown by 68.2% of the panelists ($p = 0.037$) on the treated site two days after patch removal, versus untreated and/or placebo sites. Thus, according to Castelli et al. (1998), *Ginkgo biloba*/carboxymethyl-beta-1,3-glucan formulation can mitigate against allergic contact dermatitis.

In an open, controlled study (Cesarani et al., 1998), 44 patients complaining of vertigo, dizziness or both, caused by vascular vestibular disorders were randomly treated with an extract of *Ginkgo biloba* (EGb 761), 80 mg twice daily, or with betahistine dihydrochloride, (BI) 16 mg twice daily, for three months. A complete neuro-otologic and equilibrimetric examination was performed at baseline and after three months of treatment, with evaluation of clinical findings. In the first month of therapy, vertigo and dizziness improved in 64.7% of patients treated with BI and in 65% of those who received EGb 761. Compared to baseline, no statistically significant changes were observed in cranial scans for patients with a "central" cranial pattern and for the equilibrium score in both groups. The comprehensive test battery showed that EGb 761 considerably improved visuovestibular ocular reflex. No side effects were recorded during the trial except for transient mild headache and gastric upset in two patients receiving EGb 761 and transient cyanosis of nails and lips in one patient given BI. These results suggest that EGb 761 and BI operate at different equilibrium receptor sites and show that EGb 761 can considerably improve oculomotor and visuovestibular functions.

In an open trial, *Ginkgo biloba* extract was found to be 84% effective in treating antidepressant-induced sexual dysfunction predominately caused by selective serotonin reuptake inhibitors (SSRIs, $n = 63$) (Cohen and Bartlik, 1998). Women ($n = 33$) were more responsive to the sexually enhancing effects of *Ginkgo biloba* than men ($n = 30$), with relative success rates of 91% versus 76%. *Ginkgo biloba* generally had a positive effect on all four phases of the sexual response cycle: desire, excitement (erection and lubrication), orgasm and resolution (afterglow).

This study was later confirmed by another open trial consisting of 55 patients with sexual dysfunction receiving a standardized *Ginkgo* extract, 40 or 60 mg capsules taken three to four times daily (Cohen, 1999). Forty-nine of the 55 patients reported significant response and improvement in their sexual dysfunction. There were no significant adverse effects associated with the use of *Ginkgo biloba* extract, except for one patient who reported worsening of urinary difficulties related to prostatic hypertrophy (Cohen, 1999). Dry powdered extract from leaves of *Ginkgo biloba* has synergistic effect with lyophylized roe and the mixture as such has been used for treating impotence in human males (Omar, 1998).

Studies made in 208 patients with discirculatory encephalopathy using clinical and neuropsychological tests, apparatus (REG, ECG, EEG, Doppler sonography) and laboratory (coagulogram) methods showed the prescription of drug preparations from *Ginkgo* to be a worthwhile exercise likely to benefit the patient during different stages of his/her ailment (Ivaniv, 1998). *Ginkgo* preparations showed positive time-related changes in elastic and tonic characteristics of vessels, bioelectric brain activity, emotional and behavioral, cognitive and mnestic features, especially when combined with a complex of components of the tricarboxylic acid cycle.

The effects of *Ginkgo biloba* extract EGb 761 from the points of view of hemorheology for patients of peripheral arterial occlusive diseases (PAOD) were studied by Li et al. (1998). The treatment with EGb (240 mg/d, po) and the pain-free walking distance (PFWD) were carried out for 24 PAOD patients (12 nondiabetic, ND and 12 diabetic, D) over 48 weeks. The parameters erythrocyte stiffness (ES) and relaxation time (RT), the blood plasma viscosity (eta), the plasma fibrinogen concentration (Cf) and the blood sedimentation rate (BSR), the PFWD, and maximal walking distance (MWD) were determined at six weeks before treatment (−6), at the beginning of the treatment (0), and after 6, 11, 16 and 48 weeks of treatment. At week 0, stiffness and relaxation time were significantly higher than healthy control, and the mean PFWD was only 111 m. The blood plasma viscosity value was significantly elevated and fibrinogen concentration and blood sedimentation rate were enhanced. Throughout 11 weeks of treatment, ES, RT, eta and Cf decreased gradually, and PFWD improved. Between 16 and 48 weeks, ES and RT were no longer significantly different from the controls, whereas eta and Cf decreased gradually but remained higher than normal, BSR decreased, and the PFWD improved by a factor of 3.8 times (D) and 3.3 times (ND). From these results, Li et al. (1998) concluded that EGb has therapeutic effects in PAOD patients.

The effects of *Ginkgo* leaf concentrated oral liquor (GLC) on airway hyperactivity and inflammation, clinical symptoms and pulmonary functions of asthma patients were determined in a randomized controlled trial (Li et al., 1997). In contrast to placebo group, GLC significantly reduced airway hyperactivity ($P < 0.05$) and improved clinical symptoms ($P < 0.05$) and pulmonary functions ($P < 0.05$) of the asthmatic patients. Thus, GLC is proposed as an effective drug of anti-airway inflammation.

In addition to the numerous double-blind and randomized controlled trials with *Ginkgo biloba* extracts dating back to 1984, six meta-analysis of clinical studies with *Ginkgo biloba* have been reported (Table 6.5). These analyses relate predominantly to diseases such as Alzheimer's, dementia, claudication, cerebral performance and neurophysiology of patients with organic brain syndrome where *Ginkgo biloba* has generally been found to have significant positive clinical effects.

Recently, Ernst and Pittler (1999) systematically reviewed the clinical evidence of *Ginkgo biloba* preparations as a symptomatic treatment for dementia. Computerized literature searches were performed to identify all double-blind, randomized, placebo-controlled trials assessing clinical end points of *Ginkgo biloba* extract as a treatment for dementia. Databases included Medline, Embase, Biosis and the Cochrane Library. There were no restrictions regarding the language of publication. Data were extracted in a standardized, predefined fashion, independently by both authors. Nine double-blind, randomized, placebo-controlled trials met the inclusion criteria and were reviewed. These studies of varying methodological quality collectively suggest that *Ginkgo biloba* extract is more effective for dementia than placebo. However, few, generally mild, adverse effects were reported.

Flint and van Reekum (1998) reviewed the drug treatment of Alzheimer's disease (AD) to provide guidelines for the physician on how to integrate these treatments into the overall management of this disorder and made a qualitative review of randomized, double-blind, placebo-controlled trials of medications used to treat cognitive deficits, disease progression, agitation, psychosis or depression in AD. A computerized search of Medline was used to identify relevant literature published during the period of 1968–1998. Key words used in the search were "randomized controlled trials" with "dementia" and with "Alzheimer's disease." Donepezil and *Ginkgo biloba*, two of the four agents currently available in Canada to treat the cognitive deficits of AD, were associated with a statistically significant but clinically modest improvement in cognitive function in a substantial minority of patients with mild to moderate AD. These data indicate that selected medications can be used to treat cognitive deficits, disease progression, agitation, psychosis and depression in AD. However, considerable heterogeneity was observed in patients' responses to these medications.

The effect of treatment with *Ginkgo biloba* extract on objective measures of cognitive function in patients with AD was determined by Oken et al. (1998) through a formal review of the current literature. According to these authors, only four of the 50 studies examined met all inclusion criteria of clear diagnoses of dementia and AD. In total, there were 212 subjects in each of the placebo and *Ginkgo* treatment groups. Quantitative analysis of the

TABLE 6.5. Meta-Analyses of Clinical Studies with *Ginkgo biloba*.

Authors	Type of Study	Disease/Diagnosis	Assessment
Ernst and Pittler, 1999	Computerized literature searches for randomized controlled trials to March 1998	Dementia	Nine studies out of 18 double-blind, randomized, placebo-controlled trials met the inclusion/exclusion criteria. Collectively, the trials suggest that *Ginkgo biloba* is efficacious in delaying the clinical deterioration of patients with dementia.
Flint and van Reekum, 1998	Literature review 1968–1998 of randomized double-blind placebo-controlled trials	Alzheimer's-cognitive deficits, agitation, psychosis or depression in AD	Inconclusive data. GB associated with a significant but clinically modest improvement in cognitive function.
Oken et al., 1998	Review of 50 studies	Cognitive function in Alzheimer's disease	Small but significant effect of 3–6 months. Treatment with 120–240 mg of GB extract on cognitive function in AD.
Ernst, 1996	Review of 10 controlled trials	Intermittent claudication	Studies implied that GB is an effective therapy for intermittent claudication.
Hopfenmüller, 1994	Review of 11 controlled clinical trials	Cerebrovascular insufficiency	Seven studies confirmed effectiveness of GB vs. placebo based on total scores of clinical symptoms.
Kleijnen and Knipschild, 1992	Review of 40 trials	Cerebral insufficiency	Eight well-performed trials met the inclusion/exclusion criteria and confirmed the effectiveness of GB in cerebral insufficiency

AD = Alzheimer's disease.
GB = *Ginkgo biloba* extract.

164

literature showed a small but significant effect of three- to six-month treatments with 120 to 240 mg of *Ginkgo biloba* extract on objective measures of cognitive function in AD. The drug had no significant adverse effects in formal clinical trials, but two cases reported bleeding complications (Oken et al., 1998).

Itil et al. (1998) compared the pharmacological effects of *Ginkgo biloba* extract EGb with those of tacrine, also known as tetrahydroaminoacrine (THA), one of the two currently approved drugs in the United States for the treatment of Alzheimer's disease. The authors studied 18 subjects (11 males, seven females of an average age of 67.4 years) with light to moderate dementia [Mini Mental mean score = 23.7, ranges: 15–29 (Geriatric Depression Scale mean scores = 3.7; range: 3.2–5.4)]. Each subject was randomly administered a single oral "Test-Dose" of either 40 mg of tacrine or 240 mg of EGb in two separate sessions within three- to seven-day intervals. Before drug administration and at one- and three-hour intervals after drug administration, CEEGs were recorded for a minimum of 10 minutes. The CEEGs were analyzed using Period Analysis. EGb induced pharmacological effects in the central nervous system and exhibited more therapeutic effects (compared to nonresponders) when drugs were administered chronically.

The effectiveness of *Ginkgo biloba* in the treatment of intermittent claudication was evaluated by a systematic review of Medline-search-identified ten controlled trials on the subject (Ernst, 1996). These were heterogeneous in all respects and, with only few exceptions, of poor methodological quality. All the studies implied that *Ginkgo biloba* is an effective therapy for intermittent claudication. This hypothesis was confirmed in a monocenter, randomized, placebo-controlled double-blind study with parallel-group comparison undertaken to demonstrate the efficacy of *Ginkgo biloba* special extract EGb 761 on objective and subjective parameters of the walking performance in trained patients suffering from peripheral arterial occlusive disease in Fontaine stage IIb (Blume et al., 1996). A total of 60 patients were recruited (42 men; aged 47–82 years) with angiographically proven peripheral arterial occlusive disease of the lower extremities and an intermittent claudication existing for at least six months. No improvement had been shown despite consistent walking training, and a maximum pain-free walking distance on the treadmill of less than 150 m was recorded at the beginning of the study. The therapeutic groups were treated with either *Ginkgo biloba* extract EGb 761 at a dose of three times one film-coated tablet of 40 mg per day by oral route or placebo over a duration of 24 weeks following a two-week placebo run-in phase. The main outcome measure was the difference of the walking distance between the start of treatment and after 8, 16 and 24 weeks of treatment as measured on the treadmill (walking speed 3 km/h and slope of 12%). As secondary parameters, the corresponding differences for the maximum walking distance, the relative increase of the pain-free walking distance, the Doppler index and the subjective evaluation of the patients were analyzed. The absolute changes in the pain-free walking distance in treatment weeks 8, 16 and 24 as against the treatment beginning (median values with 95% confidence interval) led to the following values for the patients treated with *Ginkgo biloba* special extract EGb 761:19 m, 34 m and 41 m. The corresponding values in the placebo group were 7 m, 12 m and 8 m. The advantage of the EGb 761-treated group as compared to the placebo group could be verified statistically at the three time points with $p < 0.0001$, $p = 0.0003$ and $p < 0.0001$. The test for the presence of a clinically relevant difference of 20% between EGb 761 and placebo also produced a statistically significant result ($p = 0.008$). The Doppler index remained unchanged in both therapeutic groups. A corresponding statistically significant advantage for the EGb 761 group was observed on a descriptive level for the other parameters tested. The tolerance of the treatment was very good. The results of this placebo-controlled study show that treatment with *Ginkgo biloba* extract EGb 761 produces a statistically highly significant and clinically relevant improvement of the walking performance in trained patients suffering from intermittent claudication with very good tolerance of the study preparation.

Eleven controlled clinical trials were evaluated in a meta-analysis of studies on the effectiveness of the *Ginkgo biloba* extract LI 1370 (Kaveri forte) in patients with cerebrovascular insufficiency in old age (Hopfenmüller, 1994). All studies were placebo-controlled randomized double-blind trials, using in most cases a daily dose of 150 mg extract. The requirements for the quality of the studies were the basic criteria for the performance of clinical drug tests analyzed from the biometrical scope. Three studies were excluded from the meta-analysis according to methodological or objective reasons. In two further studies, the evaluation of the physician or the patients was missing, therefore, the studies could not be used for the analysis of the "global effectiveness." All other studies were comparable with regard to diagnoses, inclusion and exclusion criteria as well as methodology. Therefore, a statistical meta-analysis could be performed for them, analyzing the parameters "single symptoms," total score of clinical symptoms and "global effectiveness." For all analyzed single symptoms, significant differences could be concluded, indicating the superiority of *Ginkgo biloba* in comparison to placebo. The analysis of the total score of clinical symptoms from all relevant studies indicated that seven studies confirmed the effectiveness (*Ginkgo biloba* being better compared to placebo), while only one study was inconclusive (the medications were not different).

Kleijnen and Knipschild (1992), by means of a critical review, sought evidence from controlled trials in humans on the efficacy of *Ginkgo biloba* extracts in cerebral insufficiency. The methodological quality of 40 trials on *Ginkgo* and cerebral insufficiency was assessed using a list of predefined criteria of good methodology, and the outcome of the trials was interpreted in relation to their quality. A comparison of the quality was made with trials of co-dergocrine, which is registered for the same indication. There were eight well-performed trials out of a total of 40. Shortcomings were limited numbers of patients included and incomplete description of randomization procedures, patient characteristics, effect measurement and data presentation. In no trial was double-blindness checked. Virtually all trials reported positive results. In most trials, the dosage was 120 mg *Ginkgo* extract a day, given for at least four to six weeks.

In addition to the studies described above, there is evidence of the use of *Ginkgo biloba* preparations in combination with other phytochemicals for treating patients suffering from viral diseases, especially AIDS (Beljanski, 1990). Apparently, the standardized hydrolysate of *Ginkgo biloba* extract (Bioparyl) effectively mitigates the accumulation of gamma globulins produced by HIV infection (Beljanski, 1990). *Ginkgo biloba* extract (EGb 761) has also been found to decrease vasomotor disorder of the extremities during a Himalayan mountain expedition (Roncin et al., 1996).

2.5. SAFETY

The seed and leaf of *Ginkgo biloba*, despite their medicinal properties, also contain some toxic principles. Potent allergens such as ginkgolic acids are present in the seed, pulp and leaves, and contact with the fresh seed pulp may cause dermatitis and other allergic reactions. Ginkgolic acid (125 μg) causes swelling of the lymph nodes when injected in mice (Van Beek et al., 1998). Thus, the German Monograph E allows a concentration of less than 5 ppm ginkgolic acid for the dry extract of *Ginkgo biloba* leaf. 4'-*O*-methylpyridoxine, another potentially toxic compound present in edible *Ginkgo* nuts, may cause convulsions and loss of consciousness, particularly in children, when excessive amounts of nuts are consumed. Food poisoning from the consumption of *Ginkgo biloba* seeds has been attributed to the presence of the toxic substance 4'-*O*-methylpyridoxine that antagonizes vitamin B6 and inhibits the formation of 4-aminobutyric acid in the brain (Keiji et al., 1988). Cytotoxicity due to the antivitamin B6 from *Ginkgo biloba* seed has also been reported in 70 cases in Japan (Keiji et al., 1985). This toxin has also been detected in leaves and some boiled *Ginkgo*-containing foods.

The crude alcoholic extract of the *Ginkgo biloba* leaf has been in use for several years in France for the treatment of peripheral vascular disease, and this has not been associated with any known side effects. Administration of a single dose of the ginkgolide BN 52063 was not associated with any side effects. However, since PAF seems to be involved in many natural processes, such as the implantation of the human embryo in the uterus, and is present in human amniotic fluid, it is worth remembering that these could be interfered with. Toxicological studies in animals with the PAF antagonist BN 52021 do not support such concerns (Chung and Barnes, 1988).

Guinot et al. (1988) studied the safety of BN 52063 in normal healthy volunteers to find a maximum well tolerated dose for use in the initial and multiple dose studies. Subsequent studies confirmed the PAF-antagonist activity of BN 52063 in humans and investigated basic pharmacokinetics and pharmacodynamics and a safe dose range. In a single-blind crossover trial, five healthy male volunteers took increasing single doses of BN 52063, 20 mg, 40 mg, 80 mg, 120 mg and placebo, separated by at least one week. BN 52063 had no significant effects on blood pressure, pulse, electrocardiogram and laboratory parameters at each dosage level. In a further two-week double-blind, randomized, crossover study, six healthy normal subjects were treated with 40 mg BN 52063 or placebo, three times a day for a week, the two treatment periods being separated by one week. Again, treatment with BN 52063 was found to be safe, well tolerated and significantly not different from the placebo treatment. In a separate study, dosages of BN 52063 up to 720 mg as a single dose and 240 mg/day for two weeks or even 360 mg/day for one week were well tolerated clinically and biologically by healthy volunteers (Bonvoisin and Guinot, 1989).

2.6. CONCLUSIONS

More than 280 studies have been published on *Ginkgo biloba* since the 1950s, covering areas concerning the pharmacological and therapeutic effects of *Ginkgo biloba* extract on the vascular tissue, impotence, memory dysfunction and several other pathologies. At present, extracts of *Ginkgo* leaves are widely recommended in the Asian and European medical communities and account for annual sales of approximately $500 million. In fact, during 1998, physicians in Germany wrote more prescriptions (5.4 million) for *Ginkgo biloba* extract than any other drug. It is also available in Europe and Asia as an over-the-counter drug (OTC). The product is marketed in the U.S. and Canada in dry powdered and liquid form, and in tablets and capsules for oral use.

The very complex and unique structure of the macrocyclic lactones (ginkgolides) impeded the study and function of these macromolecules for pharmaceutical uses prior to 1988 (Braquet, 1988). Lactones similar to these structures are now emerging as potentially important cancer chemotherapeutic agents and are now in several human clinical trials (Wender et al., 1999). This resurgence of interests in lactones will no doubt lead to revisiting the clinical studies performed with ginkgolides in the late 1980s. This may spur new studies that should be undertaken to reexamine the use of *Ginkgo biloba* for cancer treatment in addition to traditional use in the treatment of various diseases. The chemical, pharmacological and clinical literature on the standardized extract of *Ginkgo biloba* is very impressive. However, although use of the extract is widespread in Europe, it is only recently that it has gained acceptance in clinical studies in North America. *Ginkgo biloba* and its constituents also have proven applications in nonpharmaceutical, especially cosmetic, industries (Van Beek et al., 1998). No doubt, these opportunities will be fully explored as new studies on *Ginkgo biloba* and its constituents come to the fore. At the dawn of the third millennium, the sixth for *Ginkgo biloba*, the use of *Ginkgo biloba* appears to have no boundaries.

3. REFERENCES

Adamski, R and E. Styp-Rekowska 1971 "Stability of hypericin in juice, dry extract, and tablets from *Hypericum perforatum* plants," *Farm. Pol.* 27:237–241 (CA 75:91286k).

Ahlemeyer, B., A. Mowes and J. Krieglstei. 1999. "Inhibition of serum deprivation and staurosporine-induced neuronal apoptosis by *Ginkgo biloba* extract and some of its constituents," *Eur. J. Pharmacol.* 367·423–430.

Aizenman, B. E. 1969. "Antibiotic preparations from *Hypericum perforatum*," *Mikrobiol. Zh.* (Kiev) 31:128–133 (CA 70.118006e).

Ames, B. N. 1983. "Dietary carcinogens and anticarcinogens. Oxygen radicals and degenerative diseases," *Science.* 221·1256–1264.

Anonymous. 1997. "Antioxidant properties of a series of extracts from medicinal plants," *Biofizika* 42(2):480–483.

Anonymous 1998a. "*Hypericum perforatum*, St John's wort " In: *PDR for Herbal Medicines,* Gruenwald, J., T. Brendler and C. Jaenicke, Eds., Medical Economic Company, Inc., Montvale, New Jersey, pp 905–908.

Anonymous. 1998b. "*Ginkgo biloba, Ginkgo*." In. *PDR for Herbal Medicines,* Grueruald, J., T Brendler and C. Jaenicka, Eds., Medical Economy Company, Inc., Montvale, New Jersey, pp. 871–873.

Araya, O. S and E. J. H. Ford. 1981. "An investigation of the type of photosensitization caused by the ingestion of St. John's wort (*Hypericum perforatum*) by calves," *J. Comp Path* 91·135–141.

Baureithel, K. H , K. B. Buter, A. Engesser, W. Burkard and W. Schaffner. 1997. "Inhibition of benzodiazepine binding *in vitro* by amentoflavone, a constituent of various species of *Hypericum*," *Pharm Acta Helv.* 72(3):153–157.

Becker L. E. and G. B. Skipworth. 1975. "*Ginkgo*-tree dermatitis, stomatis, and proctitis," *JAMA.* 231:1162–1163.

Beljanski, M. 1990. "Anti-virus composition and its uses," European Patent EP 0373986. Publication date 1990-06-20.

Berghoefer, R. and J. Hoelzl. 1987. "Biflavonoids in *Hypericum perforatum*. Part 1. Isolation of 13,II8-biapigenin," *Planta Med.* 53:216–217.

Bhattacharya, S. K., A. Chakrabarti and S S Chatterjee 1998 "Activity profiles of two hyperforin-containing *hypericum* extracts in behavioral models," *Pharmacopsychiatry.* 31 (Suppl. 1): 22–9.

Biber A., H. Fischer, A. Romer and S. S Chatterjee. 1998. "Oral bioavailability of hyperforin from *hypericum* extracts in rats and human volunteers," *Pharmacopsychiatry.* 31 (Suppl. 1):36–43.

Bladt, S. and H. Wagner. 1994. "Inhibition of MAO by fractions and constituents of *Hypericum* extracts," *J. Geriatr. Psychiatry Neurol.* 7:S57–S59.

Block, G. 1992. "The data support role for antioxidants in reducing cancer risk," *Nutrition Reviews.* 50(7):207–213.

Blume, J., M. Kieser and U. Hölscher 1996 "Placebo-controlled double-blind study of the effectiveness of *Ginkgo biloba* special extract EGb 761 in trained patients with intermittent claudication," *Vasa* 25:265–274.

Bombardelli, E and P. Morazzoni. 1995. "*Hypericum perforatum*," *Fitoterapia.* 66·43–68.

Bonvoisin, B. and P. Guinot 1989. "Clinical studies of BN 52063 a specific PAF antagonist " In· Ginkgolides—Chemistry, Biology, Pharmacology and Clinical Perspectives. Volume 2, Braquet, P , Ed., J R. Prous Science Publishers, Barcelona, Spain, pp. 845–854

Bors, W , W. Heller, C. Michel and K. Stettmaier. 1996. "Flavonoids and polyphenols " In. *Chemistry and Biology Handbook of Antioxidants* New York, NY. Marcel Dekker, Inc , pp. 409–465

Braquet, P. 1988. Ginkgolides—Chemistry, Biology, Pharmacology and Clinical Perspectives. Volume 1 J. R. Prous Science Publishers, Barcelona, Spain, pp 794.

Brockmoller, J., T Reum, S. Bauer, R Kerb, W. D. Hubner and I Roots 1997 "Hypericin and pseudohypericin: Pharmacokinetics and effects on photosensitivity in humans," *Pharmacopsychiatry.* 30 (Suppl. 2).94–101.

Brondz, I. and T. Greibrokk. 1983 "n-1-alkanols of *Hypericum perforatum*," *Journal of Natural Products.* 46 940–941.

Brondz, I , T Greibrokk, P Groth and A. Aasen. 1983. "The absolute configuration of hyperforin, an antibiotic from *Hypericum perforatum* L , based on the crystal structure determination of its p- bromobenzoate ester," *Acta Chem. Scand. Ser. A* A37:263–265

Büter, B., C. Orlacchio, A Soldati and K. Berger. 1998 "Significance of genetic and environmental aspects in the field cultivation of *Hypericum perforatum*," *Planta Med.* 64(5) 431–437.

Castelli, D., L Colin, E Camel and G. Ries 1998. "Pretreatment of skin with a *Ginkgo biloba* extract/sodium carboxymethyl-beta-1,3-glucan formulation appears to inhibit the elicitation of allergic contact dermatitis in man," *Contact Dermatitis.* 38:123–126.

Cesarani, A., F. Meloni, D. Alpini, S. Barozzi, L. Verderio and P. F. Boscani. 1998. "*Ginkgo biloba* (EGb 761) in the treatment of equilibrium disorders," *Adv. Ther.* 15·291–304

Chang, H. M , P. P. H. But. 1987. *Pharmacology and Applications of Chinese Materia Medica,* World Scientific Publ. Co. Singapore. Vol. 2, pp. 1096–1101.

Chatterjee, S. S., B. L. Gabard and H. E. W. Jaggy. 1986. "Pharmaceutical compositions containing bilobalide for the treatment of neuropathies." US Patent 4,571,407 (February 18, 1986).

Chatterjee, S. S., B L. Gabard and H. E. W. Jaggy. 1990. "Pharmaceutical compositions containing bilobalide for the treatment of neuropathies," US Patent 4,892,883 (January 9, 1990).

Chatterjee, S. S., S K Bhattacharya, M. Wonnemann, A. Singer and W. E. Muller. 1998a. "Hyperforin as a possible antidepressant component of hypericum extracts," *Life Sci.* 63(6):499–510.

Chatterjee, S. S., M. Noldner, E Koch and C. Erdelmeier. 1998b. "Antidepressant activity of *Hypericum perforatum* and hyperforin: The neglected possibility," *Pharmacopsychiatry.* 31 (Suppl. 1)· 7–15.

Chavez, M. L. and P. I. Chavez. 1997. "Saint John's wort," *Hospital Pharmacy.* 32·1621–1632.

Chen, C., T. Wei, Z. Gao, B. Zhao, J. Hou, H. Xu, W. Xin and L. Packer. 1999 "Different effects of the constituents of EGb761 on apoptosis in rat cerebellar granule cells induced by hydroxyl radicals," *Biochem. Mol. Biol. Int.* 47:397–405.

Chialva, F., G. Gabri, P. Liddle and F. Ulian. 1981. "Study on the composition of the essential oil from *Hypericum perforatum* L. and Teucrium chamaedrys L.," *Riv Ital. EPPOS* 63:286–288 (CA 96:11497a).

Chung, K F. and P. J. Barnes. 1988. "Clinical perspectives of PAF-acether antagonists." In Ginkgolides—Chemistry, Biology, Pharmacology and Clinical Perspectives. Volume 1. Braquet, P, Ed, J. R. Prous Science Publishers, Barcelona, Spain, pp 333–344.

Chung, H S, A Harris, J. K. Kristinsson, T A Ciulla, C Kagemann and R Ritch. 1999. "*Ginkgo biloba* extract increases ocular blood flow velocity," *J. Ocul. Pharmacol Ther.* 15.233–240.

Clostre, F 1999 *Ginkgo biloba* extract (EGB 761) "State of knowledge in the dawn of the year 2000," *Ann. Pharm. Fr* 57·1S8–1S88.

Cohen, A. J. and B. Bartlik. 1998 "*Ginkgo biloba* for antidepressant-induced sexual dysfunction," *J Sex Marital Ther.* 24.139–143.

Cohen, A. J. 1999 "Method for treating sexual dysfunction disorders with compositions containing *Ginkgo biloba*," U S Patent 5897864. Publication date 1999-04-27.

Cook, N C. and S Samman. 1996. "Flavonoids—Chemistry, metabolism, cardioprotective effects, and dietary sources," *Nutr. Biochem.* 7:66–76

Costes, C. and T. Chantal. 1967. "Carotenoid pigments of the petals of the inflorescence of St. John's wort (*Hypericum perforatum*)," *Ann Physiol. Veg.* 9.157–177 (CA 68.66335y).

Cracchiolo, C 1998. FAQ on St. John's wort (*Hypericum perforatum* and *Hypericum* angustifolium) v.3 1 k, http://www primenet.com/-camilla/stjohns faq.

Crowell, P. L. 1999. "Prevention and therapy of cancer by dietary monoterpenes," *J. Nutr* 129.775S–778S

Curtis-Prior, P., D. Vere and P. Fray. 1999. "Therapeutic value of *Ginkgo biloba* in reducing symptoms of decline in mental function," *J. Pharm. Pharmacol.* 51:535–541.

De Smet, P. A and W. A. Nolen. 1996. "St. John's wort as an antidepressant," (Editorial), *BMJ* Aug, 3,313(7052):241–242.

DeFeudis, F. V. 1991. *Ginkgo biloba* Extract (EGb-761): Pharmacological Activities and Clinical Applications Elsevier Science Ltd., Paris, pp. 187.

Denke, A, H. Schempp, E. Mann, W. Schneider and E. F. Elstner 1999. "Biochemical activities of extracts from *Hypericum perforatum* L: 4th Communication: Influence of different cultivation methods," *Arzneimittelforschung,* 49(2):120–125.

Derbentseva, N. A. and A. S. Rabinovich. 1968. "Isolation, purification, and study of some physicochemical properties of novoimanin " In: *Novoimanin Ego Lech,* Svoistva., A.I. Solov'eva, ed., Naukova Dumka, Kiev, Ukraine, pp. 15–18.

Derbentseva, N. A., E. L. Mishenkova and O. D. Garagulia 1972. "Effect of tannins from *Hypericum perforatum* on influenza viruses," *Mikrobiol Zh (Kiev).* 34:768–772.

Dimpfel, W., F Schober and M. Mannel. 1998. "Effects of a methanolic extract and a hyperforin-enriched CO_2 extract of St. John's Wort (*Hypericum perforatum*) on intracerebral field potentials in the freely moving rat," *Pharmacopsychiatry.* 31 (Suppl. 1) 30–35.

Dorossiev, I. 1985. "Determination of flavonoids in *Hypericum perforatum*," *Pharmazie.* 40:585–586.

Duke, A. J 1992. "*Hypericum perforatum* L." In: *Handbook of Phytochemical Constituents of GRAS Herbs and Other Economic Plants,* CRC Press, Inc., Boca Raton, FL, pp. 302–303.

Ernst, E. 1996. "*Ginkgo biloba* in treatment of intermittent claudication. A systematic research based on controlled studies in the literature," *Fortschr. Med.* 114:85–87.

Ernst, E and M. H. Pittler. 1999. "*Ginkgo biloba* for dementia· A systematic review of double-blind, placebo-controlled trials," *Clinical Drug Investigation* 17:301–308.

Ferrandiz, M L and M. J Alcaraz. 1991. "Anti-inflammatory activity and inhibition of arachidonic acid metabolism by flavonoids," *Agents Actions* 32·283–288.

Flint, A. J. and R. van Reekum. 1998 "The pharmacologic treatment of Alzheimer's disease A guide for the general psychiatrist," *Can. J. Psychiatry.* 43:689–697.

Furukawa, S 1932 "Constituents of *Ginkgo biloba* L. leaves," *Sci. Papers Inst. Phys. Chem. Res. Tokyo.* 19:27–38.

Gellerman, J. L., W. H. Anderson and H. Schlenk. 1976. "6-(Pentadec-8-enyl)-2,4-dihydroxyben-zoic acid from seeds of *Ginkgo biloba*," *Phytochem.* 15:1959–1961

Giese, J. 1999. "Taste for nutraceutical products," *Food Technol.* 53(10):43

Gobbato, S., A. Griffini, E. Lolla and F. Peterlongo. 1996. "HPLC quantitative analysis of biflavones in *Ginkgo biloba* leaf extracts and their identification by thermospray liquid chromatography-mass spectrometry " *Fitoterapia* 67:152–158.

Gonsette, R. E. 1982. "Treatment of multiple sclerosis," *Bull. Soc. Belge. Ophtalmol.* 199–200 275–280

Guinot, P., C. Brambilla, J. Duchier, A. Taytard and C Summerhayes. 1988. "The clinical effects of BN 52063, a specific PAF-acether antagonist.asthma." In: *Ginkgo*lides—Chemistry, Biology, Pharmacology and Clinical Perspectives. Volume 1. Braquet, P., ed., J R. Prous Science Publishers, Barcelona, Spain, pp. 345–354.

Gulick, R M , V. McAuliffe, J. Holden-Wiltse, C. Crumpacker, L. Liebes, D S Stein, P. Meehan, S. Hussey, J. Forcht and F. T Valentine. 1999. "Phase I studies of hypericin, the active compound in St. John's Wort, as an antiretroviral agent in HIV-infected adults," *Ann. Intern. Med* 130(6):510–514.

Gurevich, A. I., V. N Dobrynin, M. N. Papovko, I. D. Ryabova, B. K. Chernov, N. A Derbentseva, B. E. Aizenman and A D. Garagulya. 1971. "Hyperforin, an antibiotic from *Hypericum perforatum*," *Antibiotiki.* 16:510–512 (CA 75.956225t).

Hippius, H. 1998. "St. John's Wort *(Hypericum perforatum)*—a herbal antidepressant," *Curr. Med. Res. Opin.* 14(3) 171–184.

Hobbs, C. 1989. "St. John's wort *(Hypericum perforatum* L.). A review," *HerbalGram.* 18/19:24–33. http://www.healthy.net/library/articles/hobbs/hypericm.htm

Hoelzl, J. and E. Ostrowski 1987. "St. John's wort *(Hypericum perforatum* L.) HPLC analysis of the main components and their variability in a population," *Dtsch. Apoth. Ztg.* 127:1227–1230 (CA 107:112686).

Hollman, P. C., M. G. L. Hertog and M. B. Katan. 1996. "Analysis and health effects of flavonoids," *Food Chem.* 57:43–46.

Hopfenmüller, W. 1994. "Evidence for a therapeutic effect of *Ginkgo biloba* special extract. Meta-analysis of 11 clinical studies in patients with cerebrovascular insufficiency in old age," *Arzneimittelforschung.* 44:1005–1013.

Hoult, J. R., M. A. Moroney and M. Paya. 1994. "Action of flavonoids and coumarin on lipoxygenase and cyclooxygenase," *Methods Enzymol.* 234:443–554.

Irie, J., M. Murata and S. Homma. 1996. "Glycerol-3-phosphate dehydrogenase inhibitors, anacardic acids, from *Ginkgo biloba*," *Biosci. Biotech. Biochem.* 60:240–243.

Itil, T. M., E Eralp, I. Ahmed, A. Kunitz and K. Z Itil. 1998. "The pharmacological effects of *Ginkgo biloba*, a plant extract, on the brain of dementia patients in comparison with tacrine," *Psychopharmacol. Bull.* 34:391–397.

Ivaniv, O. P. 1998. "The results of using different forms of a *Ginkgo biloba* extract (EGb 761) in the combined treatment of patients with circulatory encephalopathy," *Lik Sprava.* 8:123–128.

Jaggy, H. and E. Koch. 1997. "Chemistry and biology of alkylphenols from *Ginkgo biloba* L.," *Pharmazie.* 52:735–738.

Janssens, D., C. Michels, G. Guillaume, B. Cuisinier, Y. Louagie and J Remacle. 1999. "Increase in circulating endothelial cells in patients with primary chronic venous insufficiency: Protective effect of Ginkor Fort in a randomized double-blind, placebo-controlled clinical trial," *J. Cardiovasc. Pharmacol.* 33.7–11.

Joly, M , M. Hagg-Berrurier and R Anton. 1980 "La 5'-méthoxybilobétine, une biflavone extraite du *Ginkgo biloba*," *Phytochem.* 19:1999–2002

Karryev, M O. and N. F. Komissarenko. 1980. "Phytochemical study of *Hypericum* L. plants of the *Turkumenian flora*," *Izv. Akad. Nauk Turkm SSR*, Ser. Biol Nauk 1980 52–57. (CA 93 182809w).

Keiji, W., I Seikou, U Kaori, S Masakatsu and H Masanobu 1985 "An antivitamin B6, 4'-methoxypyridoxine, from the seed of *Ginkgo biloba* L ," *Chem. Pharm. Bull.* 33·3555–3557

Keiji, W., I. Seikou, U. Kaori, T Yutaka, S. Keiko, S Masakatsu and H. Masanobu 1988 "Studies on the constitution of edible and medicinal plants I. Isolation and identification of 4-O-methylpyridoxine, toxic principle from the seed of *Ginkgo biloba* L.," *Chem Pharm. Bull.* 36.1779–1782.

Kim, H. K , K. H. Son, H. W. Chang, S. S. Kang and H P. Kim. 1998 "Amentoflavone, a plant biflavone. A new potential anti-inflammatory agent," *Arch. Pharm. Res.* 21(4)·406–410

Kim, H. K , K. H. Son, H. W. Chang, S S. Kang and H P. Kim. 1999. "Inhibition of rat adjuvant-induced arthritis by ginkgetin, a biflavone from *Ginkgo biloba* leaves," *Planta Med* 65·465–467.

Kitanov, G. 1983 "Determination of the absolute configuration of catechins isolated from *Hypericum perforatum*." *Farmatsiya (Sofia).* 33:19–22 (CA 99 50290j).

Kleijnen, J. and P. Knipschild. 1990. "*Ginkgo biloba*," *Lancet.* 340:1136–1139.

Kleijnen, J. and P. Knipschild. 1992. "*Ginkgo biloba* for cerebral insufficiency," *Br J. Clin. Pharmac* 34:352–358.

Laakmann, G , C. Schule, T. Baghai and M. St. Kieser. 1998. "John's wort in mild to moderate depression: The relevance of hyperforin for the clinical efficacy," *Pharmacopsychiatry* 31 (Suppl 1). 54–59.

Lavie, G., Y. Mazur, D. Lavie and D. Meruelo. 1994. "The chemical and biological properties of hypericin—a compound with a broad spectrum of biological activities," *Med. Res. Rev.* 15·111–119

Lavie G , Y. Mazur, D. Lavie, A M. Prince, D. Pascual, L Liebes, B. Levin and D. Meruelo. 1995. "Hypericin as an inactivator of infectious viruses in blood products," *Transfusion* 35·392–400.

Lees, A. M., H. Y. I Mok, R S. Lees and M. A. McCluskey. 1977. "Plant sterols as cholesterol-lowering agents: Clinical trials in patients with hypercholesterolemia and studies of sterol balance," *Atherosclerosis.* 28.325–338

Lenoir, S , F. H. Degenring and L. Saller. 1999. "A double-blind randomized trial to investigate three different concentrations of a standardized fresh plant extract obtained from the shoot tips of *Hypericum perforatum*," *Phytomedicine.* 6(3):141–146

Li, M. H., H. L Zhang and B. Y. Yang. 1997. "Effects of *Ginkgo* leaf concentrated oral liquor in treating asthma," *Chung Kuo Chung Hsi I Chieh Ho Tsa Chih.* 17:216–218.

Li, A L., Y. D Shi, B. Landsmann, P. Schanowski-Bouvier, G. Dikta, U. Bauer and G. M. Artmann. 1998. "Hemorheology and walking of peripheral arterial occlusive diseases patients during treatment with *Ginkgo biloba* extract," *Kuo Yao Li Hsueh Pao.* 19:417–421.

Lietti, A., A. Cristoni and M Picci 1976. "Studies on *Vaccinium myrtillus* anthocyanins. I Vasoprotective and anti-inflammatory activity," *Arznein. Forsch/Drug Research* 26:829–832

Linde, K , G. Ramirez, C. D. Mulrow, A. Pauls, W. Weidenhammer and D. Melchart 1996. "St John's wort for depression—an overview and meta-analysis of randomized clinical trials," *British Med. J.* Aug 3.313(7052):253–258.

Lingaerde, O., A. R. Foreland and A. Magnusson. 1999. "Can winter depression be prevented by *Ginkgo biloba* extract? A placebo-controlled trial," *Acta Psychiatr. Scand* 100:62–66.

Lopez-Bazzocchi, I , J. B. Hudson and G. H. N. Towers. 1991. "Antiviral activity of the photoactive plant pigment hypericin," *Photochem. Photobiol.* 54:95–98.

Mackerras, D. 1995. "Antioxidants and health," *Food Australia (Supplement).* 47(11):1–23.

Maisenbacher, P and K. A Kovar. 1992. "Analysis and stability of Hyperici oleum," *Planta Med.* 58(4) 351–354

Maruyama, M., A. Terahara, Y. Itagi and K. Nakanishi. 1967a. "The ginkgolides I Isolation and characterization of the various groups," *Tetrahedron Lett.* 4:299–302.

Maruyama, M., A. Terahara, Y Itagi and K. Nakanishi. 1967b. "The ginkgolides II Derivation of partial structures," *Tetrahedron Lett.* 4·303–308.

Maruyama, M., A. Terahara, Y. Nakadaira, M. C Woods and K. Nakanishi 1967c. "The ginkgolides III Structure of the ginkgolides," *Tetrahedron Lett.* 4:309–313.

Maruyama, M., A. Terahara, Y. Nakadaira, M. C. Woods, Y. Takagi and K. Nakanishi. 1967d "The ginkgolides IV. Stereochemistry of the ginkgolides," *Tetrahedron Lett* 4.314–319.

Mathis, C. and G. Ourisson. 1963. "Etude chimio-taxonomique du genre *Hypericum*," *Phytochemistry.* 2:157–171

Mathis, C and G. Ourisson. 1964. "Etude chimio-taxonomique du genre *Hypericum* III. Repartition des carbures satures et des monoterpenes dans les huiles essentielles d'hypericum," *Phytochemistry* 3.133–137.

Mazza, G. 1997. Anthocyanins in Edible Plant Parts: A Qualitative and Quantitative Assessment. In: *Antioxidant Methodology in vivo and in vitro Concepts.* O. I. Aruoma and S L. Cuppett (Eds). Champaign, IL, AOCS Press, pp 119–140.

Meruelo, D., G. Lavie and D. Lavie. 1988. "Therapeutic agents with dramatic antiretroviral activity and little toxicity at effective doses: Aromatic polycyclic diones hypericin and pseudohypericin," *Proc. Natl. Acad. Sci., USA.* 85(14):5230–5234.

Michel, P. F. 1986. *Ginkgo biloba*: L'Arbre Qui à Vaincu Le Temps. Felin, Paris

Middleton, E. and C. Kandaswami. 1992. "Effects of flavonoids on immune and inflammatory cell functions," *Biochem. Pharmacol.* 43.1167–1179

Mori, M. 1982 "n-hexacosanol and n-octacosanol: Feeding stimulants on the larvae of the silkworm, Bombyx mori," *J. of Insect Physiology.* 28:969–973.

Muller, W. E., A. Singer, M. Wonnemann, U. Hafner, M. Rolli and C. Schafer. 1998. "Hyperforin represents the neurotransmitter reuptake inhibiting constituent of hypericum extract," *Pharmacopsychiatry.* 31 (Supp. 1): 16–21.

Nahrstedt, A. and V. Butterweck. 1997. "Biologically active and other constituents of the herb *Hypericum perforatum* L.," *Pharmacopsychiatry.* 30 (Suppl. 2): 129–134.

Nakanishi, K. 1988. "*Ginkgo*lides-Isolation and structural studies carried out in the mid 1960's." In: Ginkgolides—Chemistry, Biology, Pharmacology and Clinical Perspectives, Volume 1. Braquet, P., ed., J. R. Prous Science Publishers, Barcelona, Spain, pp. 27–36

Negrash, A K. and P. Y. Pochinok. 1972 "Comparative study of chemotherapeutic and pharmacological properties of antimicrobial preparations from common St. John's wort," Fitonotsidy, Mater Soveshch. 6th. Meeting date 1969, pp. 198–200 (CA 78:66908u).

Nielsen, H. and P. Arends. 1978 "Structure of the xanthonolignoid kielcorin." *Phytochemistry.* 17:2040–2041.

NIH. 1997. St. John's Wort Study Launched, Press Release, Oct. 1, National Institutes of Health, Bethesda, MD

Oken, B. S., D. M. Storzbach and J. A. Kaye. 1998. "The efficacy of *Ginkgo biloba* on cognitive function in Alzheimer's disease," *Arch. Neurol* 55:1409–1415.

Okpanyi, S. N. and M. L. Weischer 1987. "Experimental animal studies of the psychotropic activity of a *Hypericum* extract," *Arzneim.-Forsch.* 37.10–13.

Ollivier, B., G. Balansard, C. Maillard and E. Vidal. 1985 "Separation and identification of phenolic acids by high-performance liquid chromatography and ultraviolet spectroscopy. Application to *Parietaria officinalis* L. and to Saint-John's-wort (*Hypericum perforatum* L.)" *J Pharm. Belg.* 40:173–177.

Omar, L. I. 1998. "Medication for impotence containing lyophilized roe and a powdered extract of *Ginkgo biloba*," U S. Patent 5730987 Publication date 1998-03-24.

Orth, H C , C Rentel and P. C. Schmidt. 1999. "Isolation, purity analysis and stability of hyperforin as a standard material from *Hypericum perforatum* L.," *J. Pharm Pharmacol.* 51(2)·193–200.

Page, C. P. and D N Robertson. 1988. "PAF, airway hyperactivity and asthma· The potential of ginkgolides in the treatment of asthma " In. Ginkgolides—Chemistry, Biology, Pharmacology and Clinical Perspectives, Volume 1. Braquet, P., ed., J. R. Prous Science Publishers, Barcelona, Spain, pp. 305–312.

Peter, H., J Fisel and W. Weisser 1966 "Zur pharmakologie der wirkstoffe aus *Ginkgo biloba*," *Arzneim-Forsch/Drug Res.* 16:719–725

Rand, E E., J. Barnes and C Stevinson. 1998. "Adverse effects profile of the herbal antidepressant St. John's wort (*Hypericum perforatum* L.)," *Eur. J. Clin Pharmacol* 54(8)·589–594

Rao, S G., A L. Udupa, S L. Udupa, P G. M. Rao, G. Rao and D. R. Kulkani 1991. "*Calendula* and *Hypericum* Two homeopathic drugs promoting wound healing in rats," *Fitoterapia.* 62(6):508–510.

Razinskaite, D. 1971. "Active substances of *Hypericum perforatum* (St John's wort) 2 Flavonoids and dynamics of their content," *Liet. TSR Mokslu Akad. Darb* Ser C (1): 89–100 (CA 75 72427r).

Remacle, J , A. Houbion, I Alexandre and C. Michiels 1990 "Behavior of human endothelial cells in hyperoxia and hypoxia. Effect of *Ginkor fort*," *Phlebologie.* 43:375–386.

Rigney, U., S. Kimber and I. Hindmarch 1999. "The effects of acute doses of standardized *Ginkgo biloba* extract on memory and psychomotor performance in volunteers," *Phytother. Res.* 13:408–415

Rocha, L. 1994. "An antifungal gamma-pyrone and xanthones with monoamine oxidase inhibitory activity from *Hypericum brasiliense*," *Phytochemistry.* 36(6).1381–1385.

Rocha, L., A Marston, O Potterat, M. A. Kaplan, H. Stoeckli-Evans and K. Hostettmann. 1995. "Antibacterial phloroglucinols and flavonoids from *Hypericum brasiliense*," *Phytochemistry.* 40(5)·1447–1455.

Roncin, J. P., F. Schwartz and P. D'Arbigny. 1996. "EGb 761 in control of acute mountain sickness and vascular reactivity to cold exposure," *Aviat. Space Environ Med.* 67:445–452.

Saljic, J. 1975 "Ointment for the treatment of burns," *Ger.Offen.* 2,406,452 (CL. A61K), 21 Aug 1975 (CA 83, 197797).

Saponara, R. and E. Bosisio. 1998. "Inhibition of cAMP-phosphodiesterase by bioflavones of *Ginkgo biloba* in rat adipose tissue," *J. Nat Prod.* 61:1386–1387.

Schulz, H. and M Jobert. 1994. "Effects of hypericum extract on the sleep EEG in older volunteers," *J. Geriatr. Psychiatry Neurol.* Oct. 7 Suppl 1· S39–S43

Shellenberg, R., S. Sauer and W Dimpfel. 1998. "Pharmacodynamic effects of two different hypericum extracts in healthy volunteers measured by quantitative EEG," *Pharmacopsychiatry.* 31 (Suppl. 1). 44–53.

Shi, H. and E. Niki. 1998. "Stoichiometric and kinetic studies on *Ginkgo biloba* extract and related antioxidants," *Lipids.* 33:365–370.

Singer, A., M. Wonnemann and W E. Muller. 1999. "Hyperforin, a major antidepressant constituent of St. John's wort, inhibits serotonin uptake by elevating free intracellular Na," *J. Pharmacol. Exp. Ther.* 290(3):1363–1368.

Snider, S. R. 1984. "Octacosanol in Parkinsonism [letter]," *Ann. Neurol.* 16:723.

Spinnewyn, B., N. Blavet, F. Clostre and P. Braquet. 1988. "Protective effects of ginkgolides in cerebral post-ischemic phase in Mongolian gerbils." In Ginkgolides—Chemistry, Biology, Pharmacology and Clinical Perspectives. Volume 1. Braquet, P., ed., J. R. Prous Science Publishers, Barcelona, Spain. pp. 665–679.

Stevinson, C. and E. Edzard. 1999. "Safety of *Hypericum* in patients with depression: A comparison with conventional antidepressants," *CNS Drugs.* 11:125–132.

Sticher, O. 1977. "Plant mono-, di- and sesquiterpenoids with pharmacological or therapeutical activity." In: *New Natural Products and Plant Drugs with Pharmacological, Biological or Therapeutical Activity*, H. Wagner and P. Wolff, eds., Springer-Verlag, NY.

Stricher, O. 1993. "Quality of *Ginkgo* preparations," *Planta Med.* 59:1–11.

Suzuki, O , Y. Katsumata, M. Oya, S Blandt and P Wagner. 1984. "Inhibition of monoamine oxidase by hypericin," *Planta Medica.* 50:272–274.

Upton, R. 1997. "St. John's Wort Monograph. American Herbal Pharmacoepea and Therapeutic Compendium," *HerbalGram.* 40(5):1–32.

Van Beek, T. A., E. Bombardelli, P. Morazzoni and F. Peterlongo. 1998. "*Ginkgo biloba* L " *Fitoterapia.* 69:195–243.

Watson, D. G. and E. J Oliveira. 1999. "Solid-phase extraction and gas chromatography-mass spectrometry determination of kaempferol and quercetin in human urine after consumption of *Ginkgo biloba* tablets," *J. Chromatogr. B. Biomed. Sci. Appl.* 723:203–210.

Wender, P. A., K. W. Hinkle, M. F. T. Koebler and B. Lippa. 1999. "The rational design of potential chemotherapeutic agents: Synthesis of bryostatin analogues," *Med Res. Rev.* 19(5):388–407.

Wheatley, D. 1998. "*Hypericum* extract. Potential in the treatment of depression," *CNS Drugs.* 9:431–440.

Wichtl, M. 1994. "St. John's wort " In: *Herbal Drugs and Phytopharmaceuticals* (Translated from the German by Bissett, N.G), CRC Press, Inc., Boca Raton, FL. pp. 273–275

Zhu, L., J. Wu, H. Liao, J. Gao, X. N. Zhao and Z. X. Zhang. 1997. "Antagonistic effects of extract from leaves of *Ginkgo biloba* on glutamate neurotoxicity," *Chung Kuo Yao Li Hsueh Pao.* 18:344–347.

CHAPTER 7

Valerian, Saw Palmetto and Goldenseal as Herbal Medicines

S. DIAMOND
G. H. N. TOWERS

1. VALERIAN

1.1. INTRODUCTION

THERE are possibly 300 species of Valeriana, but only three species are used commonly as phytomedicines. These are *Valeriana officinalis* L. or valerian, a Eurasian member of the Valerianaceae, *V. wallichii* of India and Pakistan and *V. edulis* of Mexico. Clinical efficacy has been shown mainly for *V. officialis*. The Latin name is a reference to valere or well-being and officinalis to its recognized use as a medicinal plant. *V. officinalis* is a polymorphous complex subspecies with populations dispersed over temperate and subpolar Eurasian zones (Bradley, 1992). Differences in phytochemical constituents have been documented for the subspecies (Titz et al., 1982, 1983). Since we are mostly concerned with *V. officinalis* in this review, this species will henceforth be referred to as valerian.

Valerian has been in use therapeutically for more than 2,000 years and was described by Dioscorides (50 AD) as a sedative (Morazzoni and Bombardelli, 1995). It is one of the herbal drugs positively assessed and considered to be safe by the German Commission E (Blumenthal et al., 1998) and was one of the most prescribed monopreparations for psychotropic indications in Germany in 1994 (Cott, 1995). More than 5 million units of valerian extracts are sold in Germany and over 10 million units are sold in France every year (Carper, 1997). The European Scientific Cooperative on Phytotherapy (ESCOP) Monographs on the Medicinal Uses of Plant Drugs (1997) recognizes valerian preparations for treating tension, restlessness and irritability with difficulty in falling asleep (ESCOP, 1997). It is also listed as a hypotensive drug in the British Herbal Compendium, the companion to the British Herbal Pharmacopoeia (Bradley, 1992). Over 200 studies have been published on the pharmacology of valerian, mostly in Europe, within the last 30 years. The phytochemistry and biological activities of valerian and related plants have been carefully reviewed by Houghton (1988), and his review is essential reading for anyone interested in this subject. Several comprehensive reviews of the history, constituents and activity of *Valeriana* have also been published in recent years (Houghton, 1998, 1999; Volz, 1997; Hobbs, 1989, 1994; Jaspersen-Schib, 1990).

1.2. CULTIVATION

Valerian is widely cultivated in Europe, particularly in Holland, and also in the U.S. and Japan. The rhizomes and roots, with and without stolons (Valerianae radix), are harvested in autumn or spring and are used in medicine. Various factors, including time of harvest, influence the phytochemistry of valerian (Bos et al., 1998a; Wagner et al., 1972; Chapelle, 1972). Based on studies carried out between 1989 and 1993 in The Netherlands, levels of essential oil in underground parts peak in September, ranging from 1.2% to 2.1% (v/w), based on dry weight (DW), and the composition of the oil remains constant (Bos et al., 1998a). Valerenic acid and its derivatives and the valepotriates peaked in February–March, with contents of 0.7–0.9% (DW) and 1.1–1.4% (DW), respectively. Selections can be made to maximize the levels of active ingredients. Plant material harvested in September of the same year of sowing was found to have a content of 0.9% essential oil and 0.5% valerenic acid and derivatives (Bos et al., 1998a). Valerian is generally easy to grow, however, there have been recent reports of a phytoplasma-induced yellow disease in valerian grown in Alberta, Canada (Hwang et al., 1997).

1.3. USES

A very interesting and thorough review of the early history of valerian was published by Valpiani (1995). Valerian has been used as an aromatic and diuretic. It was recommended for the treatment of digestive problems, flatulence, nausea, stagnant liver and urinary tract disorders. The Greeks used valerian as an emmenagogue, antiperspirant and for vaginal yeast infections. It may also have been used very early in the treatment of neuroses (Wheelwright, 1974). In the 17th to 19th centuries, it was used extensively in European medicine in the treatment of nervous disorders and became established in German folk medicine. Eventually, the oil was considered to be the active component responsible for the sedative effects (Culbreth, 1983). The oil was first noted as the active component in the British Pharmaceutical Codex of

1923. Valerian is still maintained in the British, German, American and Indian pharmacopoeias (Valpiani, 1995) as well as in the pharmacopoeias of Austria, Belgium, the Czech Republic, Egypt, France, Greece, Hungary, Italy, The Netherlands, Romania, Russia, Switzerland and former Yugoslavia (Newall et al., 1996). From 1820 to 1942, valerian was listed in the U.S. Pharmacopoeia as a tranquilizer. It was also listed in the National Formulary, the pharmacists' guide, until 1950 (Boyle, 1991). In North America, *V. edulis* (Tobacco root) was traditionally pit-cooked for up to 48 hours and eaten, warm or cold, alone or with other roots or meat by First Nation peoples of British Columbia and Ontario (Kuhnlein and Turner, 1991). Scoulers valerian (*V. scouleri* Rydb.), Sitka valerian (*V. sitchensis* Bong.) and Mountain valerian (*V. dioca* L.) were also commonly used as food flavoring and as medicines (Turner, 1997; Pojar and Mackinnon, 1994). The Alaskan Tlingit applied the crushed roots to a mother's nipples when weaning a child, rubbed them on sore muscles or blew them onto animal traps for luck. The roots were also used as an antiseptic on wounds (Shaw and On, 1979). Valerian was an ingredient of soups and broths in medieval Europe.

In addition to its effects on humans, the plant has an euphoric effect on cats and other small mammals similar to the effects of nepetalactone, an iridoid from *Nepeta* cataria (catnip). The herb has also been traditionally used by some groups to quiet unfriendly dogs and to attract horses (Conway, 1975).

A recent study also found valerian effective as a bath treatment for generalized fibromyalgia (Ammer and Melnizky, 1999). The primary use of valerian is as a mild hypnotic to induce sleep and relieve anxiety (Wagner et al., 1998). Alcoholic extracts made from fresh roots of valerian (or other species of *Valeriana*) have a sedative effect (Houghton, 1988). This is probably due to the presence of valepotriates and valerenic acid (*European Pharmacopoeia,* 1983). Products with valerian are indicated on the label for treating: states of nervous excitation, difficulty in falling asleep and spasmodic gastrointestinal pains of nervous origin.

Dosages range from 2 to 3 grams of drug (root or underground parts) taken once to several times per day (Wichtl, 1994; Blumenthal et al., 1998). Preparations include infusions, extracts and tinctures. For use as a sleeping aid, standardized extracts are recommended at 300 to 500 mg to be taken about an hour before bedtime (Carper, 1997).

1.4. PHYTOCHEMISTRY

Early studies on valerian were concerned with chemical characterization of the volatile oil that has a characteristic penetrating and unpleasant odor, emphasized on storage because of the release of isovaleric and related short chain acids by enzymatic hydrolysis of some of the constituent terpenoid esters. Volatile oils from valerian, consisting largely of monoterpenes (includ-

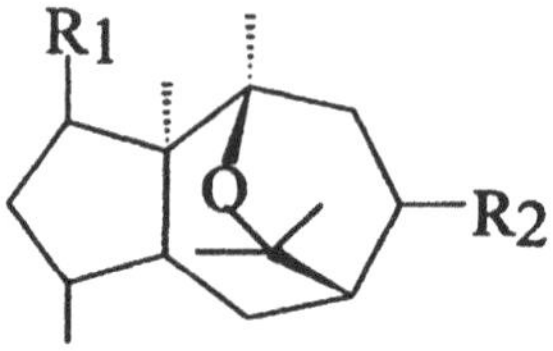

R₁	R₂	
H	H	Kessane
OH	H	Kessyl-2-ol
H	OH	Kessyl-6-ol
OH	OH	Kessyl glycol
OAc	H	Kessyl-2-ol acetate
H	OH	Kessyl-6-ol acetate
OAc	OAc	Kessyl glycol diacetate

Figure 7.1 Structures of sesquiterpenes in *Valeriana.*

ing iridoids) and sesquiterpenes, range from 0.2 to 2.8% depending on the origin of the material (Bokstaller and Schmidt, 1997). Initially, kessane (sesquiterpene) derivatives were identified in *V. officinalis* var. *latifolia,* grown in Japan (Ukita, 1944). Some of these compounds are shown in Figure 7.1. Other terpenoids in the oil include isovalerianate, bornyl acetate, bornyl formate, eugenyl acetate, eugenyl isovalerate, isoeugenyl isovalerate, pinene, terpinene, terpinolene, camphene, *p*-cymene, fenchene, limonene, myrcene, phellandrene, carvacrol, thymol, carvone (around 8%), patchouli alcohol, esters of valerianol and caryophyllene. As with other species in this family, the volatile oils are extremely complex and variable, with many other terpenoid esters being present in very small amounts. The constituents vary considerably with the cultivars and growing conditions including time of harvesting and soil conditions (Guenther, 1952). The lack of correlation between volatile oil content and biological activity led to further phytochemical studies with the isolation of the valtrate esters (epoxy-iridoid esters) from *V. wallichii* (Thies, 1966). These valepotriates, as they came to be known, emerged as target active molecules.

The valepotriates are unstable compounds decomposing rapidly under the influence of moisture, heat (> 40°C) and acidity (pH < 3). Some of these compounds and their degradation products are shown in Figure 7.2. The content of valepotriates varies greatly among species of *Valeriana* and in

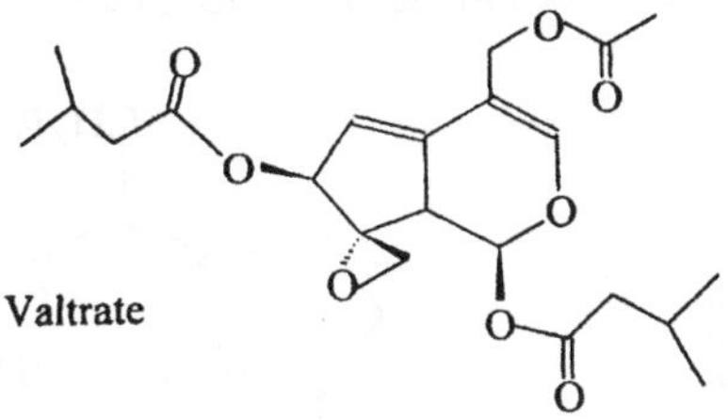

Figure 7.2 Structures of valepotriates in *Valeriana.*

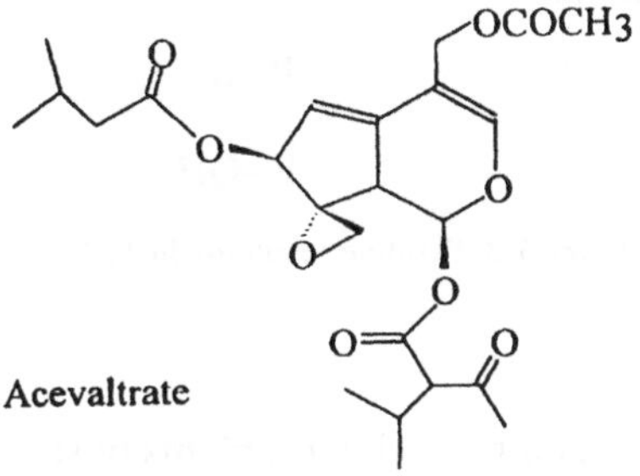

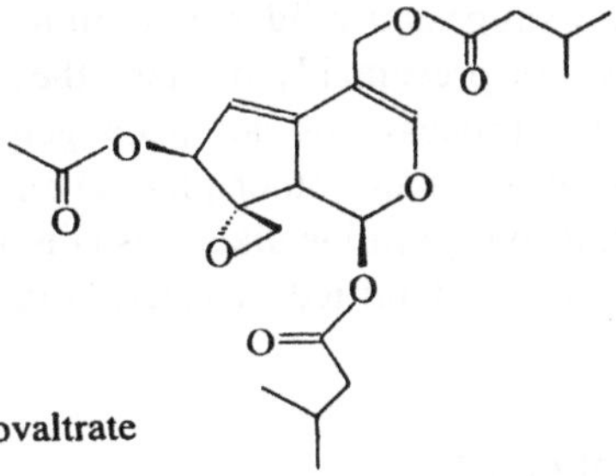

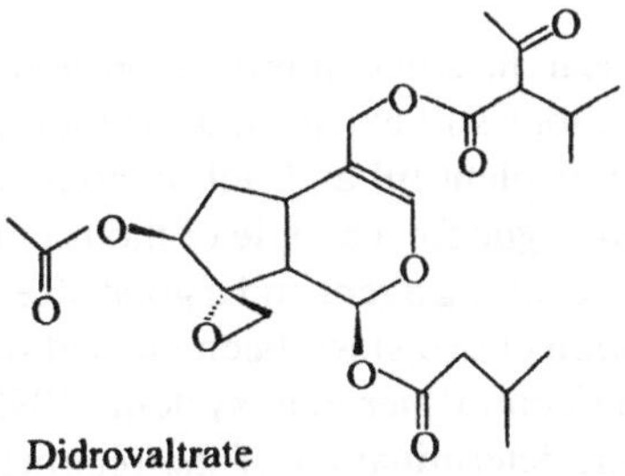

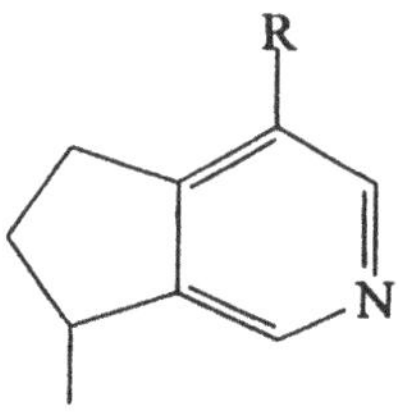

R=CH$_2$OCH$_3$ Valeranine
R=CH$_3$ Actinidine

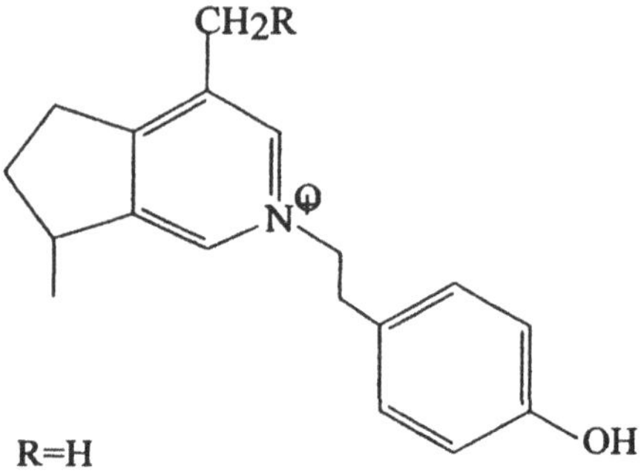

R=H

R=OH

Figure 7.3 Pyridine alkaloids in *Valeriana.*

cultivars of *V. officinalis,* emphasizing problems of standardization in preparations of traditional medicines, particularly where the active principles have not been identified with any confidence. The instability of these iridoids in aprotic solvents has suggested milder conditions of extraction, such as with supercritical fluids, but presumably because they are no longer target molecules, this expensive process has not been generally adopted for valerian preparations. In addition, many other phytochemicals, including flavonoids, chlorogenic acid and four pyridine alkaloids (Figure 7.3), present in very low amounts, have also been identified in valerian (Chevalier, 1907; Torsell and Wahlberg, 1967).

1.5. PHARMACOLOGY

In spite of extensive research, some imprecision about pharmacological activities of valerian preparations still exists, and the literature on the pharmaceutical potential of valerian is plentiful and rich in contradictions (Bruneton, 1995). The traditional use, in vogue for a couple of thousand years and possibly longer, is as a sedative, i.e., a calming agent and a good sleeping aid (Chevalier, 1907). Numerous investigations have since been carried out on the effects of valerian preparations on the central nervous system (CNS) (Morazzoni and Bombarelli, 1995). However, determination of the pharmacological activities of valerian have been carried out mostly on fractions or compounds from extracts and not with the whole plant drug. This is typical of modern studies of many traditional drugs and may partly account for the difficulty in verifying claims of efficacy. The volatile oil of valerian accounts for only one-third of its sedative activity (Gstirmer and Kind, 1951; Gstirmer and Kleinbauer, 1958) and is not correlated with the major components, i.e., the bornyl esters (Stoll et al., 1957).

The sedative effects of tinctures were demonstrated in mice (Leuschner et al., 1993), but the overall sedative effects of the plant could not be accounted for on the basis of the valepotriate content that may reach 1.2% w/w in roots (Hikino et al., 1980). The activity of aqueous rather than ethanolic extracts in humans on decreasing the time taken to fall asleep suggests that valepotriates may not have significant effects since they are only slightly soluble in water (Leathwood et al., 1982; Leathwood and Chauffard, 1985; Balderer and Borbely, 1985). Traditional preparations of valerian stored for some time, e.g., after 60 days (Adzet et al., 1975), have very low levels of valepotriates (Bounthanh et al., 1980). The valepotriates are alkylating agents by virtue of the epoxide function in these molecules, and they exhibit cytotoxicity and mutagenicity (Bounthanh et al., 1981, 1983; Von der Hude et al., 1985, 1986; Hansel, 1990, 1992; Keochanthala-Bounthanh et al., 1990, 1993). The decomposition products of the valepotriates, e.g., homobaldrinal (Wagner et al., 1980) or valtroxal (Veith et al., 1986) have been suggested to be active constituents.

Valeranone (Figure 7.4) or kessyl esters display no sedative or hypnotic effects at the normal levels found in valerian (Morazzoni and Bombardelli, 1995), although they have been found to prolong sleep in mice and rats (Takamura et al., 1975). Valeranone is one of the major constituents of the essential oil of the Indian nard (*Nardostachys jatamansi* DC) that grows wild in Nepal and is used in Asia for its CNS sedative properties (Bruneton, 1995).

Valerenic acid (Figure 7.5), a stable active ingredient found consistently in valerian extracts and often used as a marker compound for standardization, reduces motor activity and increases sleeping time in mice (Hendriks et al., 1985). Valerenic acid resembles pentobarbital, binds to the same receptors sites in the brain and prolongs sleeping time (Hendriks et al., 1985). It prevents the enzymatic breakdown of gamma-aminobutyric acid (GABA) in brain cells; this may explain the decrease in CNS activity common to valerian extracts, or, in other words, the sedative effects (Riedel et al., 1982). Aqueous root extracts of valerian (which contain valerenic acid) have been shown to inhibit the uptake and stimulate the release of gamma-aminobutyrate in brain cortical

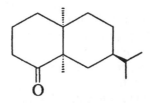

Figure 7.4 Structure of valeranone.

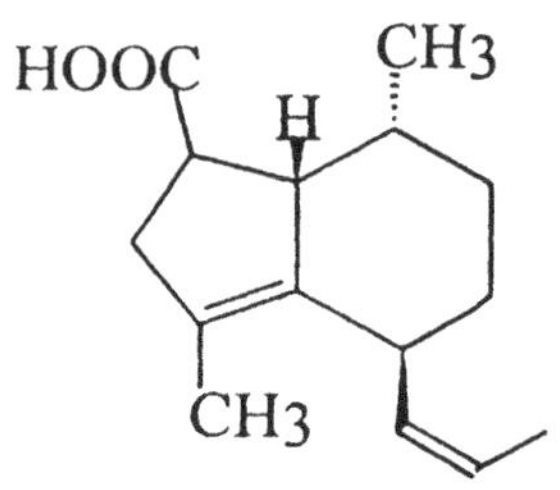

Figure 7.5 Structure of valerenic acid.

synaptosomes, causing sedation of the brain arousal systems, similar to the synthetic benzodiazepines, Valium® and Halcion® (Santos et al., 1994). The amount of GABA in aqueous extracts of valerian has been shown to be sufficient to account for its release in synaptosomes (Santos et al., 1994). Valerian's active ingredients competitively inhibit pentobarbital and benzodiazepines based on animal studies. *In vitro* studies show that valerian's active constituents mediate the release of GABA (Leuschner et al., 1993) and bind the same receptors as benzodiazepines but with less affinity and milder clinical effects (Mennini et al., 1993).

A commercially available valerian root extract (Valdipert) was found to have weak anticonvulsive properties (Leuschner et al., 1993) that may help explain one of the traditional uses of valerian for treatment of convulsions (Castleman, 1988). The essential oil from roots of Chinese valerian is rich in borneol, isoborneol and bornyl acetate and has been found to have sedative effects in mice even after treatment with caffeine (Buchbauer et al., 1992). Studies with Hokkai-kisso, the roots of Japanese valerian, show prolonged sleep in mice, reversed reserpine-induced hypothermia and several CNS calming effects (Sakamoto et al., 1992).

1.6. CLINICAL TRIALS

Between 1975 and 1992, there were four double-blind, placebo-controlled clinical trials on valerian proving its efficacy as a sedative (Schulz et al., 1997). At least six controlled trials in Europe have shown that valerian can shorten the time to fall asleep, prolong sleep time, increase deep sleep stages, increase dreaming, reduce nighttime awakenings and significantly improve sleep quality in normal sleepers and insomniacs (Carper, 1997). One of the impressive results of all of these studies is the remarkable safety of this botanical sedative compared to conventional synthetic drugs used for the same purpose. In a review of several botanical medicines including valerian, side effects were reported in less than 3% of patients (Schulz et al., 1997).

One double-blind study with crossover trial was carried out on 27 persons taking a valerian root extract preparation in tablet form, Valeriana NATT, for sleep difficulties, and it showed a significant effect (Lindahl and Lindwall, 1989). According to these authors, intake of 400 mg of valerian extract (containing primarily sesquiterpenes) before bedtime significantly improves sleep quality when compared with a placebo ($p < 0.001$). Forty-four percent of subjects reported perfect sleep, and 89% reported improved sleep from taking Valeriana NATT. No side effects were observed.

Another study on the effect of valerian on human sleep showed, on the basis of subjective sleep parameters, that valerian has significant sleep promoting action and possesses mild hypnotic action after comparing two groups of healthy young subjects at home ($N = 10$) and at a sleep laboratory ($N = 8$) with a placebo (Balderer and Borbely, 1985). It was determined that at home, both doses tested (450 and 900 mg) reduced sleep latency (time to fall asleep) and wake time after sleep onset, while no change in sleep stages or EEG spectra were seen. Nighttime motor activity was enhanced in the middle third of the night and reduced in the last third, leading to the conclusion that the effect was dose dependent. In the sleep laboratory, however, 900 mg had no significant effect when compared with the placebo (Table 7.1). No residual effect was detected in the morning after waking. Another human trial showed a significant decrease in the time taken to settle down to sleep for eight habitual insomniacs with a dosage of 450 mg of valerian aqueous extract when compared to a placebo (Leathwood and Chauffard, 1985). The higher dosage of 900 mg was seen to produce no further benefits, and side effects were not observed.

The effect of an aqueous extract of valerian root was tested on 128 people comparing a placebo, 400 mg of a valerian extract and a proprietary preparation containing valerian (Leathwood et al., 1982). The valerian extract produced a significant improvement in sleep quality based on subjective measures with a questionnaire and sleep latency scores. People with sleep difficulties (i.e., irregular sleepers, poor sleepers and smokers) benefited the most. The proprietary preparation caused sleepiness or grogginess in the morning, an uncommon side effect not reported for valerian. Only one person out of the 166 who began the study had nausea as a side effect. Leathwood and Chauffard (1983) utilized electroencephalography (EEG) and questionnaires to quantify the physiological and subjective aspects of sleep. EEG evaluations showed no significant effect of the root extract, whereas subjective criteria suggested significant sedative action. The resulting EEG pattern confirmed a shorter mean sleep latency and increased mean latency to first awakenings, although these findings were not significant (Leathwood and Chauffard, 1983). Another short-term study using EEG tests compared Kava (*Piper methysticum* G.

TABLE 7.1. Subjective and Objective Sleep Parameters and Nighttime Motor Activity After 900 mg Valeriana and Placebo. Mean Values (N = 8) with Standard Errors in Parenthesis. ADAPT: Adaptation Night; PL 1, PL 2: Placebo Nights: VAL: Valeriana Night: PL-PD: Post-Drug Placebo Night. [PL2 Corresponds to Night 3 (N = 4) or Night 5 (N = 4)].

	ADAPT	PL 1	PL 2	VAL	PL-PD
Subjective sleep parameters					
Sleep latency (min)	33 1	20 5	18 5	14.8	17 2
	(6 3)	(7 3)	(2 6)	(4 5)	(4 1)
Wake after sleep onset (min)	14 4	14 3	7 8	2 8	14 8
	(7 3)	(8 9)	(4 6)	(1 2)	(7 6)
Number of awakenings	1 4	0 9	1 1	1 2	1 1
	(0 5)	(0 3)	(0 4)	(0 3)	(0.7)
Objective sleep parameters (min)					
Total sleep time		402 9	408 6	417 4	407.8
		(14.0)	(4 9)	(2 7)	(7 3)
Wake after sleep onset		35.8	24.5	19 0	28 4
		(13 6)	(5 2)	(2 1)	(7 3)
Moving time		14 4	11 5	12 4	12 7
		(2.4)	(1.9)	(2 3)	(1.3)
Stage 1		10 9	17 3	15 3	17 9
		(2 7)	(2 4)	(3 3)	(4 9)
Stage 2		208.8	201 7	213 1	206 1
		(8 1)	(8.2)	(7 8)	(9 1)
Stage 3		26 0	28.9	30.0	36 5
		(3 6)	(3.8)	(4 9)	(4 5)
Stage 4		56 1	60.3	57 6	57 3
		(6.4)	(9 3)	(6 9)	(8 3)
REM sleep		101 1	100 5	101 4	93 9
		(10 1)	(7 7)	(10 1)	(6 9)
Latency until stage 2		27 4	26 1	19 2	22 8
		(20 7)	(8 4)	(4 6)	(12 1)
Latency until REM sleep		78 1	70 3	88.1	80 3
		(9 9)	(6 9)	(8 2)	(5.3)
Nighttime motor activity (%)					
First third of night	58 5	53 0	49 5	55 8	58.8
	(6 6)	(9.1)	(5 6)	(7 3)	(7 3)
Second third of night	55 9	61 4	58.4	66 8	51.3
	(14 5)	(4.6)	(4 2)	(3 1)	(6 7)
Third third of night	70.3	65.5	67 5	64 2	66 9
	(3 2)	(6.4)	(5 6)	(3 9)	(5 4)

Reproduced with permission from G Balderer and A A Borbely (1985)

Forster), passion flower (*Passiflora incarnata* L.), lavender (*Lavandula angustifolia*) and valerian against placebo and diazepam (Valium®). With caffeine as a stimulant, the botanicals had distinctly different effects on the EEG, except for lavender and valerian, which showed remarkably similar profiles (Schultz et al., 1997, 1998).

A preparation of valerian (160 mg) and lemon balm (80 mg) competed equally with Halcion® (0.125 mg triazolam) in reducing sleep latency and improving sleep quality in twenty people (Dressing, 1992). Unlike Halcion®, the herbal preparation did not cause loss of concentration the next day or produce morning somnolence. Another study on the effects of valerian, propanolol and their combination on activation performance and mood of healthy volunteers under social stress conditions found that valerian alone (100 mg) produced the best effects on concentration, even though this dosage did not have a significant effect on pulse frequency compared to propanolol and the combination (Kohnen and Oswald, 1988). Objective measures on the effects of valerian for reducing sleep latency also compare favorably against commonly used synthetic benzodiazepine- and barbiturate-type hypnotics including Secobarbital® (100 mg), Flurazepam® (30 mg), Triazolam® (0.5 mg), Amylobarbital® (100 mg) and Temazepam® (15 mg) (Kales et al., 1977). This study found that of the synthetic drugs, only Flurazepam® significantly reduced sleep latency. Other clinical studies, however, have questioned the effectiveness of Flurazepam® (Cirignotta et al., 1975; DiPerri and Meduri, 1975). The larger doses of Temazepam® needed to shorten sleep latency produced morning somnolence, similar to the effects of the required dosages of Nitrazepam® (5 mg) (Hindmarch, 1975). Since these synthetic drugs can have side effects including the risk of addiction or overdose, withdrawal problems, psychotic disturbances, memory loss and hallucinations, it would seem logical to utilize valerian instead, which has none of these associated risks. Valerian products are also much less expensive than synthetic sleep aids: valerian costs only 14 cents per day compared to 89 cents per day for Halcion® (French, 1996).

One clear benefit of valerian preparations over the synthetic prescription drugs is their inability to interact with or augment the negative effects of alcohol when taken at the recommended dosages (Albrecht, 1995). It has also been reported that valerian root extracts can be safely used during the day to combat anxiety and nervous conditions without reducing concentration, affecting reaction time or diminishing driving performance scores in a car (Albrecht, 1995). The tincture, however, labeled according to the German Commission E monograph (1986), must indicate the statement: "When taking this valerian tincture, normal reaction capability can be so far altered that ability for active participation in road traffic or for operation of machinery is impaired. This happens to a greater degree in conjunction with alcohol" (Bradley, 1992). The 1990 monograph, however, does not include this statement (Blumenthal et al., 1998).

1.7. CYTOTOXIC EFFECTS

Valepotriates are the most cytotoxic of valerian root constituents and tinctures when used in human small-cell lung cancer and colorectal (COLO320) cancer cell lines (Bos et al., 1998b). The cytotoxicity is likely due to the epoxide function in these molecules that have demonstrated alkylating ability (Braun et al., 1982). One of the decomposition products of the valepotriates, isovaltral (Figure 7.2), is three times more toxic than cisplatin, the cytostatic drug used as a reference material. The diene valepotriates are only slightly more toxic than cisplatin, and the monoenes are 2 to 3 times less toxic. Valepotriates, however, are most probably destroyed by gastric acidity. Valerenic acids exhibit low toxicities (100–200 mM). The freshly prepared tincture has significantly higher cytoxicity than that stored for two months, and it has been suggested that tinctures should be stored for two months prior to use. The valepotriates have no influence on tumor growth in rats and, therefore, do not appear to be useful in cancer chemotherapy (Berger et al., 1986). Studies on the cytotoxic and antitumor activities of valtrate and dihydrovaltrate isolated from the roots of *Valeriana wallichii*, as well as baldrinal, a breakdown product of valtrate, showed *in vitro* activity on cultured rat hepatoma cells (HTC line) and *in vivo* activity on female mice Krebs II ascitic tumors (Bounthanh et al., 1981).

1.8. ASSAYS

The presence of the valepotriates in the drug is indicated by the development of a blue color on treatment with hydrochloric acid, after extraction with dichloromethane; this color reaction is typical of a number of iridoids and is the result of deacylation and substitution of dichlorocyclo-penta[c] pyriliium ions (Bruneton, 1995). The test is described in detail in the German Commission E text (Blumenthal et al., 1998). Several HPLC methods have been used to measure valtrate, valerenic acid and other active ingredients of valerian (Tittel and Wagner, 1978; Perry et al., 1996; Bos et al., 1996). TLC methods for valerian extracts, including Mexican and Indian species of *Valeriana*, are described by Wagner and Bladt (1996) and Stahl and Schild (1969). The valepotriates are difficult to resolve chromatographically by TLC. A standardized extract containing three valepotriates is marketed in Germany under the trade name Valmane (Houghton, 1988). It consists of 80% didrovaltrate, 15% valtrate and 5% acylvaltrate. There are about 18 valepotriates in valerian including a chloro derivative, valechlorine (Popov et al., 1973). A new method has recently been reported and recommended based on HPLC fingerprinting and the use of marker compounds for identifying and standardizing herbal drugs including valerian (Lazarowych and Pekos, 1998).

1.9. SAFETY

Valerian is recognized as a safe traditional medicine in Canada and as a General Recognized as Safe (GRAS) herb in the U.S. when taken in typical recommended amounts. Recommended dosages rarely produce side effects, and none are noted in the German Commission E monograph (Blumenthal et al., 1998). At recommended dosages, any side effects are minor but may include nausea or headaches. Large amounts may produce headaches, restlessness, nausea, heart palpitations, insomnia and morning grogginess (O'Hara et al., 1998; Hobbs, 1989). Valerian overdose at approximately 20 times the recommended therapeutic dose in an adult reported recently appears to be benign as the mild symptoms were resolved within 24 hours (Willey et al., 1995). Valerian should not be used concomitantly with barbiturates because excessive sedation may occur (Miller, 1998). As a comparison, common side effects of Valium® (Diazepam) are drowsiness and ataxia (difficulty in coordinating muscular movements), fatigue, dizziness, nausea, blurred vision, diplopia, vertigo, headache, slurred speech, tremors, hypoactivity, dysarthria, euphoria, impairment of memory, confusion, depression, incontinence or urinary retention, constipation, skin rash, generalized exfoliative dermatitis, hypotension and changes in libido (Canadian Pharmacists Association, 1998). Also, synthetic sedatives are contraindicated for geriatric patients and on patients with emotional disorders, epileptic patients and during pregnancy.

1.10. CONCLUSIONS

Published research shows that valerian is effective as a mild sedative and sleep aid. It is also added to bath water for external application. No significant side effects are noted, although reports of gastrointestinal complaints and headache have been reported.

2. SAW PALMETTO

2.1. INTRODUCTION

Saw palmetto is a small shrubby palm of the sandy soils of the southern U.S. and appears in the states of South Carolina, Louisiana, Georgia, Alabama, Mississippi and Florida (Bennett and Hicklin, 1998; Small, 1926). The correct botanical name is *Serenoa repens* (Bartram) J. K. Small or *Serenoa serrulata* Hook. Both specific epithets seem suitable as serrulata means saw-toothed (the petioles bear spiny thorns along their margins), and repens refers to the underground creeping nature of the plant. Saw palmetto is the most common palm in the U.S. and is the only member of the genus *Serenoa* (Arecaceae, subfamily Coryphoideae, tribe Coryphae and subtribe Livistoninae). This shrubby palm commonly grows in dense thickets beneath pine trees; rarely does the stem become tall and the plant tree-like, reaching a maximum height of about 2 meters. The plant usually grows to approximately a meter tall with wide, fan-shaped leaves spanning 0.31 meter. Creamy white flowers bloom among the fans. The part of this plant used in medicine is the dark brown or black single-seeded fruit or berry that looks somewhat like a shriveled olive in shape and size and ripens from green, to yellow, to orange and finally to purplish-black. The European Pharmacopoeia requires whole, dried fruits with undamaged fat-soluble fractions (Bennett and Hicklin, 1998).

The newest edition of the United States Pharmacopeia and the National Formulary list saw palmetto berries as an accepted botanical medicine (USP 23 - NF 18, 1999). Saw palmetto berries were listed in the 19th Edition of the U.S. Dispensatory for treating chronic and subacute cystitis, chronic bronchitis, laryngitis, catarrh and enlarged prostate (Wood et al., 1907), while the 23rd edition stressed their usefulness for treating, "the enlarged prostate of old men" (Wood and Osol, 1943). The Extra Pharmacopoeia of Great Britain and the British Pharmaceutical Codex recommend saw palmetto for treating cystitis and chronic bronchitis and for stimulating the genitourinary tract and mucous membranes (Bennett and Hicklin, 1998).

Many herbal preparations used for prostate conditions are also used for kidney, bladder and urethral conditions that affect urination in men. An excellent overview of conventional, experimental and botanical treatments of nonmalignant prostate conditions is provided by Brinker (1994). The use of saw palmetto extracts for the successful treatment of benign prostatic hypertrophy (BPH) has been reviewed systematically by Wilt and colleagues (1998).

2.2. CULTIVATION

Saw palmetto is one of the most abundant plants in Florida where the berries are wild harvested to meet the worldwide demand. Saw palmetto berries have been exported to Europe from Florida since 1602 for their medicinal properties (Bennett and Hicklin, 1998). Today, the annual fruit harvest exceeds 6.8 million kg, and based on the estimated 1.4 million hectares of saw palmetto-covered lands in Florida, a shortage of this herbal drug is not expected within the coming years (Bennett and Hicklin, 1998).

2.3. USES

The berries have been used as food and medicines by First Nations dating back to precontact times with archaeological sites showing that it was important for Florida's pre-Columbian, nonhorticultural peoples (McGoun, 1993). The berries were soon utilized in a likewise manner by early settlers, although many found them unpalatable (Snow, 1996; Vogel, 1970; Hutchens, 1973). Indigenous Indians of the southern U.S. were using saw palmetto berries for treating testicular atrophy, erectile dysfunction and prostate inflammation in the early 1700s (Wilt et al., 1998; Lowe and Ku, 1996). Saw palmetto came

TABLE 7.2. **Diseases and Ailments Against Which Saw Palmetto Has Been Used.**

Anticatarrhal/anti-inflammatory	Genito-urinary stimulant
Appetitic stimulant	Hormone regulation
Asthma	Impotence, male
Breast atrophy	Kidney disease (incl Bright's)
Breast enlargement	Neuralgia
Bronchitis	Prostate gland (enlargement)
Colds	Reproductive organs
Diabetes	Respiratory congestion and infections
Diarrhea and Dysentery	Sedative
Digestive aid	Sexual stimulant
Diuretic	Testicular atrophy
Dysuria	Urinary tract infection and cystitis
Epididymitis	Whooping cough
Expectorant	

Adapted from Bennett and Hicklin (1998)

into use about a hundred years ago by the eclectic physicians for the treatment of prostatic hypertrophy and relief of bladder symptoms associated with this condition in old men (Anonymous, 1891; Bloyer, 1896; Tegarden, 1898; Showerman, 1892; Watkins, 1899). The berries were listed as official medicines in the U.S. Pharmacopoeia from 1906 to 1916 and in the National Formulary from 1926 to 1950.

2.4. PHYTOCHEMISTRY

Presently, the primary use of saw palmetto is for the treatment of benign prostatic hyperplasia (BPH). Other reported medicinal uses are listed in Table 7.2. The components of saw palmetto that have received the most attention are the lipids. The oil is made up of triacylglycerols with fatty acids of chain lengths usually less than 14 carbons and predominantly with lauric acid followed by myristic and oleic acids (Bone, 1998). The berries also contain flavonoids, terpenoids and polysaccharides (Bombardelli and Morazzoni, 1997). There are remarkably no unusual alkanes, alkenes, polyprenols, sterols or free fatty acids in the nonpolar extracts. β-Sitosterol is the major sterol found in the fruit (Bombardelli and Morazzoni, 1997). Because the hexane-soluble fraction is the material used in the symptomatic treatment of BPH, it would be interesting to substitute this material with compositions of identified constituents of the extracts in order to discover whether there are "hidden" phytochemicals with biological activities that have so far eluded detection. Minor constituents that have been overlooked may play an important role in the therapeutic value of these extracts. However, it seems that the total lipid and sterol extract is currently the drug of commerce (Bach and Ebeling, 1996;

Hohenfellner, 1996; Sultan et al., 1984; Weissrer et al., 1996). Variation in the liposterol content of the fruit, and other possible active ingredients, based on phenological patterns of the plant and geographical location of collections have yet to be reported. Solvents used for extraction of the liposterols include ethanol, hexane or supercritical CO_2. Manual cold-pressing of the berries is also done in small-scale operations, followed usually by recovery of the residual oil in the presscake using hot water and skimming the oil from the surface (Bennett and Hicklin, 1998). The fruit also contains a lipase that remains active in the berries during ripening and drying, resulting in a rancid odor and taste. A short steam heat treatment that will destroy enzymes without damaging the essential fatty acids or sterols may be useful in arresting the oxidization process.

2.5. ETIOLOGY OF BENIGN PROSTATE HYPERPLASIA

The mechanism of the pharmacological action of saw palmetto in BPH, particularly the liposterolic fraction of the plant extract, has received considerable attention.

BPH, one of the most common medical conditions in men over 50, is progressive with age (Guess, 1992). Research has shown that most men over 50 experience some extent of obstructive urinary symptoms due to BPH (Berry et al., 1994; Lytton et al., 1968). In the U.S., more than 300,000 prostatectomies are carried out annually (McConnell et al., 1994). Other frequently used therapies in the U.S. include intrusive devices, surgery, pharmaceutical drugs and, finally, herbal preparations (Barry, 1997; Oesterling, 1995). In Germany, herbal treatments are the primary approach to resolving the symptoms of BPH and are prescribed in 90% of cases (Buck, 1996). The change in urine flow following any treatment is often used as a measure of success.

The development of BPH requires that the individual has functioning testes. The testes produce the androgen, testosterone, that is reduced in the prostate to the androgen, dihydrotestosterone (DHT) by the enzyme 5-alpha-reductase. These androgens, in higher than normal levels, cause an overgrowth of the epithelial cells of the prostate (i.e., hyperplasia). In other words, BPH occurs when hormone imbalances cause a progressive thickening of the epithelial and fibromuscular structures within the prostate gland (Tenover, 1991; Lawson, 1993). This hyperplasia causes obstruction and irritation of the urethra, thus, interfering with normal urinary functions and resulting in the typical symptoms of frequent and weak urination, incomplete voidance and need for frequent nocturnal urination. In the aging process, prostatic levels of testosterone fall, whereas the DHT levels remain high. There is no significant increase in DHT or 5-alpha-reductase levels in BPH over normal prostate tissue with aging although DHT receptor content may increase in BPH. In addition to the high levels of DHT, the levels of the estrogen, estradiol, a product of testosterone, also remain high in the aging process.

Androgens and estrogens are considered to act synergistically in stimulating cell growth in BPH (Brinker, 1994). Brinker references several studies that document the role of estrogens in BPH through stromal-epithelian interactions. A study of beagle dogs showed that estradiol increases the concentration of prostatic androgen receptors and of intraprostatic DHT; castrated beagles given estrogens and androgens had greater prostatic enlargement than those given only androgens (Bluestein and Oesterling, 1993). The original treatment for menopausal symptoms in women, such as hot flashes, began with the administration of synthetic estrogens until it became clear that estrogens given alone led to the development of endometrial hyperplasia that often developed into endometrial cancer over time (Stewart, 1999). This problem of hyperplasia and increased risk of cancer was rectified with the addition of progesterone to the treatment regimen, thus, restoring the important estrogen-to-progesterone ratio cited as critical by many researchers (Zava et al., 1997, 1998; Lee, 1990a, 1990b, 1991, 1993; Siiteri and Wilson, 1970; Schilcher, 1998).

Other studies have confirmed an antiestrogenic activity for saw palmetto extracts on prostatic tissue of benign prostatic hypertrophy patients (Di Silverio et al., 1992) possibly due to the presence of sterols (Berges et al., 1995; Klippel et al., 1997).

2.6. PHARMACOLOGY

Studies show that the liposterolic fruit extracts of saw palmetto reduce the uptake of testosterone and dihydrotestosterone by more than 40% and inhibit the conversion of testosterone into DHT by inhibiting the steroid enzyme, 5-alpha-reductase. DHT binds to cell receptors in the prostate where it initiates RNA and DNA synthesis; this then increases protein synthesis and abnormal growth of the prostate gland (Robbers and Tyler, 1999; Foster and Tyler, 1999).

Phytosterols present in the fruit were originally believed to be estrogenic, but it is now understood that their action is also antiestrogenic and may explain the antiandrogenic activity of the extract (Wilt et al., 1998; Dreikorn and Richter, 1989). Sitosterol is the principal sterol in higher plants and is found in the serum and tissues of healthy humans, albeit at concentrations 800–1,000 times lower than that of endogenous cholesterol. The glucoside of β-sitosterol, sitosterolin, is also present. These compounds are reported to have a synergistic stimulatory effect on the immune system and a prophylactic effect on a variety of diseases (Pegel, 1997). Research on sitosterol shows anti-inflammatory, antiulcer, antidiabetic and anticancer activity (Pegel, 1997). A randomized, placebo-controlled, double-blind clinical trial of β-sitosterol in patients with BPH showed significant improvement in symptoms and urinary flow parameters (Berges et al., 1995). A group of 200 patients suffering from BPH were given either 20 mg of β-sitosterol (with other various mixed phytosterols) or placebo three times daily. Divergence between the placebo and treatment group did not occur until about four weeks of treatment that led the researchers to conclude that β-sitosterol had a significant positive effect in patients with symptomatic BPH, as measured by the modified Boyarsky-score after six months.

Objective measures of urine flow and a decrease of mean residual urinary volume were also improved more than in the placebo group; these parameters did not change in the placebo group ($p < 0.01$). Comparison between International Prostate Symptom Score (IPSS) questionnaires were given at three and six months that also confirmed the effectiveness of the β-sitosterol treatment. Finasteride (Proscar™), a 5-alpha-reductase inhibitor, decreased prostatic volume by up to 30% over 12 months and improved Boyarsky scores with a reduction of up to four points, which was within the same range as the β-sitosterol treatment (Finasteride Study Group, 1993). The authors note that studies with other common drugs used to treat BPH symptoms, including the alpha-receptor blocking agents doxazosine, phenoxybenzamine, prazosine, alfuzosine, are also comparable with the β-sitosterol results, but these synthetic drugs frequently cause side effects including dizziness, low blood pressure, tachycardia and orthostatic problems.

Other research on sitosterols in human nutrition have shown that saw palmetto berries, as well as other herbs (i.e., *Pygeum africanum* and pumpkin seed) frequently used for treating BPH, also contain significant levels of sitosterols (Pegel, 1997). Since 1974, sitosterol and sitosterolin, in a ratio of 100:1, have been available in capsule form as a product called Harzol in Germany for the treatment of BPH, usually at a daily dose of 60 mg plant sterols per day (approximately 44 mg sitosterol plus campesterol, campestanol, dihydrositosterol and stigmasterol) (Ebbinghaus, 1974; Pegel, 1984; Henneking and Heckers, 1983; Berges et al., 1995; Klippel et al., 1997). This product is now available in Canada under the name Moducare™ Sterinol™.

2.7. CLINICAL TRIALS

Over 20 successful human studies have been documented on the usefulness of saw palmetto berry extract for treating benign prostatic hyperplasia or BPH. An excellent review of 18 randomized controlled trials for treating BPH involving 2,939 men concluded that saw palmetto improves urologic symptoms and flow measures. Saw palmetto produced improvement in symptoms and urinary flow similar to Finasteride, the most common synthetic drug prescribed for treating BPH, and was noted to have fewer side effects (Wilt et al., 1998).

One clinical study looked at 50 men suffering from lower urinary tract symptoms (LUTS) presumed secondary to BPH. These men were given a commercially available form of saw palmetto (160 mg twice per day) for six months to assess the effects of this treatment on voiding symptoms and urodynamic parameters. The men chosen had a minimum International Prostate Symptom Score (IPSS) of 10. The mean IPSS improved from 19.5 ± 5.5 to 12.5 ± 7.0 ($p < 0.001$) among the 46 men who completed the study. Significant improvements were noted after only two months of treatment. There was no significant change of peak urinary flow rate, postvoid residual urine volume or detrusor pressure at peak flow and mean serum PSA level among patients completing the study. The authors concluded that the saw palmetto berry product was well tolerated and may significantly improve lower urinary tract symptoms in men with BPH, without significant improvements in objective measures of bladder outlet obstruction (Gerber et al., 1998).

Recently, the effectiveness of various phytopharmaceutical agents such as saw palmetto berry extracts, nettle root extracts, pumpkin seeds, pollen extracts and different phytosterols as phytotherapy of BPH was reviewed (Bracher, 1997). These plant products demonstrate a significant benefit compared to placebo treatment in several clinical trials.

A three-year prospective multicenter study on Sabal extract IDS 89 (a saw palmetto berry extract) determined the potential and limitations of this treatment for alleviating the symptoms of BPH (Bach and Ebeling, 1996). The results obtained from 435 patients suggest that saw palmetto berry extract has significant therapeutic efficacy. A marked symptomatic improvement that included a 50% reduction in residual urine and a 6.1 mL/s increase in peak urinary flow rate was observed. The need for nocturnal urination improved or normalized in 73% of the patients, daytime frequency of urination and feeling of incomplete emptying improved by 54% and 75%, respectively, and prostate swelling improved or normalized in 55% of patients.

Physicians and patients voted its efficacy as good or very good in over 80% of the cases, and, thus, the quality of life in four out of five patients was markedly improved by long-term saw palmetto IDS 89 therapy. The drug was well tolerated by 98% of the patients. The deterioration rate at the end of the three-year treatment period was significantly lower in the treated group than in the untreated BPH subjects, and thus the therapy can also reduce the need for and incidence of surgery. In comparison, the prescription drug, Proscar® (finasteride) showed only a 30% decrease in symptoms scores over three years, with only slight improvement in urine flow and almost no change in residual urine volume. Also, there was a dropout rate of 10.7% with finasteride patients due to negative side effects, compared to only 1.8% with saw palmetto (Bach et al., 1997).

Another study of the liposterolic saw palmetto extract (160 mg, twice daily) resulted in increased urinary flow rates and volume and decreased prostate size after 45 days of treatment. In a clinical trial with 505 patients, 88% of doctors and patients considered the treatment to be effective, with only 5%

of patients reporting side effects—much better than conventional treatments (Braeckman, 1994).

2.8. SAFETY

Saw palmetto extracts have been found to be well tolerated, and only one withdrawal, due to gastrointestinal upset, from a study has been reported in the literature (Tasca, 1985). Data synthesis of 18 randomized controlled trials involving 2,939 men concluded that adverse effects due to saw palmetto were generally mild and comparable with placebo (Wilt et al., 1998).

2.9. CONCLUSIONS

The available data on saw palmetto document that the liposterolic extracts possess antiandrogenic activity and prove beneficial in the treatment of benign prostatic hyperplasia. Saw palmetto improves urologic symptoms and flow measures, and compared with finasteride, saw palmetto produces similar improvement in urinary tract symptoms and urinary flow and is associated with fewer adverse treatment events (Wilt et al., 1998). In addition, the extracts possess anti-inflammatory and immunostimulant activity based on *in vitro* and *in vivo* studies (Newall et al., 1996).

3. GOLDENSEAL

3.1. INTRODUCTION

The medicinal use of the golden rhizomes and roots of *Hydrastis canadensis* L. (Goldenseal) appeared in the U.S. Pharmacopoeia in 1830 (Boyle, 1991). Goldenseal is a herbaceous perennial in the *Ranunculaceae* (buttercup) family that was once common in shady woods as an understory plant on rich moist soil in the western peninsula of Ontario and eastern U.S. from Vermont to Minnesota and south to Georgia, Alabama and Arkansas (Eichenberger and Parker, 1976). The rough-hairy plant, which grows up to 50 cm high, has one to three palmately lobed leaves up to 25 cm in diameter. The knotted, subcylindrical rhizomes are yellow-brown outside and bright yellow inside with a distinctive odor and bitter taste. Due to its popularity as a medicinal plant, it has been collected extensively, leading to its classification as an endangered species. This species was used by North American aboriginal peoples, particularly Cherokee, as an insect repellant when mixed with bear grease and for inflammations, wounds, debility and dyspepsia (Foster, 1991), as an antibiotic and as a dye for clothing (Thornton, 1998). The plant derives its name from its bright yellow color due to the presence of the yellow alkaloid, berberine, and the cuplike scars on the top of the rhizomes that resemble wax seals. It has various trivial names such as Ground Raspberry, Eye-balm and Eye-root. As a dye, it was known as Indian paint, golden root, Indian turmeric, jaundice root, wild curcuma and yellow puccoon (Snow, 1998). It was also used by eclectic physicians for inflammatory diseases of the bladder, tonsillitis, pharyngitis and myalgic tenderness (Felter, 1985). It accounts for about 6% of the herbal supplement sales in the U.S. (Richman and Witkowski, 1997). Goldenseal is used in modern phytotherapy in France, Germany, Spain, Australia and Great Britain to treat infections of the gastrointestinal, urogenital and respiratory tracts (Scassocchio et al., 1998). Monographs on goldenseal are currently found in the British Herbal Pharmacopoeia, the British Herbal Compendium (Bradley, 1992), Martindale's 30th edition (Reynolds and Martindale, 1993), and in the pharmacopoeias of Brazil, Egypt, France and Romania (Newall et al., 1996).

3.2. CULTIVATION

This small shade-loving perennial takes up to five years to produce a mature root and is difficult to cultivate. Plants may be started from seed, but most authorities recommend planting with two-year-old rhizomes purchased from specialty nurseries (Castleman, 1988). Cultivation strategies are very similar to those used for ginseng production (Ballard, 1989). Overharvesting and deforestation in the 1800s and up to this day have led to the species becoming so rare that it has been listed by CITES in June 1997 as a "threatened" species (Blumenthal, 1997) and more recently as endangered in a number of states in the U.S.

3.3. USES

Goldenseal roots and rhizomes have been used since ancient times in India and China as an antidiarrheal medication. Hobbs (1990a, 1990b) reviewed the history of goldenseal in early American medical botany and recommends the liquid extract especially for colds, flu and sinus infections as well as for lowering inflammation and cooling infections of the mucous membranes. A survey of its medical uses by Duke (1985) shows that it has been used as a diuretic, echarotic, stimulant and as a cure for boils, hemorrhoids, ringworm, ulcers, alcoholism, asthma, cancer, catarrh, chancroids, constipation, corns, deafness, dyspepsia, faintness, glossitis, impetigo, jaundice, leucorrhea, liver ailments, lumbago, sciatica and at least two dozen other ailments. None of these cures, however, are supported by scientific data, and an understanding of the bases of its "usefulness" in so many diseased conditions is lacking. Other herbs traditionally used as "bitters" have a similar broad spectrum of use (Treben, 1994). In Europe, India, China and Africa, "bitter vegetable drugs'' are considered medicinal agents and are used to stimulate appetite, aid digestion, reduce cholesterol and promote health (Blumenthal et al., 1998; Iwu, 1993; Bradley, 1992).

According to a recent review of the scientific literature available on goldenseal, the plant's extracts can be utilized as an antidiarrheal and antiseptic agent (O'Hara et al., 1998). The plant can also be used to improve digestion as a bitter tonic. Externally, an infusion of the root can be used to treat skin infections, although caution must be exercised to avoid ulceration of mucosal membranes (Newall et al., 1996). Standardized extracts of 10% alkaloids (5% hydrastine) are recommended at 250 mg per day, not to be taken for periods longer than a week (Flynn and Roest, 1995). More traditional dosage forms include the dried rhizome (0.5 to 1.0 gram), decoction, liquid extract (0.3–1 mL) or tincture (2–4 mL) taken three times daily (Newall et al., 1996).

3.4. CHEMICAL COMPOSITION AND PHARMACOLOGY

Goldenseal contains a number of isoquinoline alkaloids including berberine, hydrastin, hydrastindine, isohydrastidine, berberastine, canadine, canadaline, corylpalmine and isocorylpalmine (Messana et al., 1980; Leone et al., 1996), but only the first two have been investigated in some details. The plant does not appear to have been examined for other potentially bioactive compounds.

Berberine is a yellow quaternary ammonium protoberberine alkaloid that also occurs in species belonging to nine other botanical families (Santavy, 1970). Berberine is the most studied of the alkaloids in goldenseal and has been shown to possess fungicidal and antibacterial activities (Greathouse and Watkins, 1938; Yoshikazu, 1976; Nakamura, 1977; Pizzorno and Murray, 1985) as well as activity against protozoa, e.g., *Giardia lamblia*, *Trichomonas vaginalis* and *Entamoeba histolytica* (Kaneda et al., 1991). In a noncontrolled clinical study with children, berberine, in oral doses of 10 mg/kg/day for 10 days was found to be successful in treating giardiasis (Gupte, 1975). It significantly inhibits the secretory responses of the heat-labile enterotoxins of *Vibrio cholerae* and *Escherichia coli* in rabbits (Sack and Froehlich, 1982) and heat-stable enterotoxins of *E. coli* in pigs (Zhu and Ahrens, 1982). It also affects the development rate and survival of insects (Devitt et al., 1980).

Berberine was found to be phototoxic in near-UV light to mosquito larvae, with an LC50 for acute toxicity at 8.8 ppm compared to 250 ppm for control larvae kept in the dark (Philogene et al., 1984). In Chinese hamster ovary cells, there was a slight increase in chromosomal aberrations with berberine in near-UV light. This phototoxin was found to be a singlet oxygen generator in experiments with the chemical trap, 2,5-dimethylfuran (Philogene et al., 1984), and was two to four times more active against *Staphylococcus aureus* and the pathogenic yeast, *Candida albicans*, in UV-A compared to dark (Towers et al., 1997). Berberine was tested in a double-blind trial on 12 patients with chronic open angle glaucoma, and no changes in the bioregulation of the intraocular pressure were observed (Thumm and Tritscher, 1977). In view of its photogenotoxicity, its use in eye medications may not be advisable. It has been recommended and used in the treatment of irritable bowel syndrome (IBS) (Bone, 1998).

Hydrastine, a related isoquinoline alkaloid, induced positive inotropic and negative chronotropic effects in rat atria, potentiated electrically evoked contractions in mouse vas deferens and inhibited them in guinea pig ileum (Bartolini et al., 1990). It also has a profound intracellular effect on the proto scolex of the tapeworm. *Echinococcus granulosus*, of sheep and humans and may be a promising protoscolicide as adjuvant to hydatid surgery (Chi Sheng et al., 1991). It disrupts microtubules and affects mitochondria and lysosomes in cysts (Yao and Pao, 1989). But, it is not known if similar effects would occur in human cells, and it is not known if hydrastine is phototoxic.

Pulmery et al. (1996) recently found that the individual alkaloids from *Hydrastis* tested in rabbit aorta strips produced a dose-related inhibition of contraction responses when exposed to low doses of adrenaline. This was true only for berberine, canadine and canadaline. Hydrastine was completely inactive. At higher adrenaline concentrations, none of the isolated alkaloids inhibited contractions induced by adrenaline. However, the total extract of the root inhibited adrenaline-induced contractions at both low and high dosage levels. Hydrastinine hydrochloride, a fluorescent oxidation product of hydrastine has been used, in combination with synephrine and chlorhexidine, in eye drops to treat conjunctival hyperthermia and eye strain (Bruneton, 1995).

A recent study of the antimicrobial activity of goldenseal extract and its major isolated alkaloids, berberine, beta-hydrastine, canadaline and canadine, confirmed the plant's potency against six strains of microorganisms including *Staphylococcus aureus*, *Streptococcus sanguis*, *Escherichia coli*, *Pseudomonas aeruginosa* and *Candida albicans* (Scazzocchio et al., 1998). The antimicrobial activity was evaluated in liquid medium by contact test, and the killing time on a low-density bacterial inoculum was measured. The results (Table 7.3) demonstrate that canadaline is the most potent antimicrobial alkaloid, not the well-known berberine and hydrastine, suggesting that the opening of the C ring of benzylisoquinoline alkaloid is important for antimicrobial activity (as in canadaline) while the quaternary nitrogen group is not necessary (Scazzochio et al., 1998).

Advocates claim that goldenseal has immunostimulatory effects (Werbach and Murray, 1994), but these claims have not been adequately verified (Tyler, 1998). Recently, however, Rehman et al. (1999) investigated the antigen-specific *in vivo* immunomodulatory potential of continuous treatment with *Echinacea* and goldenseal root extract over a period of six weeks using rats that were injected with the novel antigen keyhole limpet hemocyanin (KLH) and were re-exposed to KLH after the initial exposure. Immunoglobulin pro-

TABLE 7.3. The Killing Times of *Hydrastis Canadensis* Total Standardized Extract and of the Major Alkaloids Obtained on a Low Density Bacterial Inoculum.

Tested Material		S. aureus Atcc 25923	S. aureus Atcc P6538	S. sanguis Atcc 10556	E. coli Atcc 25922	P. aeruginosa Atcc 27853	C. albicans Atcc 3153
Standard Ext.							
dir.	1 mL	15 min	15 min	4 min	2 h	15 min	15 min
Dil 1/2	0.5 mL	15 min	30 min	8 min	2 h	30 min	1 h
Dil 1/4	0.25 mL	30 min	30 min	15 min	>2 h	1 h	2 h
Canadaline							
dir	3.0 mg/mL	2 min	8 min	4 min	30 min	15 min	>2 h
Dil 1/2	1.5 mg/mL	8 min	15 min	30 min	2 h	1 h	>2 h
Canadine							
dir	3.0 mg/mL	30 min	30 min	4 min	>2 h	15 min	2 h
Dil 1/2	1.5 mg/mL	1 h	1 h	15 min	>2 h	30 min	>2 h
Berberine							
dir.	3.0 mg/mL	30 min	8 min	8 min	30 min	>2 h	1 h
Dil 1/2	1.5 mg/mL	1 h	30 min	8 min	2 h	>2 h	2 h
β-Hydrastine							
dir	3.0 mg/mL	>2 h	>2 h	>2 h	>2 h	>2 h	>2 h
Dil 1/2	1.5 mg/mL	>2 h	>2 h	>2 h	>2 h	>2 h	>2 h

Source: Scazzocchio et al. (1998). Reproduced with the permission of FITOTERAPIA.

duction was monitored via ELISA continuously over a period of six weeks. The goldenseal-treated group showed an increase in the primary IgM response during the first two weeks of treatment, suggesting that bioactive components of goldenseal may enhance immune function by increasing antigen-specific immunoglobulin production (Rehman et al., 1999).

Berberine was effective in treating diarrhea among patients with enterotoxigenic *E. coli* and *Vibrio cholerae* with a single oral dose of 400 mg (Rabbani et al., 1987). At 5 mg/kg for six days, berberine was found to be significantly better than placebo and as effective as metronidazole at 10 mg/kg (for the same duration) in treating children with giardia (Choudhry et al., 1972).

3.5. ASSAYS

Hydrastis rhizoma extracts can be distinguished, by TLC, from extracts of *Berberis* and *Mahonia,* genera that also contain berberine, by its specific spectrum of protoberberine alkaloids (Wagner and Bladt, 1996). The alkaloids can be detected, after TLC, without chemical treatment, by virtue of their fluorescence at 365 nm. The major alkaloids of *Hydrastis* have been determined by reversed-phase HPLC (Leone et al., 1996) and by colorimetry and spectrophotometry (El-Masry et al., 1980).

Recently, the technique of capillary electrophoresis-mass spectrometry (CE-MS) was applied for determination of isoquinoline alkaloids in crude methanolic extracts of several medicinal plants including *Hydratis canadensis.* For the CE separations, ammonium formate buffer solutions (70 or 100 mM, pH 3.0 or 4.0) containing 10% methanol or 20–60% acetonitrile as additives were used (Sturm and Stuppner, 1998).

3.6. SAFETY

At recommended dosages, berberine and berberine-containing plants are considered to be nontoxic (Pizzorno and Murray, 1985). High doses of hydrastine, however, are reported to cause exaggerated reflexes, convulsions, paralysis and death from respiratory failure (Genest and Hughes, 1969). At doses of 2–3 grams, goldenseal can lower heart-beat and at higher doses it can be paralyzing to the central nervous system (Flynn and Roest, 1995). Berberine and the other *Hydrastis* alkaloids can stimulate the uterus and should not be used during pregnancy as they may induce abortion (Newall et al., 1996). Goldenseal extracts are known to cause severe ulceration and, therefore, cannot be recommended as a douche (Duke, 1985). Goldenseal is contraindicated for those suffering from high blood pressure, heart disease, diabetes, glaucoma or a history of stroke (Castleman, 1988). Prolonged use of goldenseal is believed to interfere with vitamin B absorption (Newall et al., 1996).

3.7. CONCLUSIONS

There is no question that goldenseal can be used effectively as a treatment for diarrhea and as an antiseptic. Intriguing possibilities exist for the use of goldenseal as an anticancer agent and against the AIDS virus. This herb warrants more serious attention from the medical profession.

4. REFERENCES

Adzet, T., J. Iglesias, R San Martin, and M T. Torrent. 1975. "Etude de certains esters de *Centranthus ruber* et action pharmacodynamique de quelques-unes de ses preparations galeniques," *Plant Med.* 27:194–198.

Albrecht, M. 1995. "Psychopharmaceuticals and safety in traffic," *Zeits Allegmeinmed.* 71 1215–1221

Ammer, K. and P. Melnizky. 1999. "Medicinal baths for the treatment of generalized fibromyalgia," *Forschende Komplementamedizin.* 6(2):80–85

Anon 1891. "Saw palmetto," *Eclectic Med. J.* 51 229–230.

Bach, D., M. Schmitt and L. Ebeling. 1997. "Phytopharmaceutical and synthetic agents in the treatment of benign prostatic hyperplasia (BPH)," *Phytomedicine* 3/4:309–314.

Bach, D. and L. Ebeling. 1996 "Long-term drug treatment of benign prostatic hyperplasia— results of a prospective 3-year multicenter study using Sabal extract IDS 89," *Phytomedicine.* 3(2) 105–111.

Balderer, G. and A. A. Borbely. 1985. "Effect of valerian on human sleep," *Psychopharmacology.* 87 406–409

Ballard, L. 1989. "A Grower's Guide to Goldenseal," Nature's Cathedral, Norway, Iowa.

Barry, M. J. 1997. "A 73 year old man with symptomatic benign prostatic hyperplasia," *JAMA.* 278:2178–2184.

Bartolini, A., A. Giotti, S Giuliani, P. Malmberg-Aiello and R. Pattachini. 1990. "Biculline actions on isolated rat atria, mouse vas-deferns and guinea pig ileum are unrelated to GABA A receptor blockade," *General Pharmacology.* 21:277–284.

Bennett, B. C and J. R. Hicklin. 1998. "Uses of saw palmetto (*Serenoa repens*, Arecaceae) in Florida," *Economic Botany.* 52(4):381–393.

Berger, M. R., F. Garzon and D. Schmahl. 1986. "Studie zur tumorinhibierenden Wirkung von Valepotriaten an Transportieren und chemisch induziereten autochthoren Tumoren der Ratte," *Arzneimittel-Forschiung.* 36:1656–1659.

Berges, R. R., J. Windeler, H. J. Trampisch and T. H. Senge. 1995. "Randomized, placebo-controlled, double-blind clinical trial of β-sitosterol in patients with BPH," *Lancet* 345 (8964).1529–1532.

Berry, S. J., D. S. Coffey, P. C. Walsh and L. L. Ewing. 1994. "The development of human benign prostatic hyperplasia with age," *J. Urol.* 132:474–479.

Bloyer, W. E. 1896. "Saw palmetto," *Eclectic Med. J.* 56 581–582.

Bluestein, D. L. and J. E. Oesterling. 1993. "Hormonal therapy in the management of benign prostatic hyperplasia." In: *Prostate Diseases,* H. Lepor and R. K. Lawson (eds), Philadelphia, PA: W.B. Saunders Co., Chap. 16, pp. 182–197.

Blumenthal, M., W. R. Busse, A. Goldberg, J. Gruenwald, T. Hall, C. W Riggins and R. S Rister 1998. The Complete German Commission E Monographs Therapeutic Guide to Herbal Medicines. American Botanical Council, Austin, Texas. pp. 226–227, 581–582

Blumenthal, M. 1997. "Saving medicinal plants from extinction. potential threat, possible solution," *Natural Pharmacy.* July, pp. 18 and 20

Bokstaller, S. and P C. Schmidt. 1997. "A comparative study on the content of passionflower flavonoids and sesquiterpenes from valerian roots extracts in pharmaceutical preparations by HPLC," *Pharmazie.* 52(7):552–557.

Bombardelli, E and P. Morazzoni. 1997. "*Serenoa repens* (Bartram)," J K. Small. *Fitoterapia.* 68(2) 99–113.

Bone, K 1998. "Saw palmetto—A critical review," *Eur. J. Herbal Med.* 4(1)·15–24.

Bos, R., H. J. Woerdenbag, F. M. S. van Putten, H. Hendriks and J J C Scheffer. 1998a "Seasonal variations of the essential oil, valerenic acid and derivatives and valepotriates in *Valeriana officinalis* roots and rhizomes, and the selection of plants suitable for phytomedicines," *Planta Med.* 64.143–147.

Bos, R., H. Hendriks, J. J. C. Scheffer and H. J Woerdenbag. 1998b. "Cytotoxic potential of valerian constituents and valerian tinctures," *Phytomedicine.* 5(3):219–225.

Bos, R., H. J. Woerdenbag, H. Hendriks, J. H Zwavin, P. A G M De Smet, G. Tittle, H. V. Wikstrom and J J. C. Scheffer. 1996 "Analytical aspects of phytotherapeutic valerian preparations," *Phytochem. Anal.* 7(3):143–151.

Bounthanh, D , R. Misslin and R. Anton. 1980. "Activitée comparé de preparations galeniques de valeriane, *Valeriana officinalis* sur le comportement de souris," *Planta med.* 39:241–242.

Bounthanh, C., L. Richert, J. P. Beck, M. Haag-Berrurier and R Anton. 1983. "The action of valepotriates on the synthesis of DNA and proteins of cultured hepatoma cells," *Planta Med.* 49:138–142.

Bounthanh, C., L. Richert, J. P. Beck, M. Haag-Berrurier and R. Anton. 1981 "Valepotriates, a new class of cytotoxic and antitumor agents " *Planta Med.* 41:21–28.

Boyle, W 1991. *Official Herbs. Botanical Substances in the U S. Pharmacopoeias 1820–1990.* East Palestine, OH: Buckeye Naturopathic Press, pp. 1–77.

Bracher, F 1997 "Phytotherapy of benign prostatic hyperplasia," [German], *Urologe-Ausgabe A* 36(1):10–17.

Bradley, P R. 1992. *British Herbal Compendium Volume 1. A handbook of scientific information on widely used plant drugs. Valerian Root.* Bournemouth, Dorset. British Herbal Medicine Association, pp. 214–217.

Braeckman, J 1994. "The extract of *Serenoa repens* in the treatment of benign prostatic hyperplasia: a multicenter open study," *Curr. Ther. Res. Clin Exp.* 55:776–785.

Braun, R., W. Dittmar, M. Machut and S. Weickmann. 1982. "Valepotriate mit Epoxidstruktur-beatliche Alkylantien," *Deutsche Apotheker-Zeitung.* 122:1109–1113.

Brinker, F. 1994. "An overview of conventional, experimental and botanical treatments of non-malignant prostate conditions," *Brit. J. Phytotherapy.* 3:154–176.

Bruneton, J. 1995. *Pharmacognosy, Phytochemistry, Medicinal Plants.* Intercept, Ltd., UK. pp. 484, 741.

Buchbauer, G., W. Jager, L. Jirovetz, F. Meyer and H Dietrich. 1992. "Effects of valerian root oil, borneol, isoborneol, bornyl acetate and isobornyl acetate on the motility of laboratory animals (mice) after inhalation," *Pharmazie.* 47.620–622.

Buck, A. C. 1996. "Phytotherapy for the prostate." *Br. J. Urol.* 78:325–326.

Canadian Pharmacists Association, 1998. *Compendium of Pharmaceuticals and Specialties.* Gillis, M. C., Welbanks, L., Bergeron, D., Cormier-Boyd, M., Hachborn, F., Jovaisas, B., Pagotto, S., Repchinsky, C., Bisson, R., Tremblay, R., Levesque, J., Baxter, D., McIntosh, R., Danis, M. (eds.), Ottawa, Ontario: CPA, pp. 1767–1769.

Carper, J. 1997. *Miracle Cures: Dramatic New Scientific Discoveries Revealing the Healing Power of Herbs, Vitamins and Other Natural Remedies.* New York, NY. Harper Collins Publ. Inc. pp. 120–129.

Castleman, M. 1988. *The Healing Herbs. The Ultimate Guide to the Curative Power of Nature's Medicines* Emmaus, PA: Rodale Press, pp. 201–204, 362–365.

Chapelle, J. P 1972. "Seasonal changes of valeropotriates in native Valeriana procurrens Wallr.," *J. Pharm Belg* 27·570–576.

Chevalier, J. 1907. "Pharmacodynamic action of a new alkaloid contained in the roots of fresh Valerian," *Compt. Rend.* 144:154–157.

Chi Sheng, C. -K., C -H. -Y. Chi Sheng and T. C. Chung Ping. 1991. "Protoscolicidal effects of some chemical agents and drugs against *Echinococcus granulosus,*" *J. Parasitology and Parasitic Diseases.* 9:137–139.

Choudhry, V., M. Sabir and V. Bhide. 1972. "Berberine in giardiasis," *Indian J. Pediat.* 9:143–144.

Cirignotta, F., F Rasi, T. Saquegna, A. Forti and E. Lugaresi. 1975. In: Levin, P., Koella, W. P., (eds.), *Sleep.* 1974, Karger, Basel, p. 491.

Conway, D. 1975. *The Magic of Herbs.* London: Jonathon Cape

Cott, J. 1995. "Natural product formulations available in Europe for psychotropic indications," *Psychopharm. Bull* 31·745–751.

Culbreth, D. 1983. *Manual of Materia Medica.* USA: Eclectic Medical Publishers.

Devitt, B. D , B. J. R. Philogene and C. F. Kinks. 1980. "Effects of veratrine, berberine, nicotine and atropine on development characteristics and survival of the dark-sided cutworm *Euoxoa messoria* (Lepidoptera: Noctuidae)," *Phytoprotection.* 61:88–102.

DiPerri, R. and M. Meduri. 1975. In: Levin, P., Koella, W. P. (eds.) *Sleep.* Karger, Basel. p. 496.

Di Silverio F., G. D'Erarno, C. Lubrano, G. P. Flammia, A. Sciarra, E. Palma, M Caponera and F. Sciarra. 1992. "Evidence that *Serenoa repens* extract displays an antiestrogenic activity in prostatic tissue of benign prostatic hypertrophy patients," *Eur. Urol.* 21:309–314

Dreikorn, K and R. Richter. 1989. "Conservative nonhormonal treatment of patients with benign prostatic hyperplasia," In: *New Developments in Biosciences 5, Prostatic Hyperplasia.* R. Ackerman and F. H. Schroeder (eds), Berlin, Germany: Walter de Gruyter and Co., pp. 109–131.

Dressing, H 1992. "Insomnia: Are valerian/balm combinations of equal value to benzodiazepine?" *Therapiewoche.* 42:726–736.

Duke, J. A. 1985 *Handbook of Medicinal Herbs.* CRC Press Inc., Boca Raton, FL, pp. 238–239.

Ebbinghaus, K. D. 1974. "Die konservative Therapie des Prostata-Adenoms," *Munc. Med. Wschr.* 116:2209–2212.

Eichenberger, M. D. and G. R. Parker. 1976. "Goldenseal (*Hydrastis canadensis* L.) Distribution, phenology and biomass in an oak-hickory forest," *Ohio J. Sci.* 76:204–210.

El-Masry, S., M. A. Korany and A. H. Abou-Donia. 1980. "Colorimetric and spectrophotometric determinations of hydrastis alkaloids in pharmaceutical preparations," *J. Pharm. Sciences.* 69:597–598.

ESCOP. 1997. "European Scientific Cooperative on Phytotherapy, Monographs on the Medicinal Uses of Plant Drugs," ESCOP Secretariat, Argyle House, Exeter, UK.

European Pharmacopoeia. 1983. European Department for the Quality of Medicines, Council of Europe, Strasbourg. Part 2, 453-1-3.

Felter, H. W. 1985. *The Eclectic Materia Medica, Pharmacology and Therapeutics.* Portland, Oregon: Eclectic Medical Publications, pp. 417–422.

Finasteride Study Group. 1993. "Finasteride (MK-906) in the treatment of BPH," *Prostate.* 22:291–199.

Flynn, R. and M. Roest. 1995. *Your Guide to Standardized Herbal Products.* Prescott, Arizona: One World Press, pp. 38–39.

Foster, S. 1991 "Goldenseal· *Hydrastis canadensis,*" In. *American Botanical Council, Botanical Series No. 309* Austin, Texas, pp. 1–4.

Foster, S. and V. E. Tyler. 1999 *Tyler's Honest Herbal: A Sensible Guide to the Use of Herbs and Related Remedies* New York, NY: The Haworth Herbal Press, Inc, pp. 343–345.

French, M 1996. "The power of plants: An overview of herbal therapies," *Advance Nurse Practitioners* July 1996· 16–21.

Genest, K. and D. W. Hughes. 1969. "Natural Products in Canadian Pharmaceuticals IV. *Hydrastis canadensis.*" *Can. J. Pharm. Sci.* 4:41–45.

Gerber, G. S., G. P. Zagaja, G. T. Bales, G. W. Chodak and B A. Contrenas. 1998. "Saw palmetto (*Serenoa repens*) in men with lower urinary tract symptoms: Effects on urodynamic parameters and voiding symptoms," *Urology.* 51(6):1003–1007.

Greathouse, C G. and G. N. Watkins. 1938 "Berberine as a factor in the resistance of *Mahonia trifoliata* and *M sawaseyi* to *Phymatotrichum* root rot," *Am. J. Bot.* 25:743–748.

Gstirmer, F. and H. H. Kind. 1951. "Chemical and physiological examination of Valerian preparations," *Pharmazie.* 6:57–63.

Gstirmer, F. and H. H. Kind. 1958. "Zur pharmakalogische Prufung der Baldrianwurzel," *Pharmazie.* 13.415–420.

Guenther, E. 1952. *The essential oils.* Vol. VI. D. Toronto, Canada. Van Nostrand Inc., pp. 23

Guess, H A. 1992. "Benign prostatic hyperplasia antecedents and natural history," *Epidemiol. Rev.* 14·131–153.

Gupte, S. 1975. "Use of berberine in treatment of giardiasis," *Am. J. Dis. Child* 129:866.

Hansel, R. 1990. "Pflanzliche Sedativa. Informierte Vermutung zum Verstandnis ihrer Wirkweise," *Z Phytother.* 11.14–19.

Hansel, R. 1992 "Indischer Baldrian nicht empfehlenswert?" *Z Phytother.* 13:130–131.

Hendriks, H., R. Bos, H. J. Woerdenbag and A. S. Koster. 1985. "Central nervous system depressant activity of valerenic acid in the mouse," *Planta Medica.* 28:31.

Henneking, K. and H. Heckers. 1983. "Prostataadenom. Indikation zur Therapie mit sitosterinhaltigen Phytopharmaka?" *Med. Welt.* 34:625–632.

Hikino, H, Y. Hikino, H. Kobinata, A. Aizawa, C. Konno and Y. Ohizumi. 1980. "Study of the efficacy of oriental drugs 18: Sedative properties of Valeriana roots," *Shoyakugaku Zasshi.* 34:19–24.

Hindmarch, I. A. 1975. "1.4-Benzodiazeine, Temazeparn (K 3917), its effect on some psychological parameters of sleep and behaviour," *Arzneimittel-Forsch.* 25:1836–1839.

Hobbs, C. 1994. *Valerian—The Relaxing Herb.* Capitola, California· Botanica Press.

Hobbs, C 1990a. "Golden seal in early American medical botany," *Pharmacy in History.* 32:79–82.

Hobbs, C. 1990b. *Handbook for Herbal Healing. A Concise Guide to Herbal Products.* Santa Cruz, California: Botanica Press. pp. 132.

Hobbs, C. 1989. "Valerian: a literature review," *Herbalgram.* 21:19–35.

Hohenfellner, M. 1996. "Lipophilic extract from *Sabal serulata* inhibits contractions in smooth muscular tissue—Comment," *Aktuelle Urologie.* 27(3):157–158.

Houghton, P. J. 1999. "The scientific basis for the reputed activity of Valerian," *J. Pharm. Pharmacol.* 51(5):505–512.

Houghton, P. J. (ed.) 1998. "Valerian," In· *The Genus Valeriana. Medicinal and Aromatic Plants—Industrial Profiles.* Hardwood Academic Publishers.

Houghton, P. J. 1988. "The biological activity of Valerian and related plants," *J. Ethnopharmacology.* 22:121–142

Hutchens, A. R 1973. *Indian Herbology of North America.* Ontario: Merco. pp. 243–344.

Hwang, S. F., K. F. Chang, R. J. Howard and S. F Blade. 1997. "Yellows diseases of calendula (*Calendula* officinalis) and valerian (*Valeriana officinalis*) in Alberta, Canada, associated with phytoplasma infection." *J. Plant Diseases and Protection.* 104(5):452–458.

Iwu, M. M. 1993. *Handbook of African Medicinal Plants.* Boca Raton, Florida: CRC Press, Inc., pp. 167–168, 188–189.

Jaspersen-Schib, R. 1990. "Sedatifs a base de plantes," *Schweiz. Apoth. Ztg.* 128:248–251.

Kales, A, E O Bixler, J D. Kales and M. B. Scharf. 1977. "Comparative effectiveness of nine hypnotic drugs. Sleep laboratory studies," *J. Clin. Pharmacol.* 17:207–213.

Kaneda, Y., M Torii, T. Tanaka and M. Aikawa. 1991. "*In vitro* effects of berberine sulfate on the growth and structure of *Entamoeba histolytica, Giardia lamblia* and *Trichomonas vaginalis,*" *Annals Trop. Med. Parasit.* 85·417–425.

Keochanthala-Bounthanh, C, M. Haag-Berrurier, J. P. Beck and R. Anton. 1993 "Effects of two monoterpene esters, valtrate and didrovaltrate, isolated from *Valeriana officinalis,* on the ultrastructure of hepatoma cells in culture," *Phytother. Res.* 7:124–127.

Keochanthala-Bounthanh, C., M. Haag-Berrurier, J. P. Beck and R. Anton. 1990. "Effects of thiol compounds versus cytoxicity of valepotriates on cultured hepatoma cells," *Planta Med.* 56·190–192.

Klippel, K. F., D. M. Hiltl and B. Schipp. 1997. "A multicentric, placebo-controlled, double-blind clinical trial of beta-sitosterol (phytosterol) for the treatment of benign prostatic hyperplasia," *Brit. J. Urol.* 80(3).427–432.

Kohnen, R. and W. D. Oswald. 1988. "The effects of valerian, propanolol and their combination on activation performance and mood of healthy volunteers under social stress conditions," *Pharmacopsychiatry.* 21(6):447–448.

Kuhnlein, H. V. and N. J. Turner. 1991. *Traditional Plant Foods of Canadian Indigenous Peoples. Nutrition, Botany and Use.* Philadelphia, PA: Gordon and Breach Scientific Publishers. p. 265.

Lawson, R. K. 1993. "Etiology of Benign Prostatic Hyperplasia." In: *Prostate Diseases.* Lepor and Lawson (eds), Philadelphia, PA: W. B. Saunders Co., chap. 7, pp. 89–95.

Lazarowych, N. J. and P. Pekos. 1998. "Use of fingerprinting and marker compounds for identification and standardization of botanical drugs: strategies for applying pharmaceutical HPLC analysis to herbal products," *Drug Information J.* 32(2):497–512.

Leathwood, P. D. and F. Chauffard. 1985. "Aqueous extract of valerian reduces latency to fall asleep in man," *Planta Medica.* 28:144–148.

Leathwood, P. D. and F. Chauffard. 1983. "Quantifying the effects of mild sedatives," *J. Psychiatr. Res.* 17:115–122.

Leathwood, P. D., F. Chauffard, E. Heck and R. Munoz-Box. 1982. "Aqueous extract of valerian root (*Valeriana officinalis* L.) improves sleep quality in man," *Pharmacol., Biochem. Behaviour.* 17:65–71.

Lee, J. R. 1993 "Natural Progesterone. The Multiple Roles of a Remarkable Hormone," Sebastopol, CA: BLL Publishing.

Lee, J R 1991 "Dietary effects on breast cancer risk in Singapore," *Lancet.* 337:1197–1200.

Lee, J. R. 1990a "Osteoporosis reversal: The role of progesterone," *Intern. Clin. Nutr. Rev.* 10:384–91

Lee, J. R. 1990b. "Osteoporosis reversal with transdermal progesterone (letter)," *Lancet* 336:1327.

Leone, M. G., M F. Cometa, M. Palmery and L. Saso 1996 "HPLC determination of major alkaloids extracted from *Hydrastis canadensis* L.," *Phytotherapy Research* 10 545–546.

Leuschner, J., J. Muller and M. Rudmann 1993 "Characterization of the central nervous depressant activity of a commercially available valerian root extract," *Arzneimittel-Forschung.* 43 638–641

Lindahl, O and L Lindwall. 1989. "Double blind study of a valerian preparation," *Pharmacology, Biochem. Behaviour.* 32.1065–1066.

Lowe, F C. and J. C. Ku. 1996. "Phytotherapy in treatment of benign prostatic hyperplasia A critical review," *Urology.* 48:12–20.

Lytton, B., J. M Emery and B M. Howard. 1968. "The incidence of benign prostatic hypertrophy," *J. Urol.* 99.639–645.

McConnell, J D., M. J Barry and R. C. Bruskewitz 1994. "Benign prostatic hyperplasia Rockville, MD· Agency for Health Service, US Dept. of Health and Human Services Clinical Practice Guideline No 8, AHCPR publication 94-0582

McGoun, W. E. 1993. "Prehistoric peoples of south Florida Tuscaloosa, AL," The University of Alabama Press, Tuscaloosa, AL

Mennini, T, P Bernasconi, E. Bombardelli and P Morazzoni. 1993. "*In vitro* study of the interaction of extracts and pure compounds from *Valerian officinalis* roots with GABA, benzodiazepine and barbiturate receptors in rat brain," *Fitoterapia.* 54:291–300.

Messana, I., R La Bua and C Galeffi. 1980. "The alkaloids of *Hydrastis canadensis* L (*Ranunculaceae*) Two new alkaloids: hydrastidine and isohydrastinidine." *Gazz Chim. Ital.* 110 539–543.

Miller, L G. 1998. "Herbal Medicinals—Selected clinical considerations focusing on known or potential drug-herb interactions," *Archives of Internal Medicine.* 158(20).2200–2211.

Morazzoni, P. and E. Bombardelli 1995. "*Valeriana officinalis*: Traditional use and recent evaluation of activity," *Fitoterapia.* 66(2) 99–112.

Nakamura, J. 1977. "Material active against fungi. "Patent. Ger Offen., 2552630, West Germany, 37 pp.

Newall, C A., L. A Anderson and J. D. Phillipson. 1996. "Herbal Medicines. A Guide for Health-Care Professionals," Pharmaceutical Press, London, U K. pp. 260–263

Oesterling, J. E. 1995. "Benign prostatic hyperplasia," *N. Engl. J. Med.* 332:99–109.

O'Hara, M. A, D. Kiefer, K. Farrell and K. Kemper. 1998. "A review of 12 commonly used medicinal herbs," *Arch. Fam. Med.* 7:523–536.

Pegel, K. H. 1997 "The importance of sitosterol and sitosterolin in human and animal nutrition," *South African J. Science.* 93:263–268.

Pegel, K. H. 1984 "β-sitosterin-alpha-D-glucosid (sitosterolin). Eine aktive Wirksubstanz in Harzol.," *Extracta Urologica.* 7 (suppl.): 105–111.

Perry, N. B., E. J. Burgess, S. D. Lorimer and J W. VanKlink. 1996. "Fatty acid anilides as internal standards for high performance liquid chromatographic analyses of *Valeriana officinalis* L. and other medicinal plants," *Phytochem Anal* 7(5):263–268.

Philogene, B. J R., J. T. Arnason, G. H. N Towers, Z. Abramoski, F. Campos, D. Champagne and D. McLachlan. 1984. "Berberine: a naturally occurring phototoxic alkaloid," *J Chem Ecol.* 10:115–123.

Pizzorno, J. E. and M. T. Murray. 1985. "*Hydrastis canadensis*, Berberis vulgaris, Berberis aquitolium and other berberine containing plants," In: *Textbook of Natural Medicine.* Seattle, WA· John Bastyr College Publications.

Pojar, J. and A. Mackinnon (eds.) 1994. "Plants of Coastal British Columbia, including Washington, Oregon and Alaska," Vancouver, B.C , Canada: Lone Pine Publishing, p 333.

Popov, S , N. V. Handzhieva and N. Marekov. 1973. "Halogen-containing valepotriates isolated from *Valeriana officinalis* roots," *Doklady Bolgarskoi Academii Nauk.* 26.913–915.

Pulmery, M., M. F. Cometa and M. G. Leona. 1996. Further studies of the adrenolytic activity of the major alkaloids from *Hydrastis canadensis* L. on isolated rabbit aorta," *Phytotherapy Research.* 10.547–549.

Rabbani, G., T Butler, J. Knight, S. Sanyai and K. Alam. 1987. "Randomized controlled trial of berberine sulfate therapy for diarrhea due to enterotoxigenic *Escherichia coli* and *Vibrio cholerae*," *J. Infect Dis.* 155:979–984.

Rehman, J , J. M Dillow, S. M. Carter, J. Chou, B. Le and A. S. Maisel. 1999. "Increased production of antigen-specific immunoglobulins G and M following *in vivo* treatment with the medicinal plants *Echinacea angustifolia* and *Hydrastis canadensis*," *Immunol. Lett.* 68:2–3, 391–395.

Reynolds, J E. F. and W. Martindale. 1993. *The Extra Pharmacopoeia. The Universally Acclaimed Source of Drug Information,* 30th ed. London: The Pharmaceutical Press, pp. 1355–1356.

Richman, A. and J. Witkowski. 1997. "Herbs . . . by the numbers," *Whole Foods Magazine.* October, p. 22.

Riedel, E., R. Hansel and G Ehrke. 1982. "Hemmung des β-Aminobuttersaureabbaus durch Valensaurederivate," *Planta Medica.* 46:219–220.

Robbers, J. and V. E. Tyler. 1999. *Tyler's Herbs of Choice: The Therapeutic Use of Phytomedicinals.* New York, NY: The Haworth Herbal Press Inc., pp. 103–105.

Sack, R B. and J. L. Froehlich. 1982. "Berberine inhibits intestinal secretory response of *Vibrio cholerae* and *Escherichia coli* enterotoxins," *Infection and Immunity.* 35·471–475.

Sakamoto, T., Y. Mitani and K. Nakajima. 1992. "Psychotropic effects of Japanese valerian root extract," *Chem. Pharm. Bull.* 40(3):758–761.

Santavy, F. 1970. "Papaveraceae alkaloids." In: R. H F. Manske (ed.), *The alkaloids,* Vol. XII New York· Academic Press, pp. 333–454.

Santos, M. S , F. Ferreira, A. P. Cunha, A. P. Carvalho, C. F. Ribeiro and T. Macedo 1994. "Synaptosomal GABA release as influenced by valerian root extract-involvement of the GABA carrier," *Arch. Int. Pharmacodyn. Ther.* 327.220–231.

Scazzocchio, F., M F. Cometa and M. Palmery. 1998 "Antimicrobial activity of *Hydrastis canadensis* extract and its major isolated alkaloids," *Fitoterapia LXIX, Suppl.* 5:785–791.

Schilcher, H. 1998. "Herbal drugs in the treatment of benign prostatic hyperplasia." In: *Phytomedicines of Europe: Chemistry and Biological Activity* Larry D. Lawson and Rudolf Bauen (eds.) Washington, DC· American Chemical Society, pp. 62–73.

Schultz, V., W. D. Hubner and M. Ploch. 1997. "Clinical trials with phyto-psychopharmacological agents," *Phytomedicine.* 4(4):379–387.

Schultz, H., M. Jobert, and W. D. Hubner. 1998. "The quantitative EEG as a screening instrument to identify sedative or tranquillizing effects of single doses of plant extracts in comparison with diazepam," *Phytomedicine.* 5(6):449–458.

Shaw, R. J. and D. On. 1979. *Plants of Waterton-Glacier National Parks.* Missoula, MD. Moutain Press Publ. Co., p 84.

Showerman, J M. 1892. "Saw palmetto (*Sabal serrulata*)," *Eclectic Med. J.* 52:432.

Siguel, E. N. 1997. "Dietary modulation of omega-3/omega-6 polyunsaturated fatty acid ratios in patients with breast cancer," *J. Natl. Cancer Inst.* 89(15):1123–1131.

Siiteri, P. and J. D. Wilson. 1970 "Dihydrotestosterone in prostatic hypertrophy: The formation and content of dihydrotestosterone in the hypertrophic prostate of man," *J. Clin. Invest.* 49.1737–1739.

Small, J K 1926 "The saw palmetto—*Serenoa repens*," *J New York Botanical Garden.* 27.193–202

Snow, J M 1998. "*Hydrastis canadensis* L (*Ranunculaceae*)," *The Protocol Journal of Botanical Medicine* 2 25–28

Snow, J M 1996 "Monograph *Serenoa repens* Bartram (Palmae)," *The Protocol Journal of Botanical Medicine.* Winter 1996 pp. 15–16.

Stahl, E and W. Schild. 1969. "Thin layer chromatography for determination of pharmacopoeia drugs. Valeriana root, Valerian radix," *Arznemittel-Forschung* 19.314–316.

Stewart, D. 1999. Personal communication. July 4.

Stoll, A , E. Seebeck and D. Stauffacher. 1957 "New investigations on Valerian Schweiz," *Apoth. Ztg.* 95.115–120.

Sturm, S. and H. Stuppner. 1998 "Analysis of isoquinoline alkaloids in medicinal plants by capillary electrophoresis-mass spectrometry," *Electrophoresis.* 19.16–17.

Sultan, C., A Terraza and C. Devillier. 1984. "Inhibition of androgen metabolism and binding by a liposterolic extract of *Serenoa repens* in human foreskin in fibroblasts," *J. Steroid Biochem. Mol Biol.* 20.2041–2048.

Takamura, K., H. Nabata and M. Kawaguchi. 1975 "Pharmacological activity of kessylglycol 8-monoacetate," *Yakugaku Zasshi.* 93.1205–1209.

Tasca, A. 1985. "Treatment of obstruction in prostatic adenoma using an extract of *Serenoa repens.* Double-blind clinical test v placebo," *Minn Urol Nefrol* 37 87–91.

Tegarden, J. L 1898. "Saw palmetto," *Eclectic Med. J.* 58:653–654.

Tenover, J S. 1991. "Benign prostate hypertrophy," *Endocrine and Metabol. Clin. North America.* 20(4) 893–909.

Thies, P W. 1966. "Uber die Wirkstoffe des Baldrians 2. Mitteilung. Zur Konstitution der Iso Valeriansaureester Valepotriat, Acetoxyvalepotriat und Dihydrovalepotriate," *Tetrahedron Lett* 7(11).1163–1170.

Thornton, L. 1998. "The ethics of wildcrafting," *The Herb Quarterly.* Fall. 41–42

Thumm, H W. and J. Tritscher 1977. "The action of berberin-drops on the intraocular (IOP) pressure," *Klinische Monatsblatter fur Augenheikunde.* 170:119–123.

Tittel, G. and H. Wagner. 1978. "High performance liquid chromatographic separation and quantitative determination of valepotriates in Valeriana drugs and preparations," *J. Chromatography* 148.459–468.

Titz, W., J Jurenitsch, J. Gruber, I. Schabus, E. Titz and W Kubelka. 1983. "Valepotriates and essential oil of morphologically and karyologically defined types of *Valeriana officinalis.* II. Variation of some characteristic components of the essential oil," *Sci. Pharm.* 51.63–86.

Titz, W , J. Jurenitsch, E. Fitzbauer-Busch, E. Wicho and W. Kubelka. 1982. "Valepotriates and essential oil of morphologically and karyologically defined types of *Valeriana officinalis* I. Comparison of valepotriate content and composition," *Sci. Pharm.* 50.309–324.

Torsell, K. and K. Wahlberg. 1967. "Isolation, structure and synthesis of alkaloids from *Valeriana officinalis* L ," *Acta Chem. Scand* 21:53–62

Towers, G H. N., M Lefranc and Z. Abramowski. 1997. "Plant alkaloids: antibacterial and antifungal screening." (unpublished).

Treben, M 1994. *Health through God's Pharmacy* Ennsthaler Verlag, Steyr, Austria, pp. 52–59

Turner, N. J. 1997. "Food Plants of Interior First Peoples," University of British Columbia, Vancouver, B.C.: UBC Press, p. 169–170.

Tyler, V. E. 1998. "Importance of European Phytomedicinals in the American Market: An Overview," In: Phytomedicines of Europe, Chemistry and Activity. ACS Symposium Series 691. (Eds) Larry D Lawson and Rudolf Bauer. pp. 2–12

Ukita, T 1944. "Structure of Kessoglycol," *J Pharm. Soc. Japan* 64·285–294

USP 23 - NF 18. 1999. "The United States Pharmacopeia XXIII and The National Formulary XVIII," Rockville. MD. United States Pharmacopeial Convention, Inc. pp. 4454–4455.

Valpiani, C 1995. "*Valeriana officinalis,*" *ATOMS Journal.* 1(2)·57–63.

Veith, J , G Schneider, B Lemmer and M. Willems. 1986. "The effect of degradation products of valepotriates on the motor activity of light-dark synchronized mice [German]," *Planta Med.* (3):179–183.

Volz, H. P. 1997. "Efficacy of valerian extracts on sleep—Review of the literature," *Psychopharmakotherapie* 4(1) 18–22

Von der Hude, Scheutwinkel-Reich, M. and R. Brau. 1986. "Bacterial mutagenicity of the tranquillizing constituents of Valerianaceae roots," *Mutat Res.* 169:23–27.

Von der Hude, W., M. Scheutwinkel-Reich, R. Braun, and W Dittmar. 1985 "*In vitro* mutagenicity of valepotriates," *Arch. Toxicol.* 56:267–277.

Vogel, V. J. 1970. "American Indian Medicine." Norman, OK University of Oklahoma Press: pp. 365–366.

Wagner, H. and S Bladt. 1996 *Plant Drug Analysis. A Thin Layer Chromatography Atlas,* 2nd Edition Springer Verlag. Publ. pp. 42–43, 346–347.

Wagner, H., K Jurcic and R. Schaette. 1980. "Comparative studies on the sedative action of Valeriana extracts, valepotriates and their degradation products," *Planta Med.* 39 358–365.

Wagner, H , R Schaette. L. Horhammer and J. Holzl. 1972. "Dependence of the valepotriates and essential oil content in *Valeriana officinalis* L. on various exogenous and endogenous factors," *Arzneimittel-Forschung.* 22(7):1204–1209.

Wagner, J., M. L. Wagner and W. A. Hening. 1998. "Beyond benzodiazepines: alternative pharmacologic agents for the treatment of insomnia [Review]," *Ann. Pharmacother.* 32.680–691.

Watkins, L. 1899. "Saw palmetto," *Eclectic Med. J* 59 276–277

Weissrer, H , S. Tunn, B. Behnke and M. Kreig. 1996. "Effects of the *Sabal serrulata* extract IDS 89 and its subfractions on 5-alpha-reductase activity in human benign prostatic hyperplasia," *Prostate.* 28.300–306.

Werbach, M. R. and M. T. Murray. 1994. *Botanical Influences on Illness. A Sourcebook of Clinical Research.* Tarzana, CA: Third Line Press, pp. 23–24.

Wheelwright, E. G. 1974. *Medicinal Plants and Their History.* Dover, New York, USA (reprint of *The Physick Garden,* Publ. by Houghton Mifflin, Boston, U.S.A , 1935). pp. 1–288.

Wichtl, M. 1994. *Herbal Drugs and Phytopharmaceuticals. A Handbook for Practice on a Scientific Basis,* Norman Grainger Bisset (ed.) Medpharm GmbH Scientific Publishers, Birkenwaldstrasse 44, D-70191 Stuttgart, Germany: CRC Press, pp. 513–516.

Willey, L. B , S. P. Mady, D J. Cobaugh and P. M. Wax. 1995. "Valerian overdose: a case report," *Vet Human Toxicol.* 37:364–365.

Wilt, T. J., A. Ishani, G. Stark, R. MacDonald, J. Lau and C. Mulrow. 1998. "Saw palmetto extracts for treatment of benign prostatic hyperplasia. A systematic review," *JAMA.* 280 (18):1604–1609.

Wood, H. C and A Osol. 1943 *United States Dispensatory,* 23rd Edition, Philadelphia, PA: J. B. Lippincott, pp. 971–972.

Wood, G. B., J. P. Remington and S. P. Sdtler. 1907. *United States Dispensatory,* 19th Edition, Philadelphia, PA: J. B. Lippincott, pp. 1069.

Yao, C. -K. and L. -H. Pao. 1989. "The ultrastructural changes in the germinal tissues of *Echinococcus granulosus* cysts treated with d-hydrastine," *Acta Pharmacologica Sinica.* 10:185–187.

Yoshikazu, K 1976 "Organic and biological aspects of berberine alkaloids," *Heterocycles.* 4·197–219

Zava, D T, C M Dollbaum and M Blen 1998. "Estrogen and progestin bioactivity of foods, herbs and spices," *Proc Soc Exp Biol. Med* 217 369–378

Zava, D. T, M Blen and G Duwe 1997. "Estrogenic activity of natural and synthetic estrogens in human breast cancer cells in culture," *Environ. Health Perspectives* 105 637–645.

Zhu, B. and F A Ahrens 1982 "Effect of berberine on intestinal secretion mediated by *Escherichia coli* heat-stable enterotoxin in jejunum of pigs," *Am J. Veterinary Res* 43·1594–1598

CHAPTER 8

Evening Primrose Oil: Pharmacological and Clinical Applications

C. L. BROADHURST
M. WINTHER

1. INTRODUCTION

EVENING primrose oil (EPO) is currently the best available natural triacyl-glycerol source for exogenous (supplemental) gamma-linolenic acid (GLA). GLA is an 18-carbon polyunsaturated fatty acid (PUFA) with three double bonds. GLA is derived from the 18-carbon PUFA, linoleic acid (LA), with two double bonds. The additional double bond in GLA vs. LA is not a chemical triviality, but rather a difference of such fundamental biochemical significance that it forms the foundation and rationale for writing this chapter.

Sections 1 and 2 of this chapter cover the basic physiological functions and metabolism of PUFA, with an emphasis on GLA. Section 3 of this chapter outlines the common botanical sources for GLA-rich oils and summarizes some of the data that identify EPO as the most biologically active triacylglyc-erol form of GLA. Section 4 of this chapter gives the rationale and scientific basis for the use of EPO as a functional food. Five conditions, aging, alcohol-ism, atopic eczema, diabetes and hyperactivity disorders, for which GLA can be considered a conditionally essential nutrient, are discussed in detail. Dietary PUFA are discussed from an anthropological standpoint, so that a frame of reference for the type and of amounts of PUFA that humans are likely to have evolved from eating is established. Section 5 of this chapter discusses many of the medical conditions for which human research has consistently shown benefits from supplementing with EPO and (in some cases) other sources of GLA. Conditions ranging from diabetic neuropathy to rheumatoid arthritis to premenstrual syndrome are addressed. Section 6 of this chapter reiterates the value of EPO as a functional food and discusses the future of EPO for cancer treatment.

Essential fatty acids (EFA) are nutrients that must be consumed as part of the diet because they cannot be manufactured by mammals. Classically, the two EFA are restricted to linoleic acid of the n-6 series and alpha-linolenic acid (ALA) of the n-3 series. Both of these are 18-carbon *cis*-PUFA. In order to be utilized by all mammals, the double bonds in PUFA must be in the *cis* configuration. However, the 18-carbon PUFA must be further desaturated and elongated in order to form the complete set of LA and ALA PUFA derivatives required for normal metabolism. As will be discussed, the enzyme systems responsible for desaturation apparently do not function efficiently in humans; therefore, many researchers consider some of the PUFA derivatives to be EFA or at least conditionally essential. The case for essentiality is supported by the normalization of many pathological conditions upon administration of exogenous PUFA derivatives, despite no apparent deficiency in intake of LA and/or ALA. GLA administration during numerous research and clinical studies using EPO supplementation are the best documented examples of this phenomenon.

PUFA have a number of functions in the body. They are major components of all cell membranes and central nervous system tissue. The fluidity, flexibility and functionality of cell membranes are compromised without adequate provision of PUFA. PUFA also modulate the behavior of various membrane bound proteins such as receptors, enzymes and ion channels, and they are involved in the transport and disposal of cholesterol. PUFA are partially responsible for the impermeability of the skin to water and possibly for the regulation of permeability in the gut and other tissues. Finally, they are precursors for many short-lived, hormone-like substances that are known collectively as eicosanoids. These include the prostaglandins, thromboxanes and leukotrienes. PUFA can also be a source of energy through beta-oxidation or be stored as adipose tissue. With this broad spectrum of physiological functions, it is hardly possible to find an organ system in the body that cannot be significantly affected by a PUFA deficiency.

2. METABOLISM OF GAMMA-LINOLENIC ACID

The two groups of PUFA derived from LA and ALA are called the n-6 and the n-3 series, respectively. Fatty acids are frequently referred to by numerical nomenclature. The number defines the position of the first double bond in the molecule counting from the terminal methyl group. For example, GLA is also called 18:3n-6. The number before the colon gives the number of carbon atoms in the molecule (18 in this case); and the number after the colon gives the total number of double bonds (three in this case). The n-6 and n-3 series are not interconverted in mammals; both series are essential, and intake needs to be balanced in the diet. An unbalanced intake of either series is typically detrimental. It is generally believed that the enzymes operating in the PUFA metabolic pathways are the same for the n-6 and the n-3 series. In fact, sharing enzymes in the metabolic pathways results in a competition between the two series (Brenner, 1981; Huang and Nassar, 1990; Horrobin, 1990a; Horrobin and Manku, 1990; Zurier, 1993; Gerster, 1998).

2.1. DESATURATION AND ELONGATION OF POLYUNSATURATED FATTY ACIDS

Once in the body, the 18-carbon PUFA LA and ALA can be converted to their respective series of PUFA derivatives (Figure 8.1). The process involves a series of alternating desaturations (removal of two hydrogen atoms for insertion of a double bond) and elongations (addition of two carbon atoms). This eventually produces 20 and 22 carbon highly polyunsaturated fats known as long-chain PUFA (LC-PUFA). The two rate-limiting enzymes in the pathway to LC-PUFA are delta-6-desaturase (D-6-D) and delta-5-desaturase (D-5-D). The n-6 22-carbon LC-PUFA (22:5 n-6) is not considered biochemically important for humans, whereas 22:6n-3 (docosahexaenoic acid, DHA) is exceedingly important and at least conditionally essential (Crawford, 1993; Phylactos et al., 1994; Broadhurst et al., 1998; Gerster, 1998; Clandinin, 1999).

Considerable research over the last 20 years has identified factors that can affect the activity of D-6-D in particular. These include dietary, environmental and genetic factors (Brenner, 1981; Horrobin, 1990a; Horrobin and Manku, 1990). For example, magnesium, B6 and biotin are required for normal desaturation and elongation activity. Zinc is also thought to be required for D-6-D (Cunnane et al., 1982; Cunnane, 1988), but some researchers do not agree, arguing that anorexia induced by severe zinc deficiency altered the results in previous work (Elder and Kirchgessner, 1995). A high consumption of processed oils hydrogenated under high-temperature conditions and containing *trans*-fatty acids can reduce the activity of D-6-D. Stress hormones such as cortisol and adrenaline have been shown to reduce the body's ability to make GLA. Alcoholism, smoking and viral infections can alter body chemistry so that GLA production is reduced. Finally, as will be discussed in detail in Section 4 of this chapter genetic conditions such as atopy, aging and even maleness can reduce PUFA metabolic conversion and/or increase PUFA requirements.

When D-6-D is not functioning properly or efficiently, the body's requirement for PUFA changes to include derivatives such as GLA. Typical dietary sources (exogenous) of GLA can rarely provide sufficient GLA to compensate for lack of endogenous production due to defective D-6-D activity. Therefore, it becomes necessary to supplement the diet with this nutrient. EPO is one of

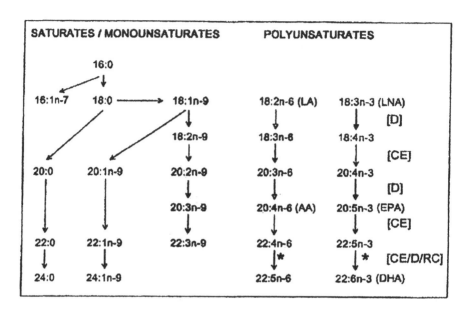

n-6 Fatty acids

Linoleic	18:2
Gamma-linolenic	18:3
Dihomogamma-linolenic	20:3
Arachidonic	20:4
Adrenic	22:4
Tetracosatetraenoic	24:4
Tetracosapentaenoic	24:5
Docosapentaenoic	22:5

Δ 6-desaturase
elongase
Δ5-desaturase
elongase
elongase
Δ6-desaturase
ß-oxidation

n-3 Fatty Acids

Alpha-linolenic	18:3
Octadecatetraenoic	18:4
Eicosatetraenoic	20:4
Eicosapentaenoic	20:5
Docosapentaenoic	22:5
Tetracosapentaenoic	24:5
Tetrahexaenoic	24:6
Docosahexaenoic	22:6

Figure 8.1 (Top) Long-chain fatty acid pathways in mammals. Palmitic acid (16:0) is the starting point for saturates (18:0 to 24.0) and monounsaturates (16·1n-7 through 24.1n-9). Linoleic acid (LA, 18·2n-6) and α-linolenic acid (LNA, 18:3n-3) are the main precursors to the polyunsaturates (n-3 and n-6 PUFA series). Pathways proceed through chain elongation [CE] and desaturation [D]. The final step in PUFA metabolism involves retroconversion [RC] from 24-carbon intermediates [ʻ] not shown. (Bottom) Detail of polyunsaturates showing desaturase enzyme system.

the richest and most biologically active sources of GLA. Decades of research have demonstrated repeatedly that dietary supplementation with EPO can correct physiological alterations resulting from a defective or inactive D-6-D. Usually, the elongation conversion of GLA to dihomo gamma linolenic acid (20:3n-6; DGLA) proceeds efficiently and rapidly.

2.2. PROSTAGLANDIN SYNTHESIS

Many of the end products of PUFA metabolism are short-lived regulatory molecules such as hydroxy fatty acids, prostaglandins and leukotrienes. In general, these substances are produced locally when and where they are required and are then destroyed following completion of their immediate function. They are involved in regulating a vast number of physiological functions and are necessary for normal growth, development and maintenance.

Prostaglandin E1 (PGE1), a metabolite of DGLA, is one of these end products that has a number of beneficial effects in the body (Figure 8.2). It is a vasodilator and is antithrombotic, hypotensive and hypocholesterolemic (Horrobin, 1990a, 1990b, 1990c; Sardesai, 1992; Zurier, 1993). In addition, it can inhibit the inflammatory response by reducing arachidonic acid (AA) mobilization and metabolism. The lipoxygenase enzymes convert AA into leukotrienes, some of which are responsible for keeping chronic inflammation running once it has been triggered (Belch, 1989; Thien and Walters, 1995;

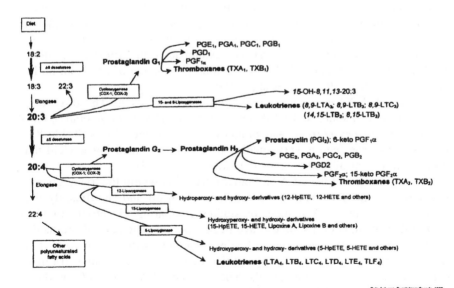

Figure 8.2 Metabolites of N-6 fatty acids.

Zurier, 1993; Rothman et al., 1995). Leukotrienes can also contract smooth muscle. Leukotrienes C4 and D4 in particular are bronchoconstrictors approximately 10^3 times more potent than histamine, and thus, play a significant role in asthma (Thien and Walters, 1995). Another DGLA end product, 15-OH-DGLA, can reduce inflammation by inhibiting 5- and 12-lipoxygenase activity. Increasing the production of PGE1 and 15-OH-DGLA is the major biochemical and clinical justification for supplementation with EPO.

3. BOTANICAL SOURCES OF GAMMA-LINOLENIC ACID RICH OILS

The occurrence of significant levels of GLA in plant sources is rare. Most higher plants contain ALA as the constituent 18-carbon PUFA. In contrast, the seed lipids of evening primrose *Oenothera biennis*, family Onagraceae, contain predominantly GLA (Mukherjee and Kiewitt, 1987; Ratnayake et al., 1989; Zygadio et al., 1994). Relatively high amounts of GLA are also found in several species of *Boraginaceae, Saxifragaceae, Scrophulariaceae, Aceraceae, Ranunculaceae* and *Anemonoideae,* but GLA is not characteristic of any genera (Muuse et al., 1988; Traitler et al., 1988; Mazza and Marshall, 1988; Galle et al., 1993; Tsevegsuren and Aitzetmuller, 1996). Fungi including *Mucor javanicus* and *Mortierella ramanniana* have been known for some time to produce GLA (Shaw, 1966; Amano et al., 1992; Kennedy et al., 1993). A recent study of over 150 zygomycetous fungi found that GLA is more common in fungi than previously reported and composes 35–65% of the total fatty acids of several species, i.e. *Mucor indicus, Circenella simplex* and *Syzygites megalocarpus* (Weete et al., 1998). The cyanobacterium *Spirulina plantensis* is also known to produce GLA, but the total percentage of lipid can be low (Grattan-Roughan, 1989; De et al., 1999).

At present, EPO is the most important source of GLA. The three other main commercial sources of GLA are borage seed (*Borago officinalis*), black currant seed (*Ribes nigrum*) and fungal (*Mucor javanicus*) oils (Table 8.1). Evening primrose seed is very small and produces a much lower yield than other commercial oilseed sources (Brandle et al., 1993).

The distribution of the GLA and other fatty acids within the triacylglycerol structure can interfere with the metabolism of an oil and significantly alter fatty acid absorption. A number of studies comparing bioactivity of GLA from different fungal oil sources have been reported. The arterial production of PGE1 in rats was measured following treatment with various oils. Rats were fed diets containing equal quantities of GLA from evening primrose, black currant, borage or fungal oils. The rats fed GLA from EPO produced the highest amount of PGE1 (Jenkins et al., 1988). EPO was also found to be more effective than borage, black currant, or fungal oils for lowering blood

TABLE 8.1 Approximate Fatty Acid Composition of Commercial GLA Containing Oils. Values are Percentages of Total Saponifiable Lipid.

Fatty Acid	EPO	Borage	Black Currant	Fungal
16 0	6 2	10 7	7 5	20 4
18 0	1.8	5 1	1 5	8 1
18·1n-9	10 9	16 3	13 6	41.3
18.2n-6	72 5	38 9	46 0	10 0
18 3n-6	8 8	22.1	15 0	19 4
18 3n-3		0 14	12 3	
18·4n-3	—		2.8	
20 0		0 25	0.1	
20.1n-9		0 4	0.2	
22 1n-9		2 5	1 0	

pressure (Engler, 1993) and improving nerve conduction velocity (Dines et al., 1996).

In contrast to EPO, borage oil can apparently provoke production of pro-inflammatory thromboxane in human platelets. Platelet aggregation in healthy men was shown to continually increase over a 43 d period of borage oil supplementation, and EPO resulted in higher production of prostacyclin than corn oil or borage oil (Horrobin and Manku, 1990; Barre et al., 1993). In a long-term feeding study, Fukushima et al. (1997) found that EPO lowered cholesterol in rats more than fungal oil. Black currant oil may be less effective than EPO for preventing cardiovascular disease (Horrobin and Manku, 1990; Charnock et al., 1994).

Compositional changes in the lipids of developing evening primrose seeds indicate that at the early stages of maturation, both ALA and GLA are synthesized by desaturation of LA (Cisowski et al., 1993). The ALA formed is channeled almost exclusively to phospholipids and glycolipids, which are constituents of the cell membrane, whereas GLA is esterified mainly to the storage lipids, i.e., triacylglycerols. Dilinoleoyl-gamma-linolenoylglycerols (DLGLA) are the major molecular species of triacylglycerols formed that contain gamma-linolenoyl moieties (Ratnayake et al., 1989).

The DLGLA molecular structure is at least partially responsible for the superior bioactivity of EPO. It appears that GLA in the DLGLA form is more bioavailable than other forms. Other oil sources of GLA contain less DLGLA (Table 8.2). These other sources also contain a number of other fatty acids such as ALA that may interfere with metabolism of GLA. Overall, research indicates that supplemental GLA sources other than EPO are less effective and may, in some cases, be detrimental.

TABLE 8.2. DLGLA Content of Commercially Important Sources of GLA.

Oil Source	Percentage of Oil
Evening primrose	15–18
Borage	10–12
Black currant	4–6
Fungal	2

4. NUTRITIONAL RELEVANCE OF EVENING PRIMROSE OIL

4.1. GLA AS A CONDITIONALLY ESSENTIAL FATTY ACID

The current definitions of LA and ALA as the only two essential fatty acids (EFA) is not in accordance with current PUFA research. In many cases, LC-PUFA derivatives must be considered the true EFA, especially docosahexaenoic acid (DHA). Broadhurst et al. (1998) and Crawford et al. (1999) have emphasized the argument that preformed dietary DHA and arachidonic acid (AA) are strictly essential for the evolution and maintenance of modern human intelligence and visual performance. Although an 18-carbon PUFA, under certain circumstances, GLA may need to be considered an EFA *sensu stricto* as well. There are pathological and genetic conditions in which the metabolism of LA into GLA appears to be so impaired that GLA cannot be produced in great enough quantity to support normal health or metabolism. If these conditions exist, it will be beneficial, if not necessary, to provide a source of preformed GLA. Some examples where GLA may be conditionally essential are as follows.

4.1.1. Aging

In the 1970s, it was first reported that an animal's ability to convert LA to GLA lessened with age. It was thus proposed that the resulting deficiency in GLA and its metabolites could be a key factor in the aging process (Dillon, 1987; Hrelia et al., 1991; Lorenzini et al., 1997). Metabolism of ALA appears to decline in older animals as well, indicating a reduction in D-6-D activity. Supplementation of elderly animals' diets with EPO increased GLA levels and unexpectedly increased the levels of ALA metabolites. Similar results have been observed in humans, with the result that EPA and DHA also increased with GLA administration; suggesting that presumably exogenous GLA activates D-6-D (Dillon, 1987; Vericel et al., 1987; Horrobin, 1990a). However, more recent research directed specifically at this issue has found that GLA administration does not augment the levels of ALA metabolites in humans (Barre et al., 1993; Brouwer et al., 1998).

Fatty acid determinations in elderly humans confirm that there is a decline in LA metabolites with age. Elderly Japanese were shown to have a significant fall in plasma DGLA, while elderly Scottish individuals had low DGLA and AA (Dillon, 1987; Vericel et al., 1987; Bamberg et al., 1997; Okuyama et al., 1997). Bolton-Smith et al. (1997) examined adipose samples from over 10,000 individuals aged 40 to 59 and found that the GLA/LA ratio decreased with age, independent of diet. A decline in D-6-D activity with age was apparent and was more pronounced in females. These results strongly suggest that an increase in dietary GLA with age is necessary to ensure that adequate amounts of LC-PUFA and DGLA metabolites are produced.

PUFA are likely to play a role in age-related osteoporosis, a condition more prevalent in females than males. Lower levels of PGE2, an AA metabolite, are associated with increased bone formation and reduced bone resorption rates (Watkins, 1998). High intakes of fish oil and, to a lesser extent, EPO tend to increase production of series 1 and 3 prostaglandins at the expense of series 2, so these PUFA may be beneficial for optimizing bone remodeling. In a pilot study, 40 elderly osteoporotic women were divided into four groups. Group 1 received 4 g EPO, group 2 received 4 g fish oil, group 3 received 4 g fish oil plus EPO and group 4 received 4 g olive oil placebo for 16 weeks (van Papendorp et al., 1995). Fish oil (group 2) increased serum calcium clearance, osteocalcin and procollagen, and it decreased alkaline phosphatase. EPO alone had no significant effects, but the positive results from the fish oil group were also seen in the fish oil plus EPO group (group 3), and EPO may have potentiated the effects of the fish oil. Doses of 2 g/d each EPO and fish oil appear reasonable and safe and may enhance bone formation, especially when used on a long-term preventive basis.

A decline in D-6-D activity may be a factor in some of the pathologies of aging such as reduced immunity, increased incidence/severity of chronic inflammatory conditions, dementia, visual decline, hypertension and lipid disorders. The possibility exists that the onset of these pathologies may be deferred or avoided by supplementation with GLA in conjunction with EPA and DHA.

4.1.2. Alcohol Metabolism and Alcoholism

Alcohol consumption influences n-6 PUFA metabolism in a number of ways. Upon initial exposure, there is an enhancement of PGE1 synthesis. However, long-term consumption can cause reduced levels of PGE1 and other eicosanoids due to inadequate supplies of DGLA. This lack of substrate DGLA may be the result of inadequate absorption of LA from the diet or insufficient conversion of LA to GLA because of reduced D-6-D activity. In addition, long-term consumption of alcohol can lead to reduced levels of LA in body tissues and an increased requirement for dietary EFA. This is likely caused by inadequate absorption of fatty acids from the diet due to hepatic and pancreatic damage. Withdrawal from alcohol may lead to a dramatic drop in PGE1 levels, partly due to depletion of the fatty acid precursor stores and partly due to the loss of alcohol's stimulating effect on PGE1 synthesis (Horrobin, 1987a; Glen et al., 1990).

EPO treatment reduced the severity of withdrawal symptoms and prevented fetal damage in pregnant animals consuming alcohol (Varma and Persaud, 1982; Duffy et al., 1992; Burdge et al., 1997). In studies of human alcoholics, EPO supplementation during the first three weeks of withdrawal from alcohol significantly reduced the patients' requirements for tranquilizers, indicating that the severity of the withdrawal symptoms were reduced. In addition, the patients' hepatic enzyme levels dropped rapidly, signifying an improved rate of return to normal liver function (Wolkin et al., 1987; Glen et al., 1990).

It may be concluded that GLA is conditionally essential in individuals who habitually consume large or perhaps even moderate amounts of alcohol. This represents a fairly large percentage of the population. The essentiality of GLA in this case follows logically when one considers that brewing and organized fermentation are but recent inventions on the time scale of human evolution, dating back perhaps 6,000 years. Many primitive societies were not exposed to alcohol until European contact several hundred years ago (Szathmary, 1994; Broadhurst, 1997a, 1997b). Under these circumstances, we cannot expect that the majority of humans have PUFA conversion systems that have fully adapted to the continual presence of high levels of ethanol. In fact, Native American groups who have been studied have been shown to have a high susceptibility to alcoholism/alcohol abuse and weak to nonexistent D-6-D and, especially, D-5-D enzyme systems (Horrobin, 1987a, 1987b; Horrobin and Manku, 1990; Lillioja et al., 1993; Pan et al., 1995; Tataranni et al., 1996).

4.1.3. Atopic Eczema

For many years, dermatologists have recognized that allergic individuals often have dry skin and follicular keratoses. Both of these conditions are also manifestations of EFA deficiency in humans. In the 1930s, Hansen successfully treated eczema with 30–60 g/d LA (Hansen et al., 1962). This was a surprisingly large quantity of the nutrient needed to produce a positive response, considering that 1–2 g/d LA is sufficient to correct skin lesions caused by PUFA deficiency. This discrepancy can now be understood since it is widely accepted that persons with atopic eczema and chronic allergic illness have genetically impaired D-6-D (Manku et al., 1982).

The case for a genetic impairment of D-6-D and probably D-5-D in atopic eczema is quite strong. Children and adults with atopic eczema are reported to have normal or elevated levels of LA and ALA but reduced levels of their derivatives in plasma, RBC and adipose tissue. In particular, LA levels in atopic eczema patients are higher, but DGLA levels are lower than in normal individuals (Manku et al., 1982, 1984; Wright, 1990; Wright and Sanders, 1991). A lack of DGLA reduces the production of anti-inflammatory PGE1 and 15-OH-DGLA, and this imbalance can account for the immunological defects and clinical symptoms experienced by allergic individuals. EPO supplies GLA directly, allowing synthesis of DGLA. Oral supplementation of EPO and borage oil has been shown to increase PGE1 and 15-OH-DGLA in the epidermis (Ziboh and Chapkin, 1988; Miller et al., 1991; Lehmann et al., 1995; Ziboh, 1996).

Infants at risk for atopy can be identified by high levels of IgE in umbilical cord blood. If these infants are breast-fed, thus receiving GLA, DGLA and AA from maternal stores, they have a lower incidence of atopy than formula-fed infants (Strachan, 1995; Chandra, 1997). However, atopic mothers have lower levels of the LA derivatives in their milk (Wright and Bolton, 1989; Horrobin, 1990a), so the protective effect of breast-feeding may be somewhat diminished, leading to the emergence of atopy in the infants regardless of feeding practice. EPO supplementation during lactation has been shown to increase the levels of GLA and DGLA in breast milk and is good insurance against atopic eczema. EPO can also be administered successfully directly to infants at risk for or presenting with atopic eczema (Fiocchi et al., 1994).

There is also evidence to suggest that EFA incorporation into cell membranes and other aspects of EFA metabolism may also be abnormal in atopic individuals. EFA-deficient animals lose excessive amounts of water through their skin due to lack of "waterproofing" sebum within the stratum corneum. Excessively dry skin is also present in eczema patients, who typically have low levels of LA-rich acyl sphingolipids. Increasing the levels of PUFA in the skin reduces water loss and improves skin condition. Compromised sebum production also increases skin itching and reduces the skin's antimicrobial action (Ziboh and Chapkin, 1988; Lehmann et al., 1995).

Supplementation with EPO in randomized, double-blind, placebo-controlled trials has been shown to normalize the levels of n-6 PUFA in eczema patients (Burton, 1990; Kerschser and Korting, 1992; Andreassi et al., 1997; Brehler et al., 1997). The increases in n-6 PUFA correspond to significant clinical improvements. EPO is particularly effective in relieving itching, reducing by more than two-thirds the requirement for topical steroid use, and in enabling patients to reduce or discontinue drugs such as oral steroids, antihistamines and antibiotics. Microscopic skin roughness in visually normal skin areas in patients with atopic eczema can also be significantly reduced.

EPO was first granted regulatory approval for the treatment of atopic eczema in the UK in October 1988 and has subsequently been approved for pharmaceutical use in more than a dozen countries. This testifies to its efficacy for atopic dermatoses: however, the role of EPO is more that of a functional food or essential nutrient than a drug. Atopic individuals have a genetic condition that apparently requires an exogenous source of GLA for normal epidermal function.

4.1.4. Diabetes

Diabetics have severely altered PUFA metabolism. The actions of D-6-D, D-5-D, and the cofactor for conversion of DGLA are abnormally poor, hence, LC-PUFA becomes chronically deficient. This chronic deficiency decreases insulin sensitivity and causes or contributes to many diabetic complications. The realization that abnormal membrane phospholipid profiles is of major significance in both Type I *and* Type II diabetes (Storlein et al., 1996; Dutta-Roy, 1994; Keen and Mattock, 1990) is of fundamental importance. It is known that insulin stimulates and glucagon inhibits the desaturase enzyme system, thus influencing the PUFA available for membrane incorporation. Conversely, the fatty acid composition of membrane lipids influences the action of insulin. Experimental data have shown that increasing membrane fluidity by feeding higher dietary levels of PUFA increases the number of insulin receptors and insulin action (Hagve, 1988; Dutta-Roy, 1994).

In Type I diabetics or chemically-induced diabetic animal models, insulin administration and EPO and fish oil supplementation can partially correct the membrane abnormalities. However, most diabetics are Type II (non-insulin-dependent diabetes mellitus, NIDDM), characterized by hyperinsulinemia or insulin resistance. Roughly 75% of Type II diabetics have elevated levels of insulin, but this insulin is not effective (Anderson, 1997). Type II diabetic membrane abnormalities thus cannot be mainly the result of insulin/glucagon influences on the desaturase system, but rather the diabetic condition may result from accumulating membrane abnormalities (Broadhurst, 1997a).

Evidence exists that such accumulating membrane abnormalities may be related to a genetic failure of D-5-D and/or D-6-D coupled with other lipid disorders. Central obesity is associated with a high incidence of NIDDM as well as cardiovascular disease, hypertension and premature mortality. Abdominally-obese individuals tend to have a diminished capability to utilize glucose peripherally or extract insulin during its first portal passage. They also have increased circulating free fatty acids (that reduces glucose metabolism) and a decline in insulin receptor number (Kissebah and Hennes, 1995). Three populations prone to central obesity, Arizona Pima Indians (Lillioja et al., 1993; Tataranni et al., 1996), South Asian Indians (McKeigue et al., 1991) and Australian Aborigines (O'Dea, 1991; Sinclair, 1992; Szathmary, 1994) appear to have a genetic predisposition for NIDDM. However, it is acknowledged that widespread health problems did not arise until these populations adopted Western food shopping and dietary habits.

Dramatic increases in calories, refined carbohydrates, total fat and n-6/n-3 PUFA ratios have affected these Native populations. In a comparison of

Australian Caucasian vs. Pima subjects, increasing n-6/n-3 PUFA ratios corre-lated positively with obesity and insulin resistance. However, the Pimas had much lower n-3 PUFA levels in their membrane phospholipids than Australian or especially Swedish subjects (Tataranni et al., 1996). In the case of the Australians vs. the Pima, Pan et al. (1995) found little evidence for greatly differing dietary intakes, however, the average DHA content of muscle phos-pholipids was 2.5 ± 0.7 vs. $1.2 \pm 0.1\%$, respectively.

This situation probably reflects decreased dietary intake and decreased desaturase/elongase activity. In the Pima studied by Pan et al. (1995), impaired insulin action and obesity were found to be independently related to reduced D-5-D activity. Vancouver Island Canadian Natives and Greenland Inuits (Eskimos), closed-society groups accustomed to traditional marine diets, have been shown to have a limited ability to convert PUFA to LC-PUFA, which is likely to be genetic (Horrobin, 1987b; Horrobin and Manku, 1990). Vancouver Island Indians are considered to have mainly reduced D-5-D activity, while Greenland Inuit have reduced D-6-D activity. The overall incidence of NIDDM in the Pima is the highest of any ethnic group known (50%), and specific genetic factors are likely to play a role (Prochazka et al., 1993).

Australian Aborigines returned to their native foods, rich in LC-PUFA but low in total fat, reduced their diabetes significantly in weeks in a dietary intervention study (Sinclair, 1992). Diabetic and nondiabetic Australian Abo-rigines were put on three diet trials: group 1—40% energy (en) from fat with 75% en from beef; group 2—20% en from fat with 80% en from coastal tropical seafood; group 3—13% from fat, with 85–87% en from kangaroo meat, freshwater fish and yams. The kangaroo meat consumed was only 1–2% fat by wet weight. Diets for groups 2 and 3 were traditional diets, low in LA and saturated fatty acids, and produced marked improvements in the metabolic abnormalities associated with diabetes and reduction in cardiovascu-lar disease risk factors, including hyperlipidemia and blood pressure. Plasma phospholipids in groups 2 and 3 had roughly equal ratios of AA, EPA and DHA. Prostacyclin activity was estimated *in vivo*, and tropical fish and kangaroo diets had evidence for high activity. High activity was not seen in comparison subjects on cold water marine fish (plasma AA roughly one-fifth of EPA plus DHA) or vegetarian (AA, DHA and, especially, EPA much lower) diets.

A tendency to store fat, release insulin and desaturate/elongate only the minimum amount of dietary 18-carbon PUFA metabolically necessary were likely positive factors for humans adapted to paleolithic foods and constant physical activity but may be maladaptive today. These are adaptions to a diet low in fat and often calories, relatively high in n-3 PUFA, raw and unprocessed and free from high-density carbohydrates; often such that dietary fat rather than carbohydrate was an absolute requirement for energy (Broadhurst, 1997b).

Most or all Native Americans descended from a small group of individuals who crossed the Bering Land Bridge between 11,000 and 14,000 years ago

(Grayson, 1993). [Recently the scanty evidence for entry circa 20,000 years was officially acknowledged by the paleoanthropological community (*c.f.* Roosevelt et al., 1996), but this has little bearing on the present argument]. "American Indians," regardless of where they live in the Americas, have more in common with high-latitude Inuit hunters than they do Caucasian Americans. Similarly, Australian Aborigines descended from an isolated small group migrating from Indonesia circa 40,000 years ago and never practiced agriculture until after Western contact (Morwood et al., 1998).

Under these circumstances, Native Americans and Australians may be but two examples of ethnic groups for which dietary LC-PUFA and GLA are strictly essential to prevent diabetes and diabetic complications. Adopting a Western diet with low GLA/LA and high AA/DHA is apparently especially detrimental to Native Americans and Australians, and EPO and fish oil may be functional foods necessary for achieving good health in these societies.

4.1.5. Hyperactivity

In 1981, the founders of the Hyperactive Children's Support Group, Vicky Colquhoun and Sally Bunday, conducted a survey of hyperactive children. Colquhoun and Bunday (1981) found that many symptoms of EFA deficiency were apparent in hyperactive children. The following points regarding these children were made:

(1) Many had colic, eczema, asthma, allergies or repeated infections. It is common for atopic children to have lower plasma levels of EFA.

(2) Hair analysis indicated many children were zinc deficient. Zinc is thought to be required for enzymatic desaturation/elongation.

(3) Some food additives and natural ingredients such as salicylates could cause rapid deterioration in behavior, presumably by reducing the conversion of EFA to prostaglandins.

(4) Nearly all the children were very thirsty. Thirst is a cardinal sign of EFA deficiency. The skin is composed primarily of lipids; if the fatty acid composition is abnormal, then the skin is unable to hold water in the body, resulting in abnormal thirst yet low urine volume.

(5) Almost all hyperactive children are male. Males require more EFA than females, thus marginal EFA deficiencies would affect males to a greater extent.

This proposed EFA deficiency syndrome is unlikely to be due to a lack of dietary PUFA, because typically only one family member is affected. The most likely cause is an intrinsic failure to convert enough LA and ALA to LC-PUFA and/or PUFA metabolites. Mitchell et al. (1987) reported that the levels of DHA, DGLA and AA were significantly lower in 48 hyperactive children than in 49 age- and sex-matched controls, providing clinical evidence to corroborate the anecdotal observations.

EFA deficiency has also been linked to attention deficit-hyperactivity disorder (ADHD). Stevens et al. (1995) compared 53 boys with ADHD to 43 control normal boys. The ADHD boys had much lower levels of PUFA in plasma and red blood cells. They also had much higher incidences of EFA deficiency symptoms, such as eczema, dermatitis, excessive thirst, frequent urination and brittle nails. Further, the ADHD boys also had a higher incidence of infections, recurrent pain and allergies. LC-PUFA were particularly deficient in 53 ADHD boys as compared to the normal subjects. Although this syndrome is multifactorial, the ADHD boys showed undeniable evidence of the vicious negative feedback cycle between inadequate nutrition, chronic allergies and hyperactive behavior.

In another study, 96 boys ages six–12 were found to have behavior, learning and health problems associated with low total PUFA, especially n-3 PUFA. EFA deficiency symptoms including thirst, frequent urination and dry skin were common in the boys, and those who were the most PUFA deficient exhibited the most learning and behavior problems. Frequent colds and antibiotic use were associated specifically with n-6 deficiency (Stevens et al., 1996).

Despite the multifactorial nature of hyperactivity disorders, the importance of EFA deficiency is underscored with the improvements seen in interventional studies. Horrobin (1982) reported that about 80% of a group of 100 hyperactive children treated with EPO seemed to respond. Interestingly, the 20% nonresponders were those with no history of atopic disorders in the family. In a double-blind, placebo-controlled crossover study of 31 children, EPO supplementation raised plasma DGLA and was associated with significant improvements on two performance tasks and parent ratings (Aman et al., 1987). Sixty-seven percent of hyperactive children showed some improvement with EPO vs. placebo in crossover case studies (Blackburn, 1990).

Dyslexia is often characterized by a visual defect that decreases the eye's ability to adapt to the dark. In a controlled research study, supplemental DHA at 480 mg per day was shown to improve this problem in dyslexics. Control subjects did not improve, except for one who was a vegan and did not consume DHA in his diet. Reading ability and behavior were also reported to improve with DHA supplementation (Stordy, 1995, 1997).

Considering the current widespread use of Ritalin for ADHD, large-scale interventional trials with a combination of EPO and fish oil are of utmost importance. Basic research and preliminary results make a good case that hyperactivity disorders are a manifestation of a genetic requirement for PUFA derivatives. GLA and LC-PUFA may be strictly essential; conversion of LA and ALA in hyperactive children is probably not adequate to maintain normal brain function, or the inadequate conversion exacerbates a preexisting brain abnormality. Providing essential nutrients and/or functional foods to help correct this abnormality is far more preferable to the use of stimulant medications.

4.2. THE ROLE OF EPO AS A FUNCTIONAL FOOD

GLA is not known to be widely distributed in Angiospermata outside of Boraginacea, however, many plants have trace amounts and still others may have GLA that has not yet been detected or reported. The number of major agricultural species consumed is but a small fraction of the known edible plants. Ethnographic studies of hunter-gatherer groups have consistently shown that such groups often consume hundreds of edible plant species, and there is no reason to think this differed in the past (Blumenschine, 1991; O'Dea, 1991; Bunn and Ezzo, 1993; Broadhurst, 1997a, 1997b). Similarly, wild game, the only meat available to hunter-gatherers, is much lower in total fat than livestock, but the relative percentages of PUFA vs. monounsaturates and especially saturates in the fat is higher (Crawford et al., 1970, 1976; O'Dea, 1991). Therefore, we assume that the dietary intake of GLA of *Homo sapiens* as he evolved from earlier humanoids was at least as high as today (sans supplements) and, more plausibly, considerably higher.

Studies of the Inuit, a traditional high-latitude hunting society, provided some of the first clues that diets based heavily on marine foods are quite healthy, despite a relatively high percent of energy and cholesterol from fat. Inuit consuming traditional diets have low incidences of arthritis, coronary artery disease, hypertension and autoimmune diseases (Bang and Dyerberg, 1980; Vidgren et al., 1997). Many of these epidemiological observations are ascribed to the high dietary intake of n-3 LC-PUFA, mainly EPA and DHA, coupled with an active lifestyle under harsh conditions. Higher latitude marine fish are rich in EPA and DHA and are used in fish oil supplements. Higher latitude and temperate marine fish are also those most commonly ingested by North Americans, Europeans and Australians, except those living in Northern Australia.

However, Inuit got more percent energy from seal meat than fish, and seal meat has a different lipid profile than fish (Ackman, 1988; Christensen et al., 1995; Ikeda et al., 1998). Seal oil has EPA and DHA primarily in the triacylglycerol sn-1,3 positions as opposed to fish and squid oils, with EPA and DHA primarily in the sn-2 position. In rats, seal oil reduced plasma triglyceride, AA contents in liver phospholipids and platelet aggregation more than equivalent doses of fish or squid oil (Ikeda et al., 1998).

Horrobin (1987b) pointed out that Inuit apparently can maintain high plasma levels of DGLA (but relatively low AA) while consuming large amounts of n-3 rich marine oils. This is likely due to a combination of dietary and genetic factors, including possibly a defect in D-6-D but not D-5-D activity. The result is an elevation of anti-inflammatory and antithrombotic PGE1 in addition to the known anti-inflammatory and antithrombotic metabolites of EPA. Westerners consuming large amounts of fish oil usually experience a sharp decline in DGLA, hence, the apparent health benefits of the traditional Inuit diet are not fully realized. However, a better approximation is made by supplementing GLA in conjunction with fish oil, which raises serum DGLA but still inhibits the conversion of DGLA to AA (Horrobin and Manku, 1990). This is an excellent example of the need for balance in the overall intakes of n-3 and n-6 PUFA.

In current Western diets, the balance is shifted too far in favor of LA, which makes up 85% or more of total PUFA intake (James et al., 1992). We know with almost complete certainty that this intake of LA is many times greater than the intake we were accustomed to eating while our biochemistry was evolving due to the prevalence of agricultural seed oils (e.g., cottonseed, peanut, soybean and sunflower) in the modern food supply. Yet, despite large and consistent intakes of LA in today's diets, the argument can be made readily that GLA is a conditionally essential fatty acid.

Overall, an efficient ability to convert LA and ALA to their respective desaturated metabolites is apparently not a feature of human metabolism; in fact, the reason we can do so at all may be a fail-safe mechanism for starvation conditions when lipid in general and PUFA in particular were hardly available, and the small amount of ALA and LA in leaves and tubers was the extent of the supply (Broadhurst, 1997b). We must consider the drastic changes in not only the absolute amounts but also the sources of dietary lipids that humans have experienced in the past 100 years or so, in addition to the preceding millenia.

In summary, three key points provide a compelling argument for increased use of EPO:

(1) The provision of GLA functions to normalize metabolism in some individuals deficient, through low D-6-D activity or by other means

(2) The increasing number of elderly individuals in society implies an increasing number of persons with a need for exogenous GLA in order to help defer/prevent some of the pathologies of aging and the associated medical costs

(3) The GLA/LA ratio in the diet of *Homo sapiens* as he evolved was higher than that of today; thus, today's ratio may not be optimal for human physiology during growth and adulthood

4.2.1. Guidelines for Daily Nutritional Supplementation

The normal endogenous production of GLA in a healthy young adult is around 250–1,000 mg/d. If one assumes that a condition such as aging or diabetes conservatively reduces the ability to convert LA to GLA by 50%, then one could estimate that persons over 60 years old or who are diabetic would require approximately 125–500 mg/day exogenous GLA. It follows that older persons with Type II diabetes would likely require 500 mg/d, at least initially. Maintenance doses may be lower after a few months at higher levels of supplementation, especially if no complications are present. This basic metabolic requirement may be one reason why EPO has emerged as one of the few successful interventional treatments for diabetic neuropathy (Section 5.2 of this chapter). Supplemental requirements for atopic individuals or those who chronically consume large amounts of alcohol may be similar.

For children with clinical hyperactivity disorders, the supplemental requirements may not be very different than those for adults because the absolute EFA requirements during pre-pubertal growth periods are greater than in adulthood. Breast-fed babies receive the equivalent of 1–5 g/day of EPO. Children aged four–12 have been given doses up to 25 mL/d for a year, and one- to four-year-olds have been given 3 g/d for six months with no adverse effects (Horrobin, 1990a).

In addition, the likelihood that PUFA will be oxidized for energy rather than metabolized to LC-PUFA is much greater during growth. PUFA are apparently preferentially oxidized during energy deprivation, undernutrition or weight cycling. In growing rats, weight cycles of 24 h fasting followed by 72 h refeeding resulted in stochastic loss of both LA and ALA (Chen et al., 1995). In weanling rats, dietary LA and ALA were shown to be mostly partitioned toward β-oxidation, and can in fact be oxidized at levels greater than intake (Cunnane and Anderson, 1997).

Supplementation of 1.5–6 g/d EPO can provide 125–500 mg GLA. Other natural oils have higher concentrations of GLA, but none of these oils has been shown to have the biological activity or therapeutic benefits that EPO has. Some reasons for this have been discussed previously, with the specific triacylglycerol structure of the GLA-containing oil likely to be important in determining the biological activity. Further, there is no other natural oil source of GLA in the marketplace that maintains the pharmaceutical and quality control standards met by the Efamol® brand of EPO.

5. THERAPEUTIC APPLICATIONS OF EVENING PRIMROSE OIL

We have already noted that the scope of physiological and metabolic functions in which PUFA play a role is very broad. We have also argued that in many cases GLA should be considered an EFA *sensu stricto* in that it must be supplied in the diet. As with essential nutrients (e.g., vitamins C and E), there are potentially a number of pathological conditions that can significantly benefit or even be resolved through supplementation of the nutrient(s) at levels beyond normal estimated metabolic requirements. The clinical evidence for the benefits of supplementation of EPO is very good, and a number of examples where the effects of EPO may be considered nutritional and pharmacological— perhaps in essence a good working definition of a nutraceutical—follows.

5.1. CARDIOVASCULAR DISEASE

The four major risk factors associated with coronary heart disease (CHD) and peripheral vascular disease are elevated cholesterol and triacylglycerol levels, hypertension and enhanced platelet aggregation. Expression of these risk factors can be explained through their association with fatty acid metabolism. These factors may be the result of an inadequate supply of LA metabolites or an imbalance in the level of n-6 and n-3 PUFA within cell membranes. It has been well documented in numerous population studies that low intake and low levels of LA in plasma and adipose tissue are associated with higher CHD risk (Horrobin and Huang, 1987; Oliver, 1989; Oliver et al., 1990). Moreover, reduced plasma and adipose tissue levels of DGLA and AA have also been reported as strong markers for CHD (Horrobin, 1993a). It is possible to explain each risk factor more specifically as follows.

5.1.1. Elevated Cholesterol and Triacylglycerols

In order for cholesterol to be eliminated from the body, it must first be esterified to a fatty acid, such as LA, ALA or one of their derivatives. If levels of these fatty acids are low, then disposal of cholesterol is hampered. To compound the problem, it appears that high cholesterol levels will inhibit the 6-desaturation of LA and ALA, hence, reducing the formation of their respective metabolites. Epidemiologically, diets higher in LA-rich vegetable oils are associated with lower levels of cholesterol than diets rich in saturated animal fats (Oliver, 1989; Horrobin and Huang, 1987; Nordoy et al., 1994).

However, from a clinical standpoint, GLA is significantly more effective than its precursor LA in lowering cholesterol (Sugano et al., 1986; Huang and Nassar, 1990; Fragoso and Skinner, 1992). EPO has been shown to lower cholesterol in patients with familial hypercholesterolemia (Horrobin, 1990a; Ishikawa et al., 1989). Guivernau et al. (1994) gave 12 hyperlipidemic men 3 g/d EPO or paraffin placebo for two four-month periods in a double-blind crossover study. GLA supplementation decreased plasma triacylglycerols by 48% and increased HDL cholesterol by 22%, while lowering total cholesterol from an average of 286 ± 18 to 194 ± 22.

Rats on high-cholesterol diets experienced lower total and VLDL + LDL cholesterol when fed wild or cultivated EPO as compared to soybean oil, palm oil or fungal oil containing GLA (Fukushima et al., 1997). The superior bioactivity of EPO vs. other GLA source oils was again demonstrated, and the efficacy of a GLA oil vs. a high LA and ALA (soybean) oil in lowering cholesterol was reaffirmed. However, the rats received 10 wt% of each oil, which translates to very large doses of EPO and GLA.

Other studies have been more equivocal regarding the ability of LA, GLA and fish oil to lower cholesterol and/or triacylglycerols (Harris, 1997; Johnson et al., 1997; Okuyama et al., 1997; van Rooyen et al., 1998). For both EPO and fish oil administration, the background diet may be the determining factor as to whether PUFA of any type are effective for dyslipidemia. For example, van Rooyen et al. (1998) divided vervet monkeys into two groups of 10. One group was fed a Western atherogenic diet, and the other was fed a high-carbohydrate low-fat diet for six weeks. Then, each diet group was subdivided into groups receiving the diet plus the free fatty acids GLA or EPA for 24 weeks. In those receiving EPA, plasma total cholesterol increased with the Western atherogenic diet but decreased with the high carbohydrate diet. In the groups receiving GLA, plasma total cholesterol increased with both diets. With GLA supplementation, HDL cholesterol decreased only slightly on the Western atherogenic diet and increased on the high carbohydrate diet.

Overall, elevated triacylglycerols are considered to respond clinically to administration of n-3 LC-PUFA in the form of fish oils or methyl esters (Connor and Connor, 1997; Harris, 1997). Combined supplementation of EPO and fish oil within an overall low-fat diet may be warranted for generalized dyslipidemia (Horrobin, 1993a).

5.1.2. Hypertension

In general, higher levels of dietary and/or plasma PUFA vs. saturated fat favorably affect hypertension, but the effects are limited (Iacono and Dougherty, 1993). Dietary LA has been shown to lower blood pressure in animals with spontaneous or stress-related hypertension, but much lower amounts of supplemental GLA and DGLA are consistently more effective than LA (Horrobin, 1990a; Deferne and Leeds, 1992). In humans with hypertension, the levels of LA and ALA derivatives are typically below normal, indicating that D-6-D activity is impaired (Singer et al., 1990). EPO supplementation has been shown to lower systolic and diastolic blood pressure in humans with mild hypertension (Venter et al., 1988; Leeds et al., 1990; Deferne and Leeds, 1992). However, the Leeds et al. (1990) study was only preliminary, and very large doses of 15 g/d EPO or sunflower oil placebo were given to only eight subjects.

EPO effectively lowered blood pressure and pressor responses to angiotensin in gestational hypertension (Broughton-Pipkin, 1990). GLA was found to be 50% more effective at reducing stress-induced rises in blood pressure in normotensive and borderline hypertensive rats than EPA (Mills and Ward,

1986, 1990; Mills et al., 1990). The same authors gave 30 normotensive young males 1.5 g/d GLA from borage oil, 1.6 g/d EPA from fish oil or olive oil placebo, and the physiological responses to stress during testing were measured. Only the GLA group showed a significant reduction in systolic blood pressure and heart rate reactivity, and the GLA group also performed better in the test (Mills et al., 1990).

5.1.3. Enhanced Platelet Aggregation

GLA, DGLA, ALA and EPA have been found to reduce platelet aggregation in animal and human studies (Horrobin, 1990a; Nordoy et al., 1994; Connor and Connor, 1997; Freese and Mutanen, 1997; Ikeda et al., 1998). This effect is mainly the result of conversion of these fatty acids to anti-aggregatory metabolites such as PGE1 from GLA and DGLA and PG13 from EPA. Guivernau et al. (1994) found that four months of EPO supplementation decreased platelet aggregation induced by adenosine diphosphate and epinephrine by 45% in humans and rats. Serum thromboxane decreased 45%, and prostacyclin nearly doubled in the rats (not measured in humans). In human subjects, one month of supplementation did not produce a significant increase in bleeding time, but after three or four months, a highly significant 40% increase in bleeding time was observed.

Freese and Mutanen (1997) supplemented ALA or fish oil for four weeks in order to increase bleeding time by one minute in both groups. Platelet aggregation stimulated by collagen and the thromboxane A2 mimic I-BOP was reduced similarly by both treatments, but aggregation stimulated by adenosine diphosphate was only reduced by fish oil.

Brister and Buchanan (1998) supplemented 60 patients scheduled for coronary artery graft bypass with 3.2 g/d of EPO, fish oil, 3:1 EPO plus fish oil mixed or placebo for four weeks prior to surgery. The four-week time period did not significantly alter bleeding times. However, the purpose of the trial was to increase blood vessel wall thromboresistance during surgery in order to attenuate acute thromobosis and chronic restenosis post-surgery. The LA metabolite 13(s)-hydroxyoctodecadienoic acid (13-HODE) is synthesized in vascular cell walls under healthy basal nonthrombogenic conditions. It was hypothesized that substantially increasing 13-HODE levels just prior to surgery could decrease platelet/vessel wall interactions during surgery, thereby leading to reduced vessel wall damage and less release of hyperplastic growth factors and vasoconstrictors after surgery. (It is not clear from the study why EPO was used instead of a simple high-LA oil, so there may be other effects from using a source of both GLA and LA).

13-HODE levels in vein and artery samples were nearly doubled over placebo by the EPO treatment, whereas fish oil and the 3:1 mix nonsignificantly lowered/raised 13-HODE, respectively. This indicates that one month presurgical supplementation of EPO does not adversely affect bleeding times and may improve the bypass surgery outcome with very little expense or risk. Fish oil given post-surgery has been shown to reduce restenosis (Bairati et al., 1992; Gapinsky et al., 1993), but high doses may be contraindicated prior to surgery. Combined but staggered fish oil and EPO supplementation may be beneficial to reduce the very severe but frequent complications of coronary artery bypass surgery.

5.2. DIABETIC NEUROPATHY

Neuropathy is the most common complication of diabetes. It produces clinical symptoms in about half of all diabetics and neurophysiological abnormalities in nearly all diabetics (Vinik et al., 1992). Neuropathy is responsible for the primary events leading to skin ulceration, amputation and impotence in diabetics. The consistent results obtained in biochemical, animal and human studies confirm that GLA is a valuable treatment for diabetic neuropathy and a likely treatment for related conditions such as retinopathy.

PUFA are important for maintenance of microcirculation around nerves. When PUFA levels are reduced, the red blood cell membranes become rigid, impairing circulation and reducing the supply of oxygen and nutrients to the nerves. Osmium tetroxide treatment of neuronal membranes from diabetic animals do not stain like membranes from non-diabetic animals. The development of the stain depends partly on the presence of double bonds in PUFA. The lack of staining within the membranes of diabetic animals suggests an abnormal fatty acid structure within the membranes. Subsequent to this observation in animals, numerous studies have shown that GLA provided from EPO can either prevent or reverse neuropathy in animals. This improvement is associated with an improvement in nerve blood flow and not with changes in glucose or sorbitol levels in nerves or in diabetic control (Julu, 1992; Tomlinson et al., 1992; Cameron and Cotter, 1994).

One of the most comprehensive human studies on any natural product was done on the effect of GLA treatment for diabetic neuropathy. Two large multicenter clinical trials were conducted in which a total of 202 diabetic patients were given 480 mg GLA per day for a year, and 202 other diabetic patients were given vegetable oil placebo. Twenty-five clinical parameters were measured. In the group that received EPO, 25 of 28 parameters improved significantly. In the placebo group, 27 of 28 parameters showed deterioration. In the second year of the study, all patients were given EPO, and the 202 that were previously in the placebo group showed improvement instead of deterioration in 23 parameters (Keen et al., 1993; Horrobin, 1997).

Alpha-lipoic acid has also shown excellent results for diabetic neuropathy in humans. In a double-blind placebo-controlled study, 328 Type II diabetics were given 1,200, 600 or 100 mg/d alpha-lipoic acid i.v. for three weeks. All doses significantly reduced pain and disability, however, the 600 mg dose was judged the best, improving symptoms 25% better than placebo (Zeigler and Gries, 1997). In other studies, 400–900 mg/d doses were shown to improve diabetic neuropathy in times as short as 14 d. It must be noted that in many cases subjects were first pre-loaded with i.v. doses, then continued on oral supplementation.

In diabetic rats, a compound with equimolar proportions of GLA and alpha-lipoic acid completely corrected nerve conduction velocity and blood flow deficits in motor and sensory nerves. GLA and alpha-lipoic acid together yielded a synergism, increasing the efficacy of diabetic neuropathy treatment by an order of magnitude (Cameron et al., 1998).

5.3. GASTROINTESTINAL DISORDERS

5.3.1. Ulcers

PUFA, in general, and GLA, in particular, have protective and potentially healing effects on stomach and duodenal ulceration. One reason for this is that DGLA and AA are substrates for prostaglandins that are cytoprotective and decrease gastric acid output (Grant et al., 1988; Das, 1998). Increased consumption of LA greatly enhances gastric prostaglandin formation and suppresses acid output. The adipose tissue of chronic duodenal ulcer patients was found to be significantly lower in LA than that of control subjects (Grant et al., 1988). In fact, it has been proposed that the steady decline in the incidence of gastric and duodenal ulcers in the past 30 years, even before the introduction of acid-blocking medications, is due to the increasing amount of PUFA in the diet (Hollander and Tarnawski, 1986).

The adverse effects of nonsteroidal anti-inflammatory drugs (NSAID) on the stomach may be related to inhibition of prostaglandin formation from DGLA. In humans, EPO raises gastric mucosal prostaglandin synthesis but does not seem to protect against aspirin-induced damage (Pritchard et al., 1988). Alternatively, in rats, EPO does not raise gastric mucosal prostaglandin synthesis, but it protects against aspirin-induced damage (Huang et al., 1987).

PUFA, in general, may protect against ulceration by inhibiting the growth of *Helicobacter pylori* and by directly healing and/or providing sufficient EFA to the gut mucosa. Various PUFA have been shown to disrupt the *H. pylori* cell membranes at 10^{-3} M concentrations *in vitro* (Thompson et al., 1994). Even when LA levels are low, provision of EPA and DHA may help heal ulceration, indicating that the role of PUFA is broader than just DGLA and AA functioning as prostaglandin substrates (Das et al., 1994). The acid blocker famotidine can apparently increase the activities of D-6-D and D-5-D, which may enhance the activity of the drug and provides a strong clue that PUFA are useful in the treatment of peptic ulcer disease (Das, 1998).

5.3.2. Cyclical Irritable Bowel Syndrome (IBS)

Some females taking EPO for mastalgia and premenstrual syndrome (PMS) symptoms reported that their IBS was significantly better with EPO. IBS is often premenstrually exacerbated; even when under control through diet during most of the month, IBS may flare up irregardless of diet during the premenstrual period. GLA has not been shown to have significant effects on noncyclical IBS.

A double-blind, placebo-controlled crossover trial of 40 women with pre-menstrually-exacerbated irritable bowel syndrome showed that half of those treated with 4 g/day EPO experienced symptom relief, while none on placebo reported improvement. Each treatment period was three menstrual cycles, with a washout phase between the two treatment periods. Symptomatic relief usually began in the second month of active treatment but sometimes was not apparent until the third (Cotterell et al., 1990).

5.3.3. Ulcerative Colitis and Crohn's Disease

A six-month placebo-controlled trial of 43 ulcerative colitis patients compar-ing fish oil and high GLA EPO found that fish oil was of no benefit, and EPO was only slightly beneficial. Patients were randomized to receive fish oils, super EPO (a combination of EPO and borage oil) or olive oil placebo. Doses were 12 capsules per day for one month followed by six capsules per day for five months. Capsules were 1 g (fish, olive oil) or 250 mg (super EPO). Super EPO significantly improved stool consistency compared to fish oil and placebo, but stool frequency, rectal bleeding, rectal histology and disease relapse remained unchanged (Greenfield et al., 1993).

An open study reported success in treating Crohn's disease with EPO (Horrobin, 1990a), but this has not been further confirmed. Fish oil has been studied fairly extensively for treatment of Crohn's disease, but there is no consensus as to whether it can maintain the disease in remission (Kim, 1996). The combined use of fish oil and EPO over the long term may be effective for ulcerative colitis and Crohn's. This remains to be tested, but certainly, supplementation poses little risk and is done informally by many suffering from these conditions.

5.4. GYNECOLOGICAL DISORDERS

5.4.1. Premenstrual Syndrome and Clinical Mastalgia

Premenstrual syndrome has a variety of psychological and physical symp-toms that occur premenstrually and are spontaneously relieved within 72 h following the onset of menstrual flow. Symptoms range from depression and food cravings to skin lesions and migraines (O'Brien and Massil, 1990;

Campbell et al., 1997). Breast pain and PMS are related conditions, since breast pain is frequently one of the symptoms of PMS. However, severe breast pain (clinical mastalgia) occurs during most or all of the menstrual period, and in one-third of cases, it is noncyclical, i.e., unrelated to the menstrual cycle (Be Lieu, 1994; Holland and Gateley, 1994).

As discussed below, patients with clinical mastalgia have been shown to have abnormal blood levels of PUFA (Gateley et al., 1992a). EPO is now considered first-line therapy for clinical mastalgia, and a prescription EPO product (Efamast®) is available for treatment. It was licensed in the UK in 1992 and has subsequently been approved in a number of other countries. EPO can be expected to provide clinically useful improvement in pain in 92% of cyclical and 64% of noncyclical mastalgia cases (Holland and Gateley, 1994). EPO doses of 3 g/d have been shown to be effective in a double-blind study (Mansel et al., 1990; Gateley et al., 1992b), and, typically, patients take three to four capsules of Efamast twice daily.

PMS and mastalgia are relieved by the natural or induced suppression of secretion of ovarian hormones, so it was long thought that these conditions were hormonally induced. However, no studies have reported consistently abnormal levels of ovarian or pituitary hormones, such as prolactin which stimulates breast tissue. An abnormal sensitivity of the breast and other tissues to normal levels of circulating hormones is a more plausible explanation for breast pain and PMS (Horrobin, 1993b).

PMS occurs at a greater frequency in women with atopic allergic disorders (Atton-Chalma et al., 1980; Chan and Hanifin, 1993; Horrobin, 1993b). This implies a link between the condition and PUFA metabolism. In addition, PMS and mastalgia are observed most frequently in populations that consume high-fat diets that contain large amounts of saturated fat (Gateley et al., 1992a; Be Lieu, 1994). The pathogenesis could be the result of this high-fat consumption or a high intake of saturated fat relative to PUFA. There are three proposed mechanisms that could explain how fatty acids may be involved (Horrobin, 1990a, 1993b):

(1) Steroid hormones, including estrogens, are esterified to fatty acids in their target tissues, and those esterified to saturated fatty acids are more potent than those esterified to PUFA. Therefore, if the ratio of saturated to polyunsaturated fatty acids is high, then the tissue response to normal circulating hormone levels will be excessive.

(2) The immediate lipid environment of the steroid receptors determines the affinity of these receptors for the steroid molecules. Saturated fatty acids are associated with a much higher receptor affinity than unsaturated fatty acids. Therefore, if steroid receptors are embedded in membranes that are relatively deficient in PUFA, then the receptors will bind the steroid

more strongly, and again, normal circulating hormone levels will produce excessive peripheral action.

(3) The peripheral action of prolactin is reduced by PGE1 derived from DGLA. Therefore, if levels of PGE1 are lower than normal, then prolactin may have exaggerated effects on breast tissue.

All of the above situations are expected to be amplified in atopic disorders since the levels of LC-PUFA are much reduced (Section 4.1.3 of this chapter). Women with PMS, cyclical breast pain and noncyclical breast pain have abnormal plasma and RBC fatty acids patterns (Brush, 1990; Gateley et al., 1992a). Overall, saturated and monounsaturated fat levels are higher than normal, while PUFA are reduced. The levels of LA can be normal, moderately elevated or slightly reduced, but the levels of LA derivatives, in particular AA, are all lower than normal. This pattern is indicative of either excessive consumption of LA or inadequate conversion of LA to GLA.

Five clinical studies have confirmed that EPO is an effective treatment for PMS and breast pain (Brush, 1990; O'Brien and Massil, 1990). For example, 46 women were given 4 g/d for their entire menstrual cycles in a three- or six-month double-blind crossover study. Patients receiving all six months of therapy improved progressively and found EPO superior to placebo. In all the studies involving PMS, there was a large placebo response that could last for three to four months, so the true effect of the treatment did not become apparent until after that time. In addition, there is a slow response to the treatment due to the time required to replace saturated fatty acids in the membranes with n-6 PUFA. However, the response to treatment is progressive and can reach its maximum in four to six months. All physical and psychological aspects of PMS symptoms may improve, but the most commonly reported improvement is a reduction in breast pain. EPO and B6 (or B vitamins in general) were the nutritional supplements most often reported to help PMS in Australian women surveyed in general practice and in previous surveys in other Western countries (Campbell et al., 1997).

5.4.2. Endometriosis

Endometriosis is a condition where endometrial tissue becomes displaced from inside the uterus and migrates to other sites in the abdominal cavity. The tissue then attaches to these new locations and undergoes the usual menstrual cycle-related changes. This results in considerable pain and inflammation at the tissue site and often necessitates surgical removal of the proliferating endometrial tissue. Up to 40% of endometriosis patients are infertile. The cyclical inflammation can be halted by hormonal therapy to reduce ovarian function or by surgical removal of the ovaries, but these are drastic treatments that prevent normal conception (Casper and MacLusky, 1990). In light of the fatty acid effects on tissue response to hormone levels described above, it is reasonable to predict that PUFA may be an effective treatment for endometriosis.

Rabbit studies of endometriosis provided initial evidence that alterations in dietary fatty acids could be of benefit in this disease condition (Covens et al., 1988). Open human studies confirmed that EPO in combination with fish oil was able to reduce the severity of symptoms in a substantial proportion of women. Subsequent placebo-controlled studies using concentrated GLA and EPA have demonstrated an improvement in 90% of the women on active treatment as compared to 10% improvement on placebo (Casper and MacLusky, 1990).

5.4.3. Gestational Hypertension

It has been recognized for many years that EFA play a major role in maintaining a healthy pregnancy. More recently, it has been recognized that there is a link between pregnancy complications and maternal and/or fetal PUFA status. For example, in Inuit communities where residents eat primarily a marine diet rich in n-3 LC-PUFA, the incidence of gestational- or pregnancy-induced hypertension (PIH) is less common (Bang and Dyerberg, 1980; Iacono and Dougherty, 1993).

PIH may occur without warning and for no apparent reason in previously normotensive women. In severe cases, it can reach morbid levels in a few days or weeks. It affects approximately one woman in ten and is the single most frequently cited cause of maternal death in the UK. PIH is associated with intense vasoconstriction, increased platelet aggregability and thrombosis. If proteinuria develops in combination with hypertension, the condition is known as preeclampsia.

Studies using fatty acid supplementation in nonpregnant hypertensive patients have been promising, and so this treatment has extended to PIH. The first reported study included 34 women within the first 37 weeks of gestation who were randomly assigned to receive a LA-rich diet (20 g/day) or a normal hospital diet. The diastolic blood pressure fell slightly following the first week of treatment, but there were no other significant differences between the two groups (Broughton-Pipkin et al., 1986).

Another study examined the effects of consuming EPO for one week within the second trimester of pregnancy (Broughton-Pipkin, 1990). Resting blood pressure, pressor response to angiotensin II, components of the renin-angiotensin system and serum electrolytes were measured before and after treatment. The diastolic pressor response to angiotensin II was significantly reduced in the treated patients as compared to those in a control untreated group. The reduction in pressor response was believed to result from increased PGE1 production. Previous reports of direct PGE1 injections in pregnant animals and pregnant humans have demonstrated diminished response to angiotensin II (Wallenburg and Bremer, 1992).

5.4.4. Preeclampsia

Preeclampsia is characterized by increased vasoconstriction, frequently associated with increased platelet aggregation, reduced uteroplacental blood flow and premature delivery. Similar to cardiovascular disease, an imbalance between thromboxane A2 and prostacyclin is considered to be involved in the development and clinical expression of preeclampsia (Moodley and Norman, 1989; Homstra, 1992). These prostaglandins are modulators of vascular smooth muscle tone and platelet aggregation.

AA levels are lower and LC-PUFA in general are higher in maternal phospholipids and cholesterol esters in normal pregnancies compared to preeclamptic pregnancies (Ogburn et al., 1984; Homstra, 1992; Velzing-Aarts et al., 1997). This suggests that a diet low in AA or one that reduces the endogenous formation of AA may be protective against preeclampsia. In addition, some studies have reported a reduced excretion of vasodilator prostaglandins including PGE1 in preeclampsia. PGE1 acts as a vasodilator on the uterine arteries as well as in systemic circulation, and it counteracts the pressor effects of norepinephrine and decreases platelet aggregation. Therefore, PGE1 might improve placental perfusion, reduce thrombotic complications and prevent or ameliorate many of the characteristic features and complications of preeclampsia.

Studies combining EPO and fish oil in preventing preeclampsia have produced promising results. One study included 150 women in their first four months of pregnancy. The subjects were randomly assigned either a mixture of EPO and fish oil, magnesium oxide or placebo. Results suggested that a mixture of EPO, fish oil and magnesium may prevent preeclampsia (D'Almeida et al., 1992).

5.5. NEUROLOGICAL DISORDERS

It is not surprising that PUFA play a role in neurological disorders. Lipids compose 60% of the solid matter in the brain; and 20% of the lipids are PUFA, almost entirely DHA and AA, with minor amounts of DGLA. LA and ALA are not considered to be essential components of brain tissue and are present at very low levels (Crawford, 1993; Phylactos et al., 1994; Clandinin, 1999; Crawford et al., 1999). Alterations in PUFA intake or metabolism may have profound effects on brain structure and function. Prostaglandin metabolites of PUFA have dramatic effects on behavior when injected into the cerebrospinal fluid. These prostaglandins are also capable of modulating nerve conduction, neurotransmitter release and postsynaptic transmitter action. Series 1 and 2 prostaglandins are found in brain tissue, and their levels change rapidly following changes in dietary PUFA intake (Horrobin, 1998a). All of these factors demonstrate a link between neurological function and fatty acid metabolism.

5.5.1. Schizophrenia

Schizophrenia has been the most extensively studied neurological disease in relation to fatty acid metabolism. The condition is associated with primary overactivity of dopaminergic systems in certain parts of the brain. PGE1, the metabolite of DGLA, has been shown to dampen the dopamine effect in many tissues. Therefore, a DGLA or PGE1 deficiency could lead to an apparent excess in dopamine and cause many symptoms of schizophrenia (Horrobin, 1998a, 1998b).

Plasma phospholipid fatty acids have been measured in schizophrenics from Ireland, England, Scotland, Japan and the U.S. Results have consistently showed lower than normal levels of LA. RBC fatty acids in these subjects contain lower than normal levels of AA and DHA (Peet et al., 1996). The phosphatidylethanolamine fraction of the left frontal cortex in schizophrenics also has an abnormal fatty acid pattern. The levels of LA, GLA, DGLA and AA are reduced and the level of DHA is elevated. This indicates either accelerated metabolism of 18-carbon to LC-PUFA, selective failure to incorporate 18 and 20 carbon fatty acids into the cell membranes or rapid removal of the 18 and 20 carbon fatty acids from the membranes. Research has confirmed rapid removal as opposed to accelerated metabolism to be the major cause of this abnormality (Bates et al., 1992).

Schizophrenia may manifest itself when at least two primary genetic abnormalities in fatty acid metabolism are simultaneously present. There is an increased rate of removal of PUFA, especially AA and DHA from phospholipid membranes, and a reduced rate of incorporation of these same PUFA in cell membranes. Either or both of these abnormalities may be produced by changes in the activity or regulation of enzymes with phospholipase 2, phospholipase C or acyltransferase activity (Horrobin, 1998a, 1998b). Schizophrenia itself is associated with defects in verbal and social skills, including dyslexia, schizotypy and attention deficits, and relatives of schizophrenics who are not schizophrenic have an increased incidence of these disorders (Fish, 1987; Richardson, 1994; Crow, 1996). Bipolar disorder, alcoholism and various personality disorders are also more common in relatives of schizophrenics (Varma et al., 1997). Horrobin (1998a, 1998b) further speculates that dyslexia and schizotypy arise when only the defect in incorporation is present, whereas bipolar disorder arises when only the defect in increased removal rate is present. It has been shown that therapeutic doses of lithium inhibit phospholipase A2, which supports the case for overactivity of this enzyme in bipolar disorder and sheds new light on the mechanisms for lithium's therapeutic action on bipolar disorder (Chang et al., 1996).

Molecular studies have confirmed that the gene coding for phospholipase A2 in schizophrenics has a promoter region unlike that found in people who are not schizophrenic. This abnormal coding is responsible for the altered configuration of the enzyme, for its modified activity and ultimately for the abnormal fatty acid profile in the cell membranes (Horrobin, 1998a, 1998b). The fatty acid composition of cell membranes is critical to their function, so one could well imagine that the function of the membranes in schizophrenics would be severely altered. The administration of particular fatty acids would not affect the expression of the aberrant gene but may compensate for the abnormal function of phospholipase A2.

The rationale for fatty acid supplementation for schizophrenia is supported by some information that was collected by the World Health Organization. They completed a survey of the incidence and outcome of schizophrenia in eight countries in Africa, Asia, Europe and the Americas. They found that the incidence of schizophrenia was similar in all locations, but the outcomes were very different. There was a positive correlation between poor outcome and the presence of saturated fat in the diet and a strong negative correlation between poor outcome and vegetable and fish in the diet (Christensen and Christensen, 1988). This clearly indicates that PUFA may attenuate the disease and that saturated fats may exacerbate the condition.

Clinical studies on schizophrenics using GLA and DGLA supplementation have provided conflicting results, probably because the EFA abnormalities are broad. Both AA and n-3 PUFA are needed as well because brain cell PUFA levels take longer to change than plasma or RBC levels (Vaddadi and Gilleard, 1990). Wolkin et al. (1986) found no benefit for 10 patients. A placebo-controlled crossover with two 16-week periods showed some benefit (Vaddadi et al., 1989). Patients were given 12 × 400 mg/d EPO or placebo. During active treatment, the patients showed significant improvement in their memory and schizophrenic symptoms but not in tardive dyskinesia. After 32 weeks, all patients were provided with supplements containing the EPO plus the cofactors zinc, pyridoxine, niacin and vitamin C for 16 weeks. The combined treatment increased the incorporation of both n-6 and n-3 PUFA in RBC. In addition, it produced significant clinical improvement in memory, schizophrenic symptoms and tardive dyskinesia (Vaddadi and Gilleard, 1990).

Tardive dyskinesia is a side effect of neuroleptic medications and consists of abnormal involuntary movements. It is frequently observed in schizophrenics and increases with age and duration of medication. Spontaneous dyskinesia is clinically indistinguishable apart from the fact that it occurs in persons never exposed to neuroleptics. Both tardive and spontaneous dyskinesia are associated with abnormal PUFA profiles, especially low AA and higher DGLA, saturates and monounsaturates (Nilsson et al., 1996; Vaddadi et al., 1996).

Higher DGLA levels may reflect decreased synthesis of PGE1. Vaddadi et al. (1989) found that prior to treatment, patients with severe dyskinesia had low RBC levels of n-6 and n-3 PUFA as compared to normals and psychiatric controls. Throughout the treatment period, n-6 PUFA levels progressively rose, approaching the levels observed in the control group but remaining substantially different than those of a normal group.

5.5.2. Alzheimer's Disease

Excessive oxidation of PUFA in neuronal cell membranes may play a role in the development of Alzheimer's and related dementias. Reduced levels of PUFA have been observed in blood samples from Alzheimer patients and those suffering from other forms of dementia (Corrigan et al., 1991a, 1991b). Corrigan et al. (1991a) assigned 30 Alzheimer's patients to an EPO, zinc, selenium and vitamin E supplement or olive oil placebo for 20 weeks. At the beginning and end of the study, they were given a battery of cognitive tests. Improvements in performance on eight of nine tests were observed in the active group, and three of eight were observed in the placebo group. The degree of improvement was consistently greater in the active group; however, the improvements in the placebo group were real and may have arisen from the extra attention and stimulation afforded to the subjects by the testing.

Studies have shown that higher blood levels of antioxidants from vitamins or plants are associated with better cognition in older persons (Morris et al., 1998; Schmidt et al., 1998). For example, a 4.3-year follow-up of 633 persons over age 65 found that 91 developed Alzheimer's disease. None of the 50 persons who supplemented vitamin C and/or E were among the 91 who developed the disease (Morris et al., 1998). Fish consumption was shown to be inversely related to the incidence of dementia and, in particular, Alzheimer's disease in a study of 5,386 persons over age 55 (Kalmijn, 1997). Therefore, prevention or treatment of Alzheimer's and related dementias may be enhanced by the simultaneous administration of antioxidants, fish oil and EPO. This would provide the PUFA lost by oxidation or other processes and provide protection for it as well. Since these disorders are strongly associated with aging, the argument for essentiality of GLA (and DHA; possibly also EPA) in older persons presented in Section 4.1.1 becomes all the more compelling.

5.6. RHEUMATOID ARTHRITIS AND SIMILAR CHRONIC INFLAMMATORY CONDITIONS

GLA has been found to be effective in a variety of *in vitro* and *in vivo* models of inflammation including adjuvant arthritis, experimental allergic encephalomyelitis, naturally occurring autoimmune inflammatory disease, salmonella-associated arthritis and urate-induced inflammation (Tate et al., 1988;

Tate and Zurier, 1994; Veale et al., 1994; Mancuso et al., 1997). Lung fibrosis in hamsters induced by an antineoplastic drug (Bleomycin) was also reduced by EPO supplementation (Ziboh et al., 1997).

Rheumatoid arthritis (RA) is particularly refractory to treatment, and the side effects from long-term use of non-steroidal anti-inflammatory drugs (NSAID) are potentially severe (Parke et al., 1996). EPO, borage oil, free GLA and fish oil have been shown to be effective treatments for RA in clinical trials, but the duration of the study should be at least six months, and the doses need to be consistently high—at least 5 g/d (Endres et al., 1995; Rothman et al., 1995). For example, 90 RA patients were enrolled in a double-blind randomized study and assigned to 6 g fish oil, 3 g fish oil plus 3 g olive oil or 6 g olive oil. Overall, the placebo group continued to deteriorate, the 3 g fish oil group improved nonsignificantly and the 6 g group improved significantly (Guesens et al., 1994). Nonetheless, any positive results afforded from PUFA supplementation are encouraging due to the safety and low cost of this treatment.

Chronic inflammation is associated with defective regulation of the arachidonic acid cascade (Figure 8.2). The production of proinflammatory series-2 prostaglandins and 4-series leukotrienes is increased. When the DGLA metabolites PGE1 and 15-OH-DGLA are present in adequate amounts, they can block the release of AA from phospholipids and the conversion of AA to proinflammatory metabolites (Belch, 1989; Horrobin, 1990a; Sardesai, 1992; Zurier, 1993, Ziboh, 1996).

Chronic inflammation is also associated with immunological abnormalities. The production of cytokine inflammatory mediators (e.g., interleukins, tumor necrosis factor) and reactive oxygen metabolites is altered. Many cytokines function in part by modulating the behavior of the AA cascade, and in turn, are modulated by it, so there may be feedback regulation of prostaglandin production/action. If this feedback is negative, it tends to keep an initial inflammatory event running, hence (albeit simplistic), acute becomes chronic. The prostaglandins also potentiate histamine- and bradykinin-induced pain and edema (Belch, 1989, 1990; Zurier, 1993; Caughey et al., 1996). It is important to realize that dietary AA is not really the culprit in chronic inflammation, but rather defective regulation of its metabolism is the culprit. Supplementation of 1.5 g/d AA in healthy humans on normal diets has been shown not to alter cytokine or lymphocyte production or have an adverse effect on immune response (Kelley et al., 1997, 1998).

PGE1 is required for a normal immune response at low concentrations but inhibits immune response at high concentrations. It has been found to be antinflammatory in a variety of *in vitro* systems and animal models. PGE1 itself is difficult to administer due to its short half-life, and it is difficult to synthesize stable analogues that mimic the desirable actions of PGE1. Therefore, the use of GLA as a precursor of both PGE1 and 15-OH-DGLA is an attractive and more realistic alternative for the treatment of chronic inflammatory conditions that often require years of daily medication to control. DGLA also has the advantage that it cannot be converted into inflammatory leukotrienes, so it should have some action similar to NSAID (Tate et al., 1988; Brzeski et al., 1991; Zurier, 1993). The pharmacological action of GLA rather than a correction of any deficiency is the mode of action in this treatment.

Four clinical trials have reported positive results with sources of GLA for RA. In a 12-week trial, patients abruptly stopped NSAIDs and switched to EPO. No patients improved, but only two of 17 deteriorated. A higher rate of deterioration would have been expected following sudden cessation of conventional drugs. It was concluded that EPO was at least as effective as NSAIDs in controlling symptoms (Hansen et al., 1983). Belch et al. (1988) completed a larger trial using patients stabilized on NSAID therapy. They were randomly assigned to EPO, a combination of EPO and fish oil or placebo. After three months of treatment, they were told to reduce or discontinue their use of NSAID. The double-blind phase continued for 12 months and was followed by a single-blind phase where all patients were assigned to placebo. In the two active groups, 90% of patients reported themselves as better as compared to 30% in the placebo group. Two-thirds of the patients in the two active groups could stop or substantially reduce NSAID as compared to one-third in the placebo group. In the single-blind phase, almost all those who had improved on active treatment deteriorated on placebo. There were no significant differences between EPO and an EPO/fish oil combination.

More recently, 37 RA patients stabilized on NSAID or corticosteroids were randomly assigned to 12 capsules of borage oil per day, providing 1.4 g GLA, or cottonseed oil placebo. After 24 weeks, patients on borage oil had no deterioration, and the number and score of tender and swollen joints were significantly reduced, whereas the placebo group showed no significant improvement and some patients deteriorated (Leventhal et al., 1993). Zurier et al. (1996) randomly administered 2.8 g/day GLA as the free fatty acid, or a sunflower oil placebo, to 56 RA patients stabilized on various medications for six months. For the following six months, all patients were given active treatment. During GLA treatment periods, there were statistically significant and clinically relevant reductions in the signs and symptoms of RA. In the group of patients taking GLA for 12 months, 16 of 21 showed meaningful improvement, and seven were able to reduce their dosage of NSAID or prednisone. Three months after the 12 months of active treatment ended, follow-ups showed that most patients experienced an exacerbation of disease activity.

The success of natural oils containing GLA for RA has led to clinical recommendations and self-use of EPO for other chronic inflammatory conditions such as osteoarthritis, tendinitis and bursitis. Sometimes there is little perceived/reported benefit because individuals are using only 500–2,000 mg/d, which is unlikely to be sufficient to produce measurable clinical improvement. Higher doses and/or increased duration of treatment should be tried before assuming that EPO is not effective, since results are not as rapid as they are for NSAID and corticosteroids.

5.7. VIRAL INFECTIONS AND IMMUNOLOGICAL DISORDERS

There is a close relationship between PUFA metabolism and immune response, including the ability to respond to viral infections (Horrobin, 1990a, 1990b; Rothman et al., 1995; Sanderson et al., 1995; Calder, 1998). Some mechanisms are as follows:

(1) PUFA are required to enable interferon to play its role in resisting viral infections. Interferon is produced by the immune system cells in response to viral infection. When interferon is synthesized in the body, it produces fever. This fever stimulates the formation of prostaglandins. Interferon must be accompanied by prostaglandins in order to exert its antiviral effects. Therefore, appropriate fatty acid precursors are necessary for the body to mount an effective antiviral response. One of the viral mechanisms to combat this host defense appears to be inhibition of D-6-D.

(2) Cell regulatory lymphokines and cytokines may also require PUFA and eicosanoids to exert their normal effects. For example, stimulation of the tumoricidal effects of macrophages by exposure to lymphokines is associated with changes in PUFA content of the macrophages. Further, exposure of macrophages to EFAs in the absence of lymphokines is as effective as lymphokine exposure in stimulating a tumoricidal response.

(3) DGLA and PGE1 appear to play a particularly important role in the regulation of T lymphocyte function. This cell-mediated immune response can be inhibited by antisera that specifically block the effects of PGE1.

(4) PUFA have direct virucidal actions against many viruses, notably those with lipid envelopes, such as HIV.

5.7.1. Viral Infections

Abnormal fatty acid levels have been well documented in patients with acute viral respiratory tract infections, Epstein-Barr infections, autoimmune deficiency syndrome (AIDS) and post-viral fatigue syndrome (PVFS). In each case, there is a pattern of elevated levels of saturated and monounsaturated fatty acids and reduced levels of LA and ALA derivatives. AIDS patients are the most severely affected with PUFA including LA, DGLA and AA drastically reduced to 50–70% of normal (Horrobin, 1990b; Palangyo et al., 1992).

PUFA deficiency, especially DGLA and AA, may weaken the body's ability to cope with viral infections. Low PUFA levels may also reduce the responsiveness of the host to therapeutic agents such as interferon. Therefore, fatty acid supplementation in patients with viral infections may be beneficial in treating symptoms associated with EFA deficiency and may enhance the efficacy of other therapeutic agents.

5.7.2. Post-Viral Fatigue Syndrome

Post-viral fatigue syndrome (PVFS) may be related to an abnormal production of cytokines induced by the viral infection that persists after the acute phase of the infection has resolved (Horrobin, 1990c, 1991). EPO has been used in combination with fish oil in a placebo-controlled trial of 70 PVFS patients (Behan et al., 1990). At baseline, the patients' RBC membranes had elevated levels of saturated and monounsaturated fatty acids and reduced levels of n-6 and n-3 PUFA. Patients were given eight capsules/d of Efamol Marine (each capsule has 35 mg GLA and 17 mg EPA) or olive oil placebo. Patients symptoms including muscle weakness, aches and pains, lack of concentration, exhaustion, memory loss, depression, dizziness and vertigo were assessed at baseline, five weeks and 15 weeks. At 15 weeks, 85% of patients on active treatment improved, compared to 17% on placebo. Patients on the active treatment showed a normalization of RBC fatty acid profiles in parallel to the improvements in symptom score (Behan and Behan, 1990).

5.7.3. Auto-Immune Deficiency Syndrome

PUFAs have direct virucidal action against viruses such as HIV that have lipid envelopes. In cell culture, GLA will selectively kill cells infected with HIV and will prevent the virus from entering uninfected cells (Horrobin, 1990a, 1990b). This observation led to the hypothesis that GLA may be an effective treatment for AIDS. A phase I study using GLA as a lithium salt was used to treat a small number of HIV-infected patients. Results confirmed that there were no major toxic effects of the drug to the patient, so further studies using this treatment are possible (Winther et al., 1997).

EPO in combination with fish oil has been used to treat some of the symptoms of AIDS. Eighteen HIV seropositive ambulatory patients were treated for 12 weeks, and the severity of their symptoms were evaluated initially, at week eight and week 12. Symptoms included fatigue, weight loss, diarrhea and skin rashes and itches. Six patients dropped out of the study throughout the course of the treatment. Following treatment for 12 weeks, there was a substantial reduction in the severity of fatigue and diarrhea. Skin rashes tended to improve, but no changes were observed in the severity of skin itching. There was an average weight gain of 2.31 kg. Due to the small size of the study group, some of these changes did not reach statistical significance. A clinical assessment of 11 patients at the end of eight weeks indicated that two were progressively deteriorating, three were stable and six had improved. There was a significant elevation of T4 lymphocyte levels following treatment (Palangyo et al., 1992).

5.7.4. Raynaud's Syndrome

Intravenous PGE1 has been used successfully to relieve digital vasospasm associated with systemic sclerosis. A placebo-controlled study has shown that EPO has a modest but significant effect in relieving the vasospasm associated with a reduction in levels of thromboxane B2, the metabolite of the vasospastic and prothrombotic thromboxane A2 (Belch et al., 1985).

5.7.5. Sjogren's Syndrome

Sjogren's syndrome is associated with the loss of secretion from exocrine glands throughout the body, particularly the salivary and lacrimal glands. Exocrine gland atrophy is a symptom of EFA deficiency, thus, it is possible that EFA abnormalities play a role in Sjogren's syndrome and other chronic inflammatory immunological connective tissue diseases. Chronic inflammatory immunological connective tissue diseases are characterized by an abnormal fatty acid composition of plasma and RBC phospholipids, with PUFA being replaced by saturated fatty acids (Manthorpe et al., 1984). Such replacement stiffens the cell membranes, which is thought to produce a reduction in fluidity, functionality and altered surface receptor expression and/or binding.

Sjogren's syndrome and other chronic inflammatory immunological connective tissue diseases are almost certainly multifactorial in origin. However, it has been observed that Sjogren's patients have plasma LA at or above normal, indicating that intake and absorption were not impaired, but levels of GLA, DGLA and its metabolites were below normal. This indicates that EPO supplementation may improve Sjogren's syndrome. Placebo-controlled crossover trials with EPO in Sjogren's patients have shown a modest improvement in lacrimal gland function and relief from lethargy that is a troublesome symptom from the patients' point of view (Oxholm et al., 1986; Manthorpe et al., 1990).

6. CONCLUSIONS AND THE ROLE OF EPO IN THE FUTURE

EPO is perhaps the "type example" of our working definition of a nutraceutical in that its effects are clearly nutritional and pharmacological, and its toxicity is virtually nonexistent. Moreover, as is the case with most nutraceuticals and functional foods, the future for EPO is at once both low-tech and high-tech. Much sophisticated research has been done or is still underway that basically serves to convince us that a PUFA known to be present in breast milk might be deficient in modern Western diets and might be of value as a supplement for a large segment of the population. That any fat could in fact be deficient in the Western diet still falls out of the realm of consideration for most nutrition scientists and laypersons alike. The very large number of studies with PUFA cited in this chapter demonstrate otherwise.

The low-tech future is that hunter-gatherer diets from the past provide a better guide for the fats we need for optimum physical and mental health today than the guidelines established by most government and private nutrition agencies. In describing and quantifying how far our modern diets have strayed, emphasizing the essentiality of a broad range of PUFA—especially GLA and LC-PUFA—needs to be at the forefront. It is not sufficient, and in fact may be gravely detrimental, to base diets for the entire human population on LA and ALA as the only EFA. Currently, the incidences of diabetes (Anderson, 1997; Broadhurst, 1997a), atopic allergic conditions (Strachan, 1995; Chandra, 1997) and hyperactivity disorders (Stevens et al., 1995, 1996), are experiencing stochastic increases that cannot be related to a simple increase in diagnoses. The population of older persons in Western countries is increasing as well. We have defined these conditions as ones where PUFA metabolism is apparently genetically abnormal or less than optimally efficient. Further, the modern diet has a surfeit of LA relative to ALA, which is apparently not of great help for these conditions. With LA currently 85% or more of total PUFA intake, Western countries are experiencing functionally low n-3/n-6 PUFA *and* GLA/ LA ratios. Given these stochastic increases, it is illogical not to recommend EPO and LC-PUFA functional foods for long-term, if not lifetime, use.

Further, many Western women are delaying childbirth and/or using lengthy fertility treatments, both of which increase the risk for pregnancy complications including preeclampsia, gestational diabetes and PIH. Delayed childbirth and/ or fewer children increase the risk and severity of PMS also. EPO as a functional food may be a low-tech solution to the ever-increasing high-tech nature of having children, but it is a *preventive* solution for higher-risk pregnancies that is safe for mother and fetus.

On the high-tech future horizon, studies *in vitro* and in animals have confirmed that some PUFA derivatives, particularly GLA, DGLA and EPA, can selectively kill cancer cells without harming normal cells (Das, 1990; van der Merwe et al., 1990; Horrobin, 1994; de Kock et al., 1996). This is very interesting, because most botanical products, be they food or medicinal plants, are only effective as antimutagenic agents. A high intake of broccoli and ginseng, for example, may help prevent cancer initiation but is not necessarily effective or even beneficial at the promotion and proliferation stages.

Cancer cells are known to contain lower than normal levels of 18-carbon PUFA and especially LC-PUFA. Many types of cancer cells have greatly reduced D-6-D activity, indicating inadequate formation of DGLA and LC-PUFA but increased destruction or failure to incorporate PUFA probably plays a role as well (Das, 1990; Horrobin, 1994). Destruction of tumor masses by chemotherapeutic agents, radiation and natural defense mechanisms (e.g., tumor necrosis factor) depends upon the production of reactive oxygen metabolites. While one generally associates reactive oxygen metabolites and lipid peroxidation with negative processes such as chronic inflammation and cardiovascular disease, in the case of cancer therapy, they are necessary to disrupt replication of the malignant cells. The lower levels of PUFA that characterize cancer cells may render them less susceptible to lipid peroxidation, hence, less susceptible to destruction. It has been observed for a long time that the degree of fatty acid saturation in a malignant cell line *in vitro* correlates positively with the degree of malignancy, but the rate of lipid peroxidation correlates negatively (Dunbar and Bailey, 1975; Bartoli and Galeotti, 1979). So, PUFA, intact and peroxidized, are abnormally low in aggressive cancer cells. In a mouse skin cancer model, EPO and fish oil were found to be effective only at the promotion stage. An increase in lipid peroxidation products coincided with the inhibition of papilloma formation seen in the promotion stage (Ramesh and Das, 1998).

Lower levels of PUFA may play a role in the development of chemotherapy resistance. As the levels of PUFA drop, the levels of ROM that chemotherapeutic agents can induce become lower. Higher doses and/or more aggressive medications serve mostly to damage normal cells, hence, side effects are intolerable with very little tumor regression achieved in exchange. Further, the oxygenated metabolites of PUFA produced by the cyclooxygenase and lipoxygenase pathways are involved in the normal control of cell division. All of these observations suggest that supplying cancer cells with 6-desaturated PUFA could slow or arrest cancer cell division.

GLA in the form of EPO inhibits the growth of transplantable as well as induced mammary tumors in rats. When the GLA was supplied from borage and black currant oils, it was significantly less effective, indicating that GLA from EPO was more biologically active in this respect as well. Similar results were also observed in nude mice consuming GLA as EPO at 5% of total food intake (Aspinall et al., 1988; Rose et al., 1995; de Bravo et al., 1994). The doses of EPO required to produce beneficial effects in animals translate into relatively large but not impossible doses in humans (30–60 g/d). An open study using these high EPO levels showed they are well tolerated by patients and are associated with some regressions of cancer. Pilot studies with various doses and routes of GLA administration have provided positive results for liver, brain, breast, bladder and colon cancers and Hodgkin's disease (Singh et al., 1987; Das, 1990; van der Merwe et al., 1990; Horrobin, 1994). However, success depends partly on the ability to achieve high doses of GLA for extended periods directly at the cancer site. Therefore, intravenous infusions of purified GLA drug preparations would be alternatives superior to dietary EPO supplementation.

Some success prolonging survival in terminal pancreatic cancer was achieved with i.v. administration of a lithium GLA salt for 10 d, followed by oral therapy of 16 capsules per day. (Fearon et al., 1996). Das et al. (1995) regressed human gliomas in 15 patients with intracerebral intratumoral injections of GLA at 1 mg/d for 10 days. The treatment was safe, nontoxic to normal cells and increased survival by 1.5–2 yr. This is but a brief introduction to the potential cancer therapeutic effects of EPO and other sources of GLA, alone or in combination with n-3 LC-PUFA, and more human studies are underway.

7. REFERENCES

Ackman, R. G. 1988. "Some possible effects on lipid biochemistry of differences in the distribution on glycerol of long chain n-3 fatty acids in the fats of marine fish and marine mammals," *Atherosclerosis*, 70·171–173

Aman, M., E Mitchell and S. Turbott 1987 "The effects of essential fatty acid supplementation by Efamol in hyperactive children," *J Abnorm Child Psychol.*, 15 75–90.

Amano, N., Y Shinmen, K. Akimoto, H Kawashima and T. Amachi. 1992 "Chemotaxonomic significance of fatty acid composition in the genus Mortierella (Zygomycetes, Mortierellaceae)," *Mycotaxon*, 44 257–265

Anderson, R A 1997. "Nutritional factors influencing the glucose/insulin system Chromium," *J Am Coll Nutr*, 16·404–410

Andreassi, M , P Forleo, A. DiLorio, S. Masci, G. Abate and P. Amerio 1997. "Efficacy of gamma-linolenic acid in the treatment of patients with atopic dermatitis," *J Int. Med. Res*, 2.266–274.

Aspinall, S , P Bos and J J. Alexander 1988 "Oral administration of Efamol G to nude mice bearing human hepatocellular carcinoma xenografts," *South Afr. J. Sci*, 84·852–854

Atton-Chamla, A , G. Faure and J-R Goudard 1980. "Premenstrual syndrome and atopy," *Pharmatherapeutica*, 2·481–486

Bairati I , L Roy and F. Meyer. 1992 "Double-blind randomized, controlled trials of fish oil supplements in prevention of recurrence of stenosis after coronary angioplasty," *Circulation*, 85 950–956

Bamberg, T , X. Pelletier, H. Blain, C. Jeandel and G. Debry 1997 "Effects of malnutrition and atherosclerosis on the fatty acid composition of plasma phospholipids in the elderly," *Ann Nutr. Metab.*, 41 166–172.

Bang, H O. and J. Dyerberg. 1980 "Lipid metabolism and ischemic heart disease in Greenland Eskimos," *Adv. Nutr Res*, 3.1–22.

Barre, D E , B J Holub and R. S. Chapkin. 1993 "The effect of borage oil supplementation on human platelet aggregation, thromboxane B2 prostaglandin E1 and E2 formation," *Nutr Res.*, 13(7):739–751.

Bartoli G. M. and T Galeotti. 1979. "Growth related lipid peroxidation in tumours, microsomal membranes and mitochondria," *Biochim. Biophys. Acta*, 574·41.

Bates, C , D. Horrobin and K. Ells. 1992. "Fatty acids in plasma phospholipids and cholesterol esters from identical twins concordant and discordant for schizophrenia," *Schizophrenia Res.*, 6 1–7.

Behan, P. and W Behan. 1990. "Essential fatty acids in the treatment of postviral fatigue syndrome," In Horrobin, D. (ed). Omega-6 essential fatty acids pathophysiology and roles in clinical medicine. New York, NY.Alan R Liss, Inc., pp 275–282

Behan, P., W. Behan and D. Horrobin. 1990. "Effect of high doses of essential fatty acids on the postviral fatigue syndrome," *Acta. Neurol. Scand*, 82 209–216.

Belch, J. 1990 "Essential fatty acids and rheumatoid arthritis," In: Horrobin, D. (ed). Omega-6 essential fatty acids: pathophysiology and roles in clinical medicine. New York, NY:Alan R Liss, Inc., pp 223–237.

Belch, J. 1989. "Eicosanoids and rheumatology. inflammatory and vascular aspects," *Prostaglandins Leukotrienes Essent. Fatty Acids*, 36(4)·219–234.

Belch, J., D. Ansell, R. Madhok, A. O'Dowd and R. Sturrock. 1988. "Effects of altering dietary essential fatty acids on requirements for non-steroidal anti-inflammatory drugs in patients with rheumatoid arthritis," *Ann. Rheum. Dis.*, 47(2):96–104.

Belch, J., B. Shaw, A. O'Dowd, A. Saniabadi, P Leiberman, R Sturrock and C. Forbes. 1985. "Evening primrose oil (Efamol) in the treatment of Raynaud's phenomenon: a double blind study," *Thromb. Haemostasis*, 54(2).490–494.

Be Lieu, R. M. 1994. "Mastodynia," *Obstet. Gynecol. Clin. North Am.*, 21(3)·461–477.

Blackburn, M. 1990. "Use of Efamol (oil of evening primrose) for depression and hyperactivity in children," In: Horrobin D. F. (ed). Omega-6 essential fatty acids: pathophysiology and roles in clinical medicine. New York, NY:Alan R. Liss, Inc., pp. 345–349.

Blumenschine, R. J. 1991. "Hominid carnivory and foraging strategies, and the socio-economic function of early archaeological sites," *Philos. Trans. R Soc. Lond. B., Biol. Sci.*, 334:211–221.

Bolton-Smith, C., M. Woodward and R. Tavendale. 1997. "Evidence for age-related differences in the fatty acid composition of human adipose tissue, independent of diet," *Eur. J. Clin. Nutr.*, 51:619–624.

Brandle, J., W. Court and R. Roy. 1993. "Heritability of seed yield, oil concentration and oil quality among wild biotypes of Ontario evening primrose," *Can. J. Plant Sci.*, 73(4):1067–1070.

Brehler, R., A. Hildebrand and T. A. Luger. 1997. "Recent developments in the treatment of atopic eczema," *J. Am. Acad. Dermatol.*, 36(6).983–994

Brenner, R. 1981. "Nutritional and hormonal factors influencing desaturation of essential fatty acids," *Prog. Lipid. Res.*, 20·41–47.

Brister S. J. and M. R. Buchanan. 1998. "Effects of linolenic acid and/or marine fish oil supplements on vessel wall thromboresistance in patients undergoing cardiac surgery," *Adv. Exp. Med. Biol.*, 433:275–278.

Broadhurst, C. L., S. C. Cunnane and M C. Crawford. 1998 "Rift Valley lake fish and shellfish provided brain-specific nutrition for early homo," *Br. J. Nutr.*, 79:3–21.

Broadhurst, C. L. 1997a. "Nutrition and non-insulin dependent diabetes from an anthropological perspective," *Alt. Med. Rev.*, 2:378–399.

Broadhurst, C. L. 1997b. "Balanced intakes of natural triglycerides for optimum nutrition: An evolutionary and phytochemical perspective," *Med. Hypotheses*, 49:47–61.

Broughton-Pipkin, F. 1990. "Essential fatty acids and pregnancy hypertension," In: Horrobin, D. (ed) Omega-6 essential fatty acids: pathophysiology and roles in clinical medicine. New York, NY:Alan R. Liss, Inc., pp. 173–186.

Broughton-Pipkin, F., R. Morrison and P. O'Brien. 1986. "The effect of dietary supplementation with linoleic and gamma-linolenic acids on the pressor and biochemical response to exogenous angiotensin II in human pregnancy," *Prog. Lipid Res.*, 25:425–429.

Brouwer, D. A. J., Y. Hettema, J. J. van Doormaal and F. A. J. Muskiet. 1998. "γ-linolenic acid does not augment long-chain polyunsaturated fatty acid w-3 status," *Prostaglandins Leukotrienes Essent Fatty Acids*, 59:329–334.

Brush, M 1990. "Biochemistry of premenstrual syndrome," In: Horrobin, D. (ed). Omega-6 essential fatty acids· pathophysiology and roles in clinical medicine New York, NY. Alan R. Liss, Inc., pp. 513–522.

Brzeski, M., R. Madhok and H. Capell 1991. "Evening primrose oil in patients with rheumatoid arthritis and side-effects of non-steroidal anti-inflammatory drugs," *Br. J. Rheumatol*, 30(5):370–372.

Bunn, H. T and J. Ezzo. 1993. "Hunting and scavenging by Plio-Pleistocene hominids. nutritional constraints, archeological patterns, and behavioral implications," *J Archeol Sci.*, 20:365–398

Burdge, G. C., S. M. Wright, J. O. Warner and A. D. Postle 1997. "Fetal brain and liver phospholipid fatty acid composition in a guinea pig model of fetal alcohol syndrome: Effect of maternal supplementation with tuna oil," *J. Nutr. Biochem.*, 8(8).438–444.

Burton, J 1990. "Essential fatty acids in atopic eczema: clinical studies," In: Horrobin, D. (ed). Omega-6 essential fatty acids. pathophysiology and roles in clinical medicine. New York, NY.Alan R. Liss, Inc., pp. 67–73.

Calder, P 1998. "Dietary fatty acids and the immune system," *Nutr Rev*, 56·S70–S83.

Cameron, N. E, M. A. Cotter, D H Horrobin and H. J. Tritschler. 1998 "Effects of alpha-lipoic acid on neurovascular function in diabetic rats: interaction with essential fatty acids," *Diabetologia*, 41:390–399.

Cameron, N E. and M. A. Cotter 1994. "Effects of evening primrose oil treatment on sciatic nerve blood flow and endoneurial oxygen tension in streptozotocin-diabetic rats," *Acta Diabetol.*, 31(4).220–225.

Campbell, E M., D. Peterkin, K. O'Grady and R. Sanson-Fischer 1997. "Premenstrual symptoms in general practice," *J. Reprod. Med*, 42.637–646

Casper, R F. and N. J MacLusky. 1990. "Dietary supplementation with polyunsaturated fatty acids in the treatment of endometriosis," In: Horrobin D.F. (ed) Omega-6 essential fatty acids: pathophysiology and roles in clinical medicine New York, NY.Alan R Liss, Inc., pp. 547–555.

Caughey, G. A., E. Mantzioris, R. A. Gibson, L G. Cleland and M J James. 1996. "The effect on human tumor necrosis factor a and interleukin 1b production of diets enriched in n-3 fatty acids from vegetable or fish oil," *Am. J. Clin. Nutr.*, 63:116–122

Chan, S. and J. Hanifin. 1993. "Immunopharmacologic aspects of atopic dermatitis," *Clin. Rev. Allergy*, 11(4) 523–541

Chandra, R. K 1997 "Food hypersensitivity and allergic disease· A selective review," *Am J. Clin. Nutr.*, 66:526S–529S.

Chang, M C. J, E. Grange, O. Rabin, J. M. Bell, D. D. Allen and S. I Rappaport. 1996. "Lithium decreases turnover of arachidonate in several brain phospholipids," *Neurosci. Lett.*, 220·171–174.

Charnock, J. S., G. L. Crozier and J Woodhouse. 1994 "Gamma-linolenic acid, black currant seed and evening primrose in the prevention of cardiac arrhythmia in aged rats," *Nutr. Res.*, 14:1089–1099.

Chen, Z. -Y, C R. Menard and S. C Cunnane. 1995. "Moderate, selective depletion of linoleate and α-linoleate in weight cycled rats," *Am. J. Physiol.*, 37·R498–R505.

Christensen, M. S., B. C. Mortimer, C. E. Hoy and T. G. Redgrave. 1995. "Clearance of chylomicrons following fish and seal oil feeding," *Nutr. Res.*, 15.359–368.

Christensen, O and E. Christensen. 1988. "Fat consumption and schizophrenia," *Acta Psychiatr. Scand.*, 78:587–591.

Cisowski, W., M. Zielinska-Stasiek, M. Luckiewicz and A. Stoyhwo. 1993. "Fatty acids and triacylgycerols of developing evening primrose (*Oenothera biennis*) seeds," *Fitoterapia*, 64 155–162

Clandinin, M T 1999. "Brain development and assessing the supply of polyunsaturated fatty acid," *Lipids,* 34 131–137

Colquhoun, I. and S. Bunday 1981 "A lack of essential fatty acids as a possible cause of hyperactivity in children," *Med. Hypotheses,* 7(5) 673–679.

Connor, S L and W. E. Connor 1997 "Are fish oils beneficial in the prevention and treatment of coronary artery disease?" *Am J. Clin. Nutr,* 66 (suppl.). 1020S–1031S.

Corrigan, F. A Van Rhijn, F. MacIntyre, E. Skinner and D Horrobin. 1991a "Dietary supplementation with zinc sulphate, sodium selenite and fatty acids in early dementia of Alzheimer's type. II: Effects on lipids," *J. Nutr. Med.,* 2(3):265–272

Corrigan, F A. Van Rhijn and D Horrobin, D 1991b "Essential fatty acids in Alzheimer's disease," *Ann NY Acad Sci.,* 540·250–252

Cotterell, J. A. Lee and J Hunter. 1990 "Double-blind cross-over trial of evening primrose oil in women with menstrually-related irritable bowel syndrome," In. Horrobin, D (ed) Omega-6 essential fatty acids: pathophysiology and roles in clinical medicine New York, NY Alan R Liss, Inc., pp. 421–426.

Covens, A. L , P. Christopher and R F. Casper. 1988. "The effect of dietary supplementation with fish oil fatty acids on surgically induced endometriosis in the rabbit," *Fertil Steril.,* 49 698.

Crawford, M A , M Bloom, C L Broadhurst, W. F. Schmidt, S C Cunnane, C Galli, K Gehbremeskel, F Linseisen, J Lloyd-Smith and J. Parkington 1999 "Evidence for the unique function of docosahexaenoic acid during the evolution of the modern hominid brain," *Lipids,* 34·S39–S47.

Crawford, M A. 1993. "The role of essential fatty acids in neural development. implications for perinatal nutrition," *Am J. Clin Nutr.,* 57.703S–710S.

Crawford, M. A , N M Casperd and A J Sinclair. 1976. "The long chain metabolites of linoleic and linolenic acids in liver and brains of herbivores and carnivores," *Comp. Biochem Physiol. B. Comp. Biochem,* 54 395–401.

Crawford, M. A., M M. Gale, M H. Woodford and N. M. Casperd 1970 "Comparative studies on fatty acid composition of wild and domestic meats," *Int. J. Biochem ,* 1 295–305.

Crow, T J 1996. "Sexual selection as the mechanism of evolution of Machiavellain intelligence: a Darwinian theory of the origins of psychosis," *J. Psychopharmacol.,* 10 77–87

Cunnane, S C. and M. J Anderson 1997. "The majority of dietary linoleate in growing rats is β-oxidized or stored in visceral fat," *J. Nutr ,* 127:146–152.

Cunnane, S. 1988. "Evidence that adverse effects of zinc deficiency on essential fatty acid composition in rats are independent of food intake," *Br. J. Nutr.,* 59(2) 273–278

Cunnane, S., Y. S. Huang, D. Horrobin and J. Davignon. 1982. "Role of zinc in linoleic acid desaturation and prostaglandin synthesis," *Prog. Lipid Res.,* 20:157–160.

D'Almeida, A , J Carter, A. Anatol and C. Prost. 1992. "Effects of a combination of evening primrose oil (gamma linolenic acid) and fish oil (eicosapentaenoic + docosahexaenoic acid) versus magnesium, and versus placebo in preventing pre-eclampsia," *Women Health,* 19:117–131.

Das, U. N. 1998. "Hypothesis. cis-unsaturated fatty acids as potential anti-peptic ulcer drugs," *Prostaglandins Leukotrienes Essent Fatty Acids,* 58:377–380.

Das, U. N., V. V. S V. Prasad and D R. Reddy 1995. "Local application of gamma-linolenic acid in the treatment of human gliomas," *Cancer Lett.,* 94.147–155.

Das, U. N., K Vijay, K. Kumar, E Ramanjaneyulu, N. Joshi and V K. Dixit. 1994. "Essential fatty acids and their metabolites in duodenal ulcer," *Med. Sci. Res ,* 33·423–425

Das, U 1990 "Gamma-linolenic acid, arachidonic acid, and eicosapentaenoic acid as potential anticancer drugs," *Nutrition*, 6 429–435

De, B K, S Chaudhury and D. K Bhattacharyya. 1999 "Effect of nitrogen sources on γ-linolenic acid accumulation in *Spirulina platensis*," *J Am. Oil Chem Soc*, 76 153–156

de Bravo, M, G Schinella, H Tournier and C Quintans. 1994. "Effects of dietary gamma and alpha linolenic acid on human lung carcinoma grown in nude mice," *Med. Sci. Res*, 22 667–668

de Kock, M, M Lottering, C. Grobler, T Viljoen, T le Roux, and J. Seegers 1996 "The induction of apoptosis in human cervical carcinoma (HeLa) cells by gamma-linolenic acid," *Prostaglandins Leukotrienes Essent. Fatty Acids*, 55 403–411.

Deferne, J L. and A Leeds 1992 "The antihypertensive effect of dietary supplementation with a 6-desaturated essential fatty acid concentrate as compared with sunflower seed oil," *J Hum Hypertens*, 6 113–119

Dillon, J. 1987. "Essential fatty acid metabolism in the elderly. effects of dietary manipulation," In Horisberger, M and U. Bracco (eds) Lipids in modern nutrition. New York Raven Press 13 93–106

Dines, K, M Cotter and N Cameron 1996 "Effectiveness of natural oils as sources of gamma-linolenic acid to correct peripheral nerve conduction velocity abnormalities in diabetic rats. modulation by thromboxane A2 inhibition," *Prostaglandins Leukotrienes Essent Fatty Acids*, 55 159–165

Duffy, O, J F Menez and B. Leonard 1992. "Attenuation of the effects of chronic ethanol administration in the brain lipid content of the developing rat by an oil enriched in gamma linolenic acid," *Drug Alcohol. Depend*, 31 85–89.

Dunbar, L M and J M Bailey 1975 "Enzyme deletions and essential fatty acid metabolism in cultured cells," *J Biol. Chem*, 250 1152–1154.

Dutta-Roy, A K 1994 "Insulin mediated processes in platelets, erythrocytes, and monocytes/macrophages: effects of essential fatty acid metabolism," *Prostaglandins Leukotrienes Essent Fatty Acids*, 51 385–399

Elder, K and M Kirchgessner. 1995 "Activities of liver microsomal fatty acid desaturases in zinc-deficient rats force fed diets with a coconut oil/safflower oil mixture or linseed oil," *Biol Trace Elem Res*, 48.215–229

Endres, S, R. De Caterina, E. B Schmidt and S. D. Kristensen. 1995. "n-3 Polyunsaturated fatty acids. update 1995," *Eur. J Clin Invest.*, 25.629–638

Engler, M M. 1993. "Comparative study of diets enriched with evening primrose, black currant, borage or fungal oils on blood pressure and pressor responses in spontaneously hypertensive rats," *Prostaglandins Leukotrienes Essent Fatty Acids*, 49 809–814

Fearon, K., J Falconer, J Ross, D. Carter, J. Hunter, P Reynolds and Q Tuffnell 1996. "An open-label phase I/II dose escalation study of the treatment of pancreatic cancer using lithium gammalinolenate," *Anticancer. Res*, 16 867–874.

Fiocchi, A, M Sala, P Signoroni, G. Banderali, C Agostoni and E Riva. 1994 "The efficacy and safety of gamma-linolenic acid in the treatment of infantile atopic dermatitis," *J. Int Med Res*, 22 24–32

Fish, B 1987 "Infant predictors of the longitudinal course of schizophrenic development," *Schizophr Bull.*, 13 395–409

Fragoso, Y. and E. Skinner. 1992 "The effect of gammalinolenic acid on the subfractions of plasma high density lipoprotein of the rabbit," *Biochem. Pharmacol.*, 44:1085–1090

Freese, R. and M Mutanen 1997 "α-Linoleic acid and marine long-chain n-3 fatty acids differ only slightly in their effects on hemostatic factors in healthy subjects," *Am. J. Clin Nutr.*, 66 591–598

Fukushima, M , T. Matsuda, K Yamagishi and M. Nakano 1997. "Comparative hypocholestero-lemic effects of six dietary oils in cholesterol-fed rats after long-term feeding," *Lipids,* 32.1069–1074.

Galle, A., M. Joseph, C. Demandre, P Guerche, J. Dubacq, A. Oursel, P. Mazliak, G. Pelletier and J Kader. 1993 "Biosynthesis of gamma-linolenic acid in developing seeds of borage (*Borago officinalis* L.)," *Biochim. Biophys Acta,* 1158.52–58.

Gapinsky, J. P., J V Van Ruiswyk, G. R. Heudebert and G. S. Schectman. 1993. "Preventing restenosis with fish oils following coronary angiopathy, A meta-analysis." *Arch. Int. Med.,* 153, 1595–1601.

Gateley, C., P. Maddox, G. Pritchard, W. Sheridan, B. Harrison, J. Pye, D. Webster, L. Hughes and R. Mansel. 1992a. "Plasma fatty acid profiles in benign breast disorders," *Br. J. Surg.,* 79.407–409.

Gateley, C., M. Miers, R. Mansel and L Hughes 1992b. "Drug treatments for mastalgia: 17 years experience in the Cardiff mastalgia clinic," *J. R. Soc. Med.,* 85:12–15.

Gerster, H. 1998. "Can adults adequately convert α-linolenic acid (18:n-3) to eicosapentaneoic acid (20:5n-3) and docosahexanoic acid (22:6n-3)?" *Internat. J. Vit. Nutr. Res.,* 68.159–173

Glen, A., E. Glen, L. MacDonell and F Skinner. 1990. "Essential fatty acids and alcoholism," In: Horrobin, D. (ed). Omega-6 essential fatty acids: pathophysiology and roles in clinical medicine. New York, NY: Alan R. Liss, Inc., pp. 321–332.

Grant, H. W. K. R. Palmer, R. W. Kellt, N. H. Wilson and J. J. Nisiewicz 1988 "Dietary linoleic acid, gastric acid, and prostaglandin secretion," *Gastroenterology,* 94:955–959

Grattan-Roughan, P. 1989. "Spirulina: a source of gamma-linolenic acid?," *J. Sci Food Agric ,* 47·85–93.

Grayson, D. K. 1993. "The desert's past: a natural prehistory of the Great Basin," Washington, DC: Smithsonian Inst. Press, 356 p

Greenfield, S., A. Green, J. Teare, A. Jenkins, N. Punchard, C. Ainley and R A Thompson. 1993 "A randomized controlled study of evening primrose oil and fish oil in ulcerative colitis," *Aliment Pharmacol. Ther.,* 7.159–166

Guesens, P , C. Wouters, J. Nijs, Y. Jiang and J. Dequeker. 1994. "Long-term effect of omega-3 fatty acid supplementation in active rheumatoid arthritis," *Arthritis Rheum ,* 6:824–829

Guivernau, M., N. Meza, P. Barja and O. Roman. 1994. "Clinical and experimental study on the long term effect of dietary gamma-linolenic acid on plasma lipids, platelet aggregation, thromboxane formation, and prostacylin production," *Prostaglandins Leukotrienes Essent Fatty Acids,* 51.311–316.

Hagve, T. -A. 1988. "Effects of unsaturated fatty acids on cell membrane functions," *Scand. J. Clin Lab. Invest ,* 48:381–388.

Hansen, T. M., A. Lerche, V Kassis, I Lorenzen and J Sondergaard. 1983. "Treatment of rheumatoid arthritis with prostaglandin E1 precursors cis-linoleic acid and gamma-linolenic acid," *Scand. J Rheumatol ,* 12·85–88

Hansen, A., R. Stewart, G Hughes and L. Soderhjelm. 1962. "The relation of linoleic acid to infant feeding," *Acta Paediatr ,* 51(suppl. 137) 1–41.

Harris, W. S. 1997. "n-3 fatty acids and serum lipoproteins: human studies," *Am. J. Clin. Nutr.,* 65 (suppl.), 1645S–1654S.

Holland, P A. and C A Gateley. 1994. "Drug therapy of mastalgia: What are the options?" *Drugs,* 48·709–716.

Hollander, D. and A Tarnawski 1986. "Dietary essential fatty acids and the decline in peptic ulcer disease a hypothesis," *Gut,* 27.239–242

Hornstra, G 1992. "Essential fatty acids, pregnancy and pregnancy complications. a roundtable discussion," In· Sinclair, A. and R. Gibson (eds). Essential fatty acids and eicosanoids: invited papers from the third international congress. Champaign, Ill. AOCS Press, pp. 177–182.

Horrobin, D. F 1998a "The membrane phospholipid hypothesis as a biochemical basis for the neurodevelopmental concept of schizophrenia," *Schizophr. Res ,* 30:193–208.

Horrobin, D. F. 1998b. "Schizophrenia. the illness that made us human," *Med. Hypotheses,* 50:269–288.

Horrobin, D. F. 1997. "Essential fatty acids in the management of impaired nerve function in diabetes," *Diabetes,* 46(suppl 2): S90–S93

Horrobin, D. F 1994 "Unsaturated lipids and cancer," In: New approaches to cancer treatment" Horrobin, D F. (ed). Edinburgh: Churchill Communications, pp 3–29

Horrobin, D F. 1993a. "Omega-6 and omega-3 essential fatty acids in atherosclerosis," *Semin Thromb. Hemostas.,* 19·129–137.

Horrobin, D. F. 1993b "The effects of gamma-linolenic acid on breast pain and diabetic neuropathy: possible non-eicosanoid mechanisms," *Prostaglandins Leukotrienes Essent. Fatty Acids,* 48:101–104.

Horrobin, D F. 1991. "Essential fatty acids and the post-viral fatigue syndrome," In: Jenkins, R. and J Mowbray (eds). Post-viral fatigue syndrome. New York, NY: John Wiley & Sons., pp 393–404.

Horrobin, D. F. 1990a. "Gamma linolenic acid," *Rev. Contemp Pharm ,* 1: 1–41

Horrobin, D. F. 1990b. "Essential fatty acids, immunity and viral infections," *J. Nutr. Med.,* 1.145–151.

Horrobin, D F. 1990c "Post-viral fatigue syndrome, viral infections in atopic eczema, and essential fatty acids," *Med Hypotheses,* 32·211–217.

Horrobin, D F. 1987a. "Essential fatty acids, prostaglandins, and alcoholism. an overview," *Alcohol Clin Exp. Res.,* 11:2–9.

Horrobin, D. F 1987b. "Low prevalence of coronary heart disease (CHD), psoriasis, asthma, and rheumatoid arthritis in Eskimos are they caused by high dietary intake of eicosapentaenoic acid (EPA), a genetic variation of essential fatty acid (EFA) metabolism or a combination of both?" *Med. Hypotheses,* 22:421–428.

Horrobin, D F 1982. "Prostaglandins, essential fatty acids and psychiatric disorders: a background review," In Horrobin, D (ed) Clinical uses of essential fatty acids, Montreal, PQ. Eden Press Inc , pp. 167–174.

Horrobin, D. F. and M. S Manku. 1990 "Clinical biochemistry of essential fatty acids," In· Horrobin D. F. (ed) Omega-6 essential fatty acids: pathophysiology and roles in clinical medicine. New York· Wiley-Liss., pp. 21–53

Horrobin, D F. and Y. S. Huang 1987. "The role of linoleic acid and its metabolites in the lowering of plasma cholesterol and the prevention of cardiovascular disease," *Int. J Cardiol.,* 17.241.

Hrelia, S., A. Bordoni, P. Motta, M. Celadon and P. Biagi. 1991. "Kinetic analysis of delta-6-desaturation in liver microsomes· influence of gamma-linolenic acid dietary supplementation to young and old rats," *Prostaglandins Leukotrienes Essent. Fatty Acids,* 44:191–194.

Huang, Y. -S. and B. Nassar. 1990. "Modulation of tissue fatty acid composition, prostaglandin production and cholesterol levels by dietary manipulation of n-3 and n-6 essential fatty acid metabolites," In: Horrobin, D. (ed) Omega-6 essential fatty acids. pathophysiology and roles in clinical medicine. New York, NY: Alan R. Liss, Inc., pp 127–144

Huang, Y. -S., R Drummond and D. Horrobin. 1987. "Protective effect of gamma-linolenic acid on aspirin-induced gastric hemorrhage in rats," *Digestion,* 36:36–41

Iacono, J and R. Dougherty 1993 "Effects of polyunsaturated fats on blood pressure," *Ann Rev Nutr.*, 13 243–260

Ikeda, I , H Yoshida, M. Tomooka, A. Yosef, K. Imaizumi, H Tsuji and A Seto 1998 "Effects of long-term feeding of marine oils with different positional distribution of eicosapentaenoic and docosahexaenoic acids on lipid metabolism, eicosanoid production, and platelet aggregation in hypercholesterolemic rats," *Lipids*, 33 897–904.

Ishikawa, T., Y Fujiyama, O Igarashi, M. Morine, N. Tada, A Kagami, T Sakamoto, M Nagano and H. Nakamura 1989. "Effects of gammalinolenic acid on plasma lipoproteins and apolipoprotein " *Atherosclerosis*, 75 95–104

James, P., K Norum and I Rosenberg. 1992. "The nutritional role of fat," (Meeting summary) *Nutr Rev.*, 50 68–70.

Jenkins, D. K , J C. Mitchell, M S Manku and D F Horrobin 1988 "Effects of different sources of gamma-linolenic acid on the formation of essential fatty acid and prostanoid metabolites," *Med. Sci Res.*, 16.525–526

Johnson, M M , D D. Swan, M. E. Surette, J Stegner, T Chilton, A N. Fonteh and F. H Chilton 1997. "Dietary supplementation with gamma-linolenic acid alters fatty acid content and eicosanoid production in healthy humans," *J Nutr.*, 127·1435–1444

Julu, P. 1992. "Responses of peripheral nerve conduction velocities to treatment with essential fatty acids in diabetic rats. possible mechanisms of action," In Horrobin, D. (ed) Treatment of diabetic neuropathy a new approach. London, UK Churchill Livingstone, pp 48–61.

Kalmijn, S. 1997 "Dietary fat intake and the risk of incident dementia in the Rotterdam Study," *Ann Neurol.*, 42.776–782

Keen, H and M D Mattock. 1990 "Complications of diabetes mellitus role of essential fatty acids," In Horrobin, D F., (ed). Omega-6 essential fatty acids pathophysiology and roles in clinical medicine. New York· Wiley-Liss, pp. 447–455

Keen, H., J Payan, J Allawi, J Walker, G Jamal, A Weir, L Henderson, E Bissessar, P Watkins, M Sampson, E Gale, J Scarpello, H. Boddie, K Hardy, P Thomas, P Misra and J. P. Halonen. 1993. "Treatment of diabetic neuropathy with gamma-linolenic acid," *Diabetes Care*, 16 8–15.

Kelley, D. S., P C. Taylor, G J Nelson and B E Mackey. 1998. "Arachidonic acid supplementation enhances synthesis of eicosanoids without suppressing immune functions in young healthy men," *Lipids*, 33·449–456

Kelley, D S., P C Taylor, G. J Nelson, P C Schmidt, B E. Mackey and D. Kyle. 1997. "Effects of dietary arachidonic acid on human immune response," *Lipids*, 32.449–456.

Kennedy, M., S. Reader and R Davies. 1993. "Fatty acid production characteristics of fungi with particular emphasis on gamma linolenic acid production," *Biotechnol. Bioeng.*, 42 625–634

Kerscher, M and H Korting. 1992. "Treatment of atopic eczema with evening primrose oil, rationale and clinical results," *Clin Invest*, 70.167–171

Kim, Y I 1996. "Can fish oil maintain Crohn's in remission?" *Nutr Rev.*, 54 248–257

Kissebah, A H and M. M I. Hennes. 1995 "Central obesity and free fatty acid metabolism," *Prostaglandins Leukotrienes Essent Fatty Acids*, 52:209–211.

Leeds, A R , I Gray and M N M A Ahmad 1990 "Effects of n-6 essential fatty acids as evening primrose oil in mild hypertension," In. Horrobin, D.F (ed). Omega-6 essential fatty acids: pathophysiology and roles in clinical medicine New York, NY:Alan R. Liss, Inc., pp. 151–171

Lehmann, B , C Huber, H. Jacobi, A. Kampf and G. Wozel 1995 "Effects of dietary gamma-linolenic acid-enriched evening primrose seed oil on the 5-lipoxygenase pathway of neutrophil leukocytes in patients with atopic dermatitis," *J Dermatol Treat.*, 6·211–218.

Leventhal, L J, E G. Boyce and R. B. Zurier. 1993 "Treatment of rheumatoid arthritis with gammalinolenic acid," *Ann. Intern Med*, 119·867–873

Lillioja, S, D Mott, M Spraul, R Ferraro, J E Foley, E Ravussin, W C. Knowler, P H Bennet and C Bogardus 1993. "Insulin resistance as precursor of non-insulin dependent diabetes mellitus," Prospective studies of Pima Indians *N. Engl. J Med*, 29·1988–1992

Lorenzini, A., A. Bordoni, C. Spane, E. Turchetto, P. Biagi and S Hrelia 1997. "Age related changes in essential fatty acid metabolism in cultured rat heart myocytes," *Prostaglandins Leukotrienes Essent. Fatty Acids*, 57·143–147

Mancuso, P, J Whelan, S J. DeMichele, C. C. Snider, J A Guszcza and M D Karlstad 1997. "Dietary fish oil and fish and borage oil suppress intrapulmonary proinflammatory eicosanoid biosynthesis and attenuate pulmonary neutrophil accumulation in endotoxic rats," *Crit Care Med*, 25:1198–1206.

Manku, M., D Horrobin, N Morse, S Wright and J Burton 1984 "Essential fatty acids in the plasma phospholipids of patients with atopic eczema," *Br J. Dermatol*, 110 643–648.

Manku, M., D. Horrobin, N. Morse, V Kyte, K Jenkins, S. Wright and J. Burton 1982 "Reduced levels of prostaglandin precursors in the blood of atopic patients· defective delta-6-desaturase function as a biochemical basis for atopy," *Prostaglandins Leukotrienes Med*, 9·615–628

Mansel, R, J. Pye and L. Hughes 1990 "Effects of essential fatty acids on cyclical mastalgia and noncyclical breast disorders," In. Horrobin, D (ed) Omega-6 essential fatty acids: pathophysiology and roles in clinical medicine. New York, NY· Alan R Liss, Inc, pp 557–566

Manthorpe, R, T Manthorpe, A. Oxholm, P Oxholm and J. Prause. 1990 "Primary Sjogren's syndrome· new concepts," In Horrobin, D. (ed). Omega-6 essential fatty acids pathophysiology and roles in clinical medicine. New York, NY Alan R Liss, Inc, pp 239–253

Manthorpe, R., S Hagen Petersen and J. Prause. 1984 "Primary Sjogren's syndrome treated with Efamol/Efavit: a double blind cross-over study," *Rheumatol Int*, 4.165–167

Mazza, G. and H. H Marshall. 1988. "*Onosmodium* seed, a potential source of γ-linolenic acid," *Can Inst Food Sci Technol. J.*, 21·558–559

McKeigue, P M, B. Shah and M G Marmot 1991 "Relation of central obesity and insulin resistance with high diabetes prevalence and cardiovascular risk in South Asians," *Lancet*, 337 382–386

Miller, C., W Tang, V Ziboh and M Fletcher. 1991. "Dietary supplementation with ethyl ester concentrates of fish oil (omega-3) and borage oil (omega-6) polyunsaturated fatty acids induces epidermal generation of local putative anti-inflammatory metabolites," *J Invest. Dermatol*, 96·98–103.

Mills, D, Y. S Huang and R. Ward. 1990. "Fatty acid metabolism in normotensive and hypertensive rats I Differential incorporation of dietary omega-3 and omega-6 fatty acids," *Nutr. Res.*, 10·663–674.

Mills, D. and R. Ward 1990. "Dietary n-6 and n-3 fatty acids and stress-induced hypertension," In Horrobin, D. (ed). Omega-6 essential fatty acids: pathophysiology and roles in clinical medicine New York, NY· Alan R Liss, Inc, pp. 145–156.

Mills, D and R Ward. 1986. "Effects of essential fatty acid administration on cardiovascular responses to stress in the rat," *Lipids*, 21:139–142.

Mitchell, E, M Aman, S. Turbott and M Manku. 1987 "Clinical characteristics and serum essential fatty acid levels in hyperactive children." *Clin. Pediatr*, 26 406–411.

Moodley, J. and R Norman 1989. "Attempts of dietary alteration of prostaglandins pathways in the management of pre-eclampsia," *Prostaglandins Leukotrienes Essent Fatty Acids*, 37.145–147

Morris, M. C., L. A. Beckett, P. A. Scherr, L E. Herbert, D. A. Bennett, T. S. Field and D. A Evans. 1998. "Vitamin E and vitamin C supplement use and risk of incident Alzheimer disease," *Alzheimer Dis. Assoc. Disord.*, 12:121–126.

Morwood, M. J., P. B. O'Sullivan, F. Aziz and A. Raza. 1998. "Fission track ages of the stone tools and fossils on the East Indonesian island of Flores," *Nature*, 392:173–176.

Mukherjee, K and I. Kiewitt. 1987. "Formation of gamma-linolenic acid in higher plant evening primrose (*Oenothera biennis* L.)," *J. Agric. Food Chem.*, 35 1009–1012.

Muuse, B., M. Essers and L. van Soest. 1988. "Oenothera species and *Borago officinalis*. sources of gamma-linolenic acid. Netherland," *J. Agric. Sci.*, 36 357–363.

Nilsson, A., D F. Horrobin, A. Rosengren, L. Waller, A. Adlerberth and L. Wilhelmsen. 1996. "Essential fatty acids and abnormal involuntary movements in the general male population. a study of men born in 1933," *Prostaglandins Leukotrienes Essent. Fatty Acids*, 55:83–87

Nordoy, A., L. F. Hatcher, S. H. Goodnight, G. A. FitzGerald and W. E. Connor. 1994. "Effects of dietary fat content, saturated fatty acids, and fish oil on eicosanoid production and hemostatic parameters in normal men," *J. Lab. Clin. Med.*, 123:914–920.

O'Brien, P. and H. Massil. 1990, "Premenstrual syndrome: clinical studies on essential fatty acids," In: Horrobin, D. (ed). Omega-6 essential fatty acids: pathophysiology and roles in clinical medicine New York, NY: Alan R. Liss, Inc., pp. 523–545.

O'Dea, K. 1991. "Traditional diet and food preferences of Australian Aboriginal hunter-gatherers," *Philos. Trans. Roy Soc. Lond. B. Biol. Sci*, 334:233–241.

Ogburn, P. L., P. P. Williams, S. B. Johnson and R. T. Holman. 1984. "Serum arachidonic acid levels in normal and pre-eclamptic pregnancies," *Am. J Obstet. Gynecol.*, 148:5–9.

Okuyama, H., T. Kobayashi and S. Wantanbe. 1997. "Dietary fatty acids the n-6/n-3 balance and chronic elderly diseases. Excess linoleic acid and relative n-3 deficiency syndrome seen in Japan," *Prog. Lip. Res.*, 35:409–457.

Oliver, M 1989. "Linoleic acid and coronary heart disease," *Diabetes Nutr. Metab.*, 2(Suppl. 1).49–54.

Oliver, M., R. Riemersma, M. Thompson, M. Fulton, R. Abraham and D. Wood. 1990. "Linoleic acid and coronary heart disease," In: Horrobin, D. (ed). Omega-6 essential fatty acids: pathophysiology and roles in clinical medicine. New York, NY: Alan R. Liss, Inc , pp. 121–126

Oxholm, P., R. Manthorpe, J. Prause and D. Horrobin. 1986. "Patients with primary Sjogren's syndrome treated for two months with evening primrose oil," *Scand. J. Rheumatol*, 15 103–108.

Palangyo, K., S. Maselle, J. Mtabaji, E. Thomas, M. Winther and D. Horrobin. 1992. "An open study of essential fatty acid therapy (Efamol Marine) in patients with AIDS," *Arch. Std./HIV Res.*, 6:7–14.

Pan, D. A., S. Lillioja, M. R. Milner, D. K. Adamandia, L. A Baur, C. Bogardus and L. H. Storlien. 1995. "Skeletal muscle membrane lipid composition is related to adiposity and insulin action," *J. Clin. Invest.*, 96:2802–2808.

Parke, A., D. Parke and F. Jones. 1996. "Diet and nutrition in rheumatoid arthritis and other chronic inflammatory diseases," *J. Clin. Biochem. Nutr.*, 20.1–26.

Peet, M., J. D. E. Laugharne, J. Mellor and C. N Ramchand. 1996. "Essential fatty acid deficiency in erythrocyte membranes from chronic schizophrenic patients, and the clinical effects of dietary supplementation," *Prostaglandins Leukotrienes Essent Fatty Acids*, 55·71–75

Phylactos, A., K. Ghebremeskel, K. Costeloe, A. Leaf, L. Harbige and M. Crawford. 1994. "Polyunsaturated fatty acids and antioxidants in early development. Possible prevention of oxygen-induced disorders," *Eur. J. Clin. Nutr.*, 48(Suppl 2):S17–S23

Pritchard, P., G. Brown, N. Bhaskar and C. Hawkey. 1988 "The effect of dietary fatty acids on the gastric production of prostaglandins and aspirin-induced injury," *Aliment. Pharmacol. Ther.*, 2.179–184.

Prochazka, M , S Lillioja, J F Tait, W C Knowler, D Mott, M Spraul, P H Bennet and C. Bogardus. 1993. "Linkage of chromosomal markers on 4q with a putative gene determining maximal insulin action in Pima Indians," *Diabetes*, 42 514–519

Ramesh, G and U Das 1998 "Effect of evening primrose and fish oils on two-stage skin carcinogenesis in mice," *Prostaglandins Leukotrienes Essent. Fatty Acids*, 59 155–161.

Ratnayake, W., D Matthews and R. Ackman. 1989. "Triacylglycerols of evening primrose (*Oenothera biennis*) seed oil," *J. Am. Oil Chem. Soc.*, 66.966–969.

Richardson, A. J 1994. "Dyslexia, handedness and syndromes of psychosis-proneness," *Int J. Psychophysiol.*, 18:251–263.

Roosevelt, A C , M. Lima de Costa, and C. Lopes Machado 1996. "Paleoindian cave dwellers in the Amazon The peopling of the Americas," *Science*, 272.373–384

Rose, D , J Connolly and X. H Liu 1995. "Effects of linoleic acid and gamma-linolenic acid on the growth and metastasis of a human breast cancer cell line in nude mice and on its growth and invasive capacity *in vitro*," *Nutr Cancer*, 24.33–45.

Rothman, D , P. DeLuca and R Zurier. 1995 "Botanical lipids: effects on inflammation, immune responses, and rheumatoid arthritis," *Semin, Arthritis Rheum.*, 25.87–96

Sanderson, P., P. Yaqoob and P. Calder 1995. "Effects of dietary lipid manipulation upon rat spleen lymphocyte functions and the expression of lymphocyte surface molecules," *J. Nutr. Environ Med*, 5:119–132.

Sardesai, V 1992. "Biochemical and nutritional aspects of eicosanoids," *J. Nutr Biochem.*, 3·562–579

Schmidt, R M. Hayn, B. Reinhart, G. Roob, H Schmidt, M. Schumacher, N Watzinger and L. J. Launer. 1998. "Plasma antioxidants and cognitive performance in middle-aged and older adults. Results of the Austrian Stroke Prevention Study," *J. Am. Geriatr. Soc*, 46·1407–1410

Shaw, R. 1966 "The fatty acids of phycomycete fungi, and the significance of the gamma-linolenic acid component," *Comp. Biochem. Physiol.*, 18·325–331

Sinclair, A 1992. "Was the hunter gather diet prothrombotic? In· Sinclair, A and R Gibson (eds). Essential fatty acids and eicosanoids. (invited papers from the third international conference), Champaign, IL: AOCS Press, pp 318–324

Singer, P , M Moritz, M. Wirth, I. Berger and D. Forster. 1990. "Blood pressure and serum lipids from SHR after diets supplemented with evening primrose, sunflower seed or fish oil," *Prostaglandins Leukotrienes Essent. Fatty Acids*, 40·17–20.

Singh, N., U Das and P. Srivastava 1987. "Essential fatty acids and cancer with particular reference to Hodgkin's disease," *J. Assoc. Physicians India*, 35·137–138.

Stevens, L J., S. S. Zentall, M. L Abate, T. Kuczek and J. R. Burgess, Sr. 1996. "Essential fatty acid metabolism in boys with attention-deficit hyperactivity disorder," *Phys. Behav.*, 59 915–920.

Stevens, L. J., S. S Zentall, J. L. Deck, M L Abate, B. A Watkins, S. R. Lipp and J. R. Burgess. 1995. "Essential fatty acid metabolism in boys with attention-deficit hyperactivity disorder," *Am. J. Clin. Nutr.*, 62 761–768.

Stordy, B. J. 1997. "Dyslexia, attention deficit hyperactivity disorder, dyspraxia—do fatty acids help?" *Dyslexia Rev*, 9:1–3.

Stordy, B J 1995. "Benefit of docosahexaenoic acid supplements to dark adaptation in dyslexics," *Lancet*, 346 385.

Storlein, L. H , D. A. Pan, A. D. Kriketos, J. O'Connor, I D. Caterson, G. J. Cooney, A B Jenkins and L A. Baur 1996 "Skeletal muscle membrane lipids and insulin resistance," *Lipids*, 31·S262–S265.

Strachan, D P 1995. "Epidemiology of hay fever towards a community diagnosis," *Clin Exp Allergy,* 25 296–303

Sugano, M , T. Ide, T Ishida and K. Yoshida. 1986. "Hypocholesterolemic effect of gamma-linolenic acid as evening primrose oil in rats," *Ann Nutr. Metab ,* 30 289–299.

Szathmary, E. J E. 1994. "Non-insulin dependent diabetes mellitus among aboriginal North Americans," *Ann Rev Anthropol ,* 23 457–482

Tataranni, P. A , L J. Baier, G Paolisso, B V Howard and E. Ravussin 1996. "Role of lipids in development of noninsulin-dependent diabetes mellitus Lessons learned from Pima Indians," *Lipids,* 31 S267–S270.

Tate, G and R. Zurier 1994 "Suppression of monosodium urate crystal-induced inflammation by black currant seed oil," *Agents Actions,* 43.35–38.

Tate, G., B. Mandell, R. Karmali, M. Laposata, D. Baker, H. J. Schumacher and R Zurier. 1988. "Suppression of monosodium urate crystal-induced acute inflammation by diets enriched with gamma-linolenic acid and eicosapentaenoic acid," *Arthritis Rheum.,* 31 1543–1551

Thien, F C K and E. H. Walters. 1995. "Eicosanoids and asthma: an update," *Prostaglandins Leukotrienes Essent. Fatty Acids,* 52 271–288.

Thompson L., A Cockayne and R C. Spiller. 1994 "Inhibitory effect of polyunsaturated fatty acids on the growth of *Helicobacter pylori.* a possible explanation of the effect of diet on peptic ulceration," *Gut,* 35 1557–1561

Tomlinson, D , N Lockett and A. Carrington 1992. "Gamma-linolenic acid and the metabolic basis of conduction disorders in experimental diabetes," In Horrobin, D. (ed) Treatment of Diabetic Neuropathy A New Approach. London, UK. Churchill Livingstone, pp. 83–97.

Traitler, H , H. Wille and A. Studer. 1988. "Fractionation of blackcurrant seed oil," *J Am. Oil Chem. Soc.,* 65.755–760

Tsevegsuren, N and K Aitzetmuller. 1996 "γ-linolenic acid and stearidonic acids in Mongolian Boraginacea," *J Am Oil Chem Soc ,* 73:1681–1684

Vaddadi, K S , C. J Gilleard, E. Soosai, A. K Polonowita, R. A Gibson and G D Burrows 1996 "Schizophrenia, tardive dyskinesia and essential fatty acids," *Schizophr. Res.,* 20.287–294.

Vaddadi, K and C. Gilleard. 1990 "Essential fatty acids, tardive dyskinesia, and schizophrenia," In Horrobin, D (ed) Omega-6 essential fatty acids: pathophysiology and roles in clinical medicine. New York, NY Alan R. Liss, Inc., pp. 332–343.

Vaddadi, K , P Courtney, C. Filleard, M Manku and D. Horrobin. 1989 "A double-blind trial of essential fatty acid supplementation in patients with tardive dyskinesia," *Psychiatr. Res.,* 27:313–323.

van der Merwe, C , J Booyens, H. Joubert and C van der Merwe 1990. "The effect of gamma-linolenic acid, an *in vitro* cytostatic substance contained in evening primrose oil, on primary liver cancer A double-blind placebo controlled trial," *Prostaglandins Leukotrienes Essent Fatty Acids,* 40 199–202

van Papendorp, D. H., H. Coetzer and M C. Kruger. 1995. "Biochemical profile of osteoporotic patients on essential fatty acid supplementation," *Nutr Res ,* 15:325–334

van Rooyen, J , S Swanevelder, J. C. Morgenthal and A. J Spinnler Benade 1998. "Diet can manipulate the metabolism of EPA and GLA in erythrocyte membrane and plasma," *Prostaglandins Leukotrienes Essent. Fatty Acids,* 59 27–38.

Varma, S L , A. M Zain and S. Singh. 1997 "Psychiatric morbidity in first-degree relatives," *Am. J. Med. Genet ,* 74 7–11.

Varma, P and T. Persaud. 1982. "Protection against ethanol-induced embryonic damage by administering gamma-linolenic and linoleic acids," *Prostaglandins Leukotrienes Med ,* 8 641–6445

Veale, D , H Torley, I Richards, A Odowd, C. Fitzsimons, J Belch and R Sturrock 1994. "A double-blind placebo-controlled trial of Efamol Marine on skin and joint symptoms of psoriatic-arthritis." *Br. J Rheumatol* , 33.954–958.

Velzing-Aarts, F. V , F. R. M. van der Klis, F. P. L van der Dijs and F. A. J. Muskiet. 1997. "Umbilical vessels of preeclamptic pregnancies have low contents of both ω3 and ω6 long chain polyunsaturated fatty acids," *Prostaglandins Leukotrienes Essent. Fatty Acids*, 57.194.

Venter, C , P. Joubert and J. Booyens. 1988. "Effects of essential fatty acids on mild to moderate essential hypertension," *Prostaglandins Leukotrienes Essent Fatty Acids*, 33:49–51.

Vericel, E., M. Lagarde, F. Mendy, M. Courpron and M. Dechavanne. 1987 "Comparative effects of linoleic acid and gamma-linolenic acid intake on plasma lipids and platelet phospholipids in elderly people," *Nutr. Res* , 7 569–580

Vidgren, H M., J J. Agren, U. Schwab, T. Rissanen, O Hanninen and M. I. J. Uusitupa. 1997 "Incorporation of n-3 fatty acids into plasma lipid fractions and erythrocyte membranes and platelets during dietary supplementation with fish, fish oil, and docosahexanenoic acid-rich oil among healthy men," *Lipids*, 32.687–695.

Vinik, A , M. Holland, J Le Beau, F. Liuzzi, K. Stansberry and L. Colen. 1992. "Diabetic neuropathies," *Diabetes Care*, 15:1926–1975.

Wallenburg, H. and H. Bremer. 1992. "Principles and applications of manipulation of prostaglandin synthesis in pregnancy," *Bailliere's Clin. Obstet. Gynaecol*, 6.859–891.

Watkins, B 1998. "Regulatory effects of polyunsaturates on bone modeling and cartilage function," *World Rev. Nutr Diet.*, 83.38–51.

Weete, J D., F. Shewmaker and S. R. Gandhi. 1998. "γ-Linolenic acid in zygomycetous fungi *Syzygites megalocarpus*," *J Am. Oil Chem. Soc.*, 75:1367–1372.

Winther, M , W Schlech, B. Conway and J Singer. 1997. "Phase I open label study of the safety, pharmacokinetics and potential efficacy of LIGLA in advanced HIV infection," *Prostaglandins Leukotrienes Essent. Fatty Acids*, 57:247.

Wolkin, A., D. Segarnick, J. Sierkierski, M. Manku, D. Horrobin and J. Rotrosen 1987 "Essential fatty acid supplementation during early alcohol abstinence." *Alcohol Clin. Exp. Res* , 11 87–92.

Wolkin, A., B Jordan, R Peselow, M. Rubinstein and J Retrosen. 1986. "Essential fatty acid supplementation in tardive dyskinesia," *Am. J. Psychiatr.*, 143:912–914.

Wright, S. and T. A. B. Sanders. 1991. "Adipose tissue essential fatty acid composition in patients with atopic eczema," *Eur. J Clin. Nutr.*, 45.501–505.

Wright, S. 1990. "Essential fatty acids and atopic eczema: biochemical and immunological studies," In· Horrobin, D (ed). Omega-6 essential fatty acids: pathophysiology and roles in clinical medicine. New York, NY. Alan R. Liss, Inc., pp. 55–65.

Wright, S and C Bolton. 1989. "Breast milk fatty acids in mothers of children with atopic eczema." *Br J. Nutr.*, 62:693–697.

Zeigler, D and F A. Gries. 1997. "α-Lipoic acid in the treatment of diabetic peripheral and autonomic neuropathy," *Diabetes*, 46 (suppl. 2):S62–S66.

Ziboh, V. 1996. "The significance of polyunsaturated fatty acids in cutaneous biology," *Lipids*, 31 (suppl). 249–253.

Ziboh, V. A , M. Yun, D. M. Hyde and S. N Giri. 1997. "Gamma-linolenic acid-containing diet attenuates Bleomycin-induced lung fibrosis in hamsters," *Lipids*, 32.759–767.

Ziboh, V. and R. Chapkin. 1988. "Metabolism and function of skin lipids," *Prog. Lipid Res* , 27.81–105.

Zurier, R. 1993 "Fatty acids, inflammation and immune responses," *Prostaglandins Leukotrienes Essent. Fatty Acids*, 48 57–62

Zurier, R. B , R. G. Rossetti, E. W. Jacobson, D. M. DeMarco, N. Y. Liu, J E. Temming, B. M. White and M Laposata. 1996. ''Gamma-linolenic acid treatment of rheumatoid arthritis,'' *Arthritis Rheum.,* 39:1808–1817.

Zygadio, J., R Morero, R. Abburra and C. Guzman. 1994. ''Fatty acid composition in seed oils of some Onagraceae,'' *J. Am. Oil Chem. Soc.,* 71:915–916.

CHAPTER 9

Tea as a Source of Dietary Antioxidants with a Potential Role in Prevention of Chronic Diseases

D. A. BALENTINE
I. PAETAU-ROBINSON

1. INTRODUCTION

Infusing leaves of the plant *Camellia sinensis* in water produces a fragrant beverage that is mildly astringent and bitter in the mouth called tea. People began enjoying tea over 2,000 years ago when in the year 2737 BC the Chinese emperor Sheng Nung discovered the drink when leaves fell into a pot of boiling water (Ukers, 1994). Today, tea is a truly global drink enjoyed hot and iced for its ability to revive, refresh and relax the body and mind. Tea was originally valued for its medicinal qualities that were first reported by Chinese scholars in a medical text, the Pen T'sao, about 20 centuries ago, and it is the potential healthful properties of tea that are gaining scientific merit today (Table 9.1) (Blofeld, 1985; Segal, 1996).

Tea was imported to continental Europe from China in the early 1600s by Dutch traders, and green tea became a trendy drink in England and North America by the mid 1600s. In America during the late 1700s, a demonstration against taxation by the English known as the "Boston Tea Party" was one of the early events of the revolution and founding of the United States of America. Black tea did not become popular in Europe and the United States until the later 1800s. The history, agriculture and chemistry of tea are reviewed in "Tea: Cultivation to Consumption" (Wilson and Clifford, 1992).

The process used to convert fresh green tea leaves to tea product results in chemical changes that determine the distinctive color, flavor and aroma of green, oolong and black tea beverages. Tea contains flavonoids and methylxanthines, bioactive compounds that are important to tea flavor and tea's potential role as a beverage that helps prevent chronic diseases such as cancer (Yang

TABLE 9.1.   Traditional Health Claims for Tea.

Traditional Claims (From Chinese Medicine)	Potential Physiological Effects (From Modern Science)
Improves blood flow	Vasodilates and decreases platelet activity
Eliminates alcohol and toxins	Induces phase I and phase II enzymes
Clears urine and improves flow	Diuretic effects
Relieves joint pain	Anti-inflammatory and antiarthritic activity
Improves disease resistance	Improves immune response, prevents cancer and cardiovascular diseases.

and Wang, 1993; Dreosti, 1996) and cardiovascular disease (Tijburg et al., 1997a). Antioxidant function is one of the properties of tea flavonoids (Wiseman et al., 1997). These compounds are bioavailable (Hollman et al., 1997) and have other physiological activities such as modulation of signal transduction pathways, anti-inflammatory effects and decreased cell proliferation at G1 phase of the cell cycle (Dong et al., 1997; McCarty, 1998). However, the relevance of the bioactivity of tea flavonoids to human health remains to be established.

2. THE CHEMISTRY OF TEA FLAVONOIDS

Flavonoids are a family of plant phenolics widely distributed in fruits, vegetables and beverages. Examples of flavonoids are the catechins in tea, anthocyanins in blueberries, purple grapes and red wines, isoflavones in soy products and the flavonols in onions, apples and tea. Flavonoids are chemically classified as *"Di-benz pyrans and pyrones and their derivatives."* These are plant phenolics with a C6-C3-C6 ring structure. Part C3 is usually a heterocyclic ring containing oxygen. Derivatives including glycosides and oxidation products that retain this ring structure are also flavonoids (Figure 9.1).

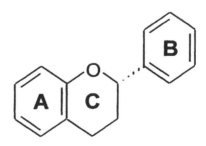

Figure 9.1 Basic flavonoid structure (dibenz pyrans and pyrones and their derivatives).

The main flavonoids found in fresh leaf are the catechins (flavan-3-ols) (Figure 9.2) and the flavonols (Figure 9.3). These flavonoids typically make up greater than 30% of the dry weight of the leaf. Epigallocatechin gallate (EGCG) is the most abundant catechin in most green, oolong and black teas. Green and oolong tea beverages typically contain 30–130 mg EGCG per cup of tea, and black tea beverages typically contain 0–70 mg EGCG per cup of tea. The flavonols such as quercetin, kaempferol, myricetin and their glycosides are present in much lower concentrations than the catechins and are found in comparable quantities in black, green and oolong tea beverages (5–15 mg/cup). Flavonol aglycones are found in tea leaf but are only minor components of tea beverages due to their poor water solubility (Balentine, 1996; Harbowy and Balentine, 1997).

The manufacturing process (Figure 9.4) employed to convert fresh green leaf to green, oolong or black teas determines the types of flavonoids found in tea beverages. During the manufacture of tea, enzymatic reactions are responsible for the development of the respective colors and flavors characteristic of each type of tea. During the manufacture of black tea, the colorless catechins are converted to flavonoids with orange-yellow to red-brown color by enzymes in tea leaves called oxidases. Oxidation of amino acids and lipids

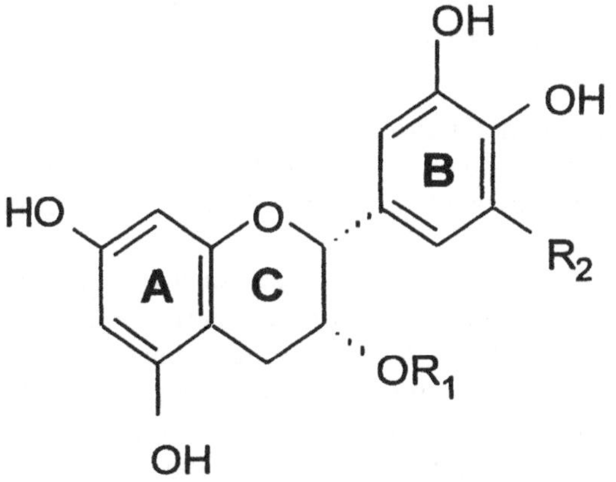

		R_1	R_2
Epicatechin	EC	H	H
Epicatechin gallate	ECG	Gallate	H
Epigallocatechin	EGC	H	OH
Epigallocatechin gallate	EGCG	Gallate	OH

Figure 9.2 Major tea catechins (flavan-3-ols)

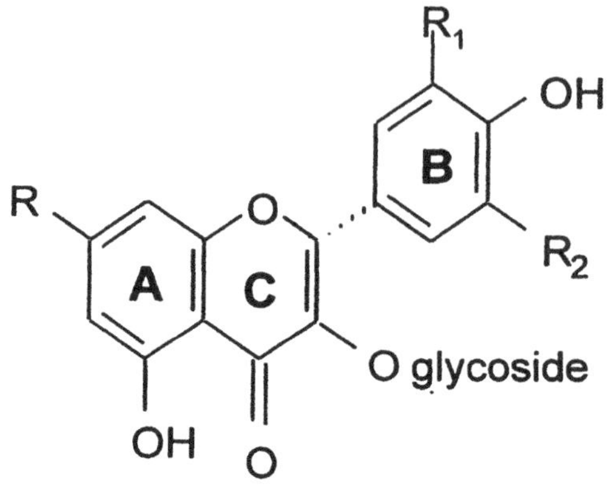

		R_1	R_2
Kaempferol glycoside	KaG	H	H
Quercetin glycoside	QuG	OH	H
Myricetin glycoside	MyG	OH	OH

Figure 9.3 Major tea flavonols.

results in the generation of numerous volatile flavor compounds. These changes are reflected in the red-amber color, reduced bitterness and increased astringency and more complex flavor of black teas (Wilson and Clifford, 1992; Balentine, 1996).

Green tea is produced using thermal processes such as steam or dry heat to inactivate enzymes that oxidize catechins to more complex flavonoids characteristic of oolong and black teas. However, some oxidation of the catechins in green teas typically occurs during withering, and 20–30% of the total flavonoids in green teas will be catechin oxidation products typical of oolong and black teas. Small quantities of a group of flavonoids characteristic of black teas called theaflavins can be found in most green teas.

Black tea contains several unique classes of flavonoids resulting from the fermentation process including theaflavins, bisflavonols (theasinensins), theaflavic acids, theaflagallins and a heterogeneous group of flavonoids called thearubigens.

2.1. THEAFLAVINS

Theaflavins are biopolymers that are bright red/orange in solutions and impart astringency and brightness to tea beverages (Obanda and Owuor, 1995).

There are four main theaflavins (Figure 9.5) and two groups of minor theaflavins called isotheaflavins and neotheaflavins (Robertson, 1992; Harbowy and Balentine, 1997). The theaflavin content of tea leaf is typically about 2% on a dry weight basis (8–40 mg/cup). The theaflavin content of tea beverages can be readily determined by HPLC analysis (Finger et al., 1992).

2.2. CONDENSED TEA FLAVONOIDS (THEARUBIGENS)

The catechin content of tea leaf is reduced by 85% during fermentation, yet only ~10% can be accounted for in the form of theaflavins and theaflavic acids. The remaining flavonoids form a heterogeneous group of water-soluble products called thearubigens that typically make 23% of black tea leaf on a dry weight basis (100–200 mg/cup) (Robertson, 1992; Balentine, 1996). These flavonoid polymers are the brown/black pigments in black tea that provide thickness and astringency to tea beverages. One group of thearubigens are proanthocyanidin polymers and two other types recently identified are theafulvin (Bailey et al., 1992) and oolongtheanin (Hashimoto et al., 1988). There are no simple methods to directly determine the thearubigen content of tea beverages.

2.3. ANTIOXIDANT PROPERTIES OF TEA FLAVONOIDS

Oxidative damage to cells and tissues by reactive oxygen and reactive nitrogen species contributes to the development of a number of chronic illnesses including cancer, cardiovascular disease and diabetes. A decreased risk of chronic disease is associated with a diet rich in plant foods. Whereas fiber,

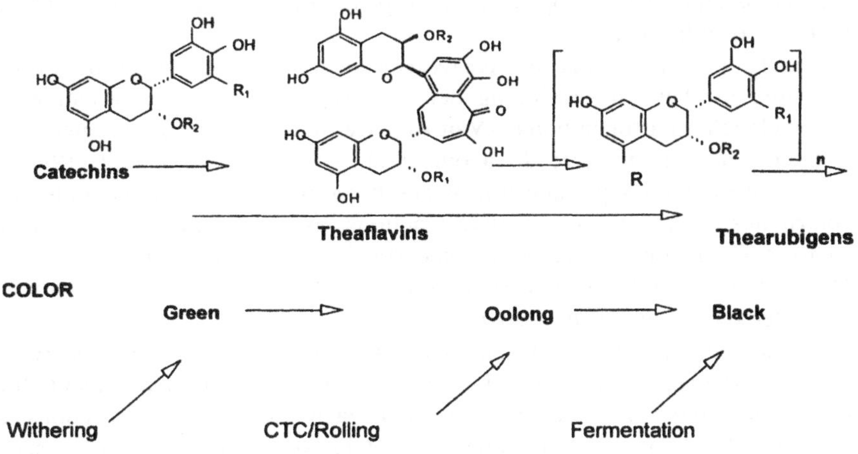

Figure 9.4 Tea processing and chemistry.

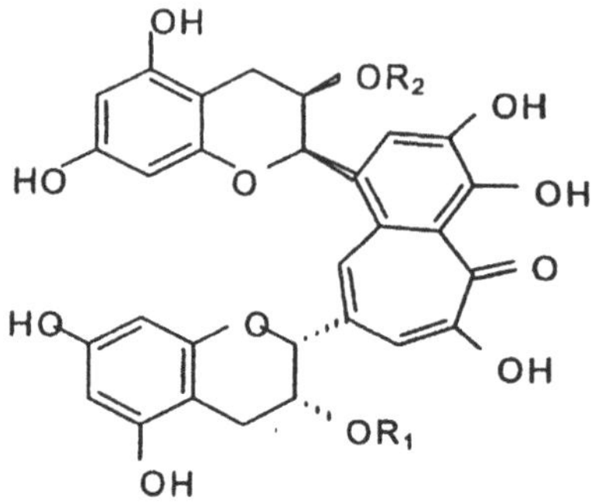

		R_1	R_2
Theaflavin	TF	H	H
Theaflavin 3-gallate	TF-3G	Gallate	H
Theaflavin 3'-gallate	TF-3'G	H	Gallate
Theaflavin di-gallate	TF-DG	Gallate	Gallate

Figure 9.5 Major theaflavins.

vitamins and minerals are important dietary nutrients that contribute to health; the exact components of plants responsible for health benefits have not been clearly established, but antioxidant phytochemicals are a class of components likely to have a role in these benefits. Tea is a plant-based beverage containing flavonoids and other biologically active phytochemicals making tea a likely part of a healthy diet.

Tea flavonoids scavenge reactive oxygen species and free radicals by several proposed mechanisms, including delocalization of electrons, formation of intramolecular hydrogen bonds (Van Acker et al., 1996a) and rearrangement of their molecular structure (Jovanovic et al., 1994; Van Acker et al., 1996b). Chelation of free copper and iron is another property of tea flavonoids that contributes to preventing oxidative reactions, as these metals are known to catalyze formation of reactive oxygen species *in vivo* (Morel et al., 1993; Miller et al., 1996b). The antioxidant function of tea flavonoids has been well characterized *in vitro* and *in vivo*.

Dietary antioxidants are substances found in foods that significantly decrease oxidative damage in humans. To be an effective dietary antioxidant, the food component must be bioavailable and reach sufficient concentrations in biological fluids or tissues to suppress or delay oxidation of substrates such as lipids, proteins or genetic material.

TABLE 9.2. Relative Antioxidant Potentials of Vitamins, Tea Beverages, Flavonoids and Carotenoids.

Antioxidant	Trolox Equivalent Antioxidant Capacity (TEAC mM)*
Vitamins	
Ascorbic acid	1 0
Vitamin E	1.0
Tea Beverages	
Green Tea (1,000 ppm tea solids)	3 8
Black Tea (1,000 ppm tea solids)	3 5
Flavonols	
Epigallocatechin	3 8
Epicatechin gallate	4 9
Epigallocatechin gallate	4.8
Flavonols	
Quercetin	4 7
Rutin	2.4
Theaflavins	
Theaflavin	2 9
Theaflavin 3-monogallate	4 7
Theaflavin 3'-monogallate	4 8
Theaflavin digallate	6 2
Carotenoids	
Lycopene	2 9
β-carotene	1 9
Lutein	1 5

* TEAC is the millimolar concentration of a Trolox Solution having the antioxidant capacity equivalent to a 1 0 mM solution of the substance under investigation
Reprinted with permission from S A Wiseman, D A Balentine and B Frei 1997 "Antioxidants in tea," *Crit Rev Food Sci Nutr* 37, 705–718 Copyright CRC Press, Boca Raton, FL

The antioxidant functions of tea extracts and purified tea flavonoids have been characterized using a number of *in vitro* model systems based on scavenging of various reactive oxygen species or stable radicals. Fenton chemistry (metal ions), UV light, radiation, pulse radiolysis and the hypoxanthine-xanthine oxidase system are typically used to generate reactive oxygen species (ROS) *in vitro*, and substrates are lipids, liposomes, micelles, ghost cell membranes or lipids such as methyl linoleate or low density lipoprotein (LDL). In one such test system, tea flavonoids have been shown to be effective scavengers in the aqueous phase of the stable radical 2,2'-azinobis-(3-ethylbenzothiazoline-6-sulfonic acid) (ABTS) (Salah et al., 1995; Miller et al., 1996b; Rice-Evans et al., 1996). Tea flavonoids are significantly more potent antioxidants than vitamin E, vitamin C, the carotenes and the xanthophylls in this test system (Table 9.2) (Miller et al., 1996a). The *in vitro* activity of tea extracts and tea flavonoids in scavenging many radicals have been well characterized, and the flavonoid, epigallocatechin gallate (EGCG), was consistently the most effective (Table 9.3). Oxidation of LDL is a factor linked to

TABLE 9.3. Scavenging of Stable Radicals by Tea Antioxidants.

Stable Radical	Reaction Phase	Detection	Relative Activity	Ref.
Fremy's salt	Polar	ESR	EGCG > ECG > EC > EGC > C > GA	Gardner et al., 1998
Fremy's salt	Polar	ESR	Green tea > Black tea	Gardner et al., 1998
Galvinoxyl	Apolar	ESR	EGCG > ECG > GA > EC > C > EGC	Gardner et al., 1998
Galvinoxyl	Apolar	ESR	Green tea > Black tea	Gardner et al., 1998
ABTS	Polar	abs 734 nm	TF-dg > TF-mg > TF	Miller et al., 1996[b]
ABTS	Polar	abs 734 nm	ECG > EGCG = Q > EGC > GA > EC	Salah et al., 1995
DPPH	Apolar	ESR	EGCG + ECG > EGC > EC >> Vit C > Vit E	Nanjo et al., 1996
DPPH	Apolar	abs 521 nm	EGCG >> ECG = GA > EC = utin > Vit E	Fourneau et al., 1996
DPPH	Apolar	abs 517 nm	(Teas) Pouchong > green > polong > black	Yen and Chen, 1995
DPPH	Apolar	abs 517 nm	EGCG > ECG > EC > GA > EGC > Vit C > trolox	Hong et al., 1994
DPPH	Apolar	abs 520 nm	ECG > EGCG > EGC > GA > EC > Vit C > Vit E	Yoshida et al., 1989
DPPH	Apolar	ESR	ECG(tet) > ECG(tri) > ECG(di) = EGCG	Hatano et al., 1989
DPPH	Apolar	ESR	ECG > EGCG	Uchida et al., 1987

Note: ABTS, 2,2'-azinobis-(3-ethylbenzothiazoline-6-sulfonic acid) radical cation; DPPH, 1,1-diphenyl-2-picrylhydrazyl radical; TF-dg, theaflavin digallate; TF-mg, theaflavin monogallate; TF, theaflavin; EGCG, epigallocatechin gallate; ECG, epigallocatechin; ECG, epicatechin gallate, EC, epicatechin; C, catechin; GA, gallic acid; Q, quercetin.

Reprinted with permission from S.A. Wiseman, D.A. Balentine and B. Frei. 1997 "Antioxidants in tea," *Crit. Rev Food Sci. Nutr. 37,* 705–718. Copyright CRC Press, Boca Raton, FL.

development of atherosclerosis (Berliner and Heinecke, 1996). Unlike native LDL, oxidized LDL is readily taken up by macrophages, inducing the formation of lipid-laden foam cells, which are precursors of the atherosclerotic fatty streak (Frei, 1995). The role of antioxidants in preventing oxidation of LDL *in vitro* and *in vivo* has been studied in depth. Tea flavonoids are very effective inhibitors of *in vitro* oxidation of LDL when oxidation is catalyzed by numerous pro-oxidants (Table 9.4). Clinical trials are required to determine the effect of tea on LDL oxidation *in vivo* as *in vivo* effects are dependent on the absorption and distribution of antioxidants into the endothelial wall of the blood vessels or to LDL particles.

3. TEA AS A DIETARY ANTIOXIDANT

There are two epidemiological studies that support the role of tea as a dietary antioxidant. It was observed that green tea consumption in smokers reduces the frequency of sister chromatid exchange (SCE) in peripheral lymphocytes to a level similar to that found in nonsmokers (Shim et al., 1995). SCE is believed to reflect smoke-induced DNA damage. Smoke contains many oxidants and free radicals that can enhance oxidative damage to DNA. Oxidative damage to DNA, in turn, can contribute to dysfunction and disease states of cells. Tea-drinking smokers have a significantly lower mean frequency of micronuclei (MNF) in lymphocytes than of non-tea-drinking smokers (Xue et al., 1992). MNF is an indicator for genotoxicity of cigarette smoke.

In vitro experiments have demonstrated the potent antioxidant activities of flavonoids as free radical scavengers and as inhibitors of LDL oxidation (Rice-Evans et al., 1995; Vinson et al., 1995a, 1995b; Miller et al., 1996b; McAnlis et al., 1998).

The oxidative hemolysis of RBC, initiated by peroxyl radicals, is inhibited by tea flavonoids *in vitro* and *in vivo* in a rat model (Zhang et al., 1997). Two other *in vitro* studies showed protective effects of green tea and black tea against lipid peroxidation of erythrocyte membranes, initiated by various inducers (Grinberg et al., 1997; Halder and Bhaduri, 1998). Tea extracts are able to inhibit lipid peroxidation and scavenge reactive oxygen species (Yen and Chen, 1995). Nonenzymatic and enzymatic lipid peroxidation is inhibited by tea flavonoids (Galvez et al., 1995; Polette et al., 1996).

Animal models support the results from *in vitro* studies (Table 9.5). Tea and tea flavonoid consumption reduces lipid peroxidation in hamsters and rats (Nanjo et al., 1993; Fremont et al., 1998; Vinson and Dabbagh, 1998). Oxidative damage to lung DNA caused by a tobacco-specific nitrosamine (NNK), which results in 8-hydroxydeoxyguanosine (8-OHdG), is inhibited in mice following green tea and EGCG consumption (Xu et al., 1992).

TABLE 9.4. Inhibition of LDL Oxidation *In Vitro* by Tea Antioxidants.

Prooxidant	Marker of Oxidative Damage	Relative Activity	Ref.
Cu^{2+}	Conjugated dienes (lag phase)	TF-dg > EGCG > TF-mg > ECG > TF > EGC > EC	Ishikawa et al., 1997
Cu^{2+}	Conjugated dienes (lag phase)	EGCG > Vit E	Ishikawa et al., 1997
Cu^{2+}	Conjugated dienes (lag phase)	GT(50 mg/mL) > BT 950 mg/mL)	Van het Hof et al., 1997
Cu^{2+}	TBARS	EGC and EC inhibitory in initiation phase, accelerative in propagation phase.	Yamanaka et al., 1997
Cu^{2+}	Conjugated dienes (lag phase)	GT = Vit E > Vit C	Luo et al., 1997
Cu^{2+}	TBARS	TF-dg > TF3-mg > TF	Miller et al., 1996b
Cu^{2+}	Conjugated dienes (lag phase)	EGCG.EC > quercetin > chlorogenic acid > rutin > vit E > genistein	Vinson et al., 1995a
Cu^{2+}	Conjugated dienes (lag phase)	Sesaminol > Q > EGCG > TF > > > myricetin > BHT	Miura et al., 1995
Cu^{2+}	Apo B fragmentation	EGCG > BHT > α-tocopherol	Miura et al., 1995
Cu^{2+}	TBARS	EGCG > ECG > EC > C > EGC	Miura et al., 1994
Cu^{2+}	TBARS	ECG > EGCG > EC > C > EGC > BHT	Miura et al., 1994
Cu^{2+}	Apo B fragmentation	EGCG > C	Miura et al., 1994
Cu^{2+}	TBARS	Flavanols most effective of the flavonoids	Vinson et al., 1995a
Cu^{2+}	TBARS	EGCG > EGC > ECG > C > BHT > vit C > vit E	Vinson et al., 1995b
Peroxynitrite	REM	ECG > GA > EC = EGC = EGC = EGCG	Pannala et al., 1997
Metmyoglobin	TBARS/REM	EGCG = ECG = EC = C > EGC > GA	Salah et al., 1995
Macrophages	TBARS	GT polyphenols inhibited oxidation	Zhenhua et al., 1991
Macrophages	LDL Uptake by receptors	GT polyphenols inhibited uptake by receptors	Zhenhua et al., 1991
Macrophages	Apo B Fluorescence/REM	GT polyphenols inhibited fragmentation	Mangiopane et al., 1992

Note: TF-dg, theaflavin digallate; TF-mg, theaflavin monogallate; TF, theaflavin; EGCG, epigallocatechin gallate; EGC, epigallocatechin; ECG, epicatechin gallate; EC, epicatechin; C, catechin; GA, gallic acid; GT, green tea; BT, black tea; Q, quercetin; BHT, butylated hydroxytoluene; TBARS, thiobarbituric acid reactive substances; REM, relative electrophoretic mobility.

Reprinted with permission from S. A. Wiseman, D. A. Balentine and B. Frei 1997 "Antioxidants in tea." *Crit. Rev. Food Sci. Nutr.* 37, 705–718. Copyright CRC Press, Boca Raton, FL.

Results from human intervention studies are limited. Preliminary data supports tea as an antioxidant *in vivo*, but there is no data on individual tea flavonoids.

Ingestion of green or black tea rapidly raises the total plasma antioxidant capacity in humans (Benzie et al., 1999; Leenan et al., 1999). A study in diabetics showed that two weeks of consuming a diet rich in flavonoids provided by tea and onions reduces the hydrogen peroxide-induced damage to lymphocyte DNA (Lean et al., 1999). Urinary and lymphocyte 8-OHdG is a marker for oxidative DNA damage caused by the attack of a hydroxyl radical on a guanosine base. Seven days of green tea consumption reduces urinary and WBC 8-OHdG in smokers and nonsmokers (Klaunig et al., 1999). In the same study population, urinary malondialdehyde (MDA) excretion, a marker for lipid oxidation *in vivo*, was reduced following green tea consumption.

4. TEA AND CARDIOVASCULAR HEALTH

Several epidemiological studies have evaluated the relationship of tea consumption with risk of heart disease, and the majority of these studies indicate a protective role of tea. A recent case-control study in men and women with no history of myocardial infarction or angina pectoris found a significantly lower risk of myocardial infarction in tea drinkers ($\geq$ 1 cup/day) compared with tea nondrinkers (Sesso et al., 1999). In a prospective cohort study, men who consumed > 4.7 cups tea/day had a 69% reduced risk of stroke compared with men who had < 2.6 cups/day (Keli et al., 1996). In the same study population, there was a significant inverse association between flavonoid intake and coronary mortality risk (Hertog et al., 1993). Black tea was the major source of dietary flavonoids in this study population, followed by onions and apples. Using a cohort of men from a follow-up of the Health Professional Study, involving 34,789 men, an inverse relationship between intake of flavonol and subsequent coronary mortality rate in subjects with previous coronary heart disease was found (Rimm et al., 1996). Black tea contributed 25% of the dietary flavonoids. A Norwegian study observed that the risk of dying from CHD was 36% lower for men consuming at least one cup of tea/day compared with men drinking less than one or no cups of tea/day (Stensvold et al., 1992). A strong inverse association of tea intake with aortic atherosclerosis was observed in a large prospective study of men and women (Geleijnse et al., 1999). The odds ratio for severe atherosclerosis decreased from 0.6 to 0.4 in subjects consuming one cup vs. > five cups of tea daily, respectively. The protective effect of tea was stronger in women than in men. In contrast, tea consumption was not associated with CHD and stroke mortality in a prospective study of 34,492 postmenopausal women (Yochum et al., 1999).

TABLE 9.5. Summary of Some *In Vivo* Studies in Animals On The Role of Polyphenols As Antioxidants.

Model	Challenge	Treatment	Biomarker	Ref.
Rats	AAPH	GTP	Red blood cell hemolysis	Zhang et al., 1997
Rats	High-fat diets: Saturated and PUFA	Tea catechins	TBARS in plasma	Nanjo et al., 1993
Rats	Diets rich in MUFA and PUFA, CuSO$_4$	Quercetin; (+)-catechin	Conjugated dienes and TBARS in VLDL + LDL	Fremont et al., 1998
Rats	2-nitropropane	EGCG; GTP	Hepatic 8-OHdG MDA	Hasegawa et al., 1995
Mice	Irradiation	Flavonols; catechins	Micronucleated reticulocytes	Shimoi et al., 1994
Mice	Irradiation	EGCG	Hepatic TBARS	Uchida et al., 1992
Mice	NNK^2	Green tea; EGCG	8-OHdG in lung tissue	Xu et al., 1992
Hamster	High cholesterol, high fat diet; CuSO$_4$	Green and black tea	Plasma TBARS; VLDL + LDL oxidizability	Vinson et al., 1998

Note: AAPH, 2,2'-azo-bis(2-amidinopropane) dihydrochloride; PUFA, polyunsaturated fatty acid; MUFA, monounsaturated fatty acid; NNK, nitrosamine 4-(methylnitrosamine)-1-(3-pyridyl)-1-butanone; GTP, green tea polyphenols; EGCG, epigallocatechin gallate; TBARS, thiobarbituric acid reactive substances; 8-OHdG, 8-hydroxy deoxyguanosine; MDA, malondialdehyde.

276

A study in Welsh men found no relation of tea consumption to ischemic heart disease, in fact, CHD mortality increased with the consumption of > two cups of tea/day compared with men drinking less tea (Hertog et al., 1997). However, the men in this population who consumed the most tea also smoked more than the other men and consumed a very high-fat diet, thereby confounding the results.

There is general consensus in the scientific community that elevated plasma cholesterol concentrations, lipid peroxidation and LDL oxidation play a pivotal role in the pathology of cardiovascular diseases. A large study in Japan observed an inverse association of green tea consumption with plasma cholesterol, LDL cholesterol, HDL cholesterol and triacyglycerols (Imai and Nakachi, 1995). In the same study, serum concentrations of lipid peroxides among heavy smokers inversely changed with the amount of green tea consumed. Results from two other studies suggest a negative association between tea consumption and serum cholesterol (Green and Harari, 1992; Stensvold et al., 1992), whereas two well-controlled studies did not find any association (Bingham et al., 1997; Tsubono and Tsugane, 1997).

In vitro studies provide a mechanistic basis for a role of tea in maintaining coronary health through antioxidant function, platelet activity and vascular reactivity. As antioxidants, flavonoids are able to scavenge free radicals and delay the onset of LDL oxidation *in vitro* (Rice-Evans et al., 1995; Vinson et al., 1995a, 1995b; Miller et al., 1996b; McAnlis et al., 1998). Other *in vitro* studies demonstrated that flavonoids efficiently inhibit lipid peroxidation, platelet aggregation and thromboxane formation (Tzeng et al., 1991; Terao et al., 1994; Polette et al., 1996).

In animal models, it has been shown that tea flavonoids can modulate lipid profiles and act as antioxidants. Elevated plasma cholesterol and triacylgycerol concentrations are risk factors for CHD. Studies on the effect of tea consumption on hypercholesterolemia are inconclusive. Plasma cholesterol concentrations of hypercholesterolemic rats and mice can be decreased by consumption of flavonoids or tea (Muramatsu et al., 1986; Matsuda et al., 1986; Igarashi and Ohmuma, 1995; Matsumoto et al., 1998; Suzuki et al., 1998). In hamsters, it was shown that consumption of black or green tea improves the lipid profile and decreases lipid peroxides and oxidizability of lipoproteins (Vinson and Dabbagh, 1998). Green tea epicatechins had a dose-dependent hypolipidemic effect in hamsters fed a high-fat diet (Chan et al., 1999). The activities of 3-hydroxy-3-methyl glutaryl coenzyme A reductase and intestinal acyl CoA: cholesterol acyltransferase are not affected by green tea epicatechins, suggesting that the hypolipidemic activity is due to interference with absorption of dietary fat and cholesterol. This was further supported by increased excretion of neutral and acidic sterols with feces in the treated group compared to the control. No effect of green or black tea on blood lipids in hypercholesterolemic rabbits was observed; however, the lag phase of *ex vivo* LDL oxidation was prolonged following the treatments (Tijburg et al., 1997b).

Results from human studies with regard to the effect of tea consumption on plasma lipids and LDL oxidation are inconclusive. Green tea or black tea intake (3 g tea solids/day) for four weeks had no effect on plasma cholesterol, LDL cholesterol, HDL cholesterol and plasma triacylglycerols in volunteers with normal plasma cholesterol and lipid profiles (Bingham et al., 1997; Ishikawa et al., 1997; Van het Hof et al., 1997; Princen et al., 1998). One study was able to demonstrate a 10–15% increase in lag time for *ex vivo* LDL oxidation after four weeks of black tea consumption (Ishikawa et al., 1997); however, two other studies were not able to show this effect (Van het Hof et al., 1997; Princen et al., 1998). Contradictory results may stem from differences in methods for LDL isolation, *ex vivo* oxidation, dietary factors, etc. Since tea flavonoids are water soluble, they are unlikely to be incorporated in the lipid core of the LDL particle but are more likely to be loosely attached to its outer surface and, as a result, tea flavonoids may be lost during the LDL isolation procedure. Applying a method that assesses LDL oxidation *in vivo* instead reflects the protective potential of tea flavonoids in the circulation following tea consumption.

Epidemiological studies suggest that tea flavonoids help reduce the risks of CHD and stroke. *In vitro* studies provide ample evidence on potential mechanisms by which tea and tea components may affect the risk of CHD. Clinical intervention trials have focused on the effect of tea drinking on blood lipid profiles. There is no clear evidence that tea drinking will reduce blood cholesterol or triacyglycerols in the subjects studied. More research remains to be done in the form of human intervention studies to demonstrate *in vivo* effects of tea consumption on risk factors of coronary events.

5. TEA AND CANCER PREVENTION

Cancer is a complex disease that occurs in tissues and organs when genetic damage to cells causes mutations in oncogenes and or tumor suppressor genes resulting in uncontrolled cell growth and metastasis (Weinberg, 1996). Many factors contribute to cancers including tobacco use, diet, lifestyle, the environment and genetics. Two-thirds of all cancers are linked to tobacco use and dietary factors. In particular, consumption of alcohol, high-fat foods and well-cooked red meats have been associated with higher risk of cancer (Trichopoulos et al., 1996). Diet is also important in lowering cancer risk. Increased consumption of plant foods and fresh vegetables and fruits has been shown to lower the risk of cancer (Anonymous, 1997), suggesting that plant-based foods contain components important to the prevention of cancer.

Epidermiological studies have shown that increased intake of plant foods reduces the risk of cancer (Steinmetz and Potter, 1991; Yu et al., 1995; Blot et al., 1996; Steinmetz and Potter, 1996; Zheng et al., 1996; Imai et al., 1997; Knekt et al., 1997; Dreosti et al., 1998). Tea and other plant foods are dietary sources of nutrients such as carotenoids, tocopherols, ascorbic acid and non-nutrient phytochemicals, especially flavonoids. In particular, tea is a significant source of flavonoid antioxidants. *In vivo* and *in vitro*, flavonoids have been shown to be physiologically active, suggesting that they may play a role in prevention of cancer (Wiseman et al., 1997; Dreosti et al., 1998).

A recent cohort study conducted among Iowa women found a significant reduction in risk of cancers of the digestive and urinary tract with increased consumption of black tea. The women who reported drinking two or more cups of black tea per day had 40–70% lower incidence rates of digestive and urinary cancer than women who never or infrequently drank tea (Zheng et al., 1996). Drinking green tea was found to significantly reduce the recurrence of stage I and stage II breast cancer among Japanese women, suggesting that green tea consumption prior to diagnosis of breast cancer improved the prognosis (Nakachi et al., 1998). There is more consistency in observational findings for a role of green tea in reduction of cancer risk and more variability in the black tea data. However, some of the variability in the black tea literature is due to insufficient data on tea intake, small differences between low and high tea intake among the populations studied and failure to properly correct for other diet and lifestyle factors. Overall, the observational findings provide support for a role of tea flavonoids in reduction of cancer risk.

Animal studies have shown that both black and green tea consistently decrease cancers of skin, lung, forestomach, esophagus, duodenum, pancreas, liver, breast and colon (Katiyar and Mukhtar, 1996; Wiseman et al., 1997). Tea flavonoids appear to reduce development of cancers in animals by several mechanisms. When tea is consumed during exposure to carcinogens, the flavonoids inhibit genetic damage by acting as an antioxidant (Xu et al., 1992) and by inducing production of enzymes required for the metabolism and elimination of cancer-causing chemicals (Katiyar and Mukhtar, 1996; Dong et al., 1997; Wiseman et al., 1997). Tea flavonoids decrease cell proliferation rates and inflammation and increase apoptosis. AP1 is a critical switch for induction of cell growth, transformation, angiogenesis and tumor invasiveness, and tea flavonoids (catechins and theaflavins) block activation of AP1 by binding to antioxidant response elements in the signal transduction pathway (Dong et al., 1997; McCarty, 1998). The animal data strongly supports a possible role for black and green tea flavonoids as dietary agents for cancer prevention.

Human trials are underway at several medical centers in the U.S., Japan and China to verify results of the animal study and to determine the efficacy of tea flavonoids in the prevention of cancer. Studies have confirmed that flavonoids in black and green tea are bioavailable, and significant levels can be found in human plasma and tissues after tea drinking (Hollman et al., 1997). Less is known on the detailed biochemistry of the absorption and metabolism of tea flavonoids but is under study in several laboratories worldwide. Tea flavonoids have been shown to function as dietary antioxidants, and in human studies, tea drinking reduces oxidative damage in smokers and nonsmokers (Wiseman et al., 1997; Klaunig et al., 1999). In addition, a recent study conducted in China found that a mixture of black and green tea flavonoids significantly reduced the severity of precancerous lesions of the mouth (Li et al., 1999). Observational and animal studies support a role for tea as part of an overall healthy diet in reducing risk of cancer. Clinical data in human studies on the role of tea in cancer prevention will emerge over the next few years.

6. BIOAVAILABILITY OF TEA FLAVONOIDS

In order to exert biological effects *in vivo*, flavonoids must be absorbed and transported to the target sites. Studies on the absorption and pharmacokinetics of tea flavonoids in humans show that plasma concentrations of individual flavonoids peak in 1.5–2.4 h after ingestion of tea (Lee et al., 1995; Van het Hof et al., 1998; Yang et al., 1998). There is some controversy that the absorption of flavonoids from tea is reduced by the addition of milk to tea due to binding of milk proteins with flavonoids, rendering them unavailable for absorption. One study by Serafini et al. (1996) provided indirect evidence for this theory by showing that the increase of total plasma antioxidant observed after consumption of black tea was eliminated by addition of milk to the tea. However, the effect of tea ingestion, with or without milk, on plasma catechin concentrations was not assessed in the study. Two other studies showed that the increase in the plasma antioxidant capacity and catechin concentrations after tea consumption are not affected by the addition of milk to tea (Van het Hof et al., 1998; Leenan et al., 1999). The majority of studies support the conclusion that addition of milk to tea does not affect the bioavailability of catechins.

A small percentage of administered flavonoids is excreted in the urine after ingestion of an acute dose of tea (Hollman et al., 1997). The majority of total urinary catechins is excreted within 8 h post-dosing (Lee et al., 1995; Yang et al., 1998). The pharmacokinetics and bioavailability of quercetin in humans have been reviewed by Graefe et al. (1999). Contrary to former belief, quercetin glycosides seem to be better absorbed than the quercetin aglycone. It appears that the relative excretion of quercetin in urine depends on the type of sugar moiety in the administered quercetin glycoside, e.g., the excretion of quercetin rutinoside after tea consumption is lower than quercetin glucoside excretion after the ingestion of onions, indicating better bioavailability of quercetin glucoside (De Vries et al., 1998). It is speculated that glucosides are absorbed by means of the glucose carrier in the proximal intestine. In plasma, quercetin occurs mainly as glucuronide and sulfate conjugates.

Little is known about the metabolism of tea flavonoids in humans after oral consumption of tea. Pietta et al. (1998) detected 4-hydroxybenzoic acid, 3,4-dihydroxybenzoic acid and 3-methoxy-4-hydroxybenzoic acid (vanillic acid) in urine of human volunteers after catechin ingestion. In recent years, more data has become available on absorption, distribution and metabolism of tea flavonoids *in vivo*. This is still an area of active research, and more data is expected to become available in the near future.

7. REFERENCES

Anonymous, 1997. "Food Nutrition and Prevention of Cancer: A Global Perspective," American Institute of Cancer Research, World Cancer Fund. Washington, DC.

Bailey, R., H. Nursten and I. McDowell. 1992. "Isolation and analysis of a polymeric thearubigen fraction from tea," *J Sci. Food Agric.* 59:365–375.

Balentine, D. A. 1996. "Tea." In: Kirk-Othmer Encyclopedia of Chemical Technology, 4th ed , John Wiley & Sons, New York, NY, pp. 747–768.

Benzie, I. F F., Y. T. Szeto, J. J. Strain and B. Tomlinson. 1999. "Consumption of green tea causes rapid increase in plasma antioxidant power in humans," *Nutr Cancer* 34:83–87.

Berliner, J A and J. W. Heinecke 1996. "The role of oxidized lipoproteins in atherogenesis," *Free Rad. Biol. Med.* 20:707–727.

Bingham, S. A., H. Vorster, J. C. Jerling, E Magee, A. Mulligan, S. A. Runswick and J H. Cummings. 1997. "Effect of black tea drinking on blood lipids, blood pressure and aspects of bowel habit," *Br. J. Nutr.* 78:41–55.

Blofeld, J. 1985. "Tea and health," *The Chinese Art of Tea* George Allen & Unwin, London. pp. 154–163.

Blot, W. J., W. H. Chow and J. K. McLaughlin. 1996. "Tea and cancer. a review of the epidemiological evidence," *Eur. J. Cancer Prev.* 5.425–438.

Chan, P T , W. P. Fong, Y. L. Cheung, Y. Huang, W. K K. Ho and Z. Y. Chen. 1999. "Jasmine green tea epicatechins are hypolipidemic in hamsters (*Mesocricetus auratus*) fed a high fat diet," *J. Nutr.* 129.1094–1101.

De Vries, J. H. M., P. C. H. Hollman, S. Meyboom, M. N C. P. Buysman, P. L. Zock, W. A. van Staveren and M. B. Katan. 1998. "Plasma concentrations and urinary excretion of the antioxidant flavonols quercetin and kaempferol as biomarkers for dietary intake," *Am. J. Clin. Nutr.* 68:60–65.

Dong, Z., W. Ma, C. Huang and C. S. Yang. 1997. "Inhibition of tumor promoter-induced AP1-activation and cell transformation by tea polyphenols, (–)-epigallocatechin gallate and theaflavins," *Cancer Res.* 57:4414–4419.

Dreosti, I. E. 1996. "Bioactive ingredients: antioxidants and polyphenols in tea," *Nutr. Rev.* 54:S51–S58.

Dreosti, I. E., M. J Wargovich and C. S. Yang. 1998. "Inhibition of carcinogenesis by tea: the evidence from experimental studies," *Crit. Rev. Food Sci. Nutr.* 37:761–770.

Finger, A., S. Kuhr and U. H. Engelhardt. 1992. "Chromatography of tea constituents," *J. Chromatogr.* 624·293–315.

Fourneau, C, A. Laurens, R. Hocquemiller and A. Cave. 1996. "Radical scavenging evaluation of green tea extracts," *Phytother. Res.* 10.529–530.

Frei, B. 1995. "Cardiovascular disease and nutrient antioxidants: role of low-density lipoprotein oxidation," *Crit. Rev. Food Sci. Nutr.* 35 83–98.

Fremont, L., M. T. Gozzelino, M. P. Franchi and A. Linard. 1998. "Dietary flavonoids reduce lipid peroxidation in rats fed polyunsaturated or monounsaturated fat diets," *J Nutr.* 128 1495–1502.

Galvez, J, J. P. de la Cruz, A. Zarzuelo and F. S. de la Cuesta. 1995. "Flavonoid inhibition of enzymatic and nonenzymatic lipid peroxidation in rat liver differs from its influence on the glutathione-related enzymes," *Pharmacology.* 51:127–133

Gardner, P T, D. B. McPhail and G C. Duthie. 1998. "Electron spin resonance spectroscopic assessment of the antioxidant potential of teas in aqueous and organic media," *J. Sci. Food Agric.* 76.257–262.

Geleijnse, J M, L. J. Launer, A. Hofman, H. A P. Pols and J. C. M. Witteman 1999. "Tea flavonoids may protect against atherosclerosis· The Rotterdam Study," *Arch. Intern. Med* 159:2170–2174

Graefe, E. U., H. Derendorf and M Veit. 1999. "Pharmacokinetics and biovailability of the flavonol quercetin in humans," *Int. J. Clin. Pharmacol. Therapeutics* 37·219–233.

Green, M. S. and G. Harari. 1992. "Association of serum lipoproteins and health-related habits with coffee and tea consumption in free-living subjects examined in the Israeli CORDIS Study," *Prev Med.* 21:532–545

Grinberg, L N., H. Newmark, N. Kitrossky, E. Rahamim, M. Chevion and E. A Rachmilewitz. 1997 "Protective effects of tea polyphenols against oxidative damage to red blood cells," *Biochem Pharmacol* 54:973–978.

Halder, J. and A. N. Bhaduri. 1998. "Protective role of black tea against oxidative damage of human red blood cells," *Biochem. Biophys. Res. Commun.* 244.903–907.

Harbowy, M. and D. A. Balentine. 1997. "Tea Chemistry," *Crit Rev. Plant Sci.* 16 415–480

Hasegawa, R., T. Chujo, K. Sai-Kato, T. Umemura, A. Tanimura and Y. Kurokawa. 1995. "Preventive effects of green tea against liver oxidative DNA damage and hepatotoxicity in rats treated with 2-nitropropane," *Food Chem. Toxicol.* 33:961–970.

Hashimoto, F., G. Nonaka and I. Nishioka. 1988. "Tannins and related compounds. LXXIX. Isolation and structure elucidation of B,B'-linked bisflavanoids, theasinensins D-G and oolong theanin from oolong tea," *Chem. Pharm. Bull.* 36:1676–1684.

Hatano, T., R. Edamatsu, M. Hiramatsu, K. Mori, Y. Fujita, T Yasuhara, T. Yoshida and T. Okuda. 1989. "Effects of the interaction of tannins with co-existing substances. VI. Effects of tannins and related polyphenols on superoxide anion radical and on 1,1-diphenyl-2-picrylhydrazyl radical," *Chem Pharm. Bull.* 37:2016–2021.

Hertog, M. G. L, E J M. Fesken, P. C. H. Hollman, M. B. Katan and D. Kromhout. 1993. "Dietary antioxidant flavonoids and risk of coronary heart disease: the Zutphen Elderly Study," *Lancet.* 342.1007–1011.

Hertog, M. G. L, P. M. Sweetnam, A. M. Fenily, P. C Elwood and D. Kromhout. 1997. "Antioxidant flavonols and ischemic heart disease in a Welsh population of men: the Caerphilly Study," *Am. J. Clin. Nutr.* 65:1489–1494.

Hollman, P. C H., L. B. M Tijburg and C. S. Yang. 1997. "Bioavailability of flavonoids from tea," *Crit. Rev. Food Sci. Nutr.* 37.719–738.

Hong, C. Y., C. P. Wang, Y. C. Lo and F. L. Hsu. 1994. "Effect of flavan-3-ol tannins purified from *Camellia sinensis* on lipid peroxidation of rat mitochondria," *Am. J. Chinese Med.* 22·285–292

Igarashi, K. and M. Ohmuma 1995. "Effects of isorhamnetin, rhamnetin, and quercetin on the concentrations of cholesterol and lipoperoxide in the serum and liver and on the blood and liver antioxidative enzyme activities of rats," *Biosci Biotech Biochem.* 59˙595–601

Imai, K and K Nakachi 1995. "Cross sectional study of effects of drinking green tea on cardiovascular and liver disease," *British Medical J.* 310˙693–696.

Imai, K , K Suga and K. Nakachi. 1997. "Cancer-preventive effects of drinking green tea among a Japanese population," *Prev. Med.* 26.769–775.

Ishikawa, T., M. Suzukawa, T. Ito and H. Yoshida. 1997. "Effect of tea flavonoid supplementation on the susceptibility of low-density lipoprotein to oxidative modification," *Am J Clin Nutr.* 66.261–266

Jovanovic, S., S Steenken, M Toasic, B. Marjanovic and M G Simic 1994 "Flavonoids as antioxidants," *J. Am. Chem. Soc.* 116.4846–4851.

Katiyar, S and H Mukhtar. 1996. "Tea in chemoprevention of cancer. epidemiological experimental studies (review)," *Int J. Oncology.* 8˙221–238.

Keli, S. O , M. G. L Hertog, E. J. M. Fesken and D. Kromhout 1996. "Dietary flavonoids, antioxidant vitamins, and incidence of stroke. The Zutphen Study," *Arch Intern Med.* 154.637–642.

Klaunig, J. E., C Han, L. M Kamendulis, J. Chen, C Heiser, M S Gordon and E. R. Mohler, 3rd. 1999. "The effect of tea consumption on oxidative stress in smokers and nonsmokers," *Proc. Soc Exp. Biol Med* 220.249–254

Knekt, P , R. Jarvinen, R Seppanen, M. Hellovaara, L Teppo, E Pukkala and A Aromaa 1997. "Dietary flavonoids and the risk of lung cancer and other malignant neoplasms," *Am J. Epidemiol.* 146˙223–230

Lean, M E , M. Noroozi, I. Kelly, J. Burns, D. Talwar, N Sattat and A Crozier. 1999. "Dietary flavonols protect diabetic human lymphocytes against oxidative damage to DNA," *Diabetes* 48˙176–181

Lee, M. J., Z. Y. Wang, H Li, L. Chen, Y. Sun, S Gobbo, D. A Balentine and C. S. Yang. 1995 "Analysis of plasma and urinary tea polyphenols in human subjects," *Cancer Epidemiol. Biomarkers Prev.* 4˙393–399.

Leenan, R , A J C. Roodenburg, L B. M. Tijburg and S. A. Wiseman 1999. "A single dose of tea with or without milk increases plasma antioxidant activity in humans," *Eur. J. Clin. Nutr.* 53.1–6.

Li, N , Z Sun, C Han and J. Chen. 1999. "The chemopreventive effects of tea on human oral precancerous mucosa lesions," *Proc. Soc. Exp. Biol. Med.* 220 218–224.

Luo, M , K. Kannar, M. L. Wahlqvist and R. C. O'Brien. 1997. "Inhibition of LDL oxidation by green tea extract," *Lancet.* 349:360–361.

Mangiopane, H., J. Thomson, A. Salter, S. Brown, G D. Bell and D. White. 1992. "The inhibition of the oxidation of low-density lipoprotein by (+)-catechin, a naturally occurring flavonoid," *Biochem Pharmacol.* 43.445–450.

Matsuda, H , T Chisaka, Y. Kubomura, J Yamahara, T. Sawada, H. Fujimura and H. Kimura. 1986. "Effects of crude drugs on experimental hypercholesterolemia I. Tea and its active principles," *J. Ethnopharmacology.* 17:213–224.

Matsumoto, N., K. Okushio and Y. Hara. 1998. "Effect of black tea polyphenols on plasma lipids in cholesterol-fed rats," *J. Nutr. Sci. Vitaminol.* 44:337–342.

McAnlis, G T , J. McEneny, J. Pearce and I. S. Young. 1998. "Black tea consumption does not protect low density lipoprotein from oxidative modification," *Eur. J. Clin. Nutr.* 52:202–206

McCarty, M F. 1998. "Polyphenol-mediated inhibition of AP-1 transactivating activity may slow cancer growth by impeding angiogenesis and tumor invasiveness," *Medical Hypotheses* 50:511–514

Miller, N. J., J. Sampson, L. P. Candeis, P. Bramley and C. Rice-Evans. 1996a. "Antioxidant activities of carotenes and xanthophylls," *FEBS Lett.* 384:240–242.

Miller, N. J., C. Castelluccio, L Tijburg and C. Rice-Evans. 1996b. "The antioxidant properties of theaflavins and their gallate esters—radical scavengers or metal chelators?" *FEBS Lett.* 392:40–44.

Miura, S., J. Watanabe, T Tomita, M. Sano and I. Tomita. 1994 "The inhibitory effects of tea polyphenols (flavan-3-ol derivatives) on Cu^{2+} mediated oxidative modification of low density lipoprotein," *Biol. Pharm. Bull.* 17:1567–1572.

Miura, S., J. Watanabe, M. Sano, T. Tomita, T. Osawa, Y. Hara and I. Tomita. 1995. "Effects of various natural antioxidants on the Cu^{2+}-mediated oxidative modification of low density lipoprotein," *Biol. Pharm. Bull.* 18:1–4.

Morel, I., G. Lescoat, P. Cogrel, O. Sergent, N. Pasdeloup, P. Brissot, P. Cillard and J. Cilliard. 1993. "Antioxidant and iron-chelating activities of the flavonoids catechin, quercetin, and diosmetin on iron-loaded rat hepatocyte cultures," *Biochem. Pharmacol.* 45:13–19.

Muramatsu, K., M. Fukuyo and Y Hara. 1986. "Effect of green tea catechins on plasma cholesterol levels in cholesterol-fed rats," *J. Nutr. Sci. Vitaminol.* 32:613–622.

Nakachi, K., K. Suemasu, K. Suga, T Takeo, K. Imai and Y. Higashi. 1998. "Influence of drinking green tea on breast cancer malignancy among Japanese patients," *Jpn. J. Cancer Res.* 89:254–261.

Nanjo, F., M. Honda, K. Okushio, N. Matsumoto, F. Ishigaki, T. Ishigami and Y. Hara. 1993. "Effects of dietary tea catechins on α-tocopherol levels, lipid peroxidation, and erythrocyte deformability in rats on high palm oil and perilla oil diets," *Biol. Pharm. Bull.* 16:1156–1159.

Nanjo, F., K. Goto, R. Seto, M. Suzuki, M. Sakai and Y. Hara. 1996. "Scavenging effects of tea catechins and their derivatives on 1,1-diphenyl-2-picrylhydrazyl radical," *Free Rad. Biol. Med.* 21:895–902.

Obanda, M. and P. O. Owuor. 1995. "Impact of shoot maturity on chlorophyll content, composition of volatile flavour compounds and plain black tea chemical quality parameter of clonal leaf," *J. Sci. Food Agric.* 69:529–534.

Pannala, A., C. Rice-Evans, B. Halliwell and S. Surinder. 1997. "Inhibition of peroxynitrite-mediated tyrosine nitration by catechin polyphenols," *Biochem. Biophys. Res. Commun.* 232:164–168.

Pietta, P. G., P. Simonetti, C. Gardana, A. Brusamolino, P. Morazzoni and E. Bombardelli. 1998. "Catechin metabolites after intake of green tea infusion," *BioFactors.* 8:111–118.

Polette, A., D. Lemaitre, M. Lagarde and E. Vericel. 1996. "N-3 fatty acid-induced lipid peroxidation in human platelets is prevented by catechin," *Thromb. Haemostasis.* 75:945–949.

Princen, H. M. G., W. van Duyvenvoorde and R. Buytenhek. 1998. "No effect of consumption of green and black tea on plasma lipid and antioxidant levels and on LDL oxidation in smokers," *Arterioscler Thromb. Vasc. Biol.* 18:833–841.

Rice-Evans, C., N. J. Miller, P. G. Bolwell, P. M. Bramley and J. B. Pridham. 1995. "The relative antioxidant activities of plant-derived polyphenolic flavonoids," *Free Rad. Res.* 22:375–383.

Rice-Evans, C. A., N. J. Miller and G. Paganga. 1996. "Structure-antioxidant activity relationships of flavonoids and phenolic acids," *Free Rad. Biol. Med.* 20:933–956.

Rimm, E. B., M. B. Katan, A. Ascherio, M. J. Stampfer and W. C. Willet. 1996. "Relation between intake of flavonoids and risk for coronary heart disease in Male Health Professionals," *Ann. Intern. Med.* 125:384–389.

Robertson, A. 1992. "The chemistry and biochemistry of black tea production: the non-volatiles," In: Tea Cultivation to Consumption. Wilson, K. C. and Clifford, M. N., Eds., Chapman and Hall, London, pp. 553–601.

Salah, N., N. Miller, G. Paganga, L. Tijburg, G. P. Bolwell and C. Rice-Evans. 1995. "Polyphenolic flavanols as scavengers of aqueous phase radicals and as chain-breaking antioxidants," *Arch. Biochem. Biophys.* 322:339–346.

Segal, M. 1996. "Tea: a story of serendipity," *FDA Consumer.* 30.22–26.

Serafini, M., A. Ghiselli and A Ferro-Luzi. 1996. "*In vivo* antioxidant effect of green and black tea in man," *Eur. J. Clin. Nutr.* 50:28–32.

Sesso, H. D., M. Gaziano, J E Buring and C. H. Hennekens 1999. "Coffee and tea intake and the risk of myocardial infarction," *Am. J Epidemiol.* 149:162–167.

Shim, J. S., M. H. Kang, Y. H. Kim, J. K. Roh, C Roberts and L. P. Lee. 1995 "Chemopreventive effect of green tea (*Camellia sinensis*) among cigarette smokers," *Cancer Epidemiol Biomarkers Prev.* 4:387–391.

Shimoi, K., S. Masuda, M. Furugori, S. Esaki and N. Kinae. 1994. "Radioprotective effect of antioxidant flavonoids in γ-ray-irradiated mice," *Carcinogenesis.* 15:2669–2672.

Steinmetz, K and J D. Potter. 1991. "A review of vegetables, fruit and cancer. I. Epidemiology," *Cancer Causes Control.* 2:325–357.

Steinmetz, K. and J. D. Potter. 1996. "Vegetables, fruits, and cancer prevention: a review," *J. Am. Diet. Assoc.* 96:1027–1039.

Stensvold, I., A. Tverdal, K. Solvoll and O. P. Foss. 1992. "Tea consumption. Relationship to cholesterol, blood pressure, and coronary and total mortality," *Prev. Med.* 21:546–553.

Suzuki, H., A. Ishigaki and Y. Hara. 1998. "Long-term effect of a trace amount of tea catechins with perilla oil on the plasma lipids in mice," *Int. J. Vitamin Nutr. Res.* 68:272–274.

Terao, J., M. Piskula and Q. Yao. 1994. "Protective effect of epicatechin, epicatechin gallate, and quercetin on lipid peroxidation in phospholipid bilayers," *Arch. Biochem. Biophys.* 308:278–284.

Tijburg, L. B. M., T. Mattern, J. D Folts, U M. Weisgerber and M. B. Katan. 1997a. "Tea flavonoids and cardiovascular diseases: A review," *Crit. Rev. Food Sci. Nutr.* 37:771–785.

Tijburg, L. B. M., S. A. Wiseman, G. W. Meijer and J. A. Westrate. 1997b. "Effects of green tea, black tea and dietary lipophilic antioxidants on LDL oxidizability and atherosclerosis in hypercholesterolaemic rabbits," *Atherosclerosis.* 135:37–47.

Trichopoulos, D., F. P. Li and D. J. Hunter. 1996. "What causes cancer?," *Scientific American.* 275:80–87.

Tsubono, Y. and S Tsugane 1997. "Green tea in relation to serum lipid levels in middle-aged Japanese men and women." *Ann Epidemiol.* 7:280–284.

Tzeng, S. H., W C. Ko, F. N. Ko and C. M. Teng. 1991. "Inhibition of platelet aggregation by some flavonoids," *Thromb. Res.* 64:91–100.

Uchida, S., R. Edamatsu, M. Hiramatsu, A. Mori, G. Nonaka, I. Nishioka, M. Niwa and M. Ozaki. 1987. "Condensed tannins scavenge active oxygen free radicals," *Med. Sci. Res.* 15:831–832.

Uchida, S., M. Ozaki, K. Suzuki and M. Shikita. 1992. "Radioprotective effects of (-)-epigallocatechin 3-O-gallate (green-tea tannin) in mice," *Life Sci.* 50:147–152.

Ukers, W. H. 1994. "Tea in the beginning," In: *All about Tea.* Hyperion Press, Westport, CT, pp. 1–22.

Van Acker, S. A. B. E., D. J. van den Berg, M. N. J. L., Tromp, D. H. Griffioen, W. P. Bennekom. W. J. F. van der Vijgh and A. Bast. 1996a. "Structural aspects of antioxidant activity of flavonoids," *Free Rad. Biol. Med.* 20:331–342.

Van Acker, S. A. B. E., M. J. de Groot, D. J. van den Berg, M. N. J. L., Tromp, G. Donné-Op de Kelder, W. J. F., van der Vijgh and A. Bast. 1996b. "A quantum chemical explanation of the antioxidant activity of flavonoids," *Chem. Res. Toxicol.* 9:1305–1312.

Van het Hof, K. H., H S. M. de Boer, S A. Wiseman, N Lien, J. A. Weststrate and L. B. M. Tijburg. 1997. "Consumption of green or black tea does not increase resistance of low-density lipoprotein to oxidation in humans," *Am. J. Clin. Nutr.* 66:1125–1132.

Van het Hof, K. H , G. A. A. Kivits, J A. Weststrate and L. B. M. Tijburg. 1998. "Bioavailability of catechins from tea: the effect of milk," *Eur. J. Clin. Nutr.* 52.356–359.

Vinson, J. A and Y. A. Dabbagh. 1998. "Effect of green and black tea supplementation on lipids, lipid oxidation and fibrinogen in the hamster: mechanisms for the epidemiological benefits of tea drinking," *FEBS Letters.* 433.44–46.

Vinson, J. A., Y. A. Dabbagh, M M. Serry and J. Jang. 1995a. "Plant flavonoids, especially tea flavonols, are powerful antioxidants using an *in vitro* oxidation model for heart disease," *J. Agric. Food Chem.* 43:2800–2802

Vinson, J. A., J. Jang, Y A Dabbagh, M. M. Serry and S. Cai 1995b. "Plant polyphenols exhibit lipoprotein bound antioxidant activity using an *in vitro* oxidation model for heart disease," *J. Agric. Food Chem.* 43:2798–2799.

Weinberg, R. A. 1996. "How cancer arises." *Scientific American.* 275.62–70.

Wilson, K. C. and M N. Clifford 1992. *Tea: Cultivation to Consumption,* Chapman and Hall, London.

Wiseman, S. A., D. A. Balentine and B. Frei 1997. "Antioxidants in tea," *Crit. Rev. Food Sci. Nutr.* 37:705–718.

Xu, Y., C. T. Ho, S. G. Amin, C. Han and F. L. Chung. 1992. "Inhibition of tobacco-specific nitrosamine-induced lung tumorigenesis in A/J mice by green tea and its major polyphenol as antioxidants," *Cancer Res.* 52 3875–3879.

Xue, K. X., S. Wang, G J. Ma, P. Zhou, P. Q. Wu, R. F. Zhang, X Zhen, W. S. Chen and Y. Q. Wang 1992. "Micronucleus formation in peripheral-blood lymphocytes from smokers and the influence of alcohol- and tea-drinking habits," *Intl. J. Cancer.* 50:702–705.

Yamanaka, N , O. Oda and S. Nagao 1997. "Green tea catechins such as (−)-epicatechin and (−)-epigallocatechin accelerate Cu^{2+}-induced low density lipoprotein oxidation in propagation phase," *FEBS Lett.* 401:230–234.

Yang, C. S. and Z -Y. Wang 1993. "Tea and cancer," *J. Natl. Cancer Inst.* 85:1038–1049.

Yang, C. S., L. Chen, M. J. Lee, D. Balentine, M. C. Kuo and S P. Schantz. 1998. "Blood and urine levels of tea catechins after ingestion of different amounts of green tea by human volunteers." *Cancer Epidemiol. Biomarkers Prev.* 7:351–354.

Yen, G. C and H. Y. Chen. 1995. "Antioxidant activity of various tea extracts in relation to their antimutagenicity," *J Agric. Food Chem.* 43:27–32.

Yochum, L., L. H. Kushi, K. Meyer and A. R. Folsom. 1999. "Dietary flavonoid intake and risk of cardiovascular disease in postmenopausal women," *Am. J. Epidemiol.* 149 943–949.

Yoshida, T., K. Mori, T. Hatano, T. Okomura, I. Uehara, K. Komagoe, Y. Fujita and T. Okuda. 1989. "Studies on inhibition mechanism of autoxidation by tannins and related polyphenols on 1,1-diphenyl-2-picrylhydrazyl radical," *Chem. Pharm. Bull.* 37:1919–1921.

Yu, G , C Hsieh, L. Wang, S Yu, X. Li and T. Jin. 1995. "Green tea consumption and risk of stomach cancer: a population-based case-control study in Shanghai, China," *Cancer Causes Control.* 6:532–538.

Zhang, A., Q. Y. Zu, Y. S. Luk, K. Y. Ho, K. P. Fung and Z. Y. Chen. 1997. "Inhibitory effects of Jasmine green tea epicatechin isomers on free radical-induced lysis of red blood cells," *Life Sci.* 61:383–394

Zheng, W., T. J. Doyle, L. H. Kushi, T A. Sellers, C.-P. Hong and A. R. Folsom. 1996. "Tea consumption and cancer incidence in a prospective cohort study of postmenopausal women," *Am J. Epidemiol.* 144:175–182.

Zhenhua, D , C Yuan, Z Mei and F Yunzhong. 1991 ''Inhibitory effect of China green tea polyphenol on the oxidative modification of low density lipoprotein by macrophages,'' *Med Sci Res* 19 767–768.

CHAPTER 10

Bilberries and Blueberries as Functional Foods and Nutraceuticals

M. E. CAMIRE

1. INTRODUCTION

1.1. HISTORICAL USE

THE bilberry (*Vaccinium myrtillus* L.) is a low-growing shrub native to northern Europe, Asia and North America. Bilberry fruit and leaves are used for a variety of medical conditions, and the small dark blue fruit is eaten fresh or made into juice and preserves. Although bilberries are often called blueberries because of their color, the "true" blueberry is a native American fruit. Three species are grown commercially: the wild or lowbush blueberry (*V. angustifolium* Ait.), the highbush blueberry (*V. corymbosum* L.) and the rabbiteye blueberry (*V. ashei* Reade), so-named because it resembles a wild rabbit's eye (Austin, 1994). American blueberries have traditionally been used as food, but there is considerable interest in the health benefits of these berries.

1.2. SYSTEMATICS AND DISTRIBUTION

Bilberries and blueberries are members of the Ericaceae (heath) family and are related to other small fruit of economic importance including the cranberry (*V. macrocarpon* Ait.) and the lingonberry (*V. vitis-idaea* L.). *V. myrtillus* appears to have arisen in Europe, then spread to far northern regions of Asia and North America. *V. angustifolium* is grown in Maine and the eastern Canadian provinces. In the U.S. New Jersey and Michigan are the major producing states (Moore, 1994), and the fruit is also grown in other states that meet the chilling requirements of 800–1,060 hours at temperatures below

7.2°C and do not have winter temperatures below −24°C (Eck, 1988). British Columbia is the largest producer of cultivated blueberries in Canada (Villata, 1998). Highbush blueberries are also successfully grown in Australia, New Zealand, Chile, Germany and throughout central Europe. *V. ashei* is cultivated primarily in Georgia, Florida, Arkansas and Texas. Harvest method, hand or mechanical, varies by species and location (Moore, 1994). The United Nations Food and Agriculture Organization (FAO) 1998 blueberry crop statistics are shown in Table 10.1.

1.3. FOOD USES

In the U.S., approximately 15% USDA grade A blueberries are sold fresh, but most of the lowbush blueberries and a significant portion of the highbush blueberries are frozen or canned (Moore, 1994). Lower quality fruit is used in jams and purees where appearance is not critical. Purees can be added to yogurt, ice cream and fruit smoothies. Combination of osmotic and air drying technology has produced shelf-stable blueberries that maintain a pleasant chewy texture (Mazza et al., 1998). These dried berries can be used as snacks, in granola, in trail mixes and in hot and cold breakfast cereals. Bakery applications for blueberries are discussed by Villata (1998). About 20% of the U.S. cultivated blueberry crop is used in food service operations (Kenyon, 1997). Markakis (1982) reviewed many applications for anthocyanins in food ingredients, but commented that berry juice and juice concentrates were exempt from toxicological testing required by the U.S. Food and Drug Administration for food colorants. Francis (1985) described several food products for which blueberries could be used as a colorant.

A fairly new application for *Vaccinium* products in the U.S. is dietary supplements, which are regulated as foods under the Dietary Supplement Health and Education Act of 1994 (DSHEA). Dietary supplements can be marketed without prior demonstration of safety and efficacy, as is required for drugs, provided the ingredients have a traditional use. Supplements are permitted to bear the same health claims as allowed for foods, but supplements

TABLE 10.1. **1998 FAO World Blueberry Production Statistics.**[a,b]

Country	Production (metric tons)
United States	95,000
Canada	33,000
Eastern Europe	17,000
Western Europe	5,900
New Zealand	1,000

[a] Source FAOSTAT Database (http //apps fao org)
[b] Australia and Chile, two nations that export blueberries, were not included in the database

have more freedom to display structure-function claims. Structure-function claims relate a compound to maintenance of normal structure or functions in the body. Foods may make such claims only for nutrients; for example, "vitamin C maintains normal capillary strength." Dietary supplements are allowed under DSHEA to make such claims for compounds for which no recommended daily intake is listed. Examples of structure-function claims on supplements containing bilberry are: "Promotes normal vision" and "Enhances eyesight" (Camire, 1997).

1.4. NONFOOD USES

Bilberries and blueberries are used in a wide variety of products, including cosmetics and pharmaceuticals. Both the fruit and leaves of *V. myrtillus* are listed in the German Commission E monographs for botanicals (Gruenwald et al., 1998). Aqueous extracts of bilberry leaves and fruit have been the most traditional form (Figure 10.1). Generally, these pharmaceutical products contain an alcohol extract of bilberry fruit, which in turn is manufactured into tinctures, pills and capsules. European products are usually standardized to contain 25% anthocyanidins, with a recommended dosage of 20–40 mg anthocyanidins (also referred to as anthocyanosides in the trade) thrice daily, or 80–160 mg of standardized bilberry extract three times per day (Murray, 1997).

Morazzoni and Bombardelli (1996) reviewed the history of medical uses for bilberries from the Middle Ages to the present. A 1994 survey found 183 products containing *V. myrtillus* extract as an ingredient (Kalt et al., 1994); however, that figure may be considerably larger today. When the survey was

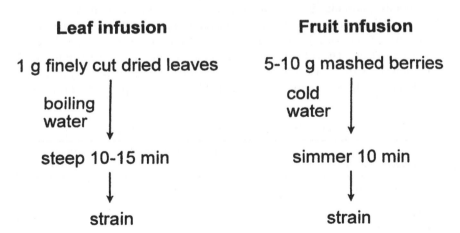

Figure 10.1 Extraction of *Vaccinium myrtillus* L. fruit and leaves. Adapted from Gruenwald et al., 1998. *PDR for Herbal Medicines,* 1st ed., Medical Economics Co., Inc., Montvale, NJ.

published, DSHEA had just been passed, and the U.S. dietary supplement market has undergone tremendous growth since then. Kurtzweil (1998) reported a 30% increase in U.S. supplement sales from 1994 to 1996. Half of the $6.5 billion sales of herbal supplements are in Europe (Anonymous: 1998).

2. CHEMICAL COMPOSITION

2.1. CHEMICAL STRUCTURES

Water is the primary constituent of the berries, followed by carbohydrates (Table 10.2). Measurement of sugars by soluble solids overestimated sugars in lowbush blueberries by 30% (Kalt and McDonald, 1996). Vitamins and minerals are fairly low in these fruits due to the low solids content. Lowbush blueberries are higher than rabbiteye blueberries in calcium, magnesium, iron and manganese, rabbiteyes are higher in potassium (Bushway et al., 1983; Eitenmiller et al., 1977).

Citric, malic and quinic acids are the major organic acids in lowbush blueberries, contributing 36%, 31% and 20% to total acidity; chlorogenic is the major phenolic acid (Kalt and McDonald, 1996). Citric (75%) and succinic (17%) acids are the most abundant acids in highbush blueberries, while rabbiteye blueberries have an acid distribution of 50% succinic, 37% malic and 10% citric (Ehlenfeldt et al., 1994). Each organic acid has its own distinctive taste, thus, these differences can contribute to real flavor differences among species.

The functional components of *Vaccinium* products appear to be the phenolic compounds, particularly the anthocyanins. Several anthocyanins are found in the fruit. The structures of the most prevalent anthocyanidins are shown in Figure 10.2. Each variation on the basic flavylium ring seems to have different biological effects. The role of the various sugar moieties is not well understood. Although the various *Vaccinium* species contain the same anthocyanins, the distribution of these compounds varies (Table 10.3). The pigments are located

TABLE 10.2. Bilberry and Blueberry Composition.[a]

Vaccinium Species	% Dry Matter	Total Soluble Solids (°Brix)	Ascorbic Acid (mg/100 g)
myrtillus	15 8	10.0	1 3[b]
angustifolium	31 3	16 5	16.4
ashei	18.4	13 2	8 4
corymbosum	23 4	14 8	10.2

[a] Source Prior et al (1998)
[b] Some losses may have occurred during shipping from Germany to Massachusetts

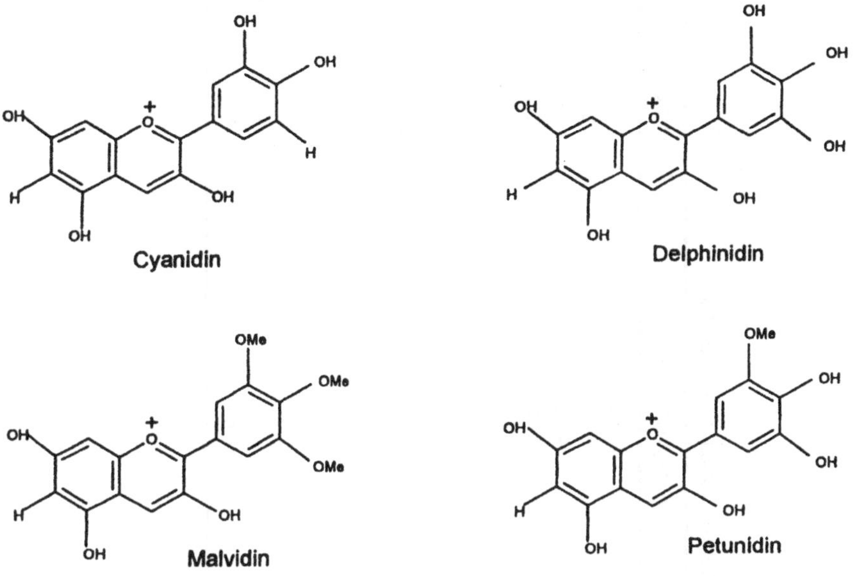

Figure 10.2 Chemical structure of anthocyaninidins found in bilberries and blueberries.

primarily in the skin of the berries, with the exception of *V. myrtillus* (Figure 10.3). Smaller blueberries have a greater surface area than large berries on an equal weight basis, and thus contain more pigments (Francis, 1985).

Other phenolic compounds are also found in these plants. Bilberry leaves contain flavonoids (quercitrin, isoquercitrin, hyperoside, avicularin, meratine, astragaline), catechin tannins, oligomeric proanthocyanidins, iridoid monoterpenes (asperuloside, monotropein), phenolic acids (chlorogenic, salicylic, gentisic) and quinolizidine alkaloids (myrtine, epimyrtine and occasionally arbutine); the fruit can contain these compounds as well as pectin (Gruenwald et al., 1998).

2.2. ANALYSIS

Total phenolics can be determined by reaction with Folin-Ciocalteau reagent, with subsequent absorbance measured at 725 nm (Singleton and Rossi, 1965). Individual compounds can be measured by high-performance liquid chromatography (HPLC), and identification of isomers is facilitated by HPLC in tandem with mass spectrometry (MS).

Analytical procedures for anthocyanins have been reviewed by Mazza and Miniati (1993) and Mazza (1997). Anthocyanins have traditionally been measured by the absorbance of the alcoholic extract at 535 nm. However, that method provides only a total value and does not permit screening for individual

TABLE 10.3. Anthocyanin Profiles of Different Vaccinium Species.[a]

Anthocyanin	V. myrtillus[b]	V. angustifolium[c]	V. corymbosum[d]
Cyanidin-3-monoarabinoside	10.46	2.62	—
Cyanidin-3-monogalactoside	8.68	4.31	5.0
Cyanidin-3-monoglucoside	9.34	4.52	5.0
Delphinidin-3-monoarabinoside	13.81	5.05	10.3
Delphinidin-3-monogalactoside	12.99	8.65	13.5
Delphinidin-3-monoglucoside	14.96	9.34	6.9
Malvidin-3-monoarabinoside	2.29	4.75	11.9
Malvidin-3-monogalactoside	3.57	8.96	12.9
Malvidin-3-monoglucoside	11.91	13.03	11.9
Peonidin-3-arabinoside	—	—	0.1
Peonidin-3-galactoside	—	1.25	0.7
Peonidin-3-glucoside	—	2.57	1.1
Petunidin-3-arabinoside	7.78	2.45	3.1
Petunidin-3-galactoside	—	4.61	8.8
Petunidin-3-glucoside	10.11	6.84	6.6

[a] % of total anthocyanin.
[b] Petri et al., (1997); averages of extracts from Italian and Polish factories.
[c] Gao and Mazza (1994); data converted to % of total anthocyanin.
[d] Kader et al. (1996); cv. Coville.

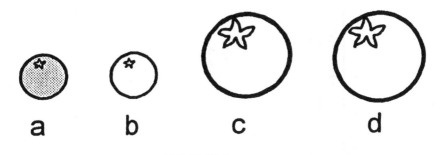

Figure 10.3 Representative size and anthocyanin distribution of bilberries and blueberries (a = *Vaccinnium myrtillus:* b = *V angustifolium.* c = *V. corymbosum:* d = *V. ashei*). Bar = 1 cm

compounds. However, it has been recommended over other techniques when a total concentration of anthocyanins is required (Petri et al., 1997). Using paper chromatography, Francis et al. (1966) first identified the presence of diglycosides in *V. angustifolium.* Reflectance colorimetry correlates poorly with anthocyanin content, presumably due to the whitish "bloom" on the berries' surface (Kalt et al., 1995).

Reversed-phase HPLC with photodiode array detectors has also been useful in separating anthocyanins (Gao and Mazza, 1994, 1995). Krawczyk and Petri (1992) encountered numerous problems while applying RP-HPLC to analysis of bilberry extract. They concluded that anthocyanin content should not be compared across methods due to self-association and copigmentation, as well as other issues. A comparison of two HPLC methods is shown in Table 10.4. An advantage of this technique is the isolation of other phenolic compounds along with the anthocyanins. Since other soluble materials in the fruit can interfere with extraction of phenolics such as chlorogenic acid, Kader et al. (1996) advised the use of ethyl acetate to remove sugars and polar compounds. Capillary electrophoresis has also shown promise for separation of closely related compounds (Bridle et al., 1996). A problem common to all analytical methods for anthocyanins is the lack of pure, readily accessible, affordable standard reference compounds.

2.3. GENETIC AND ENVIRONMENTAL EFFECTS

Lowbush blueberries, also called wild blueberries, are often hand-raked, and many cultivars, called clones, grow in the same field. Although production of single clones is practiced in Canada to a limited extent, commercial fruit is nearly always a mix of several clones. Kalt and McDonald (1996) noted many differences in composition among clones over two years, as well as significant differences during maturation.

TABLE 10.4. HPLC Methods for Detecting Individual Anthocyanins in Bilberries and Blueberries.

Reference	Gao and Mazza, 1994		Petri, Krawczyk and Kéry, 1997	
Extraction	25-30 g berries with 30 mL MeOH/formic acid/water, 70/2/28		Commercial bilberry extract	
HPLC	Waters 990 photodiode detector, 525 nm		HP10-40 diode array detector, 280 nm	
Column	Pharmacia SuperPac Pep-S, 5µM, 4 X 250 mm with a 4 X 10 mm guard column		Partisil 10-ODS	
Solvent	A. 5% formic acid in water B: methanol		A. 10% HCOOH B: MeCn	
Elution profile	0-4 min	10-12% B	0-60 min	0-8% B
	4-10 min	12-15% B	61-90 min	8% B
	10-20 min	15-20% B		
	20-23 min	20% B		
	23-32 min	20-30% B		
	32-40 min	30-35% B		
	40-48 min	35-37% B		
	48-50 min	37-70% B		
	50-53 min	70% B		
	53-55 min	70-10% B		
Flow rate	1.2 mL/min		0.5 mL/min	

296

Acylated anthocyanins were found in lowbush blueberries (*V. angustifolium*) from Nova Scotia (Gao and Mazza, 1994). Cultivar "Chignecto" contained up to 35% of its anthocyanins in the acylated form, but "Blomidon" contained no acylated pigments. The primary acylated anthocyanins were malvidin-3-acetylglucoside and malvidin-3-acetylgalactoside (Table 10.5) (Gao and Mazza, 1995). These findings suggest that lowbush and highbush blueberries could be bred to maximize the more stable acylated forms.

V. myrtillus fruit contained no flavonol glycosides, but leaves contained quercitrin, isoquercitrin and quercetin-3-arabinoside (Friedrich and Schönert, 1973). Leaf flavonol glycosides reached maximum levels by June, while phenolic acids and iridoid glucosides peaked in April and declined until harvest in August. Exploitation of the iridoid compounds could provide another product category, but excessive leaf harvest early in the season would impair fruit production. Young leaves tend to have greater concentrations of phenolic compounds (Fraisse et al., 1996).

A study examining the quality of three highbush and three rabbiteye cultivars grown at the same location in Arkansas found major cultivar and species differences (Makus and Morris, 1993). Highbush berries were larger, highly colored, and had higher titratable acidity. Rabbiteye blueberries were higher in soluble solids and seeds. Weather during the three years of the experiments affected soluble solids, dry matter, hue, firmness, titratable acidity and pH.

Linalool was the predominant volatile in ripe fruit of three rabbiteye cultivars, but each cultivar had different quantities (Tifblue > Woodward > Delite) (Horvat and Senter, 1985). Other volatile compounds varied in concentration among cultivars as the fruit ripened. Utilization of unripe fruit from processing rejects could impact the notes of cosmetic and aromatherapy products. Careful blending of green with ripe fruit is required to minimize changes in "fresh berry" scent.

TABLE 10.5. Acetylated Anthocyanin Profiles of *V. angustifolium*.[a]

Anthocyanin	As % of Total Anthocyanin
Cyanidin-3-monoarabinoside acetylated	3 61
Cyanidin-3-monogalactoside acetylated	0 35
Cyanidin-3-monoglucoside acetylated	2 38
Delphinidin-3-monogalactoside acetylated	1.25
Delphinidin-3-monoglucoside acetylated	3 68
Malvidin-3-monogalactoside acetylated	2 73
Malvidin-3-monoglucoside acetylated	6.79
Peonidin-3-galactoside acetylated	0.31
Peonidin-3-glucoside acetylated	1.49
Petunidin-3-galactoside acetylated	0 05

[a] Adapted from Gao and Mazza (1994), data for wild blueberries

2.4. PROCESSING EFFECTS

Fresh berries have a short shelf life, even when refrigerated. After two weeks of refrigerated storage, lowbush blueberries had higher levels of anthocyanins and organic acids (Kalt and McDonald, 1996), but sensory and microbial quality were not reported.

Dehydration can cause anthocyanin browning. A dried bilberry juice product, Obi-Pektin 250, contained 50% dried fruit and 50% sucrose; the anthocyanin content was 31% delphinidin, 24% cyanidin, 19% petunidin, 18% malvidin and 8% peonidin (Hong and Wrolstad, 1990). The degradation index was 1.4, polymeric color was 0.4 and contribution by tannins was 41%. The vacuum-drying process used to manufacture the powder may have prevented additional thermal degradation.

Leakage of anthocyanins from blueberries into batters has long been a problem for bakers. Although such leakage has no nutritional consequences, the resulting grayish-purple color imparted to foods is disliked by some consumers. Freezing produces microscopic cracks in berries through which the pigment-laden juice escapes (Sapers et al., 1985). Apparently, the wax coating on berries protected fresh fruit from pigment loss, but cooking hastened cell damage (Sapers and Phillips, 1985). A sodium carboxymethylcellulose coating on berries can retard bleeding (Zhang et al., 1997).

3. BIOLOGICAL AND PHARMACOLOGICAL PROPERTIES

3.1. METABOLISM

The metabolisms of individual phenolic components have been studied. Quercetin levels peaked in the plasma of human volunteers three hours after eating a quercetin-rich meal, including berries, and plasma values returned to basal levels 20 hours after the meal (Manach et al., 1998). No free quercetin was detected, but conjugated forms were recovered; the possibility of improved absorption of glycosylated forms was raised.

V. myrtillus anthocyanins administered to rats by either oral or intravenous routes were excreted in the urine and bile (Lietti and Forni, 1976). Although the liver does not appear to store anthocyanins, these compounds may adhere to capillaries, skin and kidneys, which may, in part, explain the persistence of pharmacological effects when plasma levels of anthocyanins are nil. Ring fission products were isolated from the urine of rats fed either delphinidin or malvin (malvidin-3,5-diglucoside) (Griffiths and Smith, 1972).

Rats given single doses of a mixture of 160 mg/kg cyanidin-3-glucoside (Cyg) and 20 mg/kg cyanidin-3,5-diglucoside (Cydg) had plasma levels of 907 nmol Cyg and 212 nmol Cydg thirty minutes after administration of the anthocyanins; rats given larger doses had correspondingly higher plasma levels (Miyazawa et al., 1999). Plasma and liver levels reached their peaks 15 minutes after ingestion of the mixtures. The aglycone was not detected in either type of sample. In a related human study, four volunteers given 2.7 mg/kg Cyg had plasma levels of 24 nmol, and another group had levels of 29 nmol of Cyg 60 minutes after ingesting the same dose. Neither the aglycone nor any metabolites were detected, suggesting that anthocyanins are absorbed and circulated in their native form.

3.2. ANTIOXIDANT ACTIVITY

The term "antioxidant" is common in food marketing. The Food and Nutrition Board (FNB) of the Institute of Medicine (1998) has proposed the following definition of dietary antioxidants: "A dietary antioxidant is a substance in foods that significantly decreases the adverse effects of reactive oxygen species, reactive nitrogen species, or both on normal physiological function in humans."

Although the FNB recognizes only ascorbic acid, vitamin E, β-carotene and selenium as dietary antioxidants, many compounds exhibit antioxidant activity during *in vitro* tests. However, most antioxidants can act as pro-oxidants under compound-specific conditions. No single method for verifying antioxidant activity is accepted by the medical and food science communities. The lack of a common methodology has led to discrepancies among research groups studying the same food. Measurement of the extent of oxidation *in vivo* is particularly challenging, since free radicals attack DNA, proteins and lipids (Halliwell and Aruoma, 1997).

Several studies have demonstrated the *in vitro* antioxidant activity of *Vaccinium* constituents. Chlorogenic and caffeic acids inhibited linoleic acid oxidation (Ohnishi et al., 1994). Gallic and protocatechuic acids isolated from highbush blueberries were more effective than the synthetic antioxidant BHT in delaying oxidation in lard as measured by the Rancimat test (Li et al., 1998). Crude acidified ethanol extracts from four *V. corymbosum* cultivars were studied for their antioxidant activity, but this activity was not directly related to anthocyanin or polyphenol content of the extracts (Costantino et al., 1992). Cv. Collins was the most effective inhibitor of chemically generated superoxide radicals and activity toward the enzyme xanthine oxidase (Figure 10.4), and it had the highest concentration of both anthocyanins and polyphenols.

Velioglu and coworkers (1998) compared the antioxidant activity of methanol extracts from 28 food and medicinal plants using a β-carotene bleaching assay. *V. angustifolium* cv. Fundy from Nova Scotia had higher antioxidant activity than many other products (Figure 10.5), with only 50 mg/L α-tocopherol, 200 mg/L BHT, a fibrotein, buckwheat hulls, sea buckthorn and horserad-

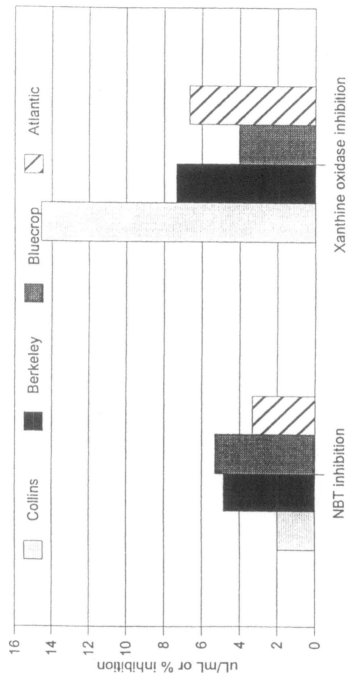

Figure 10.4 Antioxidant activity of ethanol extracts from four cultivars of *Vaccinium corymbosum*. Adapted from Costantino et al. (1992).

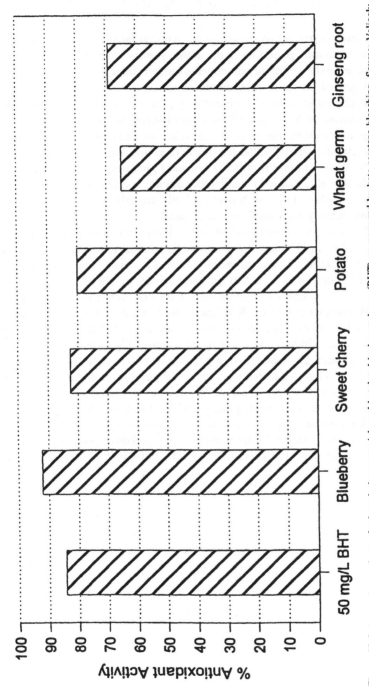

Figure 10.5 Antioxidant activity of selected plant materials and butylated hydroxytoluene (BHT) as measured by beta-carotene bleaching. Source: Velioglu et al. (1998).

ish oil having greater antioxidant activity. Antioxidant activity was not significantly related to total phenolics in the anthocyanin-rich materials, perhaps due to differences in the ability of anthocyanins to terminate free radicals compared to other types of phenolic compounds.

The oxygen radical absorbance capacity (ORAC) test has been used by the U.S. Department of Agriculture Jean Mayer Human Nutrition Research Center on Aging to evaluate the antioxidant value of acetonitrile/4% acetic acid extracts of foods. ORAC value for blueberries and bilberries were correlated with anthocyanin content ($r = 0.77$) and total phenols ($r = 0.92$) (Prior et al., 1998). ORAC values varied considerably among species and cultivars (Figure 10.6), and ascorbic acid was apparently only slightly responsible for overall antioxidant activity measured with this assay.

ORAC, anthocyanins and total phenolics increased with maturity in the two rabbiteye cultivars, Tifblue and Brightwell, studied for a period of 49 days from the formation of blue color. Highbush cv. Jersey was grown in three states, but location had no effect on composition or ORAC. Based on these findings, the researchers suggested that daily consumption of 1/2 cup (72.5 g) of blueberries would raise ORAC intake by 1–3.2 mmol. Fourteen anthocyanins tested for ORAC had greater antioxidant activity than Trolox, a water-soluble form of vitamin E (Wang et al., 1997). Cyanidin-3-glucoside and cyanidin-3-rhamnoglucoside had the greatest potency, and no clear trend emerged for a relationship between type and number of sugars and ORAC.

Human low-density lipoprotein (LDL) oxidation is believed to contribute to the formation of atherosclerotic plaques in blood vessels. Phenolic compounds found in *Vaccinium* fruit vary in their ability to retard LDL oxidation (Figure 10.7) (Meyer et al., 1998). Combinations of two and three phenolic compounds did not have the expected synergistic effects, but some additive effects were

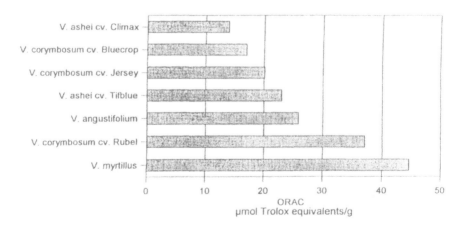

Figure 10.6 Oxygen radical absorbance capacity (ORAC) of blueberries and bilberries Adapted from Prior et al. (1998).

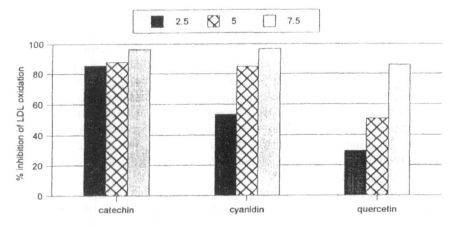

Figure 10.7 Inhibition of low-density lipoprotein (LDL) oxidation by flavonoids Adapted from Meyer et al. (1998)

observed. When these compounds were tested in a different system, the aqueous-based ABTS+ system, the antioxidant activity was reversed: quercetin > cyanidin > catechin (Rice-Evans et al., 1995).

Antioxidant activity in different systems was compared for several berries (Heinonen et al., 1998). Acetone extracts of blackberries, red raspberries and sweet cherries were more effective than highbush blueberries (cv. Jersey) in delaying hexanal formation from LDL. The high hydroxycinnamate content of blueberries was more effective in inhibiting oxidation in lecithin liposomes, while anthocyanins in the other berries were responsible for inhibiting LDL oxidation. Anthocyanins and flavan-3-ols were reduced when blueberry juice was treated with pectinase, but relative increases in extraction of other phenolic compounds may be responsible for the maintenance of antioxidant activity.

At low levels of copper (10 μM) used to induce human LDL oxidation, anthocyanins were protective in the following order: malvidin > delphinidin > cyanidin > pelargonidin; at higher copper concentration (80 μM), the order of antioxidant activity was delphinidin > cyanidin > malvidin > pelargonidin. (Satué-Gracia et al., 1997). In the liposome oxidation system, malvidin was the best antioxidant, since the others acted as pro-oxidants at the 3 μM level. These researchers concluded that a system using artificial free radical generator or a system without an oxidizable substrate may fail to fully evaluate the antioxidant potential of anthocyanins.

Aqueous extracts of *V. myrtillus* at low levels (15–20 μg/mL) retarded human LDL oxidation *in vitro* (Laplaud et al., 1997). The lag phase for production of conjugated dienes was significantly delayed by the bilberry extract, and levels of peroxides and thiobarbituric acid reactive substances (TBARS) were lower for up to seven hours after induction with copper. A net negative charge on LDL particles was also prevented. Further research is needed to evaluate dose levels and long-term consumption effects.

Although cyanidin is not one of the predominant anthocyanins in bilberries and blueberries, it has been studied the most. Cyanidin and cyanidin-3-O-β-D-glucoside (C3G) were effective antioxidants in several systems: linoleic acid autoxidation, lecithin liposome, rabbit erythrocyte membrane and rat liver microsomes (Tsuda et al., 1994). The aglycone was stronger in the liposome and rabbit cell systems. Reaction of C3G with a free radical generator (2,2'-azo-bis-2,4-dimethylvaleronitrile) produced another free radical scavenger, protocatechuic acid, and 4,6-dihydroxy-2-O-β-D-glucosyl-3-oxo-2,3-dihydro-benzofuran (Tsuda et al., 1996).

In another study, rats were fed C3G (2 g/kg diet) for 14 days (Tsuda et al., 1998). There was no difference between the control animals and the experimental group for weight gain, food intake, liver weight and serum triglycerides, phospholipids and antioxidants (α-tocopherol, ascorbic acid, reduced glutathione and uric acid). These findings support the hypothesis that anthocyanins spare nutrient antioxidants. Free and total cholesterol was lower in the C3G-fed rats. Serum was subjected to *ex vivo* oxidation by two methods: 2,2'-azobis(2-amidinopropane) hydrochloride (AAPH) and $CuSO_4$. Initially, the serum TBARS from the C3G rats were slightly but significantly lower than that of the control group, but the differences became more pronounced with time in both reaction systems. C3G did not appear to have an antioxidant effect in rat liver.

3.3. CARDIOVASCULAR PROTECTION

While cardiovascular disease (CVD) is still not completely understood, many factors contribute to its initiation and consequences. CVD is a major killer of older men and women in the U.S. and is becoming a larger health problem in other nations. Schramm and German (1998) have reviewed potential benefits of flavonoids in preventing vascular disease (Table 10.6).

Hypertension, atherosclerosis and diabetes can reduce the flexibility of arterial walls, which contributes to poor blood flow and plaque formation. Rat aortas exposed to anthocyanin-enriched blueberry extract *in vitro* exhibited relaxation caused by endothelium-generated nitric oxide (Andriambeloson et al., 1996). Delphinidin induced a maximal relaxation of 89%, which was comparable to red wine polyphenols (Andriambeloson et al., 1998). Fitzpatrick and coworkers (1995) also found that malvidin did not elicit a vasorelaxation response. Since neither malvidin nor cyanidin were effective, the number and position of hydroxyl groups, as well as any methoxylated groups, may be important for producing endothelial-dependent vasorelaxation.

A commercial *V. myrtillus* extract (Myrtocyan®) enhanced relaxation of calf aortas *in vitro* that had been exposed to adrenalin (Bettini et al., 1985). The

TABLE 10.6. Possible Mediation of Vascular Disease by Flavonoids.[a]

Risk Factor	Flavonoid Activity
Free radicals	Free radical termination and chelation of metal catalysts
Leukocyte adhesion[b]	Prevention of selection expression
Bacteria	Growth suppression
Viruses	Antiviral activity and inhibition of viral replication
AGE[c]	Inhibition of carbohydrate autoxidation, glycooxidation and glycation (Maillard reactions)
Menopause	Estrogenic properties
Proteases	Enzyme inhibition

[a] Adapted from Schramm and German (1998)
[b] Not affected by all flavonoids
[c] Advanced glycation end products

dose response was nearly linear as the extract concentration was increased from 25–100 μg/mL, and the researchers proposed that the relaxation of the blood vessels was due to inhibition of catechol-O-methyltransferases (COMT). Anthocyanins and ascorbic acid inhibited contractile responses of calf aortas in the presence of histamine or angiotensin II (Bettini et al., 1987). This effect was not observed when indomethacin or lysine acetylsalicylate were added. This study suggests that anthocyanins may have little benefit for persons taking medications that inhibit the enzymes that cause prostaglandin synthesis. Bilberry anthocyanins also reduce platelet aggregation *in vitro*, comparing favorably with other drugs (Zaragozá et al., 1985).

Anthocyanin extracts inhibited porcine elastase *in vitro*, and Lineweaver-Burk plots indicated that the inhibition was not competitive (Jonadet et al., 1983). Vascular protection in rats was measured by retention of Evans blue dye in serum and appeared to be dose dependent. Grape (*Vitis vinifera*) anthocyanins were better at protecting blood vessels than *V. myrtillus* extracts at the lowest dose, 50 mg/kg/I.P. Extracts from both fruits were significantly different from the control treatment at doses of 100 and 200 mg/kg/I.P.

Capillary strength is another important element in cardiovascular health. Easily damaged or porous capillaries contribute to electrolyte imbalances and lead to edema and other dysfunctions. The commercial *V. myrtillus* preparation Difrarel was well tolerated by patients with chronic illnesses affecting blood vessels, even though patient improvement was mixed (Amouretti, 1972). Coget and coworkers (1968) described the progress of 27 patients treated with Difrarel 20 and concluded that the drug was an effective vascular protective agent. Exposure to radiation, either therapeutic or accidental, promotes weakened capillaries. Interest in protective effects of *V. myrtillus* against such damage has been high in eastern Europe.

Oral doses of *V. myrtillus* extract were slightly less effective in increasing capillary resistance to permeability than were intraperitoneal administrations to rabbits and rats (Lietti et al., 1976). However, hamsters given oral doses (10 mg/10 g body weight) of a commercial product containing 36% bilberry anthocyanosides for two or four weeks exhibited better capillary perfusion and fewer sticking leukocytes in the capillaries that had been claimed to induce ischemia (Bertuglia et al., 1995).

A commercial *V. myrtillus* extract, known as Myrtocyan® with 25% anthocyanidins, was effective in reducing capillary permeability and increasing capillary resistance (Cristoni and Magistretti, 1987). The extract was administered intraperitoneally to rats starved for 16 h, then Evans blue dye was given intravenously, followed by an intradermal dose of histamine. Rats given doses of 50, 100, 200 or 400 mg/kg of Myrtocyan had reductions in permeability, as measured by color in skin flaps, of 32, 36, 47 and 54%, respectively. Wistar rats fed a flavonoid-free chow for three weeks exhibited more capillary resistance when the bilberry extract was given by intraperitoneal injection (Figure 10.8).

3.4. ANTIDIABETIC PROPERTIES

The leaves and fruit of *V. myrtillus* are used in Europe to treat many conditions resulting from diabetes. The high levels of glucose in the blood of diabetics trigger many deteriorative events in the body. Medical complications of diabetes include microangiopathy, cataracts, blindness due to retinopathy, neuropathy, decreased resistance to infections and hyperlipidemia.

The capillary walls in diabetic patients thicken due to collagen and glycoprotein deposits. The thickened capillaries are less flexible and more susceptible to blockage, leading to atherosclerosis. Rat aorta smooth muscle cells incorporated less radio-labeled amino acids when cultured with *V. myrtillus* anthocya-

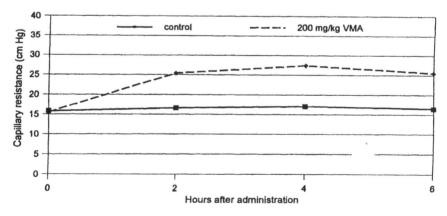

Figure 10.8 Rat capillary resistance after administration of *Vaccinium myrtillus* extract or placebo. Adapted from Cristoni and Magistretti (1987).

nins (VMA) (Boniface et al., 1986), suggesting one mechanism by which bilberry maintains normal capillary structure. Similar results were obtained for collagen content in cells from diabetic rats, and VMA reduced collagen content more than did insulin. Flavonoids inhibit aldose reductase an enzyme that converts sugars to sugar alcohols and is implicated with diabetic complications such as neuropathy, retinopathy and heart disease (Varma, 1986).

Boniface and colleagues (1986) also reported positive results for diabetic patients treated with 500–600 mg of VMA for eight to 33 months. Improvements in capillaries and microaneurysms were found for many, but not all, patients. Although there was no change in soluble collagen due to VMA, insoluble collagen levels returned to levels similar to those in normal subjects, and structural glycoproteins in diabetics were reduced by 30%. Based upon this series of experiments, the French scientists recommended that diabetic patients be given 500 mg VMA daily, split in two doses, for a period of at least several months. Sporadic doses of lower quantities were not believed to be of no value.

An aqueous alcohol extract of *V. myrtillus* leaves produced a 26% reduction in plasma glucose levels in rats made diabetic by the drug streptozotocin (Cignarella et al., 1996). Plasma triglycerides decreased in proportion with the amount of bilberry leaf extract given to rats (1.2 or 3.0 g/kg body weight) fed a hyperlipidemic diet, but the reduction was less than that obtained for rats treated with ciprofibrate (10 or 20 mg/kg). Ciprofibrate also reduced plasma levels of free fatty acids, the precursors for triglycerides, while leaf extract affected only triglycerides. Increased tendency to form blood clots or thrombi represents another diabetes-related risk for cardiovascular problems. Thrombus weight and protein content was lower in diabetic rats given ciprofibrate, but not for those rats given bilberry leaf extract. Although the German monograph on bilberry leaves describes their use for diabetes, the leaves are not recommended for therapy.

As part of a continuing cohort project known as the Nurses' Health Study, food frequency questionnaires were collected from 65,173 women in the U.S. Dietary fiber intake, especially from whole grains, in combination with a low glycemic load diet may reduce the risk for non-insulin-dependent (type II) diabetes in women (Salmerón et al., 1997). Dietary fiber from fruit did not have a significant role. However, soluble fiber, such as pectin found in blueberries, can retard the digestion of carbohydrates and subsequent absorption of sugars, lowering the glycemic index. Consumption of blueberries with breakfast cereals and other whole grain foods could add another level of protection against the onset of diabetes.

3.5. VISION IMPROVEMENT

Bilberry extract is believed to improve eyesight, particularly night vision. This health benefit is the primary reason for the product's popularity in Japan and Korea, where it is used to relieve eyestrain caused by excessive computer use (Kalt and Dufour, 1997). Since carotenoids with vitamin A activity are found in *Vaccinium* species, some of the benefits pertaining to vision are attributable to these compounds. However, a double-blind, placebo-controlled study showed that oral doses of anthocyanins are important for regeneration of visual purple (Alfieri and Sole, 1966). Adapto-electroretinograms (AERG) of two sets of six subjects were made before treatment, and one and three hours postadministration. Subjects given the bilberry adapted to the light within 6.5 minutes, compared with 9 minutes for the control group.

Scialdone (1966) used a Comberg nictometer to demonstrate that a dose of 500 mg *V. myrtillus* anthocyanins with 25 mg β-carotene improved vision in poor light due to improved mesoptic vision and reduced sensibility at direct glare. Bilberry extract slightly increased retinal sensitivity in 16 normal subjects (Magnasco and Zingirian, 1966). In a study of 20–25-year-olds with good vision, the group ($N = 30$) receiving bilberry extract had slightly, but significantly, improved adaption related to night vision compared to the group receiving a placebo (Jayle et al., 1965).

Since Maillard reactions are involved with formation of cataracts, inhibition of aldose reductase by *Vaccinium* flavonoids should aid in minimizing lens damage. Quercetin decreased sugars in sugar alcohols in lens of diabetic Peruvian rodents; quercitrin inhibited xylitol synthesis in rat lens (Varma, 1986). Diabetic hyperopia was controlled in rabbits given flavonoids before or after administration of alloxan to induce diabetes.

Bilberry extracts appear to benefit vision in several ways: improved night vision by enhanced regeneration of retinal pigments, increased circulation within the capillaries of the retina, inhibition of Maillard reactions in the lens to reduce cataract formation and protection from ultraviolet light (Figure 10.9).

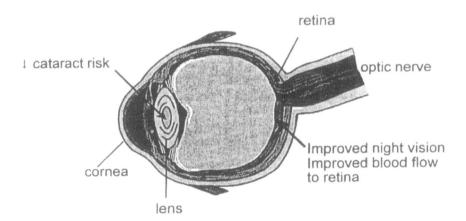

Figure 10.9 Mechanisms for improved vision due to *Vaccinium myrtillus* extracts.

The antioxidant properties of *V. myrtillus* extracts may be responsible for these health benefits. Antioxidants have been suggested to retard oxidation in the lens and slow retinal angiopathy that occurs in age-related macular degeneration and diabetic retinopathy (Trevithick and Mitton, 1999). No studies have yet been published that describe the use of blueberry extracts to treat visual problems, although these products are being used as dietary supplements for this purpose. The dose of anthocyanins required to improve or maintain vision is not yet fully understood, and the lower concentration of the pigments in blueberries suggest that they may be less effective.

3.6. INHIBITION OF CARCINOGENESIS AND MUTAGENESIS

Fruit and vegetable consumption reduces risks of certain forms of cancer, especially cancers of the gastrointestinal tract (Block et al., 1992; World Cancer Fund/American Institute for Cancer Research, 1997). Berry consumption tends to be largely seasonal with the exception of dietary supplements and frozen or canned products. Although no epidemiological studies have associated bilberry or blueberry consumption with reduced cancer risks, some recent *in vitro* studies suggest that the fruit may have a protective effect. Despite its long use in Europe, no reviews of bilberry mention its use for cancer prevention. Another important point to remember is that some chemicals may cause cancer in certain situations, yet prevent cancer in other cases (Glusker and Rossi, 1986; MacGregor, 1986).

Blueberry juice was one of the most antimutagenic fruit and vegetable products tested in a German study using *Salmonella typhimurium* TA 98 and TA100 in the Ames test (Edenharder et al., 1994) (Figure 10.10). Unfortunately the blueberry species and cultivar were not specified. Blueberry juice was effective in reducing mutagenicity caused by the polycyclic aromatic hydrocarbons IQ, MeIQ and MeIQx, even when heated to 95°C for 10 minutes. Blueberry juice was more effective than apple, grape and citrus juices, and comparable to blackberry, cherry, plum, honeydew melon and strawberry juices.

In another *in vitro* experiment, acidified methanol extracts of four *Vaccinium* species were compared for anticancer activity (Bomser et al., 1996). These crude extracts were ineffective in inducing the detoxifying enzyme quinone reductase but were able to inhibit induction of ornithine decarboxylase that is needed for the carcinogenic activity of phorbol. The experiments indicated that the proanthocyanidin fraction of *V. angustifolium, V. macrocarpon* and *V. vitis idea,* and the hexane-chloroform fraction of *V. myrtillus* showed the most promise in preventing cancer. However, commercial applications of this experiment may be delayed until safe extraction procedures are developed.

The ethyl acetate extracts of *V. corymbosum* were studied for their anticarcinogenic activity *in vitro* (Li et al., 1998). Ursolic acid and its derivative 3-

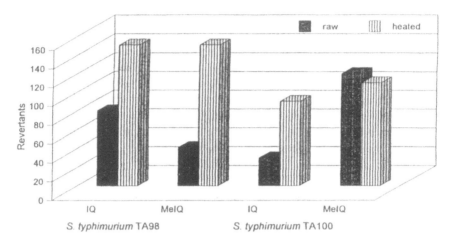

Figure 10.10 Reduction in mutagenicity of 2-amino-3-methylimidazol[4,5-*f*]quinoline, 2-amino-3,4-dimethylimidazo[4,5-*f*]quinoline (IQ) and 2-amino-3,8-dimethylimidazo[4,5-*f*]quinoxaline (MeIQ) to two types of salmonella in the Ames test Adapted from Edenharder et al. (1994).

β,19-α-dihydroxy-urs-12-en-28-oic acid were toxic to human leukemia HL-60 cells, with IC50 values of 1.5 and 1.0 ppm, respectively. This work raises the possibility that several compounds in these fruit have cancer-preventative value.

Plant cell cultures of *V. myrtillus* were found to produce a peroxidase with human Lewis determinant (Melo et al., 1997). Sialyl Lewis recognition determinants are typically found in animal cells, particularly on the surface of digestive system tumors. Since these determinants are believed to be involved with metastasis, Lewis determinants from plant sources could block sites on cancer cells and prevent spread of the disease.

3.7. ANTIMICROBIAL ACTIVITY

Rapid healing may protect against infection. Topical doses of *V. myrtillus* extract (0.5, 1.2%) reduced wound size compared with prednisone-treated controls during a three-day study, and the 2% dose resulted in smaller wounds on each day than in the control with no added compounds (Cristoni and Magistretti, 1987).

The German therapeutic monograph on bilberry fruit recommends doses of 20–60 g per day to treat acute diarrhea and mild inflammations of the mouth and throat. Tannins in bilberry and blueberry leaves are believed to be responsible for their ability to stop diarrhea (Tyler, 1994). Gallic acid, an astringent, was identified in bilberry leaves, as well as an unusual phenolic acid, melilotic acid (Dombrowicz et al., 1991).

One mechanism by which anthocyanins and other compounds inhibit microbial growth is by interference with microbial enzymes. Anthocyanins (1–2 μM) inhibited glucose oxidation by *Escherichia coli* K-12, *Staphylococcus aureus, Lactobacillus casei* 7469, *Salmonella enteriditis, S. typhosa, Proteus vulgaris* and *Aerobacter aerogenes* (Somaatmadja and Powers, 1964). At concentrations of 0.1% and 0.5%, the monoglucosides of delphinidin, petunidin and malvidin inhibited glycerol dehydrogenase, malate dehydrogenase and hexokinase, but stimulated activity for alpha-glucan phosphorylase and glutamic acid decarboxylase (Carpenter et al., 1967).

Anthocyanins are believed to protect plants from radiation and other stresses. Aflatoxin B_1 production by *Aspergillus flavus* was inhibited by anthocyanidins in the following order: delphinidin > pelargonidin > peonidin > cyanidin (Norton, 1999). Glycosides were less effective in toxin suppression than the aglycones. It is not known whether these compounds retard toxins formed by mold species found on *Vaccinium* fruit.

3.8. URINARY TRACT INFECTIONS

Cranberry (*V. macrocarpon* Ait.) juice has long enjoyed a folk reputation as a treatment for urinary tract infections (UTI). Although the low pH of the fruit was at first believed to be the antimicrobial agent, fructose and high molecular weight phenolic compounds have been found to prevent the adhesion of *Escherichia coli* cells *in vitro* (Ofek et al., 1991). These compounds have also been identified in blueberry juice. Fructose and polyphenols prevent mannose-resistant adhesions on certain P-fimbriated *E. coli* isolates from attaching to epithelial tissues in the urinary tract (Ofek et al., 1996). Purified cranberry proanthocyanidins were reported to possess antiadherence properties in an *in vitro* assay (Howell et al., 1998). Curiously, all of these researchers reported that blueberry juice was similarly active, yet only data for cranberries was published. To date, no clinical or animal studies using bilberries or blueberries to prevent or treat UTI have been published.

3.9. GASTRIC PROTECTION

Reports of the ability of *V. myrtillus* extracts (VMA) to speed wound healing inspired Cristoni and Magistretti (1987) to undertake a series of experiments to evaluate the effectiveness of such extracts against gastric ulcers. Ulcers were induced by several means: pyloric ligature, reserpine, phenylbutazone, acetic acid and restraint. VMA lowered the ulcer index in each experiment, and gastric protection was dose dependent, with 100 mg/kg oral doses most effective. Gastric juice production was unaffected by VMA, and the researchers concluded that the protective effect was due to increased mucus production. Although today we know that most gastric ulcers result from infection with the bacterium *Heliobacter pylori,* the extent to which antimicrobial activity contributed to the success of bilberry extract in this experiment is not known.

Cyanidin, given as the drug IdB 1027 at a dose of 1,200 mg/day, appeared to protect the gastric mucosa of seven male volunteers given a 1,000 mg dose of aspirin via a nasogastric tube (Barzaghi et al., 1991). Intragastric doses of IdB 1027 increased the efficiency of the gastric mucosal barrier in rats (Cristoni et al., 1989). Blueberry-coated aspirin tablets may appeal to consumers who must take aspirin despite the gastric distress the drug can produce.

3.10. EFFECTS ON THE BRAIN

Brain functions such as balance, coordination, short-term memory and information retrieval can be impaired with aging. A lifetime of oxidative stress may induce damages to lipids and proteins that lead to cellular damage. Impaired blood flow to brain tissues may also be responsible for losses in brain function. An animal study recently provided some evidence that dietary antioxidants can protect the brain from oxygen-induced damage (Joseph et al., 1998). Six-to eight-month-old F344 rats were fed a control diet or diets containing antioxidants (500 IU/kg vitamin E, 10 g/kg dried aqueous blueberry extract, 9.4 g/kg dried strawberry extract or 6.7 g/kg dried spinach extract). After eight weeks on the diet, the rats were subjected to 48 hours of 100% O_2 to induce damage similar to that found in aged rats. All antioxidant diets prevented decreases in nerve growth factor in the basal forebrain, and other adverse effects were reduced.

The thyroid hormones, thyroxine (T4) and triidothyronine (T3), regulate temperature and metabolism. Rats given intraperitoneal injections of bilberry anthocyanins (200 mg/kg/day) for five days had significantly more T3 in their brains than rats given only the solvent (26% alcohol) (Saija et al., 1990). T3 enters the brain via specific transport in the capillaries, therefore, anthocyanins may mediate T3 transport at the capillary level. The specific portions of the brain that contained more T3 in the bilberry-treated animals were those responsible for memory, vision and control of sensory input—the frontal, temporoparietal and occipital cortexes and the components of the limbic system (hippocampus, thalamus and hypothalamus).

3.11. DIAGNOSTIC APPLICATIONS

Manganese provides contrast during magnetic resonance imaging (MRI) for many ailments. Blueberries contain varying amounts of this metal, and blueberry juice is most likely more palatable than a manganese salt solution. Radiologists in Japan found that the optimal dose of manganese in blueberry juice was 3–4 mg/dL (Hiraishi et al., 1995). Spiking of 400 mL blueberry juice with an imaging agent facilitated exoscopic and virtual endoscopic 3D imaging of the stomach in three volunteers and three patients (Schmid et al., 1999).

3.12. OPTIMAL CONSUMPTION LEVELS

No recommended levels for bilberry or blueberry consumption have been made, although consumers may want to include a daily serving as part of the recommended five or more servings of fruits and vegetables. European suggested doses are 20–60 g/day of leaf or fruit extracts of *V. myrtillus* (Gruenwald et al., 1998). Bilberry supplements sold in the U.S., standardized to 10–36% anthocyanins, suggest that two or three tablets be taken daily, but no rationale for this dose has been given.

3.13. TOXICOLOGICAL RISKS

A review of bilberry studies indicated that doses of bilberry extracts as high as 400 mg per kg were nontoxic to rats (Murray, 1997). Petri et al. (1997) mentioned that prolonged or excessive use of the decoction made from *V. myrtillus* leaves can lead to hydroquinone poisoning, but the symptoms of such poisoning were not described.

Markakis (1982, p. 246) stated: "Anthocyanins are apparently harmless to health." As of October 20, 1998, there had been only five reported adverse events for dietary supplements containing bilberry (FDA, 1998). One report complained of stomach problems; a bilberry supplement and another product were taken by the patient. The other reports involved products containing mixtures of bilberry with several other botanicals and nutrients.

4. IMPROVED NUTRITIONAL VALUE

4.1. DIETARY FIBER

Bilberries have a wide range of fiber values on a fresh weight basis: insoluble, soluble and total dietary fiber IDF, 1.9–3.2%; soluble and total dietary fiber SF, 0.4–1.1%; soluble and total dietary fiber TDF, 2.3–3.9% (Plaami et al., 1992). The USDA Nutrient Database (1998) lists a value of 2.7 g/100 g total dietary fiber for blueberries, with no species or cultivar noted. Although Austin (1994) reported that rabbiteye blueberries contain about 30% fiber (d.b.), of which 0.48–87% is pectin, this high value has not been cited elsewhere. One lowbush blueberry processor cites a value of 3.07 for fiber, with a range of 2.80–3.6 g/100 g (Jasper Wyman & Sons, 1999).

4.2. FAT REDUCTION

Pectin has additional value as a replacement for fat in some food products. Although a diet rich in fruits and vegetables will generally tend to be low in fat, fat reduction of high-calorie foods may be helpful for some people. Lowbush blueberry puree used to replace oil on an equal volume basis in chocolate and spice cakes was less acceptable than full-fat versions, but consumers were not aware of the fat and calorie reduction in the cakes made with blueberries (Camire et al., 1997).

5. CONCLUSIONS

5.1. RECOMMENDATIONS FOR FUTURE RESEARCH

Certainly, more studies with human subjects are needed to verify the health benefits of bilberries and blueberries. Although double-blind, randomized, placebo-controlled experiments are preferable, the color and flavor of the fruit can limit the practical aspects of this experimental design. Powdered and/or purified materials can be put into opaque capsules to mask their identity, but beverages and other food items will be difficult to disguise. Healthy individuals may respond differently to these fruits than may individuals with diabetes or other illness. Gender and age differences must also be considered. *In vitro* tests can help better define interactions among chemicals in bilberries and blueberries, but the lack of a universal test for antioxidant activity will hinder consensus among researchers.

6. REFERENCES

Alfieri, R. and P. Sole. 1966. "Influence des anthocyanosides administrés par voie oro-perlinguale sur l'adapto-électrorétinogramme (AERG) en lumiére rouge chez l'Homme," *Comptes Rendus des Séances de la Société de Biologie et des Ses Filiales.* 160(8)·1590–1593.

Amouretti, M. 1972. "Intérêt thérapeutique des anthocyanosides de *Vaccinium myrtillus* dans un service de médecine interne," *Thérapeutique.* 48(9).579–581.

Andriambeloson, E., C. Magnier, G. Haan-Archipoff, A. Lobstein, R. Anton, A. Beretz, J. C. Stoclet and R. Andriantsitohaina. 1998. "Natural dietary polyphenolic compounds cause endothelium-dependent vasorelaxation in rat thoracic aorta," *J. Nutr.* 128(12)·2324–2333.

Andriambeloson, E., J. C. Stoclet and R. Andriantsitohaina. 1996. "Effects vasculaires d'extraits végétaux contenant des dérivés polyphénoliques " in *Polyphenols Communications 96,* J. Vercauteren, C. Chèze, M. C. Dumon and J. F. Weber. 2:421–422. Groupe Polyphénols, Bordeaux, France.

Anonymous. 1998. "News," Council for Responsible Nutrition, Nov. 1998.

Austin, M. E. 1994. *Rabbiteye Blueberries,* Agscience Inc., Auburndale, FL.

Barzaghi, N., G. Gatti, F. Crema and E. Perucca. 1991. "Protective effect of cyanidin (IdB 1027) against aspirin-induced fall in gastric transmucosal potential difference in normal subjects," *Ital. J. Gastroenterol.* 23(5)·249–252.

Bertuglia, S., S Malandrino and A. Colantuoni. 1995 "Effect of *Vaccinium myrtillus* anthocyanosides on ischaemia reperfusion injury in hamster cheek pouch microcirculation," *Pharmacol. Res.* 31(3/4):183–187.

Bettini, V, R Martino, V. Tegazzin and P Ton. 1987 "Risposte contrattili di segmenti di arterie coronarie all'istamina e all'angiotensina II in presenza di antocianosidi del mirtillo," *Cardiologia* 32(10).1155–1159.

Bettini, V, A. Fiori, R Martino, F Mayellaro and P Ton 1985 "Study of the mechanism whereby anthocyanosides potentiate the effect of catecholamines on coronary vessels." *Fitoterapia.* 56(2).67–72.

Block, G., B Patterson and A. Subar. 1992 "Fruit, vegetables, and cancer prevention· a review of the epidemiological evidence," *Nutr. Cancer.* 18 1–29

Bomser, J, D L. Madhavi, K. Singletary and M. A L Smith 1996. "*In vitro* anticancer activity of fruit extracts from *Vaccinium* species," *Planta Med* 62(3)·212–216

Boniface, R., M. Miskulin, L Robert and A M. Robert 1986. "Pharmacological properties of *Myrtillus* anthocyanosides: correlation with results of treatment of diabetic microangiopathy," in *Flavonoids and Bioflavonoids, 1985,* L Farkas, M. Gabor and F. Kallay, eds., Amsterdam: Elsevier, pp. 293–301.

Bridle, P, C Garcia-Viguera and F A. Tomás-Barberán 1996. "Analysis of anthocyanins by capillary zone electrophoresis," *J. Liq. Chrom & Rel Technol.* 19(4):537–545.

Bushway, R. J., D. F. McGann, W P Cook and A A. Bushway. 1983 "Mineral and vitamin content of lowbush blueberries (*Vaccinium angustifolium* Ait.)," *J Food Sci.* 48(12).1878, 1880.

Camire, M E. 1997. "U.S. changes in health claims and nutritional labeling for foods and dietary supplements," *Acta Hort* 446·205–209.

Camire, M. E., I. Surjawan and T. M. Work 1997 "Lowbush blueberry puree and applesauce for oil replacement in cake systems," *Cereal Foods World* 42(5):405–408.

Carpenter, J A., Y. P Wang and J. J. Powers. 1967. "Effect of anthocyanin pigments on certain enzymes," *PSEBM.* 702–706

Cignarella, A., M. Nastasi, E. Cavalli and L Puglisi 1996. "Novel lipid-lowering properties of *Vaccinium myrtillus* L. leaves, a traditional antidiabetic treatment, in several models of rat dyslipidaemia: a comparison with ciprofibrate," *Thrombosis Res* 84(5).311–322

Coget, J. and J. F. Merlen. 1968. "Étude clinique d'un nouvel agent de protection vasculaire le Difrarel 20, composé d'anthocyanosides extrait du vaccinum myrtillus," *Phlàebologie.* 21(2).221–228.

Costantino, L., Albasini, A, Rastelli, G and S Benvenuti 1992. "Activity of polyphenolic crude extracts as scavengers of superoxide radicals and inhibitors of xanthine oxidase," *Planta Medica* 58 342–344.

Cristoni, A and M. J. Magistretti. 1987. "Antiulcer and healing activity of *Vaccinium myrtillus* anthocyanosides," *Farmaco.* 42(2):29–43.

Cristoni, A., S. Malandrino and M. J. Magistretti. 1989. "Effect of a natural flavonoid on gastric mucosal barrier," *Arzneimittelforschung.* 39(5):590–592.

Dombrowicz, E., R. Zadernowski and L. Swiatek. 1991. "Phenolic acids in leaves of *Arctostaphylos uva ursi* L., *Vaccinium* vitis idea L. and *Vaccinium myrtillus* L ," *Pharmazie.* 46(9)·680–681.

Eck, P 1988. *Blueberry Science,* Rutgers Univ. Press, New Brunswick, NJ.

Edenharder, R., P. Kurz, K. John, S Burgard and K. Seeger 1994. "*In vitro* effect of vegetable and fruit juices on the mutagencity of 2-amino-3-methylimidazo[4,5-*f*]quinoline. 2-amino-3,4-dimethylimidazo[4,5-*f*]quinoline and 2-amino-3,8-dimethylimidazo[4,5-*f*]quinoxaline," *Food Chem. Toxicol.* 32(5):443–459

Ehlenfeldt, M K., F. I. Meredith and J. R Ballington. 1994 "Unique organic acid profile of rabbiteye vs. highbush blueberries," *HortSci.* 29(4):321–323.

Eitenmiller, R. R., R. F. Kuhl and C. J. B. Smit. 1977. "Mineral and water-soluble vitamin content of rabbiteye blueberries," *J. Food Sci.* 42(5).1311–1315.

Fitzpatrick, D. F., S. L. Hirschfield, T. Ricci, P. Jantzen and R. G. Coffey. 1995. "Endothelium-dependent vasorelaxation caused by various plant extracts," *J. Cardiovascular Pharmacol.* 26(1):90–95.

Food and Drug Administration. 1998. "Special Nutritionals Adverse Events Monitoring System," http.//vm.fda.gov.

Food and Nutrition Board. 1998. "Proposed definition of dietary antioxidants," http://www2.nas.edu/fnb/21a2.html.

Fraisse, D., A. Carnat and J. L. Lamaison. 1996. "Polyphenolic composition of the leaf of bilberry," *Ann. Pharmaceut Fr.* 54(6):280–283.

Francis, F. J. 1985. "Blueberries as a colorant ingredient in food products," *J. Food Sci.* 50:754–756.

Francis, F. J., J. B. Harborne and W. G. Barker. 1966 "Anthocyanins in the lowbush blueberry, *Vaccinium angustifolium*," *J. Food Sci.* 31:583–587.

Friedrich, V. H. and J. Schönert. 1973 "Phytochemical investigation of leaves and fruits of *Vaccinium myrtillus*," *Plant Medica.* 24(1):90–100.

Gao, L. and G. Mazza. 1995. "Characterization of acetylated anthocyanins in lowbush blueberries." *J. Liq. Chromat.* 18(2):245–259.

Gao, L. and G. Mazza. 1994. "Quantitation and distribution of simple and acylated anthocyanins and other phenolics in blueberries," *J. Food Sci.* (5):1057–1059.

Glusker, J. P. and M. Rossi. 1986. "Molecular aspects of chemical carcinogens and bioflavonoids," in *Plant Flavonoids in Biology and Medicine: Biochemical Pharmacological, and Structure-Activity Relationships,* Alan R. Liss, Inc., New York, N.Y. pp. 395–410.

Griffiths, L. A. and G. E. Smith. 1972. "Metabolism of myricetin and related compounds in the rat metabolite formation *in vivo* and by the intestinal microflora *in vitro*," *Biochem. J.* 130:141–151.

Gruenwald, J., T. Brendler and C. Jaenicke. 1998. *"Vaccinium myrtillus,"* Physicians Desk Reference for Herbal Medicines, 1st ed., Medical Economics Co., Inc , Montvale, NJ, pp. 1201–1202.

Halliwell, B. and O. I. Aruoma. 1997. "Free radicals and antioxidants: the need for *in vivo* markers of oxidative stress," Ch. 1 in *Antioxidant Methodology: In Vivo and In Vitro Concepts,* O. I. Aruoma and S. L. Cuppett, Eds., AOCS Press, Champaign, IL, pp. 1–22.

Heinonen, I. M., A. S. Meyer and E. N. Frankel. 1998. "Antioxidant activity of berry phenolics on human low-density lipoprotein and liposome oxidation," *J. Agric. Food Chem.* 46(10):4107–4112.

Hiraishi, K., I. Narabayashi, O. Fujita, K. Yamamoto, A. Sagami, Y. Hisada, Y. Saika, I. Adachi and H. Hasegawa. 1995. "Blueberry juice: preliminary evaluation as an oral contrast agent in gastrointestinal MR imaging," *Radiology.* 194(1):119–123.

Hong, V. and R. E. Wrolstad. 1990. "Characterization of anthocyanin-containing colorants and fruit juices by HPLC/photodiode array detection," *J. Agric. Food Chem.* 38(3):698–708.

Horvat, R. J. and S. D. Senter. 1985. "Comparison of the volatile constituents from rabbiteye blueberries (*Vaccinium ashei*) during ripening." *J. Food Sci.* 50:429–431. 436.

Howell, A. B., A. Der Marderosian and L. Y. Foo. 1998. "Inhibition of the adherence of P-fimbriated *Escherichia coli* to uroepithelial-cell surfaces by proanthocyanidin extracts from cranberries," *New Engl. J. Med* 339(15):1085–1086.

Jasper Wyman & Sons. 1999. Home page, http://www.wymans.com/nutri.html.

Jayle, G. E., M. Aubry, H. Gavini, G. Braccini and C. de la Baume. 1965. "Étude concernant l'action sur la vision nocturne," *Ann. Oculistique* 198(6).556–562.

Jonadet, M., M. T. Meunier, J. Bastide and P. Bastide. 1983 "Anthocyanosides extraits de *Vitis vinifera*, de *Vaccinium myrtillus* et de *Pinus maritimus*. I. Activités inhibitrices vis-á-vis de l'élastase *in vitro*: II. Activités angioprotectrices comparées *in vivo*," *J. Pharm. Belg.* 38(1):41–46.

Joseph, J. A., N. Denisova, D. Fisher, B. Shukitt-Hale, R. Prior and G. Cao. 1998. "Membrane and receptor modifications of oxidative stress vulnerability in aging. Nutritional considerations," *Ann. NY Acad. Sci.* 854(11):268–276.

Kader, F., B. Rovel, M. Giradin and M Metche. 1996. "Fractionation and identification of the phenolic compounds of highbush blueberries (*Vaccinium corymbosum*, L.)," *Food Chem.* 55(1):35–40.

Kalt, W. and D. Dufour. 1997. "Health functionality of blueberries," *HortTechnol.* 7(3):216–222.

Kalt, W. and J. E. McDonald. 1996. "Chemical composition of lowbush blueberry cultivars," *J. Amer. Soc. Hort. Sci.* 121(1):142–146.

Kalt, W., S. MacEwan and G. Miner. 1994. "*Vaccinium* Extract in Pharmaceutical Products." Report to the Wild Blueberry Association of North America and the Wild Blueberry Producers Association of Nova Scotia.

Kalt, W., K. B. McRae and L. C. Hamilton. 1995. "Relationship between surface color and other maturity indices in wild lowbush blueberries." *Can. J. Plant Sci.* 75:485–490.

Kenyon, N. 1997. "Cultivated blueberries. The good-for-you blue food," *Nutr. Today.* 32(3):122–124.

Krawczyk, U. and G. Petri. 1992. "Application of RP-HPLC and spectrophotometry in standardization of bilberry anthocyanin extract," *Arch Pharm.* 325(3):147–149.

Kurtzweil, P. 1998. "An FDA guide to dietary supplements," *FDA Consumer,* http://www.fda.gov/fdac/features/1998/598 guid.html.

Laplaud, P. M., A. Lelubre and M. J. Chapman. 1997. "Antioxidant action of *Vaccinium myrtillus* extract on human low density lipoproteins *in vitro*: initial observations," *Fundam. Clin. Pharmacol.* 11(1):35–40.

Li, J., M. Wang, C. T. Ho, Y. Shao, M. T. Huang and R. Rosen 1998. "Studies on blueberry, a fruit of potential antioxidative and anticarcinogenic activities," Institute of Food Technologists Annual Meeting Abstract 34B-49.

Lietti, A. and G. Forni. 1976. "Studies on *Vaccinium myrtillus* anthocyanosides. II. Aspects of anthocyanins pharmacokinetics in the rat," *Arzneim.-Forsch.* 26(5):832–835.

Lietti, A., A. Cristoni and M. Picci. 1976. "Studies on *Vaccinium myrtillus* anthocyanosides. I. Vasoprotective and antiinflammatory activity," *Arzneim.-Forsch.* 26(5):829–832.

MacGregor, J. T. 1986. "Mutagenic and carcinogenic effects of flavonoids," in *Plant Flavonoids in Biology and Medicine: Biochemical, Pharmacological, and Structure-Activity Relationships,* Alan R. Liss, Inc., New York, N.Y. pp. 411–424.

Magnasco, A. and M. Zingirian. 1966. "Influenza degli antocianosidi sulla soglia retinica differenziale mesopica." *Ann. Ottalmologia clin. oculistica.* 92(3):188–193.

Makus, D. J. and J. R. Morris. 1993. "A comparison of fruit of highbush and rabbiteye blueberry cultivars," *J. Food Quality.* 16(6):417–428.

Manach, C., C. Morand, V. Crespy, C. Demigné, O. Texier, F. Régérat and C. Rémésy. 1998. "Quercetin is recovered in human plasma as conjugated derivatives which retain antioxidant properties," *FEBS Lett.* 426.331–336.

Markakis, P. 1982. "Anthocyanins as food additives," in *Anthocyanins as Food Colors.* Academic Press, Inc., New York, pp. 245–253.

Mazza, G. 1997. "Anthocyanins in edible plant parts. a qualitative and quantitative assessment," in *Antioxidant Methodology: In Vivo and In Vitro Concepts*, O I. Aruoma and S. L. Cuppett, eds. AOCS, Champaign, IL, Ch. 8. pp 119–140.

Mazza, G. and E. Miniati. 1993. *Anthocyanins in Fruits, Vegetables, and Grains.* CRC Press, Inc., Boca Raton, FL pp. 85–130.

Mazza, G., L. Yu and D. S. Jayas 1998. "Dehydration of fruit products. Influence of drying technique on energy requirements and product quality. Presented at the International Symposium on Drying," Dallas, TX, April 1–3.

Melo, N. S., M. Nimtz, H. S. Conradt, P. S Fevereiro and J. Costa. 1997 "Identification of the human Lewis[a] carbohydrate motif in a secretory peroxidase from a plant cell suspension culture (*Vaccinium myrtillus* L.)," *FEBS Letters.* 415·186–191

Meyer, A. S., M. Heinonen and E. N. Frankel. 1998 "Antioxidant interactions of catechin, cyanidin, caffeic acid, quercetin, and ellagic acid on human LDL oxidation," *Food Chem.* 61(1):71–75.

Miyazawa, T., K Nakagawa, M Kudo, K Muraishi and K. Someya. 1999. "Direct intestinal absorption of red fruit anthocyanins, cyanidin-3-glucoside and cyanidin-3,5-diglucoside, into rats and humans," *J. Agric. Food Chem.* 47(3):1083–1091.

Moore, J. N 1994 "The blueberry industry of North America," *HortTechol.* 4(2):96–102.

Morazzoni, P. and E. Bombardelli. 1996 "*Vaccinium myrtillus* L," *Fitoterapia.* 67(1).3–29.

Murray, M. T. 1997 "Bilberry (*Vaccinium myrtillus*)," *Am. J. Nat. Med.* 4(1):18–22.

Norton, R. A. 1999. "Inhibition of aflatoxin B₁ biosynthesis in *Aspergillus flavus* by anthocyaninidins and related flavonoids," *J. Agric Food Chem.* 47(3):1230–1235.

Ofek, I., J. Goldhar and N. Sharon. 1996. "Anti-*Escherichia coli* adhesin activity of cranberry and blueberry juices," in *Toward Anti-Adhesion Therapy for Microbial Diseases*, Kahane and Ofek, eds. Plenum Press, New York, pp. 179–183.

Ofek, I , J. Goldhar, D Zafriri, H. Lis, R. Adar and N. Sharon 1991 "Anti-*Escherichia coli* adhesin activity of cranberry and blueberry juices," *New Engl. J. Med.* 324(22)·1599.

Ohnishi, M., H. Morishita, H. Iwahashi, S. Toda, Y. Shirataki, M. Kimura and R. Kido 1994 "Inhibitory effects of chlorogenic acids on linoleic acid peroxidation and haemolysis," *Phytochem.* 36(3):579–583.

Petri, G., U. Krawczyk and A. Kéry. 1997. "Spectrophotometric and chromatographic investigation of bilberry anthocyanins for quantification purposes," *Microchem. J.* 55(1):12–23

Plaami, S. P., J. T. Kumpulainen and R. L. Tahvonen. 1992. "Total dietary fibre contents in vegetables, fruits and berries consumed in Finland," *J. Sci. Food Agric.* 59(4):545–549.

Prior, R. L., G. Cao, A. Martin, E. Sofic, J. McEwen, C. O'Brien, N. Lischner, M. Ehlenfeldt, W. Kalt, G Krewer and C. M. Mainland. 1998. "Antioxidant capacity as influenced by total phenolic and anthocyanin content, maturity, and variety of *Vaccinium* species," *J Agric Food Chem.* 46(7):2686–2693.

Rice-Evans, C A., N. J Miller, P. G. Bolwell, P. M. Bramley and J B. Pridham. 1995. "The relative antioxidant activities of plant-derived polyphenolic flavonoids," *Free Rad Res.* 22(4):375–383.

Saija, A., P. Princi, N. D'Amico, R. DePasquale and G Costa. 1990 "Effect of *Vaccinium myrtillus* anthocyanins on triiodothyronine transport into brain in the rat," *Pharmacol. Res.* 22(1):59–60.

Salméron, J., J. E. Manson, M. J. Stamfer, G. A. Colditz, A. L. Wing and W. C. Willett. 1997. "Dietary fiber, glycemic load, and risk of non-insulin-dependent diabetes mellitus in women," *JAMA.* 277(6)·472–477.

Sapers, G. M. and J. G. Phillips. 1985. "Leakage of anthocyanins from skin of raw and cooked highbush blueberries (*Vaccinium corymbosum* L.)," *J. Food Sci.* 50:437–439, 443.

Sapers, G M , S B. Jones and J G. Phillips 1985 "Leakage of anthocyanins from skin of thawed, frozen highbush blueberries (*Vaccinium corymbosum* L)," *J Food Sci.* 50 432–436.

Satué-Gracia, M. T., M Heinonen and E. N Frankel 1997 "Anthocyanins as antioxidants on human low-density lipoprotein and lecithin-liposome systems," *J Agric Food Chem.* 45(9).3362–3367

Schmid, M R., T F Hany, L. Knesplova, R. Schlumpf and J. F Debatin. 1999. "3D MR gastrography. exoscopic and endoscopic analysis of the stomach," *Eur. Radiol.* 9(1)·73–77.

Schramm, D D. and J. B German 1998. "Potential effects of flavonoids on the etiology of vascular disease," *J Nutr Biochem* 9(10):560–566.

Scialdone, D. 1966 "L'azione delle antocianine sul senso luminoso," *Ann Ottalmologia Clin. Oculistica.* 92(1) 43–51.

Singleton, V L. and J. A. Rossi 1965. "Colorimetry of total phenolics with phosphomolybdic-phosphotungstic acid reagents," *Am. J. Enol Vitic* 16.144–158

Somaatmadja, D. and J. J. Powers. 1964. "Anthocyanins. V The influence of anthocyanins and related compounds on glucose oxidation by bacteria," *J Food Sci.* 129.644–654.

Trevithick, J. R. and K P. Mitton 1999 "Antioxidants and diseases of the eye," Ch. 24 in *Antioxidant Status, Diet, Nutrition, and Health,* A. M. Pappas, ed CRC Press, Boca Raton FL, pp. 545–565.

Tsuda, T , F. Horio and T. Osawa. 1998. "Dietary cyanidin 3-*O*-β-D-glucoside increases *ex vivo* oxidation resistance of serum in rats," *Lipids* 33(6) 583–587

Tsuda, T , K. Ohshima, S. Kawakishi and T Osawa 1996. "Oxidation products of cyanidin 3-*O*-β-D-glucoside with a free radical initiator," *Lipids* 31(12).1259–1263.

Tsuda, T., M Watanabe, K Ohshima, S. Norinobu, S W. Choi, S Kawakishi and T Osawa. 1994 "Antioxidative activity of the anthocyanin pigments cyanidin 3-*O*-β-D-glucoside and cyanidin," *J. Agric Food Chem.* 42(11).2407–2410.

Tyler, V E 1994. *Herbs of Choice,* Haworth Press, Binghampton, NY, pp. 51–54.

U.S Dept of Agriculture, Agricultural Research Service. 1998. USDA Nutrient Database for Standard Reference, Release 12. Nutrient Data Laboratory Home Page, http.//www nal usda.-gov/fnic/foodcomp

Varma, S. D. 1986. "Inhibition of aldose reductase by flavonoids possible attentuation of diabetic complications," in *Plant Flavonoids in Biology and Medicine. Biochemical, Pharmacological, and Structure-Activity Relationships,* Alan R Liss, Inc. pp. 343–358

Velioglu, Y. S., G. Mazza, L Gao and B. D. Oomah. 1998. "Antioxidant activity and total phenolics in selected fruits, vegetables, and grain products," *J. Agric Food Chem.* 46(10):4113–4117.

Villata, M. 1998. "Cultivated blueberries a true-blue baking ingredient," *Cereal Foods World.* 43(3) 128–130.

Wang, H., G Cao and R. L Prior. 1997. "Oxygen radical absorbing capacity of anthocyanins," *J Agric. Food Chem* 45(2) 304–309.

World Cancer Research Fund and American Institute for Cancer Research. 1997. *Food, Nutrition and the Prevention of Cancer: A Global Perspective.* AICR, Washington, DC.

Zaragozá, F , I Iglesias and J. Benedi. 1985. "Estudio comparativo de los efectos antiagregantes de los antocianósidos y otros agentes," *Archiv. Farmacol Toxicol* 11(3).183–188.

Zhang, H. L , A A. Bushway, T. Work, M. E Camire and R. Work 1997. "Prevention of anthocyanin leakage of individually quick frozen (IQF) lowbush blueberries in blueberry muffins," *Acta Hort* 446·211–217.

CHAPTER 11

Licorice in Foods and Herbal Drugs: Chemistry, Pharmacology, Toxicology and Uses

Z. Y. WANG
M. ATHAR
D. R. BICKERS

1. HISTORY, PRODUCTION AND CONSUMPTION

In the plant kingdom, the Leguminosae family occupies an economically important position. Traditionally, the family has been divided into three subfamilies, Fabaceae, Fabales and Faboideae. Licorice (liquorice) is the name applied to the roots and stolons of some *Glycyrrhiza* species, which belong to Fabaceae. It is collected in spring and autumn, sliced after the removal of residual stem and rootlets, dried in the sun and used as such without further processing or stir-baked with honey. Figure 11.1 shows the leaves, flowers, fruits and root of this plant. Licorice was known to many ancient civilizations. The belt where it is found or cultivated runs roughly between the 30th and 45th parallels of north latitude and passes through Spain, Italy, Greece, Russia, Turkey, Syria, Iraq, Iran and China. However, relatively small amounts also grow in South Africa. In addition, it also grows in Australia (New South Wales) and in New Zealand. Neither South nor Central America seem to cultivate licorice. However, there are scattered patches of wild licorice growing in several places in the United States (Huseman, 1944).

The licorice plant grows to a height of about one meter. It sends down a tap root, which may be 6–8 meters long. Its botanical name, *Glycyrrhiza*, is derived from the Greek words meaning sweet root. Its Chinese name is "Gancao," which also means "Sweet grass." The genus *Glycyrrhiza* consists of 30 species (Stake, 1986), many of which are found in the Northeast and Northwest of China such as *Glycyrrhiza glabra* L., *Glycyrrhiza kansuensis* Chang et Peng and *Glycyrrhiza inflata* Bayal (Jiansu New Medical School, 1979). At least 13 of them have been characterized chemically.

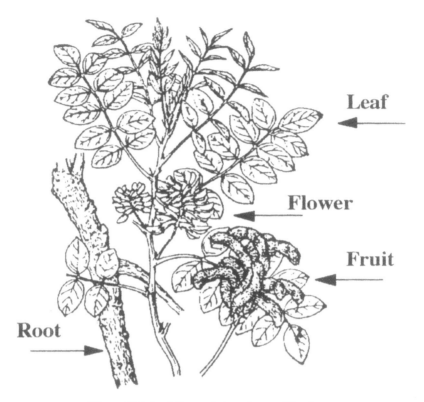

Figure 11.1 Leaf, flower, fruit and root of licorice.

Licorice root is one of the oldest and most frequently used medicinal herbs in the history of Chinese medicine. The earliest written document indicating the utilization of licorice is Codex Hammurabi dating back to 2100 BC (Gibson, 1978; Fenwick et al., 1990). References to the effectiveness of licorice are contained in the "Shen Nong Ben Cao Jing," the first Chinese dispensary book with original anonymous volumes that probably appeared by the end of third century. Licorice is recommended for prolonging life, improving health and curing injury or swelling. It is also known for its detoxification effects (Jiansu New Medical School, 1979; Nomura and Fukai, 1998). Out of 110 prescriptions recorded in the ancient Chinese book of medicine, "Shang Han Lun," seventy include licorice (Nomura and Fukai, 1998). A large quantity of licorice was also discovered during excavation of a famous Pharaoh's tomb in Egypt in 1923 (Reid, 1995). In the Mediterranean countries, licorice was well known even in the time of Theophrastus. The Greek physician seems to have been the first to describe the utility of licorice in Greek literature, in the "Enquiry into plants" (270 BC) (Huseman, 1944).

Licorice is widely used as a flavoring and sweetening agent in tobaccos, chewing gums, candies, toothpastes, beverages, etc. Licorice is also used in cosmetic and pharmaceutical products, eg., anti-ulcer drugs (Aspalon®, Caved-S®, etc.) (Gibson, 1978; Food and Drug Administration, 1985; Fenwick et al., 1990; Nomura and Fukai, 1998). In the U.S., 90% of the licorice supply is used by the tobacco industry, the rest being shared equally between the food and pharmaceutical industries (Food and Drug Administration, 1985). It is difficult to estimate the mean yearly intake of licorice products. However, rough estimates from the Nordic countries based on the import and production data on liquorice sweets, suggest yearly consumption of about 1.5 kg/person. This value is in good agreement with the data reported from the UK, U.S. and Belgium (Spinks and Fenwick, 1990). Studies on the pattern of licorice intake have also been carried out (Ibsen, 1981; Simpson and Currie, 1982). In a survey of 939 Danish schoolchildren (ages seven to 18 years), 60% of them consumed licorice regularly, and 5% of them ingested more than 100 g licorice a week (Ibsen, 1981). An investigation from New Zealand on licorice intake among 603 high-school students (Simpson and Currie, 1982) showed that licorice was eaten regularly by 29% of the girls and 17% of the boys. At least 200 g was consumed per week by 5.9% of the girls and 4.9% of the boys, and at least 500 g by 1.8% of the girls and 1% of the boys. The intake in two students exceeded 1,000 g licorice per week.

2. CHEMISTRY OF LICORICE

Licorice root extract contains essential oils, alkaloids, polysaccharides, poly-amines, triterpenoids and flavonoids (Fenwick et al., 1990). The main sensory characteristics of licorice have been identified in a concentrate of the volatile oil fraction containing estragole, anethole, eugenol, indole, cumic aldehyde and g-nonalactone (Fenwick et al., 1990).

2.1. TRITERPENOID SAPONINS

Saponins are complex molecules consisting of sugars linked to triterpenoid or steroid or steroidal alkaloids. Triterpenoid saponins occur widely throughout the plant kingdom including human foods, especially beans, spinach, tomatoes and potatoes. The most important constituent of licorice is glycyrrhizin (GL), which is present in quantities ranging between 3.63–13.06% in dried roots (Jiansu New Medical School, 1979). GL is a saponin with a pentacyclic triterpene structure of β-amyrine type and is used as a tool for recognizing the *Glycyrrhiza* species (Figure 11.2). GL is found in root and stolon, but not in seed, leaf or stem of licorice. On the other hand, soyasaponins are detected in seed, hypocotyl and rootlet. GL is nearly 170 times as sweet as sucrose,

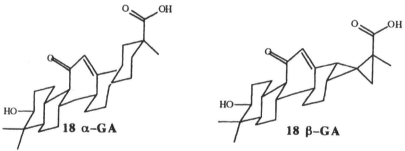

Figure 11.2 Chemical structure of GL, GA, 18 α-GA and 18 β-GA.

and its sweetness is much more persistent. However, its taste is completely lost when the saponins are hydrolyzed to the aglycone derivative (Nomura and Fukai, 1998). GL has also been shown to be partly hydrolyzed by glucuronidase to its aglycone, glycyrrhetinic acid (GA), that exists as two stereoisomers (Agarwal et al., 1991), 18 α-glycyrrhetinic acid (α-GA) and 18 β-glycyrrhetinic acid (β-GA) (Figure 11.2). At least fifty other saponins have also been isolated from *Glycyrrhiza* species, which include liquoric acid, 24-hydroxy-11-deoxoglycyrrhetinic acid, glycyrrhetol and uralenic acid (Segal et al., 1970; Zhang et al., 1986; Zeng et al., 1991a, 1991b; Cai et al., 1991; Kitagawa et al., 1993a, 1993b, 1993c, 1993d; Morris, 1976).

β-GA produces a typical type I spectra by interacting with oxidized cytochrome P-450, while α-GA shows almost no interaction. β-GA was more effective than α-GA in inhibiting monooxygenase activity, which may be due to stereochemical differences between α-GA (trans conformation of the D/E-ring) and β-GA (*cis* conformation of the D/E-ring) (Agarwal et al., 1991).

About three hundred polyphenols ranging in quantities between 1 and 5% have so far been isolated from the dried root of *Glycyrrhiza* species (Yang et al., 1990; Zhou and Zhang, 1990; Liu and Liu, 1990; Kusano et al., 1991; Kuo et al., 1992; Lee et al., 1996; Vaya et al., 1997). The main phenolic compounds are liquiritin and liquiritigin and their chalcone derivatives known as isoliquiritigin and isoliquiritin, respectively (Figure 11.3). Apparently, isomerization of isoliquiritin leads to the formation of liquiritin because the flavone glucoside is only detectable on boiling the alcoholic extract or on sun drying or storage of the roots (Nomura and Fukai, 1998). Some licorice polyphenols with isoprenylated substituted structures such as glycyrrihisoflavone, licoricidin, isoangustone A and kanzonol G as shown in Figure 11.3, may be of pharmaceutical importance (Nomura and Fukai, 1998; Vaya et al., 1997; Reiners, 1966; Liu and Liu, 1989; Kiuchi et al., 1990; Hatano et al., 1991a, 1991b; Kitagawa et al., 1994).

Glycyrrhisoflavone resembles genistein structurally, but contains an isoprenoid group on the B-ring. These polyphenols exhibit pharmacological properties as phytoestrogens and monoamine oxidase inhibitors. The concentrations of licocoumarone, glycyrrhisoflavone and genistein required for 50% inhibition of monoamine oxidase activity are 6.0×10^{-5}, 9.5×10^{-5} and 9.5×10^{-5} M, respectively (Hatano et al., 1991b).

2.2. LICORICE POLYSACCHARIDES

Licorice contains a group of polysaccharides reported to have immuno-modulating effects that include enhancing the phagocytic property of macrophages, inducting IL-1 secretion and IFN release from macrophages and enhancing natural killer cell activities (Yang and Yu, 1990).

Three polysaccharides, namely, glycyrrhizans UA, UB and UC, were isolated from the licorice root. They were homogeneous on electrophoresis and gel chromatography. *Glycyrrhiza*n UA is composed of L-arabinose, D-galactose, L-rhamnose and D-galacturonic acid in a molar ratio of 20:14:1:3, and glycyrrhizan UB is composed of L-arabinose, D-galactose, D-glucose, L-rhamnose and D-galacturonic acid in a molar ratio of 12:10:1:10:20, in addition to small amounts of *O*-acetyl groups and peptide moieties, respectively. About 10% (glycyrrhizan UA) and 35% (glycyrrhizan UB) of the D-galacturonic acid residues exist as methyl esters. *Glycyrrhiza*n UC with a molecular mass of approximately 69,000, is composed of L-arabinose, D-galactose, D-glucose and L-rhamnose in a molar ratio of 10:30:27:1. Methylation analyses, carbon-13 nuclear magnetic resonance and periodate oxidation studies indicate that it is an arabino-3, 6-galacto-glucan-type polysaccharide (Tomoda et al., 1990; Shimizu et al., 1990). Methylation analyses of primary, secondary and tertiary

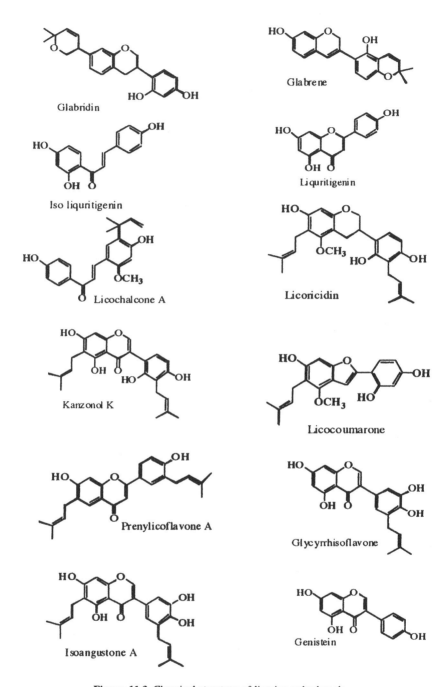

Figure 11.3 Chemical structure of licorice polyphenols.

Smith degradation products and of the limited hydrolysis products indicate that glycyrrhizan UA has a backbone chain composed of β-1,3-linked D-galactose. All of the galactose units in the backbone carry side chains composed mainly of α-1,5-linked L-arabino-β-1,6- or 1,3-linked D-galactose residues at position 6. Removal of the arabinosyl side chains substantially reduces immunological activity (Shimizu et al., 1992).

Two anticomplementary polysaccharide fractions (GR-2IIa and GR-2IIb), isolated from the roots of *Glycyrrhiza uralensis*, each show five anticomplementary polysaccharides (GR-2IIa-1-5 and GR-2IIb-1-5) on HPLC. Similarly, two anticomplementary and mitogenic polysaccharides of GR-2IIc (GR-2IIc-1-2A and -2IIc-2) can be isolated by gel filtration and HPLC. GR-2IIc-1-2A showed the most potent anticomplementary activity. GR-2IIa-1-5 and GR-2IIb-1-5 contained 40–85% and 50–90% of GalA, respectively, in addition to Rha, Ara and Gal. GR-2IIc-1-2A and -2IIc-2 mainly comprised Glc, Gal, GalA and GlcA in addition to Rha, Fuc, Xyl, Ara and Man. The large fractions from GR-2IIa-2IIc showed more potent anticomplementary activities than the original polysaccharide fractions, whereas the intermediate fractions and oligogalacturonides were inactive (Zhao et al., 1991).

3. CANCER CHEMOPREVENTIVE EFFECT OF LICORICE

3.1. ANTIMUTAGENIC EFFECT

Our previous studies have shown the antimutagenic effect of GA (Wang et al., 1991). In the presence of Araclor 1254-induced rat liver S-9 as an enzyme source for metabolic activation, β-GA was more effective than α-GA in inhibiting the mutagenicity of benzo(a)pyrene (BaP), 2-aminoflurene (2-AF) and aflatoxin B_1 (AFB_1) in *Salmonella typhimurium* strains TA100 and TA98. At a concentration of 0.5 mg/plate, β-GA and α-GA inhibited AFB_1-induced mutagenesis in TA98 by 77% and 23%, respectively (Wang et al., 1991). Licorice extract also blocks mutagenesis of furapromidum, zhengdingmycin hydrochloride, *N*-methyl-*N*-nitro-*N*-nitrosoguanidine and methylmethanesulfonate in strains of *Salmonella typhimurium* (TA 97, TA 98, TA 100, TA 102) (Shi, 1992). In addition, DNA binding of mutagens is reduced by these agents (Leibman, 1971; Abdugafurova et al., 1990; Ngo et al., 1992). We also observed that β-GA inhibits rat epidermal aryl hydrocarbon hydroxylase (AHH), 7-ethoxycoumarin-*O*-deethylase (ECD) and 7-ethoxyresorufin-*O*-deethylase (ERD) activities, in a dose-dependent manner, whereas α-GA was less effective in this regard (Wang et al., 1991). Micronuclei formation in mouse bone marrow cells induced by cytoxan is also diminished by licorice extract (Fu et al., 1995).

3.2. ANTICARCINOGENESIS EFFECT

Oral administration of disodium glycyrrhizinate in drinking water to male B6C3F1 mice for 96 wk showed evidence of chronic toxicity or tumorigenicity when compared to the control group treated with water alone (Kobuke et al., 1985).

In a two-stage skin tumorigenesis assay using 7, 12-dimethylbenz(a)anthracene (DMBA) as a tumor initiator and 12-O-tetradecanoylphorbol-13-acetate (TPA) as a tumor promotor, we have shown that oral administration of 0.05% GL in water as the sole source of drinking fluid to Sencar mice either prior to DMBA application or post-DMBA application protects against skin tumor formation (Agarwal et al., 1991). Furthermore, we also compared α-GA and β-GA on DMBA/TPA skin tumorigenesis in mice. In this study, we observed that both α-GA and β-GA inhibited the skin tumor initiating activity of DMBA. However, β-GA exerted stronger inhibitory effects as compared to α-GA (Wang et al., 1991). α-GA and β-GA possessed comparable inhibitory activity as inhibitors of TPA-induced skin tumor promotion in mice (Wang et al., 1991). Our studies also indicate that oral feeding of GL to Sencar mice causes 36% and 41% inhibition of epidermal DNA binding of [³H]-BP and [³H]-DMBA, respectively (Agarwal et al., 1991). Similarly, topical application of α-GA and β-GA resulted in 41% and 59% inhibition, respectively, of binding of [³H]-BP to epidermal DNA and 30% and 48% inhibition, respectively, of binding of [³H]-DMBA to epidermal DNA (Wang et al., 1991). These results suggest that protection against DNA damage induced by tumor initiating agents may explain their antitumor initiation activity. We observed that both α-GA and β-GA inhibit lipoxygenase activity and TPA-induced cutaneous ornithine decarboxylase (ODC) activity (Wang et al., 1991). Nishino et al. (Nishino et al., 1986, 1988; Yasukawa et al., 1988; Nishino et al., 1993) reported that GA markedly suppressed the tumor promoting effect of two different tumor promoters, TPA and teleocidin on skin tumor formation in DMBA-initiated mice. GA appears to bind to the TPA receptor in epidermis, inhibits protein kinase C and ODC induction and also down-regulates the epidermal growth factor receptor (Okamoto et al., 1983; Kitagawa et al., 1986; Nishino et al., 1988; O'Brian et al., 1990; Shibata, 1994). Licochalcone A isolated from licorice showed an *in vitro* inhibitory activity against TPA-mediated phosphorylation of phospholipids in HeLa cells. It also inhibited TPA-induced skin tumor promotion (Shibata, 1994).

In our previous studies, we used licorice water extract as a test agent given to female A/J mice treated with BaP (100 mg/kg administered p.o. at two-week intervals for eight weeks) or N-nitrosodiethylamine (NDEA) (20 mg/kg administered p.o. once weekly for eight weeks). Oral feeding of 1.0% water extract of licorice in drinking water during BaP treatment (initiation period) causes a 20% and 60% reduction in lung tumor incidence and multiplicity, respectively, and also 33% and 24% reduction in forestomach tumor incidence and multiplicity, respectively (Wang et al., 1992). Oral administration of 1.0% water extract of licorice in drinking water causes 26% and 55% reduction in NDEA-induced lung tumor incidence and tumor, respectively, and 45% and 68% reduction in forestomach tumor incidence and multiplicity, respectively (Wang et al., 1992). Licorice extract reduces N-methyl-N'-nitro-N-nitrosoguanidine (MNNG)-induced stomach tumors by about 60% (Shi, 1992).

Hepatocellular carcinoma (HCC) occurs in patients with hepatitis C virus-RNA positive chronic liver disease. A retrospective study was conducted to evaluate the long-term preventive effect of Stronger Neo-Minophagen C (SNMC), whose active component is GL on HCC development (Arase et al., 1997). Patients diagnosed with chronic hepatitis C retrospectively in the hospital were divided into two groups. In Group A, 84 patients were treated with SNMC at a dose of 100 mL daily for eight weeks, then two to seven times a week for two to 16 years (median, 10.1 years). Group B comprising 109 patients were treated with placebo or interferon for extended periods (median, 9.2 years) as the control group. The patients were retrospectively monitored, and the cumulative incidence of HCC and risk factors for HCC were examined. The 10-year rates of cumulative HCC incidence for Groups A and B were 7% and 12%, respectively, and the 15-year rates were 12% and 25%, respectively. These results suggested that long-term administration of SNMC to patients with chronic hepatitis C was able to prevent liver carcinogenesis (Arase et al., 1997).

In another study, GL was reported to inhibit mouse lung and liver tumorigenesis (Watari et al., 1976; Nishino, 1992). The multiplicity of spontaneous hepatomas in C3H/He mice decreased to 50% after treatment with 5 mg GL per 100 mL drinking water. The same dose of GL in the drinking water of the carcinogen 4-nitroquinoline-1-oxide (4NQO)-initiated/glycerol-promoted ddY mice reduced lung tumor multiplicity by 75% (Nishino, 1992).

The individual and combined effects of dietary toasted soybean meal (3.13–25%) and dietary licorice root extract (0.38–3.0%) on selected liver and intestinal enzyme levels and on clinical chemistry and histopathological parameters were evaluated in male F344 rats (Webb et al., 1992). Histopathological evaluation of organs and tissues yielded no significant strain-related changes. However, hepatic glutathione transferase, catalase and protein kinase C activity showed significant induction (up to 50%) in response to increasing doses of soybean meal and licorice extract, providing evidence for marginal interaction between the two additives (Webb et al., 1992).

The Chemoprevention Branch and Agent Development Committee of the National Cancer Institute (NCI) sponsored studies for β-GA and carbenoxolone. In these studies, β-GA was effective in inhibiting rat mammary gland and mouse colon carcinogenesis (Kelloff et al., 1994).

In the carcinogen, N-methyl-N-nitrosourea (MNU)-induced rat mammary gland carcinogensis model, β-GA (1 g/kg diet or ca. 0.1 mmol/kg-bw/day) reduced carcinoma multiplicity at a lower dose than carbenoxolone (2 g/kg diet or ca. 0.2 mmol/kg-bw/day). However, both agents inhibited premalignant lesions in the DMBA-induced mouse mammary organ culture *in vitro* assay system at the same concentration (0.001 plvl). Carbenoxolone has also been tested in combination with other agents in the MNU-induced rat mammary cancer model employing a relatively smaller dose (Kelloff et al., 1994). For example, the dose (1,750 mg/kg diet or ca. 0.15 mmol/kg-bw/day), which was ineffective alone, significantly inhibited tumor incidence when administered with D,L-alpha-difluoromethylornithine (DFMO), oltipraz and DFMO plus tamoxifen. These studies suggest that carbenoxolone at low doses may be employed with other chemotherapeutic agents to enhance their pharmacological action (Kelloff et al., 1994). β-GA inhibited BaP-induced morphological changes and ornithine decarboxylase (ODC) induction in rat tracheal epithelial cells. Both β-GA and carbenoxolone were effective in suppression against anchorage-independent growth of human lung tumor A427 cells.

In summary, studies in laboratory animals suggest that licorice water extract and major terpenoid derivatives and polyphenols may have cancer chemopreventive effects.

4. PHARMACOLOGY OF LICORICE

Licorice has numerous pharmacological effects including detoxification, antioxidant, antiulcer, anti-inflammatory and antiviral properties (Kelloff et al., 1994). The water-soluble 3β-O-hemisuccinate derivative of carbenoxolone has been used clinically in Europe, Australia and Japan for treatment of gastric ulcer and esophagitis (Reynolds and Prasad, 1982; Akimoto et al., 1986; Altman, 1989; Eason et al., 1990). However, in this review, we will focus on the following pharmacological and physiological activities of licorice.

4.1. DETOXIFICATION EFFECT

Licorice is well-known for enhancing detoxification. Licorice and its constituents protect against toxin-induced hepatotoxicity both *in vitro* and *in vivo* (Nakamura et al., 1985; Shibayama, 1989; Matsuda et al., 1991; Huang, 1991; Lee et al., 1991; Wang and Han, 1993; Liu et al., 1994: Nose et al., 1994; Lin et al., 1997; Wang et al., 1998). Treatment with GL 20 h before carbon tetrachloride (CCl$_4$) administration protects against development of pericentral hepatocellular necrosis (Shibayama, 1989). GL treatment two hours prior to the administration of allyl formate also inhibits the development of periportal hepatocellular necrosis. However, GL does not protect against the development of endotoxin-induced focal and random hepatocellular necrosis. These results suggest that GL has no protective effect on hepatic injury related to sinusoidal circulatory disturbances as observed in the case of endotoxin exposure but can protect against hepatotoxicity induced by chemicals directly acting on hepatocytes (Kiso et al., 1984; Shibayama, 1989). Subcutaneous administration of GL, β-GA or α-GA for three days at 200 μmol/kg decreased acetaminophen (500 mg/kg, ip)-induced liver injury in mice (Liu et al., 1994). The oral administration of β-GA before D-galactosamine (GalN)-treatment significantly diminishes acute liver damage (Matsuda et al., 1991; Nose et al., 1994). Pretreatment of licorice polyphenols also inhibited CCl_4-induced necrosis in mice (Wang and Han, 1993).

GL can prevent the cancer chemotherapeutic agent-induced toxicity in tumor-bearing mice and in cancer patients (Akimoto et al., 1986; Hall et al., 1989). Fifty-seven patients with postoperative breast cancer received mitomycin-methotrexate-futraful (MMF) therapy with GL and 60 were given MMF alone. There was only one patient (1.7%) in the GL plus MMF treatment group in whom the MMF alone was discontinued because of liver dysfunction; however, 15 patients (25%) in the latter group had to stop the therapy because of its side effects (Akimoto et al., 1986).

The mechanism for hepatoprotection by licorice is not fully understood. It is known that many hepatotoxicants require metabolic activation, particularly, through hepatic cytochrome P-450-dependent systems. The effect of licorice on microsomal drug metabolism (e.g., benzo(a)pyrene hydroxylase (AHH), 7-ethoxycoumarin O-deethylase (ECD) and erythromycin D (ErD) monooxygenase activity) and mutagen-DNA binding have been studied (Leibman, 1971; Abdugafurova et al., 1990; Ngo et al., 1992). Recent studies indicate that prolonged intake of licorice extracts or GL results in the induction of hepatic CYP3A- and, to a lesser extent, 2B1- and 1A2-dependent microsomal monooxygenase activities in male and female mice, in addition to 6β-(mainly associated to CYP3A), 2α-, 6α-(CYP2A1, 2B1), 7α-, 16α-(CYP2B9) and 16β-testosterone hydroxylase activities (Paolini et al., 1998). The induction of cytochrome P450-dependent activities by the prolonged intake of licorice extracts or GL may result in the accelerated metabolism of coadministered drugs with important implications for their disposition (Paolini et al., 1998).

4.2. ANTIULCER AND ANTI-INFLAMMATORY EFFECT

Licorice and its constituents have long been known to promote the healing of gastrointestinal ulcers (Rainsford and Whitehouse, 1977; Jiansu New Medical School, 1979; Eason et al., 1990; Dehpour et al., 1994; Kelloff et al., 1994; Farina et al., 1998). For example, aspirin coated with licorice reduces the ulcer index from 1.5 ± 0.12 to 0.5 ± 0.12 and its incidence from 96% to 46%. However, coatings with derivatives such as carbenoxolone and enoxolone were less effective (ulcer index, 0.70–0.94, incidence 62–76%) (Dehpour et al., 1994). In a double-blind trial, 43 patients with an endoscopically confirmed, symptomatic duodenal ulcer were randomly allocated to treatment with either carbenoxolone sodium or placebo, both provided in identical "positioned-release" capsules (Paolini et al., 1998). The 40 patients who satisfactorily completed the trial were evenly distributed between the two treatment groups. The groups were well matched with regard to clinical features and initial ulcer size. Endoscopic review of ulcer healing after six weeks of treatment showed that 12 patients (60%) receiving carbenoxolone had healed ulcers, compared with five (25%) receiving placebo ($p < 0.05$). Symptomatic remission occurred by the fourth week in 17 patients (85%) receiving carbenoxolone, compared with six (30%) receiving placebo ($p < 0.001$). In addition, a relationship between serum carbenoxolone levels and the occurrence of ulcer healing was observed (Young et al., 1979).

The most widely observed pharmacological action of GA is the hydrocortisone-like anti-inflammatory effect in animal models of edema and arthritis (Tangri et al., 1965; Nikitina, 1966; Murav'ev et al., 1983; Inoue et al., 1989). GA and its derivatives applied 30 min before treatment with arachidonic acid or 12-O-tetradecanoylphorbol-13-acetate (TPA) inhibited edema in murine ear or paw (Inoue et al., 1988, 1989, 1993). Oral administration of GA and its derivatives inhibited the ear edema induced by topical application of capsaicin in mice (Inoue et al., 1996). Licorice polyphenols were also reported to inhibit croton oil-induced murine ear edema (Fu et al., 1995) and TPA-induced skin inflammation and tumor promotion (Shibata, 1994).

In the 1950s, licorice was reported to have anti-inflammatory activity and to cause fluid retention suggesting that licorice might exhibit properties of glucocorticoids or mineralocorticoids in humans. Prostaglandins are thought to promote healing of ulcers by stimulating mucous secretion and cell proliferation. Thus, local increase in prostaglandin concentration, due to the inhibition of cyclooxygenase and lipoxygenase activity by licorice-derived components, promotes healing of ulcers (Ren and Wang, 1988; Baker, 1994). 11β-hydroxysteroid dehydrogenase is one of the major factors involved in the regulation of the levels of circulating cortisol. Cortisone does not bind to the mineralocorticoid or glucocorticoid receptor. Licorice-derived compounds inhibit 11β-hydroxysteroid dehydrogenase activity (Monder et al., 1989; Baker, 1994; Zhang et al., 1994; Baker, 1995; Soro et al., 1997). When given orally to rats, it partially inhibits renal 11β-dehydrogenase activity (Monder et al., 1989).

Renal proximal tubular preparations, kidney homogenates and renal microsomes readily convert corticosterone to 11β-dehydrocorticosterone. However, GL and its synthetic analog, carbenoxolone, inhibit this reaction in a dose-dependent manner (Monder et al., 1989). The mineralocorticoid effects of GL may be mediated by direct binding of glycyrrhetic acid to the mineralocorticoid receptor (Armanini et al., 1983, 1989). The affinity of GA for the receptor was about 3,000 and 10,000 times less than that of aldosterone in human leukocytes and rat kidney, respectively (Armanini et al., 1983, 1989). However, the anti-inflammatory effect of GL may be through the suppression of phospholipase A2 and its effect on platelet aggregation. GL was found to inhibit phospholipase A2-induced carboxyfluorescein release from the D, L-dipalmitoyl phosphatidylcholine liposomes. Part of this inhibitory effect of GL is accounted for by the physical interaction between the substrate and liposome membranes (Okimasu et al., 1983).

4.3. ANTIHEPATITIS VIRUS EFFECT

About 200 million people worldwide are chronically infected with hepatitis B virus (HBV). This disease causes acute and chronic hepatitis, cirrhosis and hepatocellular carcinoma. GL has been widely used to treat chronic hepatitis B in China and Japan. Studies on the biological effects of GL show that it normalizes elevated serum transaminases levels such as alanine and aspartate aminotransferase and improves liver histology in patients with chronic hepatitis B, resulting in a complete recovery from chronic HBV infection (Suzuki et al., 1983; Eisenburg, 1992). GL is also used to supplement interferon in the treatment of patients with chronic HBV infection and to treat nonresponders to interferon therapy (Fujisawa et al., 1973; Hayashi et al., 1991). Pompei et al. (1979) reported that GL inhibited the growth and cytopathology of several unrelated DNA and RNA viruses and inactivated herpes simplex virus particles irreversibly. Further studies indicated that GL may directly interact with HBV (Takahara et al., 1994; Sato et al., 1996). GL suppressed the secretion of hepatitis B surface antigen (HBsAg) resulting in its accumulation in cytoplasmic vacuoles in the Golgi area. The secreted HBsAg was modified by *N*-linked and *O*-linked glycans, but its sialylation was dose-dependently inhibited by GL. Thus, GL suppressed the intracellular transport of HBsAg at the trans-Golgi area after *O*-linked glycosylation and before its sialylation. HBsAg particles were observed mainly on the cell surface in the GL-treated culture but not in the untreated culture. This suggests that asialylation of HBsAg particles resulted in the novel surface nature of GL-treated HBsAg particles. However, desialylated HBsAg is known to have enhanced immunogenicity (humoral antibody response and cell-mediated immunity) compared to sialylated HBsAg (Takahara et al., 1994; Sato et al., 1996). GL and/or combination therapy with interferon is also useful in improving hepatitis A (HAV)-(Crance et al., 1994, 1995; Cuyck-Gandre et al., 1995) or C virus (HCV)-induced chronic hepatitis (Fujisawa, 1991; Acharya et al., 1993; Abe et al., 1994; Da Nagao et al., 1996; Arase et al., 1997).

GL or its derivatives exert antiviral action on other viruses too, including simplex virus, varicella-zoster virus, epstein barr virus, parainfluenza virus and influenza virus A2, which act by different mechanisms (Pompei et al., 1980; Dargan and Subak-Sharpe, 1985, 1986; Baba and Shigeta, 1987; Chang et al., 1989; Agarwal et al., 1990; Flavin-Koenig, 1996; Badam, 1997; Utsonomiya et al., 1997).

4.4. ANTI-AIDS EFFECT

The anti-HIV activity of licorice extracts and its various derivatives has been studied in an *in vitro* system using various strains of HIV (Nakashima et al., 1987; Ito et al., 1987, 1988; De Clercq, 1988; Tochikura et al., 1989; Hatano et al., 1988; Hirabayashi et al., 1991; Pliasunova et al., 1992; Vodovozova et al., 1996). GL was investigated for its antiviral action on HIV (HTLV-III/LAV), using the cytopathic effect and plaque-forming assay system in MT-4 cells (aHTLV-I-carrying cell line). Cloned Molt-4 cells (clone No. 8), which are sensitive to HIV and fuse to form giant cells after infection, were also used to assess the cytopathic effect of HIV. GL completely inhibited HIV-induced plaque formation in MT-4 cells at a concentration of 0.6 mM (IC_{50}:0.15 mM). Furthermore, GL inhibited giant cell formation of HIV-infected Molt-4 clone No. 8 cells. GL showed no direct effect on the reverse transcriptase of HIV. However, the exact mechanism of anti-HIV action observed in these studies remains to be elucidated (Ito et al., 1987). Mechanistic studies show that GL exhibits a dose-dependent reduction in protein kinase C (PKC) activity in MOLT-4 (clone No. 8) cells. The PKC inhibitor, 1-(5-isoquinolinesulfonyl)-2-methylpiperazine dihydrochloride (H-7), also inhibited HIV-1 replication in MOLT-4 (clone No. 8) cells, suggesting that PKC inhibition may be one of the mechanisms of action of GL. In addition, GL may owe its anti-HIV-1 activity, at least in part, to an interference with virus-cell binding, since this agent at 1.2 mM concentration partially inhibited the adsorption of radiolabeled HIV-1 particles to MT-4 cells. At this concentration, GL also suppressed giant cell formation induced by co-culturing MOLT-4 (clone No. 8) cells with MOLT-4/HTLV-IIIB cells, whereas the PKC inhibitor H-7 failed to do so (Ito et al., 1988).

In MT-4 cells after HIV infection, the virus-induced cytopathic effect and the expression of viral antigens were inhibited by 0.184 mM of glycyrrhizin sulfate (GLS). Moreover, GLS completely inhibited HIV-induced plaque formation in MT-4 cells at a concentration of 1 mg/mL (736 μM), the 50% inhibitory dose being 0.055 mg/mL (40 μM). GLS was found to be an efficient inhibitor of reverse transcriptase. The effect of GLS was four times stronger than the parent compound GL (Nakashima et al., 1987).

The anti-HIV activity of licorice extract and its various derivatives were also studied in a murine AIDS model (MAIDS) system to evaluate the effects of GL and other derivatives *in vivo* (Watanbe et al., 1996). C57BL/6 mice were inoculated with LP-BM5 murine leukemia virus to cause MAIDS. GL supplement was administered at day zero or four weeks after virus inoculation, three times a week for 19 weeks. Immunological abnormalities were monitored with respect to the surface phenotype identified by two-color staining for CD3 and IL-2 receptor beta-chain. All mice infected with the virus alone developed MAIDS and died by 14 weeks after infection. Mice receiving the treatment at day zero or four weeks after infection survived three weeks longer. It appears that GL supplement may possibly suppress the progression of the disease (Watanbe et al., 1996).

Human intervention studies also proved that licorice products might have anti-AIDS effects (Watanbe et al., 1996; Hattori et al., 1989; Mori et al., 1989a, 1989b; Mori et al., 1990; Lu, 1993). The effect of GL-containing supplement in 42 hemophilia patients with HIV infection was reported (Mori et al., 1990). The dose was 100–200 mL of GL supplement in 21 patients and 400–800 mL in 21 other patients administered. The patients were divided into an asymptomatic carrier (AC) group and an AIDS-related-complex (ARC)/ AIDS group. GL supplement was administered intravenously daily for the first three weeks and every second day for the following eight weeks to the 42 HIV-infected hemophilia patients. The CD4/CD8 ratio and CD4 positive lymphocyte counts did not change during the treatment period. However, significant improvement was noted in some cases. A slight increase in mito-genic responsiveness to phytohemagglutinin, concanavalin A and pokeweed mitogen was noted in most patients of the two groups. A very significant improvement was seen in the AC group receiving over 400 mL of GL supple-ment. Furthermore, complete improvement was noted in liver dysfunction, which is one of the major problems for hemophiliacs treated with blood products. Thus, prophylactic administration of high-dose GL supplement to HIV-positive hemophiliacs who have impaired immunological ability and liver dysfunction was considered to be effective in preventing the transition from AC/ARC to AIDS (Mori et al., 1990). GL has not only an inhibitory effect on HIV replication but also exhibits interferon-inducing and natural killer (NK)-enhancing effects and improves liver function (Mori et al., 1989b). Thus, large doses of GL (200–800 mg/day) administered intravenously for more than eight weeks to nine hemophilia-A patients with HIV infection (asymptomatic carrier, AC) resulted in increased lymphocyte counts in all nine cases, elevated OKT4/OKT8 ratio in six out of the nine cases and increased OKT4-positive lymphocytes in eight out of the nine cases. In four cases, liver dysfunction noted was clearly improved. Serum electrolytes, protein, lipids and renal function were within the normal levels, and no serious side effects were observed during treatment. On the other hand, in three cases of hemophilia without HIV infection, the number of OKT4 lymphocytes was not significantly altered during treatment (Mori et al., 1989b).

4.5. ANTIATHEROGENIC EFFECT

Macrophage-mediated oxidation of LDL, a hallmark in early atherosclerosis, depends on the oxidative state of LDL and that of the macrophage function

(Aviram, 1996; Aviram and Fuhrman, 1998). Animal studies with experimental hyperlipidemia and atherosclerosis have shown that pentocyclic triterpenoids and polyphenols from herbs including licorice are capable of reducing the level of cholesterol, b-lipoproteins and triglycerides in blood and the cholesterol level in aortic tissue (Nakamura et al., 1966; Vasilenko et al., 1982, 1993; Fuhrman et al., 1997). Ammonium salt of GA (AGA) and 18-dehydroglycyrretic acid (18-DHGA) effectively reduce the concentration of total cholesterol, triglycerides and atherogenic lipoproteins in serum of rabbits with modeled cholesterol atherosclerosis. These agents reduced significantly the content of malonaldehyde (MDA) and augmented the activity of tissue superoxide dismutase. It is possible that due to their powerful hypolipidemic and antioxidant effects, AGA and 18-DHGA significantly reduce atherosclerotic changes on the aortic surface (Zakirov and Abdullaev, 1996).

Dietary supplementation with licorice (200 µg/day) or pure glabridin (20 µg/day) to the E^0 mouse for six weeks resulted in a substantial reduction in the susceptibility of LDL for oxidation along with a reduction in atherosclerotic lesion area (Fuhrman et al., 1997). Sixty percent of the placebo-treated mice showed well-defined lesions (grade 2 lesions (30%), grade 3 lesions (20%) and advanced grade 4 lesion (10%) reaching the dimension of 50,000 µm²). In contrast, only two of the 10 licorice-treated mice showed grade 3 lesions with maximum dimensions of 18,000 µm². The remaining eight licorice-treated mice showed minimal histopathologic changes (grade 1 only) (Fuhrman et al., 1997). In an *ex vivo* study, LDL isolated from the plasma of 10 normolipidemic subjects orally supplemented for two weeks with 100 mg licorice/d, was more resistant to oxidation than the LDL isolated before licorice supplementation (Fuhrman et al., 1997). These results could be related to the absorption and binding of glabridin to LDL particles and subsequent protection of LDL from oxidation (Fuhrman et al., 1997).

4.6. ANTIOXIDANT EFFECT

Anti-inflammatory, antimutagenic, anticarcinogenic and antiatherogenic effects of licorice and its derivatives may be due to its antioxidant capacity (Demizu et al., 1988; Ju et al., 1989; Kuo et al., 1992; Vaya et al., 1997; Belinky et al., 1998a, 1998b; Haraguchi et al., 1998). Crude licorice extract and its derivatives have been shown to increase the resistance toward peroxidation *in vivo* (Vaya et al., 1997; Fuhrman et al., 1997). Administration of licorice extracts for two weeks before irradiation and seven days after irradiation at a dose of 2 Gy inhibits lipid peroxidation in rat lung (Palagina et al., 1995). It is likely that licorice polyphenols contribute to the antioxidant activity of licorice. Licochalcones B and D strongly inhibit superoxide anion production by the xanthine/xanthine oxidase, 1,1-diphenyl-2-picryl-hydrazyl (DPPH) radical and Fe(III)-ADP/NAPDH-mediated lipid peroxidation (Haraguchi et al.,

1998). The incubation of mitochondrial lipid peroxidation by Fe(III)-ADP/ NADH was sensitive to these retrochalcones and was completely inhibited at 10 µg/mL concentration of all retrochalcones tested. Furthermore, these retrochalcones protected red cells against oxidative hemolysis (Haraguchi et al., 1998).

The oxidation of LDL-induced by copper ions or mediated by macrophages was inhibited by licorice isoflavans, such as glabridin (4'-O-methylglabridin, hispaglabridin A and hispaglabridin B). These agents successfully inhibited the formation of conjugated dienes, thiobarbituric acid reactive substances (TBARS) and lipid peroxides. They also inhibited the electrophoretic mobility of LDL under oxidation. The isoflavene glabrene was the most active agent of all flavonoid derivatives tested. None of the isoflavan derivatives or the isoflavene compounds were able to chelate iron or copper ions. It has been suggested that the antioxidant effect of glabridin on LDL oxidation appears to reside mainly in the 2'hydroxyl group, and that the hydrophobic moiety of the isoflavan is essential for this function. It was also demonstrated that the position of the hydroxyl group on the B ring significantly affects the inhibitory efficiency of the isoflavan derivatives on LDL oxidation, but does not influence their ability to donate an electron to DPPH or their peak potential values. In macrophage-mediated LDL oxidation, TBARS formation was inhibited by these isoflavans (Vaya et al., 1997; Belinky et al., 1998a, 1998b). The natural isoflavan glabridin, the most abundant flavonoid in licorice extract (11.6%, w/w), was shown to be a potent antioxidant that inhibits 2,2'-azobis (2-amidino-propane) dihydrochloride (AAPH)-induced LDL oxidation (at 30 µM). Furthermore, 5 µM of glabridin caused 62% inhibition in cholesteryl linoleate hydroperoxide (CLOOH) formation in the LDL particle during its oxidation. It also results in a 64% inhibition in the consumption of cholesteryl linoleate (Vaya et al., 1997). Nearly 80% glabridin binds to LDL particles through hydrophobic interactions. However, under similar experimental conditions, quercetin and catechin bind to LDL by only 0.7 and 2%, respectively (Hayashi et al., 1991). This dramatic difference could relate to differences in their lipophilicity. Glabridin, with only two phenolic hydroxyl groups, is much more lipophilic. Although glabridin and quercetin are significantly different in their lipophilic characteristics, their ability to bind with LDL and their capacity to chelate metal ions, they exhibit comparable inhibitory activity toward LDL oxidation. These observations may indicate that the mechanisms of antioxidant action of these compounds in this system are different. While quercetin conserved the vitamin E content of LDL during oxidation, glabridin failed to protect it. The observed differences in action between the two agents toward vitamin E related to the greater ability of glabridin to penetrate into the LDL core than quercetin (Belinky et al., 1998b).

5. PHARMACOKINETICS AND PHARMACODYNAMICS

5.1. METABOLISM OF LICORICE

The biotransformation of GL to the active substance GA is believed to take place in the intestines due to glucuronidase activity in gastrointestinal bacteria (Stormer et al., 1993; Akao et al., 1994; Akao, 1997a). After oral ingestion in rat, GL and its main metabolites GA-3-glucuronide (GA-3-G), GA-3-O-glucuronide (GA-3-O-G) and GA-3-sulfate (GA-3-S) are detectable in bile (Hosoya, 1988). However, in blood, the concentration of GA was very high whereas GL was very low. Similar results were observed in humans (Hosoya, 1988). The biotransformation of GL mostly occurs in the large intestine (T_{max}: 13 h) (Gunnarsdottir and Johannesson, 1997).

β-GA is converted to 3-oxo-18 β-GA (3-oxoGA) by rat liver homogenates containing NADP+ whereas in presence of NADPH, it is converted to two other metabolites 22α and 24-hydroxy-18β-GA, respectively. 3-OxoGA and 3-epi-18β-GA (3-epi-GA) seem to be hydroxylated at C-22 and C-24 positions. A metabolite of 3-oxoGA, showing a lower Rf value, was identified as 22 α-hydroxy-3-oxo-18 β-GA. It is also known that for 22 α-hydroxylation, the best substrate is 3-oxoGA, followed by GA and 3-epi-GA. On the other hand, for 24-hydroxylation, the best substrate is GA, followed by 3-oxoGA, and 3-epi-GA. However, α-GA was a poor substrate for both 22α- and 24-hydroxylation reactions (Akao et al., 1990). Rat gastrointestinal bacteria can hydrolyze GL to 3β-hydroxy-GA by GL β-D-glucuronidase and to oxide 3β-hydroxy-GA and 3α-hydroxy-GA to 3-oxoGA by 3β-hydroxy-glycyrrhetinate dehydrogenase and 3α-hydroxyglycyrrhetinate dehydrogenase, respectively. Tissue pH also plays an important role in the metabolism of GA. Metabolites, 3β-hydroxy-GA, 3-oxoGA and 3α-hydroxy-GA were obtained from the metabolism of GL at pH 8. In intestines (pH 6–7), 3β-hydroxy-GA was the main metabolite of GL. However, in stomach content (pH 4.2), the concentration of 3β-hydroxy-GA was lower. It is unknown whether GL can be metabolized to 3β-hydroxy-GA in the stomach (Akao, 1997b).

5.2. PHARMACOKINETICS

Dose-dependent pharmacokinetics of GL was investigated in rat by measuring its disappearance from plasma and excretion in bile (Ishida et al., 1992, 1994; Takeda et al., 1996; Wang et al., 1996). The decline in plasma concentration was biexponential after intravenous (i.v.) administration of 5, 10, 20 or 50 mg GL per kg. Dosage, however, had a marked effect on the pharmacokinetics, with a greater-than-proportional increase in area under the plasma concentration curve (AUC) at doses of 20 and 50 mg/kg, whereas the increase was proportional at the lower doses of 5 and 10 mg/kg. There was also a significant increase of the steady-state distribution volume and decrease in the total body and biliary clearances, at 20 and 50 mg/kg as compared to 5–10 and 5–20 mg/kg, respectively. The AUC, steady-state distribution volume and renal clearance at a given dose showed no significant difference between rats with and without bile fistula. The plasma unbound fraction increased with increasing plasma GL concentration over the observed range (2–900 micrograms/mL). No significant change in distribution volume for unbound GL was observed at the various doses, indicating that the distribution of GL into tissues is not changed by altered dose. On the other hand, a dose dependency in total body clearances for the unbound GL was observed that was attributed to dose-dependent-biliary clearances for unbound GL (Ishida et al., 1992). Enterohepatic circulation of GA in the rat was determined by kinetic analysis (Kawakami et al., 1993). When GA (2, 5, 10 and 20 mg/kg) was administered i.v. to bile duct-cannulated rats, excretion of unchanged GA in bile was about 1%, the acid-hydrolyzed products was 14–16%, and GA-3-*O*-glucuronide was only 1–2% of the administrated dose. In control rats, a secondary GA peak was observed 12 h after i.v. administration of GA (20 mg/kg). The enterohepatic circulation of GA was confirmed by the linked-rat method in which bile of the donor rat after i.v. administration of GA (20 mg/kg) was allowed to flow directly into the duodenum of the recipient rat. GA in plasma of the recipient rat could be observed after six hours and its concentration reached its maximum (approximately 0.5 μg/mL) eight to 12 hours after dosing the donor rat (Kawakami et al., 1993).

In a Phase I clinical trial, plasma levels of GA were detectable (0.9 μg/mL) after a single dose of 500 mg/m^2 (ca. < 0.03 mmol/kg-bw), however, the plasma level peaked at 4.6 μg/mL after four hours (Akao et al., 1994). The mean concentration of GA in plasma of five young volunteers was found to increase sharply from approximately 120 ng/mL at six hours to a peak value of approximately 570 ng/mL at 10 h after licorice ingestion. Four of the subjects were relatively homogeneous as far as the three parameters were concerned (AUC 4,500–7,800 ng.h/mL, T_{max} 9–12 h, and C_{max} 460–1,000 ng/mL). One person was out of range in T_{max} and another person was out of range in AUC and C_{max} (Gunnarsdottir and Johannesson, 1997). Similar results were obtained by Abe et al. (1994) in experiments on six volunteers after ingestion of Shosaikoto, a herbal medicine containing GA. Raggi et al. (1994a, 1994b, 1995) and Cantelli-Forti et al. (1994) have demonstrated that the bioavailability of glycyrrhetic acid in human plasma is greater after ingestion of pure glycyrrhizic acid than after ingestion of similar amounts contained in licorice. Limited multidose pharmacokinetics data from Phase I trial suggest approximately two fold drug accumulation over two to four months of treatment with β-GA based on AUC values (Akimoto et al., 1986).

5.3. PHARMACODYNAMICS

Kelloff et al. (1994) showed that the minimal effective chemopreventive doses in rats were 0.1 mmol/kg-bw/day for GA and 0.08 mmol/kg-bw/day for carbenoxolone. These doses are approximately 20-fold lower than the one-year no-observed-effect-level (NOEL) in these animals (1,000 mg/kg-bw/day or 2.1 mmol/kg-bw/day). Based on this ratio, it appears that an effective human dose without hypokalemia may be attainable. However, the therapeutic dose of carbenoxolone (300 mg qd or ca. 0.008 mmol/kg-bw/day), which is 10-fold lower than the dose that inhibits rat colon carcinogenesis, produces hypokalemia. Therefore, rat may not be the best model for predicting the human pharmacokinetics and safety of either agent. In rat plasma, GA shows a slow second elimination phase as is observed in humans, but hypokalemia was not observed in preclinical toxicity assays. In contrast to humans, the enterohepatic circulation was observed to be negligible in rat. Hepatic metabolism and excretion may also differ between the two species. In humans, carbenoxolone is absorbed intact, excreted into bile as glucuronide and undergoes enterohepatic circulation. In rat, however, carbenoxolone is hydrolyzed to GA before absorption, and therefore does not appear to accumulate.

6. TOXICOLOGY/SAFETY ASSESSMENT OF LICORICE

6.1. PRECLINICAL SAFETY STUDIES

In the literature of ancient Chinese medicine, licorcie is listed as a "nontoxic" herb. Our previous studies indicated that oral administration of 1% licorice water extracts for 42 weeks produces no significant changes in body weight, behavior and appearance in A/J mice (Wang et al., 1992). In a one-year NCI-sponsored study, the administration of 100, 300 and 1,000 mg/kg-bw/day (0.2, 0.6 and 2.1 mmol/kg bw/day, ig), did not produce significant changes in clinical chemistry or histology (Kelloff et al., 1994). Oral administration of disodium glycyrrhizinate in drinking water to male B6C3F1 mice for 96 wk yielded no evidence of chronic toxicity or tumorigenicity when compared to a control group receiving water alone (Kobuke et al., 1985). Following administration of GL (21.3–679.9 mg/kg-bw/day) in drinking water to female rats on seven to 17 days of gestation, a dose-related increase in embryo lethality and minor anomalies were observed (Mantovani et al., 1988).

6.2. CLINICAL SAFETY STUDIES

Recently, while evaluating the clinical safety of GA, various adverse effects were observed. These included electrolyte imbalance (hypernatremia and hypokalemia), edema, increased blood pressure and suppression of the renin angiotensin-aldosterone system (Conn et al., 1968; Bannister et al., 1977; Epstein et al., 1977a, 1977b; Werner et al., 1979; Blachley and Knochel, 1980; Toner and Ramsey, 1985; Stormer et al., 1993).

An NCI-sponsored Phase-I clinical trial of GA in patients has been conducted. Patients with a prior history of breast cancer receiving a single dose of 300–700 mg GA/m^2 (ca. 0.02–0.04 mmol/kg-bw) showed no evidence of significant toxicity (Vogel et al., 1992). Serum potassium levels decreased from baseline at four to six hours post-dosing; however, they began to return to baseline at eight hours. In a multidose study, patients ($n = 6$) receiving 400 and 500 mg/m^2 qd (ca. 0.02 and 0.03 mmol/kg-bw qd) experienced hypertension or hypokalemia that necessitated dose reduction or discontinuance, and only two patients completed the 16-week treatment protocol.

On the basis of existing data, it is not possible to determine the minimum nontoxic level of GL in humans. There is great individual variation in susceptibility to GA toxicity (Stormer et al., 1993). In the most sensitive individuals, a daily intake of 100 mg GL, which corresponds to 50 g licorice sweets (assuming a content of 0.2% GL), produces adverse effects. Most individuals who consume 400 mg GL daily have some adverse effects. Considering that a regular intake of 100 mg GL/day is the lowest-observed-adverse-effect level and using a safety factor of 10, a daily intake of 10 mg GL represents a safe dose for most healthy adults. A daily intake of 1–10 mg GL/person has been estimated in several countries. However, an uneven consumption pattern suggests that a considerable number of individuals who consume large amounts of licorice sweets are exposed to the risk of developing adverse effects (Stormer et al., 1993).

The effects of GA and related compounds resemble those of the mineralocorticoid aldosterone. The mechanism appears to be inhibition of 11β-hydroxysteroid dehydrogenase (Stewart et al., 1990), which normally inactivates corticosterone and cortisol. These glucocorticoids and aldosterone share affinity for the mineralocorticoid receptor that regulates sodium and potassium transport in the kidney, and their combined effect on the receptor results in sodium retention, edema and hypertension (Baker and Fanestil, 1991).

7. SOME TYPICAL HERBAL FORMULATIONS AND PRESCRIPTIONS CONTAINING LICORICE FROM THE CHINESE PHARMACOPOEIA

Most Chinese herbal formulations are complex mixtures that are developed using a "K-M-A-S-based principle." In this principle, the "K" stands for "King," which is the principal herb. Sometimes there are two principal herbs designated as King and Queen. The "M" stands for "Minister." The minister herbs have properties similar to the principal herb(s) that provide complementary therapeutic benefits. The ''A'' stands for ''Assistant,'' which is usually added to counteract any potential adverse effects of the King or Queen or Minister herbs. The ''S'' stands for ''Servant.'' These herbs are thought to ''harmonize'' the actions of the other ingredients and promote their rapid absorption into the bloodstream and organs. Servant herbs are sometimes included to provide swift, symptomatic relief of pain and discomfort.

Licorice is thought to be an excellent coordinator and to provide servant function in Chinese herbal formulations. Therefore, out of 250 recipes reported in the ancient Chinese Pharmacopoeia ''Shang Han Lun'' and ''Jin Gui Yao Lue,'' 120 recipes contain licorice. Licorice may enhance absorption, balance metabolism, reduce toxicity and greatly enhance the flavor of the herbal formulation.

Representative recipes in use for thousands of years in Chinese medical history (Jiansu New Medical School, 1979; Hosoya et al., 1988; Reid, 1995) are described below. Most of them are prepared as decoction or ''Tang.'' For this purpose, herbs are placed in a clean nonmetallic pot and five cups of pure water are added. The mixture is brought to a boil and the boiling continues until the broth is reduced to one cup. After filtration through a muslin cloth, the residual sediments are again mixed with two cups of water and boiled again to reduce to one cup. The two portions are mixed and divided into two equal aliquots. Usually, a warm preparation is taken on an empty stomach.

Licorice is not recommended for use in formulations that contain knoxia root (*Knoxia valerianoides* Threl et Pitard), kansui root (*Euphorbia kansuiroot* Liou et Wang) or *genkawa flower* (*Daphne genkawa* Sieb. et Zucc.). The reasons for these recommendations are unclear.

7.1. SHAOYAO GANCAO TANG

- Symptoms: spasm and pain of extremities and inflammation
- Composition: licorice (*Glycyrrhiza glabra* L.) and Peony root (*Paeonia lactiflora* Pall.)

7.2. XIAO CHAI HU TANG

- Symptoms: inflammation and alternate spells of chills and fever, feeling of fullness and discomfort in the chest and hypochondrium, bitter taste in the mouth, restlessness and vomiting
- Composition: bupleurum root (*Bupleurum falcatum* Linne), Pinellia tuber (*Pinellia ternata* Breitenbach), Scutellaria root (*Scutellaria baicalensis* Georgi), Chinese date fruit (*Zizyphus jujuba* Miller var.), Ginseng root (*Panax ginseng* C.A. Meyer), Ginger rhizome (*Zingiber officinale* Rosc.) and Licorice (*Glycyrrhiza glabra* L.)

7.3. GE GEN TANG

- Symptoms: cold and influenza with fever, headache, aching shoulder and upper arm muscles, nervous discomfort
- Composition: pueraria root (*Pueraia lobata Wild.*, Ohwi), ephedra (*Ephedra intermrdia* Schrenk et Mey.), cinnamon (*Cinnamomum cassia* Blume), peony root (*Paeonia lactiflora* Pall.), Chinese date fruit (*Zizyphus jujuba* Miller var.), ginger rhizome (*Zingiber officinale* Rosc.) and licorice (*Glycyrrhiza glabra* L.)

7.4. MA XING SHI GAN TANG

- Symptoms: cold and influenza with high fever, cough and dyspnea
- Composition: ephedra (*Ephedra intermrdia* Schrenk et Mey.), gypsum, apricot kernel (*Prunus armeniaca* L.) and licorice (*Glycyrrhiza glabra* L.)

7.5. SI JUNZI TANG

- Symptoms: lassitude, shortness of breath, poor appetite and loose stool
- Composition: ginseng root (*Panax ginseng* C.A. Meyer), white atractylodes rhizome (*Atractylodes lanca,* DC.), poria (*Poria cocos* Wolf) and licorice (*Glycyrrhiza glabra* L.)

7.6. SHI QUAN DA BU TANG

- Symptoms: shortness of breath, spontaneous sweating, susceptibility to common cold and after-illness manifestations of tiredness and lassitude
- Composition: astragalus root (*Astragalus membrananaceus* Bunge), cinnamon bark (*Cinnamomum cassia* Blume), rehmannia root (*Rehmannia glutinosa* Libosch.), peony root (*Paeonia lactiflora* Pall.), chuanxiong rhizome (*Ligusticum wallichii* Franch.), atractylodes lancea rhizome (*Atractylodes lancea,* DC.), angellica root (*Angellica acutiloba* Kitagawa), ginseng root (*Panax ginseng* C.A. Meyer), poria (*Poria cocos* Wolf) and licorice (*Glycyrrhiza glabra* L.)

8. CONCLUSION

Licorice is a dietary herb with therapeutic and chemopreventive properties. Laboratory studies provide evidence that licorice possesses anticarcinogenic properties due to its content of triterpenoids, particularly β-GA. Licorice polyphenols, such as isoflavones, may also contribute to these effects. Licorice may, therefore, be a potentially attractive cancer chemopreventive agent for humans. We propose that it is time to develop carefully controlled human clinical trials in high cancer-risk populations. Since licorice is readily available, we suggest that standardized water or ethanol extracts of licorice should be employed in human intervention studies. Studies on combination therapy or prevention are also recommended. For example, we suspect that the combination of licorice and green tea extracts may be synergistic in human cancer chemoprevention studies.

A special strategy is needed to reduce the possible adverse effects of licorice. The toxicity of GA is significant with chronic intake and accumulation. Before human phase-II clinical trials, additional multidose pharmacokinetic studies of this herb should be performed. In addition, there is also a problem with the procurement of uniform standardized herbal preparations.

9. REFERENCES

Abdugafurova, M. A., Li, V. S., Sherstnev, M. P., Atanaev, T. B., Isamukhamedov, A. and Bachmanova, G. I 1990. "Antioxidative properties of glycyrrhyzic acid salts and their effect on the liver monooxygenase system," *Vopr. Med. Khim.* 36:29–31.

Abe, Y., Ueda, T., Kato, T. and Kohli, Y. 1994. "Effectiveness of interferon, glycyrrhizin combination therapy in patients with chronic hepatitis C," *Nippon Rinsho.* 52 1817–1822.

Acharya, S. K., Dasarathy, S , Tandon, A., Joshi, Y. K. and Tandon, B. N. 1993. "A preliminary open trial on interferon stimulator (SNMC) derived from *Glycyrrhiza glabra* in the treatment of subacute hepatic failure," *Indian J. Med. Res.* 98 69–74.

Agarwal, A. K., Tusie-Luna, M. T., Monder, C. and White, P. C. 1990. "Expression of 11 beta-hydroxysteroid dehydrogenase using recombinant vaccinia virus," *Mol. Endocrinol.* 4·1827–1832.

Agarwal, R., Wang, Z. Y. and Mukhtar, H. 1991. "Inhibition of mouse skin tumor-initiating activity of DMBA by chronic oral feeding of glycyrrhizin in drinking water," *Nutr. Cancer.* 15:187–193.

Akao, T. 1997a. "Hydrolysis of glycyrrhetyl mono-glucuronide to glycyrrhetic acid by glycyrrhetyl mono-glucuronide beta-D-glucuronidase of *Eubacterium sp.* GLH," *Biol. Pharm. Bull.* 20:1245–1249.

Akao, T. 1997b. "Localization of enzymes involved in metabolism of glycyrrhizin in contents of rat gastrointestinal tract," *Biol. Pharm Bull.* 20:122–126.

Akao, T., Aoyama, M., Hattori, M., Imai, Y., Namba, T., Tezuka, Y., Kikuchi, T. and Kobashi, K. 1990. "Metabolism of glycyrrhetic acid by rat liver microsomes-II. 22 alpha- and 24-hydroxylation," *Biochem. Pharmacol.* 40:291–296.

Akao, T., Hayashi, T., Kobashi, K., Kanaoka, M., Kato, H., Kobayashi, M., Takeda, S. and Oyama, T. 1994. "Intestinal bacterial hydrolysis is indispensable to absorption of 18 beta-glycyrrhetic acid after oral administration of glycyrrhizin in rats," *J. Pharm. Pharmacol.* 46:135–137.

Akimoto, M., Kimura, M., Sawano, A , Iwasaki, H., Nakajima, Y., Matano, S. and Kasai, M. 1986. "Prevention of cancer chemotherapeutic agent-induced toxicity in postoperative breast cancer patients with glycyrrhizin (SNMC)," *Gan. No Rinsho.* 32:869–872.

Altman, D. F. 1989 "Drugs used in gastrointestinal diseases," In *Basic and Clinical Pharmacology*, ed., B. G. Katzung, 4th ed., Appleton and Lange, Norwalk, Connecticut pp. 793–801.

Arase, Y, Ikeda, K., Murashima, N., Chayama, K , Tsubota, A., Koida, I , Suzuki, Y., Saitoh, S , Kobayashi, M. and Kumada, H 1997. "The long term efficacy of glycyrrhizin in chronic hepatitis C patients," *Cancer.* 79:1494–1500.

Armanini, D , Karbowiak, I. and Funder, J. W. 1983. "Affinity of liquorice derivatives for mineralocorticoid and glucocorticoid receptors," *Clin. Endocrinol.* 19·609–612

Armanini, D., Scali, M., Zennaro, M. C., Karbowiak, I., Wallace, C., Lewicka, S., Vecsei, P and Mantero, F. 1989. "The pathogenesis of pseudohyperaldosteronism from carbenoxolone," *J Endocrinol. Invest.* 12 337–341.

Aviram, M. 1996. "Interaction of oxidized low density lipoprotein with macrophages in atherosclerosis, and the antiatherogenicity of antioxidants," *Eur J. Clin. Chem. Clin. Biochem.* 34:599–608.

Aviram, M. and Fuhrman, B. 1998. "Polyphenolic flavonoids inhibit macrophage-mediated oxidation of LDL and attenuate atherogenesis," *Atherosclerosis 137 Suppl.* S45–50.

Baba, M. and Shigeta, S. 1987. "Antiviral activity of glycyrrhizin against varicella-zoster virus *in vitro*," *Antiviral Res.* 7·99–107.

Badam, L. 1997. "*In vitro* antiviral activity of indigenous glycyrrhizin, licorice and glycyrrhizic acid (Sigma) on Japanese encephalitis virus," *J. Commun Dis.* 29:91–99.

Baker, M. E. 1994. "Licorice and enzymes other than 11 beta-hydroxysteroid dehydrogenase: an evolutionary perspective," *Steroids.* 59:136–141.

Baker, M. E. 1995. "Endocrine activity of plant-derived compounds. an evolutionary perspective," *Proc. Soc. Exp. Biol. Med.* 208:131–138.

Baker, M E. and Fanestil, D. D. 1991. "Licorice, computer-based analyses of dehydrogenase sequences, and the regulation of steroid and prostaglandin action," *Mol. Cell Endocrinol.* 78·C99–102.

Bannister, B., Ginsburg, R. and Shneerson, J. 1977. "Cardiac arrest due to liquorice induced hypokalaemia," *Br. Med. J* 2:738–739.

Belinky, P. A., Aviram, M., Fuhrman, B., Rosenblat, M and Vaya, J. 1998a. "The antioxidative effects of the isoflavan glabridin on endogenous constituents of LDL during its oxidation," *Atherosclerosis* 137:49–61.

Belinky, P. A., Aviram, M., Mahmood, S. and Vaya, J. 1998b. "Structural aspects of the inhibitory effect of glabridin on LDL oxidation," *Free Radic Biol. Med.* 24:1419–1429.

Blachley, J. D. and Knochel, J. P. 1980. "Tobacco chewer's hypokalemia: licorice revisited," *N. Engl. J. Med.* 302:784–785.

Cai, L. N., Zhang, R. Y., Zhang, Z. L., Wang, B., Qiao, L., Huang, L. R. and Cheng, J. R. 1991. "The structure of glyeurysaponin," *Yao Hsueh Hsueh Pao.* 26·447–450.

Cantelli-Forti, G., Maffei, F., Hrelia, P., Bugamelli, F., Bernardi, M., D'Intino, P., Maranesi, M. and Raggi, M. A. 1994. "Interaction of licorice on glycyrrhizin pharmacokinetics," *Environ. Health Perspect.* 102 Suppl 9·65–68.

Chang, Y. P., Bi, W. X. and Yang, G. Z 1989. "Studies on the anti-virus effect of *Glycyrrhiza uralensis* Fish. polysaccharide," *Chung Kuo Chung Yao Tsa Chih.* 14:236–238, 255–256.

Conn, J. W., Rovner, D. R. and Cohen, E. L. 1968. "Licorice-induced pseudoaldosteronism. Hypertension, hypokalemia, aldosteronopenia, and suppressed plasma renin activity," *JAMA.* 205:492–496.

Crance, J. M., Leveque, F., Biziagos, E , van Cuyck-Gandre, H., Jouan, A and Deloince, R. 1994. "Studies on mechanism of action of glycyrrhizin against hepatitis A virus replication *in vitro*," *Antiviral Res.* 23:63–76.

Crance, J. M., Leveque, F., Chousterman, S , Jouan, A , Trepo, C. and Deloince, R. 1995. "Antiviral activity of recombinant interferon-alpha on hepatitis A virus replication in human liver cells," *Antiviral Res.* 28.69–80.

Cuyck-Gandre, H V , Job, A., Burckhart, M. F., Girond, S. and Crance, J. M. 1995. "Use of digoxigenin-labeled RNA probe to test hepatitis A virus antiviral drugs," *Pathol. Biol.* 43:411–415.

Da Nagao, Y., Sata, M., Suzuki, H., Tanikawa, K., Itoh, K. and Kameyama, T. 1996. "Effectiveness of glycyrrhizin for oral lichen planus in patients with chronic HCV infection," *J. Gastroenterol* 31 691–695.

Dargan, D. J and Subak-Sharpe, J. H. 1985. "The effect of triterpenoid compounds on uninfected and herpes simplex virus-infected cells in culture I. Effect on cell growth, virus particles and virus replication," *J. Gen Virol.* 66:1771–1784.

Dargan, D. J. and Subak-Sharpe, J. H. 1986. "The effect of triterpenoid compounds on uninfected and herpes simplex virus-infected cells in culture. II. DNA and protein synthesis, polypeptide processing and transport," *J. Gen. Virol.* 67:1831–1850.

De Clercq, E. 1988. "Perspectives for the chemotherapy of AIDS," *Chemioterapia.* 7:357–364.

Dehpour, A. R , Zolfaghari, M. E., Samadian, T. and Vahedi, Y. 1994. "The protective effect of liquorice components and their derivatives against gastric ulcer induced by aspirin in rats," *J. Pharm. Pharmacol.* 46:148–149

Demizu, S., Kajiyama, K., Takahashi, K , Hiraga, Y., Yamamoto, S., Tamura, Y., Okada, K. and Kinoshita, T. 1988. "Antioxidant and antimicrobial constituents of licorice: isolation and structure elucidation of a new benzofuran derivative," *Chem. Pharm. Bull.* 36:3474–3479.

Eason, C. T., Bonner, F. W. and Parke, D. V. 1990. "The importance of pharmacokinetic and receptor studies in drug safety evaluation," *Regul. Toxicol. Pharmacol.* 11:288–307.

Eisenburg, J. 1992. "Treatment of chronic hepatitis B. Part 2: Effect of glycyrrhizic acid on the course of illness," *Fortschr. Med.* 110 395–398

Epstein, M. T., Espiner, E. A., Donald, R. A. and Hughes, H. 1977a "Effect of eating liquorice on the renin-angiotensin aldosterone axis in normal subjects," *Br. Med J.* 1:488–90.

Epstein, M T., Espiner, E. A., Donald, R. A. and Hughes, H. 1977b. "Liquorice toxicity and the renin-angiotensin-aldosterone axis in man," *Br. Med. J* 1.209–210.

Farina, C., Pinza, M. and Pifferi, G. 1998. "Synthesis and anti-ulcer activity of new derivatives of glycyrrhetic, oleanolic and ursolic acids," *Farmaco.* 53.22–32.

Fenwick, G. R., Lutomski, J. and Nieman, C 1990. "Liquorice, *Glycyrrhiza glabra* L-Composition, uses and analysis," *Food Chem.* 38:119–143.

Flavin-Koenig, D. F. 1996. "The reversal of Epstein Barr virus induced hepatosplenomegaly in 24 hours with inhibitors of xanthine oxidase and nitric oxide synthase," *N.Z. Med. J.* 109:106–107.

Food and Drug Administration. 1985. "GRAS status of licorice (glycyrrhiza), ammoniated glycyrrhizin, and monoammonium glycyrrhizinate," *Fed. Reg.* 50:52104321045.

Fu, N., Liu, Z. and Zhang, R. 1995. "Anti-promoting and anti-mutagenic actions of G9315." *Chung Kuo I Hsueh Ko Hsueh Yuan Hsueh Pao.* 17:349–352.

Fuhrman, B., Buch, S., Vaya, J., Belinky, P. A., Coleman, R., Hayek, T. and Aviram, M. 1997. "Licorice extract and its major polyphenol glabridin protect low-density lipoprotein against lipid peroxidation: *in vitro* and *ex vivo* studies in humans and in atherosclerotic apolipoprotein E-deficient mice," *Am. J. Clin. Nutr.* 66:267–275.

Fujisawa, K. 1991. "Interferon therapy in hepatitis C virus (HCV) induced chronic hepatitis: clinical significance of pretreatment with glycyrrhizin," *Trop. Gastroenterol.* 12:176–179.

Fujisawa, K, Watanabe, H. and Kimata, K 1973. "Therapeutic approach to chronic active hepatitis with glycyrrhizin," *Asian Med. J.* 23:745–756

Gibson, M R. 1978. "*Glycyrrhiza* in old and new perspectives," *Lloydia* 41·348–354

Gunnarsdottir, S and Johannesson, T 1997. "Glycyrrhetic acid in human blood after ingestion of glycyrrhizic acid in licorice," *Pharmacol Toxicol* 81:300–302

Hall, I. H., Grippo, A A, Holbrook, D J., Roberts, G, Lin, H. C., Kim, H. L. and Lee, K H 1989. "Role of thiol agents in protecting against the toxicity of helenalin in tumor-bearing mice," *Planta Med.* 55·513–517.

Haraguchi, H., Ishikawa, H, Mizutani, K, Tamura, Y and Kinoshita, T. 1998. "Antioxidative and superoxide scavenging activities of retrochalcones in *Glycyrrhiza inflata*," *Bioorg. Med. Chem* 6 339–347.

Hatano, T., Yasuhara, T., Miyamoto, K. and Okuda, T. 1988. "Anti-human immunodeficiency virus phenolics from licorice." *Chem. Pharm. Bull.* 36.2286–2288.

Hatano, T., Fukuda, T., Liu, Y Z., Noro, T. and Okuda, T. 1991a "Phenolic constituents of licorice IV. Correlation of phenolic constituents and licorice specimens from various sources, and inhibitory effects of licorice extracts on xanthine oxidase and monoamine oxidase," *Yakugaku Zasshi* 111·311–321.

Hatano, T., Fukuda, T., Miyase, T., Noro, T and Okuda, T. 1991b. "Phenolic constituents of licorice. III Structures of glicoricone and licofuranone, and inhibitory effects of licorice constituents on monoamine oxidase," *Chem. Pharm Bull.* 39.1238–1243

Hattori, T., Ikematsu, S., Koito, A., Matsushita, S, Maeda, Y., Hada, M., Fujimaki, M. and Takatsuki, K. 1989 "Preliminary evidence for inhibitory effect of glycyrrhizin on HIV replication in patients with AIDS," *Antiviral Res.* 11:255–261

Hayashi, J., Kajiyama, W., Noguchi, A., Nakashima, K, Hirata, M., Hayashi, S. and Kashiwagi, S. 1991. "Glycyrrhizin withdrawal followed by human lymphoblastoid interferon in the treatment of chronic hepatitis B," *Gastroenterol. Jpn* 26 742–746.

Hirabayashi, K, Iwata, S., Matsumoto, H., Mori, T, Shibata, S, Baba, M, Ito, M., Shigeta, S, Nakashima, H. and Yamamoto, N. 1991. "Antiviral activities of glycyrrhizin and its modified compounds against human immunodeficiency virus type 1 (HIV-1) and herpes simplex virus type 1 (HSV-1) *in vitro*," *Chem. Pharm. Bull.* 39·112–115.

Hosoya, E. 1988. "Scientific reevaluation of Kampo prescriptions using modern technology," in *Recent Advances in the Pharmacology of KAMPO (Japanese Herbal) Medicines*, eds., E. Hosoya, E. Yamamura, A. Kumagai, Y. Otsuka, and H. Takagi, H., Elsevier Science Publishers, Tokyo, B V., pp. 17–29.

Hosoya, E., Yamamura, E., Kumagai, A., Otsuka, Y. and Takagi, H. 1988. *Recent Advances in the Pharmacology of KAMPO (Japanese Herbal) Medicines*, Elsevier Science Publishers, Tokyo.

Huang, W T. 1991. "Effect of olean-9(11), 12-diene-3 beta, 30-diol 3 beta, o-hemisuccinate Na salt, a glycyrrhetinic acid derivative, on peroxidation in CCl_4 induced mouse acute hepatitis," *Am. J. Chem. Med.* 19:115–120.

Huseman, P. A. 1944. *Licorice: Putting a weed to work*, Heffer & Sons, Ltd., London

Ibsen, K. K. 1981. "Liquorice consumption and its influence on blood pressure in Danish schoolchildren," *Dan. Med. Bull.* 28:124–126.

Inoue, H., Mori, T., Shibata, S. and Koshihara, Y. 1988. "Inhibitory effect of glycyrrhetinic acid derivatives on arachidonic acid-induced mouse ear oedema," *J. Pharm. Pharmacol.* 40.272–277.

Inoue, H, Mori, T., Shibata, S. and Koshihara, Y. 1989. "Modulation by glycyrrhetinic acid derivatives of TPA-induced mouse ear oedema," *Br. J. Pharmacol.* 96.204–210.

Inoue, H., Inoue, K., Takeuchi, T., Nagata, N. and Shibata, S. 1993. "Inhibition of rat acute inflammatory paw oedema by dihemiphthalate of glycyrrhetinic acid derivatives: comparison with glycyrrhetinic acid," *J. Pharm. Pharmacol.* 45:1067–1071.

Inoue, H., Nagata, N., Shibata, S. and Koshihara, Y. 1996. "Inhibitory effect of glycyrrhetinic acid derivatives on capsaicin-induced ear edema in mice," *Jpn. J. Pharmacol.* 71:281–289.

Ishida, S., Sakiya, Y., Ichikawa, T. and Taira, Z. 1992. "Dose-dependent pharmacokinetics of glycyrrhizin in rats," *Chem. Pharm. Bull.* 40:1917–1920.

Ishida, S., Sakiya, Y. and Taira, Z. 1994. "Disposition of glycyrrhizin in the perfused liver of rats," *Biol. Pharm. Bull.* 17:960–969.

Ito, M., Nakashima, H., Baba, M., Pauwels, R., De Clercq, E., Shigeta, S. and Yamamoto, N. 1987. "Inhibitory effect of glycyrrhizin on the *in vitro* infectivity and cytopathic activity of the human immunodeficiency virus [HIV (HTLV-III/LAV)]," *Antiviral Res.* 7:127–137

Ito, M., Sato, A., Hirabayashi, K., Tanabe, F., Shigeta, S., Baba, M., De Clercq, E., Nakashima, H. and Yamamoto, N. 1988. "Mechanism of inhibitory effect of glycyrrhizin on replication of human immunodeficiency virus (HIV)," *Antiviral Res.* 10:289–298.

Jiansu New Medical School, J. N. M. 1979. *Cyclopedia of the Chinese Traditional Medicine,* Shanghai Science and Technology Press, Shanghai, pp. 235–237.

Ju, H. S., Li, X. J., Zhao, B. L., Han, Z. W. and Xin, W. J. 1989. "Effects of glycyrrhiza flavonoid on lipid peroxidation and active oxygen radicals," *Yao Hsueh Hsueh Pao.* 24:807–812.

Kawakami, J., Yamamura, Y., Santa, T., Kotaki, H., Uchino, K., Sawada, Y. and Iga, T. 1993. "Kinetic analysis of glycyrrhetic acid, an active metabolite of glycyrrhizin, in rats: role of enterohepatic circulation," *J Pharm. Sci.* 82:301–305.

Kelloff, G. J., Crowell, J. A., Boone, C. W., Steele, V. E., Lubet, R. A., Greenwald, P., Alberts, D. S., Covey, J. M., Doody, L. A., Knapp, G. G. et al. 1994. "Clinical development plan: 18 beta-glycyrrhetinic acid," *J. Cell Biochem. Suppl.* 20:166–175.

Kiso, Y., Tohkin, M., Hikino, H., Hattori, M., Sakamoto, T. and Namba, T. 1984. "Mechanism of antihepatotoxic activity of glycyrrhizin. I: Effect on free radical generation and lipid peroxidation," *Planta Med.* 50:298–302.

Kitagawa, K., Nishino, H. and Iwashima, A. 1986. "Inhibition of the specific binding of 12-O-tetradecanoylphorbol-13-acetate to mouse epidermal membrane fractions by glycyrrhetic acid," *Oncology.* 43:127–130.

Kitagawa, I., Hori, K., Uchida, E., Chen, W. Z., Yoshikawa, M. and Ren, J. 1993a. "Saponin and sapogenol. L. On the constituents of the roots of *Glycyrrhiza uralensis* Fischer from Xinjiang, China Chemical structures of licorice-saponin L3 and isoliquiritin apioside," *Chem. Pharm. Bull.* 41:1567–1572.

Kitagawa, I., Hori, K., Sakagami, M., Zhou, J. L. and Yoshikawa, M. 1993b. "Saponin and sapogenol. XLVIII. On the constituents of the roots of *Glycyrrhiza uralensis* Fischer from northeastern China. (2). Licorice-saponins D3, E2, F3, G2, H2, J2, and K2," *Chem. Pharm. Bull.* 41:1337–1345.

Kitagawa, I., Hori, K., Taniyama, T., Zhou, J. L. and Yoshikawa, M. 1993c. "Saponin and sapogenol. XLVII. On the constituents of the roots of *Glycyrrhiza uralensis* Fischer from northeastern China. (1). Licorice-saponins A3, B2, and C2," *Chem. Pharm. Bull.* 41:43–49.

Kitagawa, I., Hori, K., Sakagami, M., Hashiuchi, F., Yoshikawa, M. and Ren, J. 1993d. "Saponin and sapogenol. XLIX. On the constituents of the roots of *Glycyrrhiza inflata* Batalin from Xinjiang, China. Characterization of two sweet oleanane-type triterpene oligoglycosides, apio-glycyrrhizin and araboglycyrrhizin," *Chem. Pharm. Bull.* 41:1350–1357.

Kitagawa, I., Chen, W. Z., Hori, K., Harada, E., Yasuda, N., Yoshikawa, M. and Ren, J. 1994. "Chemical studies of Chinese licorice-roots. I. Elucidation of five new flavonoid constituents from the roots of *Glycyrrhiza glabra* L. collected in Xinjiang,'' *Chem. Pharm. Bull.* 42.1056–1062.

Kiuchi, F., Chen, X. and Tsuda, Y. 1990. ''Four new phenolic constituents from licorice (Root of *Glycyrrhiza* Sp.),'' *Heterocycles.* 31:629–632.

Kobuke, T., Inai, K., Nambu, S., Ohe, K , Takemoto, T., Matsuki, K., Nishina, H., Huang, I. B. and Tokuoka, S. 1985 ''Tumorigenicity study of disodium glycyrrhizinate administered orally to mice,'' *Food Chem. Toxicol.* 23:979–983.

Kuo, S., Shankel, D. M., Telikepalli, H. and Mitscher, L A. 1992. ''*Glycyrrhiza glabra* extract as an effector of interception in *Escherichia coli* K12+,'' *Mutat. Res.* 282.93–98

Kusano, A., Nikaido, T., Kuge, T., Ohmoto, T., Delle Monache, G., Botta, B., Botta, M. and Saitoh, T. 1991 ''Inhibition of adenosine 3',5'-cyclic monophosphate phosphodiesterase by flavonoids from licorice roots and 4-arylcoumarins,'' *Chem. Pharm. Bull.* 39:930–933.

Lee, E , Miki, Y., Furukawa, Y., Simizu, H. and Kariya, K. 1991. ''Selective release of glutathione transferase subunits from primary cultured rat hepatocytes by carbon tetrachloride and deoxycholic acid,'' *Toxicology.* 67:237–248.

Lee, Y. S., Lorenzo, B. J., Koufis, T. and Reidenberg, M. M. 1996. ''Grapefruit juice and its flavonoids inhibit 11 beta-hydroxysteroid dehydrogenase,'' *Clin. Pharmacol. Ther.* 59:62–71.

Leibman, K. C. 1971. ''Studies on modifiers of microsomal drug oxidation,'' *Chem. Biol. Interact.* 3·289–290.

Lin, C. C., Shieh, D. E. and Yen, M. H. 1997. ''Hepatoprotective effect of the fractions of Banzhi-lian on experimental liver injuries in rats,'' *J. Ethnopharmacol.* 56:193–200.

Liu, Q. and Liu, Y. 1990. ''Application of 13C NMR to structural identification of the flavonoid glycosides,'' *Chung Kuo I Hsueh Ko Hsueh Yuan Hsueh Pao.* 12:359–364

Liu, Q. and Liu, Y. L 1989. ''Studies on chemical constituents of *Glycyrrhiza* eurycarpa P. C. Li.,'' *Yao Hsueh Hsueh Pao.* 24:525–531.

Liu, J , Liu, Y., Mao, Q. and Klassen, C. D. 1994. ''The effects of 10 triterpenoid compounds on experimental liver injury in mice,'' *Fundam. Appl. Toxicol.* 22:34–40.

Lu, W. B 1993. ''Treatment of 60 cases of HIV-infected patients with glyke,'' *Chung Kuo Chung Hsi I Chieh Ho Tsa Chih.* 13·340–342.

Mantovani, A., Ricciardi, C., Stazi, A. V., Macri, C., Piccioni, A., Badellino, E., De Vincenzi, M., Caiola, S. and Patriarca, M. 1988. ''Teratogenicity study of ammonium glycyrrhizinate in the Sprague-Dawley rat,'' *Food Chem. Toxicol.* 26:435–440.

Matsuda, H., Samukawa, K. and Kubo, M. 1991. ''Anti-hepatitic activity of ginsenoside Ro.,'' *Planta Med.* 57:523–526.

Monder, C., Stewart, P. M., Lakshmi, V., Valentino, R., Burt, D. and Edwards, C. R. 1989. ''Licorice inhibits corticosteroid 11 beta-dehydrogenase of rat kidney and liver: *in vivo* and *in vitro* studies,'' *Endocrinology.* 125:1046–1053.

Mori, K., Sakai, H., Suzuki, S., Akutsu, Y., Ishikawa, M., Aihara, M , Yokoyama, M., Sato, Y., Okaniwa, S. and Endo, Y. 1989a. ''The present status in prophylaxis and treatment of HIV infected patients with hemophilia in Japan,'' *Rinsho. Byori.* 37:1200–1208.

Mori, K., Sakai, H., Suzuki, S , Sugai, K., Akutsu, Y., Ishikawa, M., Seino, Y., Ishida, N., Uchida, T and Kariyone, S. 1989b. ''Effects of glycyrrhizin (SNMC· stronger Neo-Minophagen C) in hemophilia patients with HIV infection,'' *Tohoku. J. Exp. Med.* 158:25–35.

Mori, K., Sakai, H., Suzuki, S., Akutsu, Y., Ishikawa, M , Imaizumi, M., Tada, K., Aihara, M., Sawada, Y. and Yokoyama, M. 1990. ''Effects of glycyrrhizin (SNMC: Stronger Neo-Minophagen C) in hemophilia patients with HIV-I infection,'' *Tohoku. J. Exp. Med.* 162:183–193.

Morris, J. A 1976. ''Sweetening agents from natural sources,'' *Lloydia* 39:25–38.

Murav'ev, I A., Mar'iasis, E. D , Krasova, T. G., Chebotarev, V. V and Starokozhko, L. E 1983. "Corticoid-like action of liniments based on licorice root preparations," *Farmakol Toksikol.* 46:59–62.

Nakamura, M , Ishihara, Y , Sata, T., Torii, S., Sumiyoshi, A and Tanaka, K 1966. "Effects of dietary magnesium and glycyrrhizin on experimental atheromatosis of rats (long-term experiment)," *Jpn. Heart J.* 7:474–486.

Nakamura, T., Fujii, T and Ichihara, A. 1985. "Enzyme leakage due to change of membrane permeability of primary cultured rat hepatocytes treated with various hepatotoxins and its prevention by glycyrrhizin," *Cell Biol. Toxicol.* 1:285–295.

Nakashima, H., Matsui, T., Yoshida, O., Isowa, Y., Kido, Y., Motoki, Y., Ito, M., Shigeta, S , Mori, T. and Yamamoto, N 1987. "A new anti-human immunodeficiency virus substance, glycyrrhizin sulfate. endowment of glycyrrhizin with reverse transcriptase-inhibitory activity by chemical modification," *Jpn. J. Cancer Res.* 78:767–771.

Ngo, H. N., Teel, R. W. and Lau, B. H. S. 1992. "Modulation of mutagenesis, DNA binding, and metabolism of aflatoxin B1 by licorice compounds," *Nutr. Res.* 12:247–257

Nikitina, S. S. 1966. "Some data on the mechanism of anti-inflammatory action of glycyrrhizic and glycyrrhetic acids isolated from Glicirrhiza L.," *Farmakol. Toksikol.* 29:67–70.

Nishino, H., Yoshioka, K., Iwashima, A., Takizawa, H., Konishi, S., Okamoto, H., Okabe, H., Shibata, S , Fujiki, H. and Sugimura, T. 1986. "Glycyrrhetic acid inhibits tumor-promoting activity of teleocidin and 12-O-tetradecanoylphorbol-13-acetate in two-stage mouse skin carcinogenesis," *Jpn. J. Cancer Res.* 77:33–38.

Nishino, H., Nishino, A., Takayasu, J., Hasegawa, T., Iwashima, A., Hirabayashi, K., Iwata, S. and Shibata, S. 1988. "Inhibition of the tumor-promoting action of 12-O-tetradecanoylphorbol-13-acetate by some oleanane-type triterpenoid compounds," *Cancer Res.* 48:5210–5215.

Nishino, H 1992. "Antitumor-promoting activity of glycyrrhetinic acid and its related compounds," in *Cancer Chemoprevention*, eds., L. Wattenberg, M. Lipkin, C. W. Boone and G. J. Kelloff, CRC Press, Boca Raton, LA, pp. 457–467.

Nishino, H., Hayashi, T , Arisawa, M., Satomi, Y. and Iwashima, A. 1993. "Antitumor-promoting activity of scopadulcic acid B, isolated from the medicinal plant *Scoparia dulcis* L," *Oncology.* 50:100–103.

Nomura, T. and Fukai, T. 1998. "Phenolic constituents of licorice (*Glycyrrhiza* species)," *Fortschr. Chem. Org Naturst.* 73:1–158.

Nose, M., Ito, M., Kamimura, K., Shimizu, M. and Ogihara, Y. 1994. "A comparison of the antihepatotoxic activity between glycyrrhizin and glycyrrhetinic acid," *Planta Med* 60˙136–139.

O'Brian, C. A., Ward, N. E. and Vogel, V. G. 1990. "Inhibition of protein kinase C by the 12-O-tetradecanoylphorbol-13-acetate antagonist glycyrrhetic acid," *Cancer Lett.* 49:9–12.

Okamoto, H., Yoshida, D., Saito, Y. and Mizusaki, S. 1983. "Inhibition of 12-O-tetradecanoylphorbol-13-acetate-induced ornithine decarboxylase activity in mouse epidermis by sweetening agents and related compounds," *Cancer Lett.* 21:29–35.

Okimasu, E., Moromizato, Y., Watanabe, S., Sasaki, J., Shiraishi, N., Morimoto, Y. M., Miyahara, M. and Utsumi, K. 1983. "Inhibition of phospholipase A2 and platelet aggregation by glycyrrhizin, an antiinflammation drug," *Acta. Med. Okayama.* 37:385–391.

Palagina, M. V., Khasina, M. A., Gel'tser, B. I. and Deviatov, A. L. 1995. "Antioxidant effect of a Ural licorice product in acute impairment of pulmonary surfactant by total gamma irradiation," *Vopr. Med. Khim.* 41:32–34.

Paolini, M., Pozzetti, L., Sapone, A. and Cantelli-Forti, G. 1998. "Effect of licorice and glycyrrhizin on murine liver CYP-dependent monooxygenases," *Life Sci.* 62:571–582.

Pliasunova, O A., Egoricheva, I N , Fediuk, N. V., Pokrovskii, A. G., Baltina, L A., Murinov lu, I and Tolstikov, G A 1992 "The anti-HIV activity of beta-glycyrrhizic acid," *Vopr Virusol* 37 235–238

Pompei, R , Flore, O , Marccialis, M A., Pani, A. and Loddo, B. 1979. "Glycyrrhizic acid inhibits virus growth and inactivates virus particles," *Nature* 281:689–690

Pompei, R , Pani, A., Flore, O , Marcialis, M. A. and Loddo, B. 1980. "Antiviral activity of glycyrrhizic acid," *Experientia*. 36:304.

Raggi, M. A., Maffei, F , Bugamelli, F. and Cantelli Forti, G. 1994a. "Bioavailability of glycyrrhizin and licorice extract in rat and human plasma as detected by a HPLC method." *Pharmazie* 49.269–272.

Raggi, M. A., Bugamelli, F , Nobile, L., Schiavone, P and Cantelli-Forti, G. 1994b "HPLC determination of glycyrrhizin and glycyrrhetic acid in biological fluids, after licorice extract administration to humans and rats," *Boll. Chim. Farm* 133.704–708

Raggi, M A , Bugamelli, F., Nobile, L , Curcelli, V., Mandrioli, R., Rossetti, A and Cantelli Forti, G. 1995. "The choleretic effects of licorice: identification and determination of the pharmacologically active components of *Glycyrrhiza glabra*," *Boll. Chim Farm* 134 634–638

Rainsford, K D and Whitehouse, M W. 1977. "Non-steroid anti-inflammatory drugs: combined assay for anti-edemic potency and gastric ulcerogenesis in the same animal," *Life Sci.* 21 371–377.

Reid, D. 1995. *A handbook of Chinese healing herbs*, Shambhala Publications, Inc., Boston.

Reiners, W 1966 "7-hydroxy-4'-methoxy-isoflavone (formononetin) from liquorice root. On substances contained in liquorice root II," *Experientia* 22·359.

Ren, J. and Wang, Z. G. 1988. "Pharmacological research on the effect of licorice," *J Tradit Chin. Med.* 8.307–309

Reynolds, J E. F. and Prasad, A. B. 1982 *Martindale. The Extra Pharmacopoeia*, 28th ed., The Pharmaceutical Press, London.

Sato, H., Goto, W , Yamamura, J., Kurokawa, M., Kageyama, S , Takahara, T , Watanabe, A and Shiraki, K. 1996. "Therapeutic basis of glycyrrhizin on chronic hepatitis B," *Antiviral Res.* 30:171–177.

Segal, R , Milo-Goldzweig, I , Schupper, H and Zaitschek, D V. 1970 "Effect of ester groups on the hemolytic action of sapogenins. II Esterification with bifunctional acids," *Biochem. Pharmacol.* 19:2501–2507.

Shi, G Z. 1992. "Blockage of glycyrrhiza uralensis and chelidonium majus in MNNG induced cancer and mutagenesis," *Chung Hua Yu Fang I Hsueh Tsa Chih.* 26 165–167.

Shibata, S. 1994. "Anti-tumorigenic chalcones," *Stem Cells.* 12:44–52.

Shibayama, Y. 1989. "Prevention of hepatotoxic responses to chemicals by glycyrrhizin in rats," *Exp Mol Pathol.* 51·48–55.

Shimizu, N., Tomoda, M., Kanari, M., Gonda, R , Satoh, A. and Satoh, N. 1990. "A novel neutral polysaccharide having activity on the reticuloendothelial system from the root of *Glycyrrhiza uralensis*," *Chem. Pharm. Bull.* 38:3069–3071

Shimizu, N , Tomoda, M., Takada, K. and Gonda, R. 1992. "The core structure and immunological activities of glycyrrhizan UA, the main polysaccharide from the root of *Glycyrrhiza uralensis*," *Chem. Pharm. Bull.* 40·2125–2128.

Simpson, F. O. and Currie, I J. 1982. "Licorice consumption among high school students," *N. Z. Med. J* 95.31–33.

Soro, A., Panarelli, M., Holloway, C. D , Fraser, R. and Kenyon, C. J 1997 "*In vivo* and *in vitro* effects of carbenoxolone on glucocorticoid receptor binding and glucocorticoid activity," *Steroids.* 62·388–394.

Spinks, E. A and Fenwick, G. R. 1990. "The determination of glycyrrhizin in selected UK liquorice products," *Food Addict. Contamin* 7:769–778.

Stake, M. 1986 *"Glycyrrhiza* plants," Records of Lecture in 2nd Meeting on Crude Drugs, Licorice. Tokyo Soc. Pharmacognosy, Tokyo. p 1.

Stewart, P. M., Wallace, A M., Atherden, S. M., Shearing, C. H. and Edwards, C R 1990. "Mineralocorticoid activity of carbenoxolone: contrasting effects of carbenoxolone and liquorice on 11 beta-hydroxysteroid dehydrogenase activity in man," *Clin. Sci.* 78:49–54.

Stormer, F. C., Reistad, R. and Alexander, J. 1993. "Glycyrrhizic acid in liquorice-evaluation of health hazard," *Food Chem. Toxicol.* 31:303–312.

Suzuki, H , Ohta, Y., Takino, T., Fuisawa, K. and Hirayama, C. 1983 "Effects of glycyrrhizin on biological tests in patients with chronic hepatitis. A double-blind trial," *Asian Med J.* 26:423–438.

Takahara, T , Watanabe, A. and Shiraki, K. 1994. "Effects of glycyrrhizin on hepatitis B surface antigen: a biochemical and morphological study," *J. Hepatol* 21.601–609.

Takeda, S., Ishthara, K., Wakui, Y., Amagaya, S , Maruno, M., Akao, T. and Kobashi, K. 1996. "Bioavailability study of glycyrrhetic acid after oral administration of glycyrrhizin in rats; relevance to the intestinal bacterial hydrolysis," *J. Pharm. Pharmacol.* 48·902–905.

Tangri, K. K., Seth, P. K., Parmar, S. S. and Bhargava, K. P. 1965. "Biochemical study of anti-inflammatory and anti-arthritic properties of glycyrrhetic acid," *Biochem. Pharmacol.* 14:1277–1281.

Tochikura, T S., Nakashima, H. and Yamamoto, N. 1989. "Antiviral agents with activity against human retroviruses," *J. Acquir. Immune Defic Syndr.* 2.441–447.

Tomoda, M., Shimizu, N., Kanari, M., Gonda, R., Arai, S. and Okuda, Y. 1990. "Characterization of two polysaccharides having activity on the reticuloendothelial system from the root of *Glycyrrhiza uralensis*," *Chem. Pharm. Bull* 38·1667–1671

Toner, J M. and Ramsey, L E. 1985. "Liquorice can damage your health," *Practitioner.* 229.858, 860.

Utsunomiya, T., Kobayashi, M., Pollard, R. B. and Suzuki, F. 1997. "Glycyrrhizin, an active component of licorice roots, reduces morbidity and mortality of mice infected with lethal doses of influenza virus," *Antimicrob. Agents Chemother.* 41:551–556

Vasilenko I K , Lisevitskaia, L. I., Frolova, L. M., Parfent'eva, E P. and Skul'te, I. V 1982. "Hypolipidemic properties of triterpenoids," *Farmakol. Toksikol.* 45:66–70.

Vasilenko I. K., Semenchenko, V. F., Frolova, L. M., Konopleva, G. E , Parfent'eva, E. P. and Skul'te, I. V. 1993. "The pharmacological properties of the triterpenoids from birch bark," *Eksp Klin. Farmakol.* 56:53–55.

Vaya, J., Belinky, P. A. and Aviram, M. 1997. "Antioxidant constituents from licorice roots: isolation, structure elucidation and antioxidative capacity toward LDL oxidation," *Free Radic. Biol. Med.* 23:302–313.

Vodovozova, E. L., Pavlova I. B., Polushkina, M. A., Rzhaninova, A. A. Garaev, M. M. and Molotkovskii I. G. 1996. "New phospholipid-inhibitors of human immunodeficiency virus reproduction. Synthesis and antiviral activity," *Bioorg. Khim.* 22.451–457.

Vogel, V. G., Newman, R. A., Ainslie, N. and Winn, R J. 1992. "Phase I pharmacology and toxicity study of glycyrrhetinic acid as a chemopreventive drug," *Proc. Am Assoc. Cancer Res.* 33.208.

Wang, Z Y., Agarwal, R., Zhou, Z. C., Bickers, D. R. and Mukhtar, H. 1991. "Inhibition of mutagenicity in *Salmonella typhimurium* and skin tumor initiating and tumor promoting activities in SENCAR mice by glycyrrhetinic acid: comparison of 18 alpha- and 18 beta-stereoisomers," *Carcinogenesis.* 12:187–192.

Wang, Z Y., Agarwal, R., Khan, W. A. and Mukhtar, H. 1992. "Protection against benzo[a]pyrene-and N-nitrosodiethylamine-induced lung and forestomach tumorigenesis in A/J mice by water extracts of green tea and licorice," *Carcinogenesis.* 13:1491–1494.

Wang, G. S. and Han, Z. W. 1993. "The protective action of glycyrrhiza flavonoids against carbon tetrachloride hepatotoxicity in mice." *Yao Hsueh Hsueh Pao.* 28.572–576.

Wang, Z., Okamoto, M., Kurosaki, Y., Nakayama, T. and Kimura, T. 1996 "Pharmacokinetics of glycyrrhizin in rats with D-galactosamine-induced hepatic disease," *Biol. Pharm. Bull.* 19 901–904.

Wang, J Y., Guo, J S., Li, H , Liu, S. L. and Zern, M. A. 1998. "Inhibitory effect of glycyrrhizin on NF-kappaB binding activity in CCl4- plus ethanol-induced liver cirrhosis in rats," *Liver.* 18.180–185.

Watanbe, H , Miyaji, C , Makino, M. and Abo, T 1996. "Therapeutic effects of glycyrrhizin in mice infected with LP-BM5 murine retrovirus and mechanisms involved in the prevention of disease progression," *Biotherapy.* 9:209–220.

Watari, N., Torizawa, K., Kanai, M., Mabuchi, Y. and Suzuki, Y. 1976. "Ultrastructural studies on the protective effect of glycyrrhizin for liver injury induced by a carcinogen (3′-Me-DAB) (The second report)," *J Clin Electron Microsc.* 9:394–395.

Webb, T. E., Stromberg, P. C., Abou-Issa, H., Curley, R. W., Jr. and Moeschberger, M 1992. "Effect of dietary soybean and licorice on the male F344 rat· an integrated study of some parameters relevant to cancer chemoprevention," *Nutr Cancer.* 18:215–230.

Werner, S., Brismar, K and Olsson, S 1979. "Hypeprolactinaemia and liquorice," *Lancet.* 1 319.

Yang, G. and Yu, Y. 1990. "Immunopotentiating effect of traditional Chinese drugs—ginsenoside and glycyrrhiza polysaccharide," *Proc. Chin. Acad. Med. Sci. Peking. Union Med. Coll.* 5 188–193.

Yang, L , Liu, Y. L. and Lin, S. Q 1990. "HPLC analysis of flavonoids in the root of six *Glycyrrhiza* species," *Yao Hsueh Hsueh Pao* 25:840–848.

Yasukawa, K., Takido, M , Takeuchi, M and Nakagawa, S. 1988. "Inhibitory effect of glycyrrhizin and caffeine on two-stage carcinogenesis in mice," *Yakugaku Zasshi.* 108:794–796.

Young, G P , St John, D J. and Coventry, D. A. 1979. "Treatment of duodenal ulcer with carbenoxolone sodium: a double-masked endoscopic trial," *Med. J. Aust.* 1·2–5

Zakirov, U. B and Abdullaev, A. 1996. "The hypolipidemic and antiatherosclerotic properties of the ammonium salt of glycyrrhetic acid and of 18-dehydroglycyrrhetic acid," *Eksp. Klin. Farmakol.* 59.53–55.

Zeng, L , Lou, Z. C. and Zhang, R. Y. 1991a. "Quality evaluation of Chinese licorice," *Yao Hsueh Hsueh Pao.* 26:788–793.

Zeng, L., Zhang, R. Y. and Lou, Z. C. 1991b. "Separation and quantitative determination of three saponins in licorice root by high performance liquid chromatography," *Yao Hsueh Hsueh Pao* 26:53–58

Zhang, R. Y., Zhang, J. H. and Wang, M. T. 1986. "Studies on the saponins from the root of *Glycyrrhiza uralensis* Fisch," *Yao Hsueh Hsueh Pao.* 21·510–515.

Zhang, Y. D., Lorenzo, B. and Reidenberg, M M. 1994 "Inhibition of 11 beta-hydroxysteroid dehydrogenase obtained from guinea pig kidney by furosemide, naringenin and some other compounds." *J. Steroid Biochem Mol. Biol.* 49:81–85.

Zhao, J. F., Kiyohara, H., Yamada, H., Takemoto, N. and Kawamura, H. 1991 "Heterogeneity and characterization of mitogenic and anti-complementary pectic polysaccharides from the roots of *Glycyrrhiza uralensis* Fisch *et* D.C.," *Carbohydr. Res* 219:149–172.

Zhou, Y. and Zhang, J. 1990. "Effects of baicalin and liquid extract of licorice on sorbitol level in red blood cells of diabetic rats," *Chung Kuo Chung Yao Tsa Chih.* 15.433–448.

CHAPTER 12

Regulation of Herbal and Tea Products: International Perspectives

P. V. HEGARTY

1. SHOULD HERBAL AND TEA PRODUCTS BE REGULATED?

"No herb ever cures anything, it is only said to cure something. This is always based on the testimony of somebody called Cuthbert who died in 1678. No one ever says what he died of."

—*Miles Kingston*[1]

"Is there no Latin word for Tea? Upon my soul, if I had known that I would have let the vulgar stuff alone."

—*Hilaire Belloc*[1]

"I don't want to know what the law is, I want to know who the judge is."

—*Roy M. Cohn*[1]

FOOD laws and regulations are enacted to protect the health of the consumer, to prevent economic fraud and to ensure the essential quality and wholesomeness of foods. The quotations above summarize many peoples' attitudes to herbal products and teas. These products are complex and the market is large and growing, as we shall see below. Hence, the necessity for countries

---

[1]Quotations are from The Oxford Dictionary of Humorous Quotations (N Sherrin, ed), Oxford University Press, Oxford, U K., 1996.

to have reliable, effective and harmonized laws for these products. Laws are necessary for consumer protection because of some confusion and willful distortion of claims for the health-giving properties of some herbs and teas. Laws are required also to ensure the safe production and distribution of herbal and tea products. Thus, it is necessary to operate within legal definitions and demarcations of terms for herbal and tea products, standardized methods of production and processing and regulated health claims. Finally, it is important to know the initiators and implementers of food laws not only in one's own country, but also in other countries. Food laws and regulations frequently differ between countries; an international context will be given, where possible, in this chapter. Paraphrasing the last quotation above, it is important to know both the laws and those who regulate them.

1.1. DEFINITIONS AND CLASSIFICATIONS OF HERBAL AND TEA PRODUCTS

Laws regulating herbs and teas are complex, partly because these products have a duality as dietary ingredients and medicinal compounds. Hence, the context in which the word "herb" is used is important in their regulation. Tyler (1994) deals with nomenclature distinctions in the opening lines of his book *Herbs of Choice: The Therapeutic Use of Phytomedicals.* Herb in the botanical literature means non-woody seed-producing plants that die back at the end of the growing season. Herbs used in context of the culinary arts are vegetable products that add flavor or aroma to food. Herbs are advocated by dietitians and nutritionists as flavorful and healthful to salt and fat in the diet. Tyler emphasizes that herbs used in medicine are most accurately defined as crude drugs of vegetable origin. They are used to treat various disease states, often of a chronic nature. In recent years, herbs have had the added medical dimension of attaining or maintaining a condition of improved health. Phytomedicinals (plant medicines) are produced when various solvents extract pharmacological components of herbs.

The term "botanical ingredients" must be defined accurately because of its growing significance. The Office of Dietary Supplements, National Institutes of Health (1998) defines them "to include all plant-derived materials whether fresh, preserved, or dried full plants, plant parts, plant species mixtures, plant extracts, and compounds found in such materials." Thus, items that are commonly termed "herbs" or "herbal products," regardless of whether they meet the dictionary definition of herb (a flowering plant whose stem aboveground does not become woody), or that "are comprised of parts, extracts, or preparations of woody plants" will be included as botanical ingredients. The highest uses of botanical products are as foods. In North America, staples such as corn, wheat, rice, potatoes, leafy greens and other vegetables come under the term "botanical products" and so do the many plants used as spices and flavorings. FDA regulations list approximately 250 botanical ingredients (and their essential oils and extracts) that are generally recognized as safe (GRAS) for use in foods as spices and flavorings, essential oils and natural extractives. In addition, more than 100 botanical ingredients are listed as approved flavoring agents for use as natural flavorings in foods and beverages.

Regulatory complications are potentially caused by a multiplicity of names for some herbs. There are thousands of botanicals in use today. Thus, it is important to distinguish between the various species of plants. For example, there are many common names for different botanicals ("boneset," "feverwort" and "thoughtwort" are three of many nicknames for *Eupatorium perfoliatum*). Furthermore, plants can have multiple genus/species synonyms. An example is the accepted binomial of *Sernoa repens* for saw palmetto. But, this plant is referred to frequently as *Sabal serrulata* and less frequently as *Corypha repens* or *Brahea serrulata*. Resolution of these nomenclature problems was urged by Dr. Bernadette Marriott, Director of the Office of Dietary Supplements, at the 1998 International Workshop to Evaluate Research Needs on the Use and Safety of Medicinal Herbs. She recommended that medicinal herbs should be referred to by their correct, accepted Latin binomials. Furthermore, titles and keywords of medical articles should use the Latin binomials to ensure greater ease and reliability in retrieval by citation search engines.

1.2. INTERNATIONAL DIFFERENCES IN THE REGULATION OF HERBAL PRODUCTS

Countries differ in their approaches to the regulation of herbal products. Furthermore, laws and regulations covering these products are in a constant state of flux within many countries. This is due in part to heightened awareness of real or perceived health benefits from herbal products. For example, herbal, vitamin and mineral supplements were classified as drugs in Canada. The government response to the Report of the Standing Committee on Health (Health Canada, 1999) accepted the recommendation that herbal medicines be treated as a new category distinct from either foods or drugs. In the United States they are dietary supplements. Japan, China and other Asian countries have long histories in the use of herbs and other natural products. These products are generally regulated as drugs in these countries but not in the United States. Further caution is needed in these comparisons because the Japanese and Chinese nomenclature and classifications are different from those in the United States. Thus, there is a need for international coordination of the interpretation and application of regulations by authorities in different countries. The organization charged to achieve these objectives is the Codex Alimentarius Commission. In 1995, the Codex Committee on Nutrition and Foods for Special Dietary Uses considered a proposal from the German delegation, supported by other member nations, that the Committee consider the development of guidelines for dietary supplements of vitamins and minerals. Many amendments were considered; some were proposed by the United States. The guidelines were modified in many respects, but by 1997, were not ratified. Recent activities by this Codex Committee include a proposal that the guidelines for dietary supplements of vitamins and minerals do not include other dietary supplements containing herbs, botanicals or other substances.

The Codex Committee on Nutrition and Foods for Special Dietary Uses in 1996 addressed the issue of standards for potentially harmful herbs and botanical preparations sold as foods, including dietary supplements. A discussion paper prepared by the Canadian government proposed that the committee consider developing a list of herbs and botanicals that would not be used as food because that use may be unsafe. Several countries, including the United States, objected because this activity was not within the committee's scope of work. They also stressed that there was no consensus on the scientific risk assessment procedures available to develop such a list of potentially harmful herbs. The committee finally agreed with this position. They asked the Commission to delete this topic from further consideration. The Commission agreed to the requested deletion at their June 1997 meeting. Under the heading "Potentially Harmful Herbs and Botanical Preparations Sold as Foods" the following statement was issued: "The Commission concurred with the view of the Committee on Nutrition and Foods for Special Dietary Uses that no further action was needed concerning these products as this was a matter for national authorities to address, especially as regulations and practices in this area differed from one country to another. The matter was deleted from the Commission's Work Program." (U.S. Food and Drug Administration, 1997). Thus, international guidelines for the regulations of herbs and botanicals remain in limbo.

1.3. SIZE AND GROWTH OF THE HERBAL INDUSTRY

Explosive growth of the herbal and tea market is a further compelling argument for laws and regulations to protect consumers and to regulate producers. The global market for herbal supplements presently exceeds $15 billion. This breaks down to a $7 billion market in Europe, $2.4 billion market in Japan, $2.7 billion market in the rest of Asia and about a $3 billion market in North America (Glaser, 1999). In the United States, the herbs and botanicals industry was 15% of the $23.2 billion nutrition industry in 1997. Projections for 1999 predict growth of 45% for herbals, 28% for supplements and 25% and 2% for food and beverages, respectively. About 30% of all U.S. adults use some type of herbal product. Some 1,500 to 1,800 botanicals are sold in the United States as dietary supplements or ethnic traditional medicines. The top 10 botanical products sold at selected health food stores in 1995 were

Echinacea, garlic, goldenseal, ginseng, ginkgo, saw palmetto, aloe, ma huang, Siberian ginseng and cranberry (Commission on Dietary Supplement Labels, 1997). The market is spreading rapidly from specialty and health food stores into pharmacies and grocery stores. Most U.S. botanical product users are white, college-educated, middle-aged women (Anonymous, 1998).

It is instructive to check information for medicinal herbs sold in Canada (http://www.agr.ca/pfra/sidcpub/herbs.htm) for the following reasons. Herbs come from many countries and have price ranges depending on the following classifications: "Organic," "Non-certified medicinal," and "Non-medicinal or medicinal" (Clark, 1999). European data shows Germany is the largest herbal market (44%), with France having 28% and Italy 11%.

Worldwide, about four-fifths of all people rely to a great extent on traditional medicines based on plants and their components. Hence, herbal medicine is coming under increased scrutiny (Senior, 1998). Harmful side effects from some treatments have been reported. Mohamed Farah at the WHO Drug Monitoring Center, Uppsala, Sweden, perhaps best summarizes the problem with some herbal products. He regards herbal preparations as a lottery, saying that there is confusion in classifying the original plants, and that the final composition of an extract will vary, even if the plant is correctly identified. Fifteen million Americans take high-dose vitamin or herbal preparations along with prescription drugs, thereby risking adverse effects from unknown interactions.

The pharmaceutical companies will soon compete for FDA-approved herbal-pharmaceutical products (Glaser, 1999), which is a growth market. There are potential regulatory problems for the Food and Drug Administration (FDA) that is set up to deal with "traditional" drugs that usually have only one active ingredient. The active ingredients in herbal medications are often not known. Some people in the business expect the FDA to make changes in regulations to allow herbal remedies to be sold at least as over-the-counter (OTC) drugs. Furthermore, the FDA is willing to accept data from European trials showing that some of the better-known herbs are safe, as well as documented results of U.S. human studies using dietary supplements.

In summary, the laws and regulations governing herbal and tea products is in transition. There is an urgent need for clear regulatory guidelines for producers, distributors and consumers of herbal and tea products.

2. A BRIEF HISTORY OF HERBAL PRODUCT REGULATION IN THE UNITED STATES

Herbs have been a part of the human diet from the earliest of times, especially among the Greek, Roman and Hebrew peoples of antiquity (Toussaint-Samat, 1994). Healing properties of some herbs were recognized by our earliest ancestors (Tannahill, 1973). Medicinal advantages conferred by herbal consumption were based almost entirely on anecdotal evidence. This is sometimes the situation today also and is a significant reason for the continuing evolution in the regulation of herbs and teas. The latest legislation in the United States to include herbal products is the Dietary Supplement Health and Education Act (DSHEA) enacted in 1994. A brief history of previous legislation covering herbal products will place DSHEA in regulatory context. It is hoped that this information, and the reference to more detailed information, will provide an overview for readers interested in dietary supplement commerce in the United States. For readers interested in the regulation of herbal products in other countries, this brief overview may allow them to compare and contrast the development of regulations in their own country.

2.1. FEDERAL FOOD, DRUG, AND COSMETIC ACT OF 1938 (FDCA) IN THE UNITED STATES

Earlier regulations of herbal and tea products in the United States are given in detail elsewhere (Tyler, 1994; Pendergast, 1997). The Federal Food, Drug, and Cosmetic Act of 1938 (FDCA) established the principles for the regulation of food, drugs, cosmetics and medical devices. This included the regulation of any product containing a herbal ingredient. One major component of the FDCA applicable to vitamins, minerals and herbal products is section 403j. This section covers the misbranding of a food when it purports to be "for special dietary uses" unless its label has information about its vitamin, mineral and "other dietary properties" as the FDA requires by regulation. The intent of 403j was to ensure that vitamin and mineral products were regulated closely and that consumers knew the actual nutritional value of a product. The FDA began application of this section with emphasis on the nutritional qualities of a food; minimal attention was given to fraudulent claims. The final version of sections 403j published in 1940 and 1941 did not address the regulation of high-potency vitamin and mineral supplements and claims for foods for special dietary uses. There was no discussion on what herbs and other ingredients might qualify as "nutritional properties" and be within the regulations in 403j. These regulations were general rather than specific. Thus, there were almost no regulations for the protection of the public from deception in the sale of herbal products. The FDA applied FDCA to products such as herbs by litigating against false or misleading claims rather than against product content (Pendergast, 1997).

The issue of the regulation of vitamins, minerals and herbs in tablets or pills as a drug or a food continued unresolved for decades. A product is defined as a drug if its seller intends it for the treatment, prevention or mitigation of any disease. Thus, a number of drug cases involved herbal products that would normally be termed food ingredients (sage, fennel, ginseng). FDA took a strong stand against apparently harmless medical claims for garlic tablets. Herbs were beginning to establish a bad reputation in regulatory circles because of the lack of a clear boundary between the food and the medicinal components of herbal products. Fraudulent and misleading claims were beginning to grow for herbal products, which did not help matters.

Producers need not classify vitamin, mineral and herbal products as drugs if there are no intended therapeutic uses. Producers could also avoid misbranding when there were no misleading or false claims on the label. The legal loophole was to make claims for therapeutic uses and misleading or false claims in accompanying books, pamphlets and magazines. Legal protection was thus provided by the First Amendment. The FDA acted quickly to broaden the definition of labeling to include books and pamphlets. This control spread quickly to include lectures, newsletters and radio broadcasts in which therapeutic claims were made for dietary supplements. Pendergast (1997) gives the example of an herbal product widely described in lectures by health professionals as having beneficial health effects against a disease. The seller of the herb could be sanctioned if the FDA had never approved the product for that purpose.

2.2. FOOD ADDITIVE AMENDMENTS (1958)

The U.S. Congress passed the Food Additive Amendment in 1958. All ingredients added to foods required FDA approval if they were not in the generally recognized as safe (GRAS) category. Herbs were one group that caused problems in applying the Amendments. Are herbs food additives? The Amendments defined a food additive as either a component of a food or otherwise added to it. FDA argued that a single herbal ingredient food in capsule form was an additive, even though it was not added to anything. The courts ruled against the FDA. Legal experts argue that FDA's efforts to control the sale of herbal supplements by their interpretation of the definition of a "food additive" precipitated a movement that resulted in DSHEA several years later. FDA also attempted to exclude foreign-origin herbs from the U.S. dietary supplement market. The courts ruled against this order also. These, and other rulings, set the stage for dissatisfaction and concern during the 1960s and 1970s about the regulation of dietary supplements, including herbal products.

Various attempts to clarify the regulations failed until the Proxmire Amendments attempted to give more precise legal definitions to dietary supplements. Receiving the most attention were vitamin and mineral preparations sold in pills, tablets, capsules or liquid drops and marketed for "special dietary uses." This term was applied broadly to dietary needs brought on by a physical, physiological, pathological "or other condition" (e.g., pregnancy, obesity, disease). The amendments had direct relevance to herbal products in that the

FDA had no control over the combinations of safe vitamins, minerals and "other ingredients" that could be in a single product. This allowed manufacturers to add herbs and other "nonnutrients" to vitamin and mineral preparations.

Public interest and concern about the nutritional value of food continued to grow and was responded to by the food industry. Health claims were becoming more prevalent, but the FDA was not responding with guidelines for health claims for dietary supplements. A 1987 statement by FDA drew a distinction between claims for foods and those for dietary supplements. The regulatory agency continued to litigate against manufacturers who sold dietary supplements as unapproved drugs due to alleged drug claims. This caused discrepancies in the regulation of the food industry and the dietary supplement industry. This was one reason for the enactment of the Nutrition Labeling and Education Act of 1990 (NLEA) that became the precursor of DSHEA.

2.3. NUTRITION LABELING AND EDUCATION ACT 1990 (NLEA)

NLEA permitted food health claims. These covered the relationship of nutrients to a disease or health-related condition. NLEA further specified that FDA must establish separate procedures for health claims and standards for dietary supplements. Congress concurred with the dietary supplement industry in its complaints against the stringency and applicability of the standards. FDA maintained that the scientific criteria for all health claims should be similar for traditional foods and dietary supplements, including herbal products. They were concerned that little scientific data existed on the safety of herbs and about previous problems with herb safety. FDA published a series of documents and requests for comments on the regulation of dietary supplements. These actions convinced the dietary supplement industry and some members of Congress that the agency would persist in its two-tier approach to the regulation of health claims for food and for dietary supplements, including herbal products. DSHEA was the outcome of this phase of regulatory history. Many members of Congress commented that they received more mail, phone calls and constituent pressure to reduce the regulation burdens on dietary supplements than on any other issue, including health care reform, abortion or the deficit (McNamara, 1995).

3. DIETARY SUPPLEMENT HEALTH AND EDUCATION ACT OF 1994 (DSHEA)

A quick definition of dietary supplements, as defined by the Dietary Supplement Health and Education Act, 1994 (DSHEA) in the United States is as follows:

- product (other than tobacco) intended to supplement the diet that bears or contains one or more of the following dietary ingredients: a vitamin, mineral, amino acid, herb or other botanical; OR

- a dietary substance for use to supplement the diet by increasing the total dietary intake; OR
- a concentrate, metabolite, constituent, extract or combination of any ingredient described above; AND
- intended for ingestion in the form of a capsule, powder, softgel or gelcap, and not represented as a conventional food or as a sole item of a meal or the diet

Details are provided at http://odp.od.nih.gov/whatare/whatare.html, Office of Dietary Supplements, National Institutes of Health (1998). This office was created to coordinate research on dietary supplements as a result of DSHEA.

DSHEA is the subject of much critical analysis and comment. An early assessment was given by McNamara (1995) in which he concluded that substantial changes would occur in the way that FDA regulated dietary supplements. Major changes in legislation included the creation of a broad, new definition of "dietary supplement" products. There would be a moderation of the regulatory burdens on the use of dietary ingredients in dietary supplements. This included changes in both the safety standard for an ingredient and the regulatory procedure that changed from FDA preclearance to one of FDA policing. A further change was that additional promotional material on the nutritional benefits of dietary supplements was permissible. This included labels containing "statements of nutritional support," and references to books, articles and other publications for additional health-related information.

DSHEA deletes the "food additive" requirement for preclearance of dietary ingredients not considered by the FDA to be GRAS. However, FDA was given substantial new policing authority to stop the distribution of a dietary supplement if its safety is questionable (McNamara, 1996).

Stores in the United States can now sell herbal and nutrient concoctions with labels that have claims about relieving pain, "energizing" and "detoxifying" the body or providing "guaranteed results" (Kurtzweil, 1999). DSHEA permits dietary supplement manufacturers to market more products (including herbal and tea products) as dietary supplements. This is done to provide consumers greater access to and information on products as dietary supplements. FDA notes that under DSHEA, it requires less premarket review of dietary supplements than other products it regulates (drugs and many additives used in conventional foods). Thus, manufacturers and consumers now have the responsibility for checking the safety of dietary supplements and determining if label claims are true.

3.1. DIETARY SUPPLEMENT LABEL REQUIREMENTS

Kurtzweil (1999) of the FDA's public affairs staff summarizes the information required on labels for herbal products and other dietary supplements:

- statement of identity (e.g., "ginseng")

- net quantity of contents (e.g., 60 capsules)
- structure-function claim and the statement, "This statement has not been evaluated by the Food and Drug Administration. This product is not intended to diagnose, treat, cure, or prevent any disease."
- directions for use (e.g., "Take one capsule daily")
- supplement facts panel (lists serving size, amount and active ingredient)
- other ingredients in descending order of predominance and by common name or proprietary blend
- name and place of business of manufacture, packer or distributor— this is the address to write to for more product information

DSHEA takes the traditional concept of dietary supplements as one or more of the essential nutrients such as vitamins, minerals and protein, and extends it to include, with some exceptions, any product intended for ingestion as a supplement to the diet. This now includes herbs and botanicals. Dietary supplements are not drugs. FDA does not authorize or test dietary supplements. Under DSHEA and previous food labeling laws, supplement manufacturers may use, when appropriate, three types of claims: nutrient-content claims, disease claims and nutrition support claims, which include "structure-function claims." Further details on permissible claims and fraudulent products are given by Kurtzweil (1999). The FDA's proposal "Regulations on Statements made for Dietary Supplements Concerning the Effect of the Product on the Structure or Function of the Body" (FDA Fact Sheet, 1998) has many details pertinent to the manufacturers and consumers of herbal products and teas. If this proposed rule becomes final, unacceptable structure/function claims will have to be removed for labeling, or the product must be approved as a drug under the Federal Food, Drug and Cosmetic Act. Since the passage of DSHEA in 1994 and Spring 1998, FDA received approximately 2,300 notifications from manufacturers about structure/function claims for their products. The agency informed manufacturers that about 150 of these claims were problematic, and it estimated that another 60 claims would not be permitted under the proposed criteria. Clearly, this is an area demanding the attention of everybody involved with the sale and purchase of herbal products. The Commission on Dietary Supplement Labels (1997) recognized that, under DSHEA, botanical products should continue to be marketed as dietary supplements when properly labeled. The Commission made a strong recommendation that FDA promptly establish a review panel for OTC claims for botanical products that are proposed by manufacturers for drug use. More studies were recommended to establish an alternative system for regulating botanical products that are used for purposes other than to supplement the diet but that cannot meet OTC drug requirements. The study should include the types of disclaimers that might apply and the appropriateness of such a system within the regulatory framework of the United States. The Commission recommended a comprehensive evaluation of regulatory systems for botanical remedies in other countries. Consideration should be given to the scope of the products covered, the means of assuring safety and preventing deception, the effect of such systems on overall medical care, the definition of appropriate drug uses of products and the appropriateness and applicability of the different types of disclaimers. Thus, we will now take a brief survey of how dietary supplements, with specific reference to botanical products, are regulated in other countries.

4. INTERNATIONAL COMPARISONS OF THE REGULATION OF BOTANICALS

Four-fifths of all people in the world still rely to a great extent on traditional medicines based on plants and their components (American Medical Association, 1997). The use of herbs in medicine is ancient in its origins, but only recently have systematic attempts been made to codify them into acceptable regulations. Botanical pharmacopeias exist for a number of countries including the United States, United Kingdom, Japan, France and Germany. Regulations for the therapeutic use of botanical remedies are established in these countries, in Canada and some other industrialized countries. Countries and international agencies may differ slightly in statements related to the use of herbal products. Statements for the use of ginger are given in Table 12.1. The Commission on Dietary Supplement Labels (1997) considered that there are instances when statements concerning treatment such as those found in the WHO model monograph may be more informative to consumers than the less specific language used in some of the statements of nutritional support.

Some countries require clinical evidence to support recommended use. Other countries regard traditional use as sufficient to provide the basis for a limited therapeutic claim, but a disclaimer may be required (some examples are given in Table 12.2). WHO is providing guidelines for the regulation of traditional medicines, including botanical remedies. The regulatory system for dietary supplements, including herbs, is described above under DSHEA. A brief overview of the regulatory situation in other countries follows.

4.1. CANADA

A good example of the recent evolution of laws to cover herbs and botanicals comes from Canada. "Natural Health Products: A New Vision" was published in November 1998 (Report of the Standing Committee on Health, 1998). Terms of reference for this Standing Committee on Health investigating natural health products (NHPs) included: "consult, analyze and make recommendations regarding the legislative and regulatory regime governing traditional

TABLE 12.1. Rhizoma *Zingiberis*.

WHO Model Monograph

"11.1 Uses supported by clinical data

The principal clinical use of ginger is for the prophylaxis of nausea and vomiting associated with motion sickness, postoperative nausea, *hyperemesis gravidarum*,[1] and sea sickness.

11 2 Uses described in pharmacopoeias and in traditional systems of medicine

Ginger is also indicated for the treatment of dyspepsia, flatulence, colic, vomiting, diarrhea, spasms and other stomach complaints. Powdered ginger is further employed in the treatment of colds and flu, to stimulate the appetite, as a narcotic antagonist, and as an anti-inflammatory agent in the treatment of migraine headache, and rheumatic and muscular disorders

11 3 Uses described in folk medicine, not supported by experimental or clinical data.

Other medical uses for ginger include the treatment of cataracts, toothache, longevity, insomnia, baldness and hemorrhoids.

[1]Although ginger appears to be clinically effective in the treatment of *hyperemesis gravidarum*, it is currently not recommended for use in morning sickness during pregnancy "

Statements of Nutritional Support from Notification Letters to FDA

"Stimulates digestion. Ginger is an aromatic bitter herb that stimulates digestion."

"Ginger is one of the world's most popular spices, and a well researched herb for a healthy lifestyle The pungent taste of ginger, prized in international cuisine, has been linked to beneficial compounds which warm and soothe the stomach Ginger has been a favorite of travellers since ancient mariners discovered it in the exotic Orient "

"Ginger root is a soothing and warming herb for the stomach and may help maintain a calm stomach while travelling "

"Eases the discomfort associated with travelling Ginger is an aromatic bitter herb that eases the discomfort associated with travelling and stimulates digestion to promote gastrointestinal comfort."

From Commission on Dietary Supplement Labels (1997)

medicine (including, but not limited to, traditional herbal remedies, traditional Chinese, Ayurvedic and Native North American medicines), homeopathic preparations and vitamin and mineral supplements.'' NHPs are defined as ''substances or combinations of substances consisting of molecules and elements found in nature, and homeopathic preparations, sold in dosage form for the purpose of maintaining or improving health and treating or preventing diseases/conditions.'' Herbal remedies come under these descriptions of NHPs. Herbal remedies and other complementary medicines are considered to be drugs because they meet the following definition in the Food and Drug

TABLE 12.2. **Examples of Disclaimers Used in Other Countries.**

Country	Disclaimer
Belgium	"traditionally used in . ., even though its activity has not been established according to the actual criteria of evaluation of medicines "[1]
Canada	"traditional medicines"[1]
France	"traditionally used for .." or "used in . . ."[1]
Germany	"Traditionally used (e g) for preventive purposes This product is not intended for the cure of mitigation of illness, physical deficiencies or ailments Anyone who has such illness or ailment should consult a physician. This product is used traditionally and it cannot be deduced therefrom whether the product is generally useful "[2]
Greece	Wording frequently used· "possibly effective" and "traditionally used"[1]
Ireland	"The wording on the labeling is mandatory and states the following· i) Do not take in connection with other medications without having consulted a physician ii) Do not use for longer than two weeks The drug safety cannot be guaranteed for a prolonged period of use. iii) Should the condition not improve, consult a physician iv) Allergic reactions are possible v) Traditional herbal remedy for short-term treatment of slight discomfort and that should . . not be used for extended periods without the advice of a physician "[1]
United Kingdom	"a traditional remedy for the symptomatic relief of . " and "if symptoms persist, consult your doctor "[1]

[1] Gericke, N 1995 The regulation and control of traditional herbal medicines an international overview with recommendations for the development of a South African approach Working draft document Cape Town, South Africa Traditional Medicines Programme, University of Cape Town
[2] Nozari, F 1994 Dietary supplements Report to Congress LL94-3 Washington, DC
From Commission on Dietary Supplements Labels (1997)

Act: "'drug' includes any substance or mixture of substances manufactured, sold or represented for use in: (a) the diagnosis, treatment, mitigation or prevention of a disease, disorder, abnormal physical state, or its symptoms, in human beings or animals, or (b) restoring, correcting or modifying organic functions in human beings or animals'' (Health Canada, 1997a 1997b p. 5).

A survey by the Canada Health Monitor in 1997 found that 56% of Canadians reported taking one or more NHPs in the past six months. Among them, 20% had taken herbal remedies and teas. Some herbal products are sold as foods, while others are marketed as drugs (certain plants are even controlled under the Controlled Drugs and Substances Act). Health Canada's Information Letter

No. 771 (1990) on traditional herbal medicines was issued to all manufacturers and importers. It illustrates how the department determines whether an herbal product is a food or a drug: "The pharmacological activity of the ingredients, the purpose for which the product is intended; and the representations made regarding its use, including directions for use are the most important factors in determining whether a herbal product should be considered a food or a drug. Other factors, such as precise dosage form, may be considered in the determination of the product status. Herbal ingredients consumed more or less as desired due to the absence of pharmacological properties are appropriate as foods".

The Standing Committee on Health, referred to above, recommended that Health Canada, in conjunction with a new separate NHP Expert Advisory Committee, set out an appropriate definition of NHPs and amend the Food and Drugs Act accordingly. It was also recommended that Health Canada, in conjunction with the new NHP Expert Advisory Committee, examine the status of bulk herbs for legislative purposes. "Natural Health Products: A New Vision" and the government's response ("Government Response to the Report of the Standing Committee on Health") (Health Canada, 1999) are important documents for all Canadians interested in NHPs, including herbal products. For non-Canadians, it is a useful document demonstrating the complexities involved in the regulation of these products.

If a manufacturer sells a medicinal herb or markets an herb with drug claims, it requires a Drug Identification Number (DIN) or General Public (GP). If sold without a DIN or GP, it is not in compliance with Canadian Law (Health Canada, 1997a). The DIN and GP numbers are located on the label of any drug product approved for sale. The number indicates that a product has undergone and passed a review of its formulation, labeling and instructions for use. Bulk herbs, which are sold for further processing, have different restrictions than for finished products. It is the responsibility of the importer to ensure that all herbal products coming into Canada comply with Canadian requirements.

If a natural health product has been used for centuries, can it be assumed to be safe? Not necessarily, according to Health Canada (1997c). For example, Belladonna, an herbal ingredient used over the centuries as a sedative and for asthma, is also referred to as Deadly Nightshade. Consuming as few as three of the black berries of this invasive garden weed can be fatal. Recently, both *Stephania tetranda* and *Magnolia officinalis* caused severe kidney damage in Belgium where these products were illegally available. This kidney damage has resulted in over 50 patients requiring either chronic dialysis or kidney transplant. Import and compliance alerts have been issued in Canada to ensure that these products are not illegally imported and sold.

In summary, there is much activity in Canada to formalize the laws and regulations for herbal products.

4.2. EUROPEAN UNION

Herbal medicines are only weakly regulated in Europe, but more stringent regulations are being developed. American herbalists are challenging these developments because it would interfere with their sales into Europe (MacKenzie, 1998). DSHEA in the United States regulates herbal products as dietary supplements, and these products are subject to less regulation than their European counterparts. The sale of herbal preparations to treat a disease is subject to the same quality, safety and efficacy controls as any medicine sold in the European Union (EU). Herbal medicines used for genuine disease cannot evade European laws by simply omitting any mention of disease on the package. DSHEA includes no such restrictions. The current situation is perhaps best described by MacKenzie (1998) in that if the European industry can live with regulation, one wonders what, apart from greed, makes it necessary for American companies to insist that people have the right to choose their own poison.

In a recent development, herbal products will have common quality control rules in all countries of the EU. The herbal producers, including some pharmaceutical companies, were the creators of these regulations. They formed the European Scientific Cooperative for Phytotherapy (ESCOP) to advance the state of herbal medicine. ESCOP is under the auspices of the European Commission and publishes a series of plant species monographs for European Union marketing organizations.

A European Commission report on unconventional medicine includes a call for more research to protect the public from harm (Matthews, 1998). Publication of scientific evidence for different therapies is among the recommendations. The report draws attention to claims that are often anecdotal, exaggerated or unsubstantiated.

Some specific issues relating to herbal products in individual countries of the EU are worthy of mention. Germany is Europe's leading importer of herbal medicinal products and has well-established laws and regulations. The German government published the Commission E monographs that have recently been translated in English. They define quality standards and potency tests for over 350 single-plant drugs. France has recognized more than 200 medical plants and provided specifications governing their sale.

The Ministry of Agriculture, Fisheries and Food (MAFF) in the United Kingdom issues information bulletins periodically. Some of these bulletins concern the regulation of herbal products include the "Report on Health Effects from Traditional Remedies and Dietary Supplements" (Ministry of Agriculture, Fisheries and Food, 1996). Satisfaction is expressed, in general, because no significant health problems were associated with most types of traditional remedies or dietary supplements. A potential concern was raised on the use of some unlicensed Chinese herbal medicines. Liver damage was detected in some users. Some traditional remedies from the Indian subcontinent were also of concern because of high levels of heavy metals. The Medicines Control Agency, which has responsibility for the overall safety of medical products, extended its monitoring of adverse reactions for licensed and unlicensed medicines.

An MAFF survey that included herbal teas found that four biologically active principles (BAPs) (coumarin, safrole, isosafrole and pulegone) were generally consistent with the statutory limits. Another MAFF survey showed little evidence of significant aflatoxin contamination in retail herbs and spices. In summary, perusal of a number of Food Surveillance Information Sheets published by MAFF's Joint Food Safety and Standards Group indicated that the majority of herbal and tea products were within regulatory guidelines for components that would compromise food safety. The irradiation of herbs and spices is licensed in the United Kingdom. This is an additional safety assurance for herbal products.

The medicines control agency (MCA) is an agency of the Department of Health in the United Kingdom. In accordance with European Union and United Kingdom law, it licenses medicines with the primary objective of safeguarding the public health. The agency also has a duty to determine whether or not a product on the borderline between medicines and food is a medicine, and therefore subject to licensing. An MCA determination of this kind is subject to review by the courts, by way of judicial review or action in the civil courts, or in the course of a criminal prosecution by the MCA of a noncompliant manufacturer or trader. This situation has given rise to concerns for two reasons. First, food supplements have no definition under current United Kingdom law. Second, some consumer groups expressed concerns that proposals to amend the Medicines for Human Use (Marketing Authorizations, etc.) Regulations, 1994, would bring food supplements within the remit of the MCA or medicines legislation. Officials at the Ministry of Health state that proposed legislation will not bring food supplements within the remit of the MCA or Medicines Legislation (Hayman, 1999).

In Ireland, the Irish Medicines Board decided recently to apply the same regulations for herbal food supplements and alternative medicinal products as they apply to mainstream pharmaceutical products (Timmins, 1999). Under the new guidelines, products will be deemed to be medicinal when their labeling or accompanying or associated literature makes any preventive, curative or remedial claim. This met with resistance from the consumers for health choice who stressed that everyday food supplements such as garlic, evening primrose, fish oils and vitamin C tablets would disappear from the market.

The changes and proposed changes in legislation referred to above have prompted the European Union to prepare legislation to harmonize the differing regulations under which food supplements and alternative medicines are sold.

Biotechnology is a major concern with European regulators, consumers, food retailers, the food processing industry and politicians. Much attention is paid to genetically modified soy, corn, tomatoes and potatoes, but now genetically modified herbs are included also. At the time of writing, genetically modified (GM) chicory was undergoing regulatory review under the terms of the EU Novel Foods Regulation. This stipulates that before GM chicory can be marketed in the EU, it must be approved by all other member states. The genetic modification of herbs is an area deserving attention in the future. In conclusion, it seems that more will be heard in the near future about regulation of herbal products in Europe.

4.3. JAPAN

Japan licenses only physicians practicing Western medicine, yet the growing popularity of kampo (Japanese herbal medicine in the Chinese tradition) has challenged the medical system. Kampo was the primary form of medicine in Japan up to the mid 1800s. It is less targeted against disease than Western medicine. It does not assign names to diseases. Yet, scientific studies confirm the effectiveness of some kampo remedies. Its growing popularity, especially among younger doctors, resulted in the Japanese Association of Medical Sciences announcing that medical schools would grant degrees in the practice of kampo.

Herbs for nonmedicinal use were relatively unknown in the Japanese market until recently. Dried herbal tea was the first of these products to be introduced into Japan in 1969. Since then, other products, including herbal candy, were introduced. An increasing number of farmers are growing herbs, but concern has been expressed about the extensive use of pesticides in the growing of herbs.

Japanese regulations do not distinguish between herbs and spices. Herbs are defined as fragrant plants that grow primarily in warm climates. Importation of spices and herbs are subject to the provisions of the Food Sanitation and the Plant Protection Laws. In recent years, more stringent criteria have been adopted for food additives, aflatoxin, radioactivity and residual pesticides. Some of these criteria can apply to the importation of herbs. Labeling of herbal products is required under the Food Sanitation and Measurement Laws. Further details can be obtained at the following web sites: "Spice" at http://www.jetro.org/database/mtp/mtp93112.htm and "Marketing Guide Spices" at http://www.jetro.org/database/mkg/mkg1%2D17.htm.

The importation of teas must conform to the Food Sanitation Law. Black tea imported and sold in Japan must list specific items on the label. Further details are given under "Marketing Guide Black Tea" at http://www.jetro.org/database/mkg1%2D4.htm.

4.4. CHINA

More than 6,000 natural products are used in traditional Chinese medicine. About 500 are most commonly used, and of these, about 82% are derived from plants. Most of these products are regulated as drugs. The People's Republic of China bans the marketing of unregistered "health foods." The manufacture of foods purported to have special health effects is inspected.

5. QUALITY AND GOOD MANUFACTURING PRACTICES

Consumers are concerned about the cleanliness, purity and potency of herbal products. Good Manufacturing Practices (GMP) are internationally accepted standards to ensure appropriate quality at all stages up to the final point of sale. GMP standards apply to the premises, equipment, personnel, raw material (identification), finished product testing, sanitation/cleanliness and record keeping. A brief review of the use of GMP for herbal products in different countries is in order.

Australia adapted its drug GMP guidelines to include herbal products. These products are monitored before and after they are marketed. This is done by GMP inspections and through a random testing program for lower safety products. In the United States, drug manufacturers must meet GMP standards. But, since herbal products are considered to be dietary supplements and not drugs, they are not covered by drug GMP regulations. The Food and Drug Administration (FDA) has the authority to establish GMPs specific to dietary supplements. Germany requires herbal products sold as drugs to be manufactured according to GMP. The United Kingdom requires all manufacturers of drug products to have a license and to comply with GMPs. Canadian GMPs are among the highest in the world. This is one reason why Canadian NHPs gain international recognition because of their high quality (Report of the Standing Committee on Health, 1998).

Many organizations are developing guidelines for GMPs and quality control (QC) and of standard and information monographs for herbals. This is due to DSHEA and to the rapid growth in the use of herbal products. These organizations include the U.S. Pharmacopeial Convention (USP) and the American Herbal Pharmacopeia (AHP). The USP anticipates an eventual complement of 25 to 50 botanicals. The AHP is planning to have a collection numbering in the hundreds. Awang (1997) states that all monographs intend to provide a sound scientific basis for ensuring proper identity, purity and strength of medicinal plant material and finished products. However, emphasis varies among these different parameters, particularly between analytical and therapeutic aspects. Awang (1997) attributes the greatest impediment to general improvement of herbal products to the lack of effective enforcement of QC and manufacturing standards. This is especially true in North America, where such measures are virtually nonexistent.

6. ADVERTISING REGULATIONS FOR HERBAL PRODUCTS

This section will discuss the advertising regulations for dietary supplements within the United States only. The Federal Trade Commission (FTC), a law enforcement agency, has control over advertising. The FTC and FDA have overlapping jurisdiction to regulate the advertising, labeling and promotion of foods, over-the-counter drugs, cosmetics and devices (Peeler and Cohn, 1995). The division of regulatory labor is for FTC to deal with advertising issues and FDA to rule on labeling issues. A publication helpful to all interested parties concerned with the sale and purchase of herbal products in the United States is "Dietary Supplements: An Advertising Guide for Industry" (http://www.ftc.gov/bcp/confine/pubs/buspubs/dietsupp.htm). Another relevant FTC publication was written by Murphy et al. (1998).

7. CONCLUSIONS

Growing public interest in many countries in the use of herbal products for real or supposed health benefits brings the need for regulation. Different countries take differing approaches, but all are wrestling to improve their regulatory processes. It must seem incongruous to consumers that the same botanical is regarded as a drug in one country and as a dietary supplement in another. Editorials in medical journals address critical issues in alternative medicine (Angell and Kassirer, 1998). Foremost in their analysis is the issue of regulation of herbal medicine. Particular criticism is reserved for DSHEA in the United States. They argue that the only legal requirement in the sale of herbal remedies is that they are not promoted as preventing or treating disease. The editorial charges that to comply with this stipulation, manufacturers use labeling that has risen to an art form of doublespeak.

There is much work to be done to refine, and perhaps redefine, the laws and regulations governing herbs, botanicals and teas. This must be done at the national and international levels because of the rapid growth in international trade in herbal products. It is perhaps appropriate to end as we began this chapter by revisiting the three quotations. There are people who believe no herb will cure anything, while others have total faith in the healing and health-giving properties of herbal products. How food laws can protect both remains somewhat of an international enigma. Hilaire Belloc's dilemma on terminology remains both for herbs and their medicinal products. The last quotation has a hint of disrespect for the law. Such should not be the case with the regulations on herbs; there is too much at stake for consumers and for the manufacturers and distributors of herbal products. Experiences with DSHEA especially makes some people ponder on who the judge is. In short, much work remains to be done.

8. REFERENCES

American Medical Association. 1997. CSA Reports. Report 12 of the Council on Scientific Affairs. "Alternative Medicine." (http://www.ama-assn.org/med-sci/csa/1997/r12full.htm).

Angell, M. and Kassirer, J. P. 1998. "Alternative medicine. The risks of untested and unregulated remedies," *New England J. of Medicine,* 339, 839.

Anonymous. 1998. "Herbal health," *Environmental Health Perspectives.* 106 (12) (http.//ehpnetl.-niehs.nih.gov/docs/1998/106-12/niehsnews.html).

Awang, D. V. C 1997. "Quality control and good manufacturing practices: Safety and efficacy of commercial herbals," *Food and Drug Law J* 52. 341.

Clark, H. 1999. *Medicinal Herbs.* Agriculture and Agri-Food Canada, PFRA, Outlook, SK. pp. 1–4. (http://www.agr.ca/pfra/sidc.pub/herbs.htm).

Commission on Dietary Supplement Labels. 1997. "Report." Superintendent of Documents, U S. Government Printing Office, Washington, DC. (http://web.health.gov/dietshupp/).

FDA Fact Sheet. 1998. (April 27). "FDA's dietary supplement proposal." (http:// vm cfsan.fda.gov/-dms/ds-fact2.html).

Glaser, V. 1999. "Billion-dollar market blossoms as botanicals take root," *Nature Biotechnology.* 17, 17.

Hayman, B. 1999 "Response from the Baroness Hayman [Parliamentary under Secretary of State (Lords), Department of Health] to concerns expressed by readers of the society for the promotion of nutritional therapy's journal, about proposals to amend the medicines for Human Use (Marketing Authorizations Etc.) Regulations 1994." (http://www.doh.gov.uk/mix249.htm).

Health Canada. 1990. "Information Letter" No. 771, January 5, pp. 1–2.

Health Canada. 1997a. "Facts about the importation and sale of herbal products." (http://www.hc-sc.gc.ca/main/hc/web/english/archives/96-97/herbsale.htm).

Health Canada. 1997b. "Facts at a glance Regulation of herbal medicines." (http://www.hc-sc.gc.ca/main/hc/web/english/archives/96-97/glancee.htm).

Health Canada. 1997c. "Natural health remedies." (http://www.hc-gc.ca/main/hc/web/english/ archives/96-97/herbnae.htm).

Health Canada. 1999. "Government Response to the Report of the Standing Committee on Health." (http://www.hc-sc.gc.ca/english/archives/releases/9946ebk1.htm).

Kurtzweil, P. 1999. "An FDA Guide to Dietary Supplements." (http://vm.cfsan.fda.gov/-dms/ fdsupp.html).

MacKenzie, D. 1998. "First, do no harm," *New Scientist.* p. 53, October 24.

Matthews, H. 1998. "EC calls for more testing of alternative remedies," *British Medical J.* 317, 1270.

McNamara, S. H. 1995. "Dietary supplements of botanical and other substances: A new era of regulation," *Food and Drug Law J.* 50, 341.

McNamara, S. H. 1996. "FDA regulation of ingredients in dietary supplements after passage of the Dietary Supplement Health and Education Act of 1994: An update," *Food and Drug Law J.* 51, 313.

Ministry of Agriculture, Fisheries and Food 1996. "Report on Herbal Effects from Traditional Remedies and Dietary Supplements." (http.//www.maff.gov.uk/inf/newsrel/1996/ 960930a.htm).

Murphy, D , Hoppock, T. H. and Rusk, M K 1998. "Generic Copy Test of Food Health Claims in Advertising " (http·//www.ftc.gov/as1998/9811/foodhealrep htm).

Office of Dietary Supplements, National Institutes of Health. 1998. "Merging Quality Science with Supplement Research. A Strategic Plan for the Office of Dietary Supplements," p. 8.

Peeler, C. L and Cohn, S 1995. "The Federal Trade Commission's regulation of advertising claims for dietary supplements," *Food and Drug Law J* 50, 349.

Pendergast, W. R 1997. "Dietary supplements," in *Fundamentals of Law and Regulation.* Volume 1, eds., R P. Brady, R. M. Cooper and R. S. Silverman, FDLI, Washington D.C.

Report of the Standing Committee on Health. 1998 "Natural Health Products: A New Vision " (http·//www.parl gc ca/InfocomDoc/heal/studies/reports/healrp02-e.htm).

Senior, K. 1998. "Herbal medicine under scrutiny," *Lancet* 352, 1040.

Tannahill, R 1973 *Food in History* Crown Publishers, Inc , New York.

Timmins, E. 1999. "Guidelines threaten herbal products, says lobby group," *The Irish Times,* September 21. (http //www.ireland com/newspaper/ireland/1999/0921/hom14.htm).

Toussaint-Samat, M. 1994. *History of Food* Blackwell Publishers Ltd., Oxford, UK.

Tyler, V 1994. *Herbs of Choice: The Therapeutic Use of Phytomedicinals.* Pharmaceutical Products Press, New York.

U.S. Food and Drug Administration 1997 Information Paper· "Codex Alimentarius Commission Committee on Nutrition and Foods for Special Dietary Uses Proposed Guidelines for Dietary Vitamin and Mineral Dietary Supplements." (http://vm.cfsan.fda.gov/-dms/codex html).

CHAPTER 13

Quality Assurance and Control for the Herbal and Tea Industry

P. FEDEC
P. P. KOLODZIEJCZYK

1. INTRODUCTION

IN this closing chapter, an attempt will be made to bring into perspective the desires of the end user for a consistent product lot after lot and the safety of the product for consumption. The importance of this industry is considerable, and its value is difficult to relate to price and efficacy. A review of the terminology and meanings of quality assurance and quality control will be made (frequently known as QA/QC) from the perspective of understanding and appreciating how they apply to this industry and how they may be implemented. An overview of the most important manufacturing practices that are key to delivering a useful and safe ingredient or product will assist members of the industry to fully appreciate the commitment necessary to fulfill customer needs. A small section will be introduced exploring continuous improvement, the ultimate goal of the quality assurance system. Aspects of quality control will be discussed, with the aid of examples, along with reference to some of the difficulties and limitations inherent to this industry, in particular, as it applies to the desire for identity standardization of active components. New analytical tools and methods will be discussed in relation to sophisticated differentiation techniques of the future.

2. FACTORS CONTRIBUTING TO HERBAL INDUSTRY ECONOMICS AND THE NEED FOR QUALITY ASSURANCE

Plants and plant products are an important part of human nutrition, medical treatment and health care. The World Health Organization (WHO) estimates that 70–80% of the world's population relies on traditional forms of medicine based on plants and their derivatives. The Natural Business Communications (NBC) indicates that the natural product market was about $21 billion in 1997 for North America, of which supplements comprised $6.5 billion. Although vitamins are in the majority, herbal products and botanicals make up 31% of this market. Globally, herbals and botanicals accounted for sales of about $16.5 billion in 1997 (Yuan, 1998).

The amount of money flowing into the natural products market is increasing dramatically. There are several factors contributing to this phenomena, some rational, others based on public perception, market forces expecting quick returns, social changes and the integration of food, pharmaceutical industry and agriculture. With the general public becoming more focused on healthy aspects of food through education, aggressive advertising, rising medicare costs and an emphasis on prevention rather than treatment, alternative medicine has created a growing demand for natural products.

Today, we are witnessing the emergence of a worldwide nutraceutical industry (which includes herbal teas, extracts and other herbal products) with a huge economic potential and opportunity, accompanied however, by many regulatory and legal problems. This situation could change current herbal product markets. The quality of new products, along with validated, approved and harmonized methods for analysis will play an important role in this process.

Unfortunately, this "explosive" growth of herbal drugs in North America has brought to the marketplace not only legitimate and serious companies but also dozens of "get rich quick" companies involved in "buy cheap-sell high" operations. The latter source low-priced raw materials from less vigorously controlled parts of the world, climb on fashionable bandwagons or dump "hot" look-alike products onto the market that are not standardized or tested for possible impurities and contamination, not to mention efficacy. The average consumer has difficulty distinguishing between good and worthless (ineffective) products. Therefore, it is imperative that the herbal industry implement quality standards based on scientifically accepted analytical methods for quality assurance and control. A more in-depth treatise on guidelines for assessment of herbal medicines can be found in a work by Akerele (1993).

3. DEFINITIONS

3.1. QUALITY

Quality is or should be, for most companies, everyone's business (Townsend and Gebhardt, 1992). The concept of quality is centuries old, but one that conjures up different features and characteristics to different human disciplines

(Juran and Gryna, 1988). Technological characteristics such as properties of materials (e.g., hardness) may be suitable measures related to the ability of a good to endure wear or treatment. Psychological characteristics may include taste and beauty and confer status. Functional characteristics may be expressed by texture, composition and viscosity. Contractual quality characteristics include reliability and promptness and sometimes guarantee provisions. Ethical considerations are very important to this industry as its consumer often feels that quality is nebulous. Although quality can be defined using many of the characteristics listed above, the most universal short definition of the word is given by Juran and Gryna (1988), specifically "fitness for use." From the perspective of the food, cosmetic and drug industry, a more meaningful definition may be quality equals conformance to specifications.

3.2. SPECIFICATION

A specification is the range that is acceptable for a certain characteristic. As an example, a specification for a tea may be that the finished cut must fall between 4–10 mesh US, otherwise the product is unacceptable or "out of spec." Conformance to specifications means that your product meets all of the specifications that have been agreed upon by the manufacturer and buyer and within any regulatory requirements applicable. By conforming to specifications, the product is consistent from lot to lot, time and again. This is a key element for product quality. However, quality does not mean that a product should cost more.

3.3. QUALITY ASSURANCE

Quality assurance is the system of monitoring, inspecting and auditing justifying that the product has been made according to the protocols and standard operating procedures (SOPs) from conception of the good to its final delivery at the point of sale.

3.4. QUALITY CONTROL

Quality monitoring and control is the key to detecting "out of spec" product and minimizing waste. Detection and prevention encompass the primary objectives of quality control (DeCicco, 1997).

4. QUALITY ASSURANCE

4.1. QUALITY ASSURANCE DEFINITION

Quality assurance can be regarded as "the activity of providing the evidence needed to establish confidence, among all concerned, that the quality function is being effectively performed'' (Juran and Gryna, 1988). It is the system that is building into your process and product the assurance that each time a product is produced, it will be of the same quality. As a functioning unit within a business, Quality Assurance (QA) is concerned with quality-related activities such as quality planning, quality control, quality improvement, quality audit and reliability. This group is often responsible for cradle-to-grave planning. Depending on the size of the company and its product line, quality control and quality assurance units may be separate or combined and staffed appropriately. Where a quality assurance unit is separate, it is typically responsible for monitoring activities to assure company management that the facilities, personnel, methods, practices, records and control are in conformance to the appropriate (Current) Good Manufacturing Practices (cGMPs). In addition, the unit must also be fully conversant with the appropriate regulatory requirements governing the activities of the business. A number of these documents are available on websites (Giese, 1996).

4.2. QUALITY ASSURANCE PERSONNEL

Depending on the size of the operation, the QA personnel vary in number. Typically they need to have knowledge and expertise in the technology used to produce the product, training ability, people skills and auditing and investigative abilities. QA/QC staff must have adequate resources to perform their task (Stoker, 1997). In Canada, a manufacturer of herbal drugs must have staff in the QC/QA Department with a university or equivalent degree in a science related to the work being carried out (Matsalla, 1992). This expertise must be able to identify starting materials from a botanical perspective and have an appreciation and understanding of herbal sources, cultivation practices, age of plant material, environmental factors affecting quality (e.g., pesticide residues) and certification procedures for intercountry trade.

4.3. QUALITY ASSURANCE/QUALITY CONTROL MANAGER RESPONSIBILITIES AND SKILL REQUIREMENTS

The role of the QA/QC manager is to provide leadership for compliance to standards and to facilitate continuous improvement and product quality (Kieffer et al., 1998). They will have responsibilities falling into two categories: regulatory/technical and managerial. Kieffer et al. (1998) detail the necessary skill sets to include business knowledge, leadership, management, communication, processes, quality design, validation, manufacturing, chemistry, microbiology (and in the herbal trade—botany), audits, supplier capabilities and customer awareness.

5. QUALITY/MANUFACTURING PLANNING

All segments of the company are involved in the analysis of a business opportunity. After the market analysis, planning for manufacturing is a critical step. Depending on the market positioning and the label claims that will be made, the manufacturing practices, controls and assurances will differ slightly. Regulations governing the production of herbals are under strong debate in the United States and Canada. They fall somewhere between foods and drugs, depending on whether associations or actual claims are made indicating that they influence health and cure disease. The stronger the claim, the more stringent the GMPs to the point that the production criteria will be similar to that of pharmaceuticals. Oliver (1999) suggests that although the cGMPs are voluntary at this time, legislation will be passed to regulate the industry somewhere between food and drugs, possibly under as much control as drugs themselves.

A review of the planning for quality should demonstrate that good manufacturing and quality practices have included properly documented instructions and change controls that have been reviewed and approved by Quality Assurance. Describing a manufacturing process that has been designed, developed and demonstrated is finalized into a standard operating procedure (SOP). Format and suggestions on preparing these types of documents can be obtained from many literature sources (e.g., Uys, 1994; De Sain and Sutton, 1996). Test runs must be made and compliance to product specification and quality must be demonstrated as achievable. Problems that surface must be investigated and corrected. Plans for auditing process segments and the total operation must be devised and implemented. The final outcome should keep in mind that the concept of continuous improvement underlies a successful business.

6. QUALITY MANUAL

One of the key components of the quality system is a documented manual containing the related procedures and protocols. The segments are designed to address issues such as policies, parameters, responsibilities, SOPs, validations, records and traceability and auditing (including corrective actions and improvements).

Keeping the above issues in mind, the manual is structured to address the following major sections of the Canadian GMP Guidelines (1998):

- plant and premises
- transportation and storage
- equipment
- personnel

- sanitation and pest control
- recalls

GMPs require written procedures for production and process controls and objective evidence of records (McCaleb, 1997).

Briefly, key features of each section will be highlighted. Details can be found in the GMP Guidelines published by the Food and Drug jurisdictions as appropriate to herbals, nutraceuticals and dietary supplements.

6.1. PLANT AND PREMISES

The production facility should be designed and constructed to be easily maintained and adequately cleaned and designed to prevent contamination. It should lend itself to orderly flow patterns under clean conditions. It may be simpler to construct new and appropriately designed facilities rather than address the deficiencies faced in refurbishing old buildings. Maintenance SOPs should be available that address cleaning of ceilings, overhead structures and equipment. Good design criteria should consider flows, space, separation (to prevent cross-contamination), location, materials, containment (of dust) and environmental factors (including air exchanges). Provisions should include layouts for quarantine and sampling in addition to segregated storage. Air quality, humidity and temperatures should be considered. Adequate and segregated washup and toilet facilities need to be provided. Utilities, especially water and sewer, and air filtration or air conditioning need to be qualified and verified. If analytical facilities (quality control) are included on the premises, they must be in an area acceptable to their use.

6.2. EQUIPMENT

Attention must be focused on keeping materials isolated, preventing mix-ups or contamination. It is important that the production equipment be designed, constructed, maintained and operated to enable effective cleaning and prevention of contamination while functioning properly and safely. The design and construction must allow parts to be accessed and cleaned where such parts are not fully sealed. Extraneous materials must be excluded and lubricants must be of an approved grade. Multifunctional equipment must have a more stringent cleaning procedure that is validated not only against microbiological contamination but for other prior contaminants as well. Service records must be kept to log prior use and work performed in order to establish the existence of a functioning and effective preventative maintenance program.

6.3. PERSONNEL

A production facility must have adequate staffing of qualified personnel including well-defined position/job descriptions outlining specific responsibilities, designated authority and job tasks. Employees should have or receive adequate training for their particular function and in particular GMPs, SOPs and record keeping. In addition, information for handling of infectious and toxic materials must be supplied (MSDSs) and conveyed. Organizational charts and training records need to be maintained. Specific training requirements and skills as related to QA/QC are mentioned elsewhere, but the importance of the function is summarized appropriately by Cremer (1996) as ''a quality professional's true job is to learn, drive the quality transition, and take the organization where it cannot take itself''.

6.4. SANITATION

Depending on the regulatory jurisdiction, sanitation may be focused only on sanitary programs as related only to the facility and its equipment or it may include personnel hygiene. In the food GMP program, personnel hygiene is typically part of the personnel segment.

Sanitation programs will require written SOPs that provide defined cleaning procedures, a monitoring system evaluating the performance of the program and verification of its effectiveness. Cleaning personnel must be conversant with the need and importance of sanitation to the particular business, proper methods to be used, chemical handling procedures and their respective responsibilities. Records need to be maintained showing that approved chemicals are used and when the cleaning was performed. In addition, the sanitation programs need to address the existence of a pest control program that shows monitoring, control and records. Although pest control programs can be executed by competently trained in-house staff, it is preferable to use the services of a licensed exterminator who maintains the records as evidence for food and drug inspectors, often according to an ISO certified program.

Personnel must also obey stringent hygiene guidelines. Dress codes should be appropriate to the area and activity. Adverse medical conditions (i.e., infectious diseases) should be reported and clearance to return to work should be obtained. Unsanitary practices such as smoking, eating and drinking are not allowed and the presence of plants, food and drink are not permitted in the production area. Cosmetics and jewelry are also not permitted. Garments should be maintained by a laundry program. Adequate training and re-training programs are required.

6.5. RECALL PROGRAM

A process facility must have a written program showing how a recall of unsuitable product could be executed. Traceability is important, and the program needs to address handling of rework and returns. Agencies of jurisdiction must be notified about the recall order, and the degree of potential hazard to the health of the consumer needs to be classified. Additional information on specific compliances to be addressed can be obtained by detailed review of the respective cGMP guidelines published by the Food and Drug regulatory agencies. Some commonsense tips to developing GMP programs can be obtained from a very practical source such as the one written by Immel (1997).

7. ELEMENTS OF A SUCCESSFUL QUALITY PROGRAM

There are a number of critical ingredients and actions that need to come together for a successful quality program that delivers improvements to product, production and compliance quality.

(1) Management must be strongly committed to making it happen. There must be leadership, more than just verbal acknowledgement.

(2) There must be a well-defined focus and alignment to a visible strategic plan. Management needs to ensure that adequate resources are assigned. In particular, QA/QC staff must have their own adequate resources (capable, experienced and possessing a high degree of integrity) to help their companies achieve the quality objectives (Stoker, 1997). Set policies and standards should be defined by QA/QC and senior management that describe the company's philosophy, approach and commitment to quality.

(3) Standards need to be defined reflecting the applicable regulatory requirements so that consistency of product can be assured. One of the most useful steps in planning a quality assurance program is the preparation of detailed product specifications (Morrissey, 1993) so that there is little doubt as to what is required. They must be practical and achievable, yet flexible enough to take improvements into account. Specifications should address raw materials, processing, packaging, finished product, shelf life, transport and government and trade association standards. Requirements of Certificates of Analysis for raw materials may reduce the need for expensive in-house testing.

(4) Adherence to and implementation of hazard analysis critical control point (HACCP) principles is very important in production of safe edible goods (Rault, 1995). Detailed steps to understanding and implementing the seven HACCP principles can be obtained in texts such as that written by Pierson and Corlett (1992). These principles can be extended and applied to evaluation of product quality in addition to food safety.

(5) Training programs for all employees of the company need to be prepared and implemented with provision for re-training with new and amended information.

(6) A system of measurement and reporting needs to be established. These parameters need to be specific, controllable and understandable so that

product, production and compliance to quality can be reviewed from many viewpoints—customer complaints, audits, recalls, internal and external benchmarks (Stoker, 1997). Data gives feedback to management so that actions and resources can be deployed to correct and enact improvements.

A major requirement for a quality improvement program to succeed is that it must be well received by those who will be most impacted, i.e., managers, supervisors and the workforce (Juran and Gryna, 1988). The success of past initiatives can have significant positive or negative effects. Cardinal sins such as lack of upper management involvement, a disorganized approach, significant increases to workload and little or no provision for rewarding efforts diminish chances for quality improvement. To avoid some of these difficulties, quality management approaches are using structured systems such as those achieved through the ISO 9000 programs. The elements to be addressed within the ISO system parallel the GMP components very closely.

8. QUALITY CONTROL

8.1. QUALITY CONTROL DEFINITION

Quality control may also have multiple meanings, but citing the guru on the subject, "quality control is the regulatory process through which we measure actual quality performance, compare it with quality goals and act on the difference" (Juran and Gryna, 1988). Other definitions may include a part of the regulatory process such as product inspection or the tools, skills or techniques through which some or all of the quality function is carried out. In this industry, quality control refers to the tools the technical group uses to measure the accuracy and precision of the methods, procedures and compliance to specifications and is generally a very objective process.

8.2. QUALITY CONTROL PERSONNEL

In the simplest sense, the quality control person or quality control technician will typically require training in microbiology because this factor (biological contamination) is a significant health risk factor. A combined chemistry background is also essential. Larger units with broad analytical capabilities will consist of scientists and technicians skilled in the appropriate sciences related to the products being generated and knowledgeable in analytical methods, equipment and methods development. Small-scale or emerging herbal product manufacturers may be operating at a level where the QC expertise and equipment sophistication is beyond the economics for low-volume operation. In that case, a QA person may be all that is required, and the control analyses may be secured through a competent and established (preferably local) analytical laboratory (Turner, 1997; Kolodziejczyk and Fedec, 1999).

9. CONTROLLING QUALITY IN CRUDE HERBAL AND COMMERCIAL PRODUCTS

9.1. DEFINING PARAMETERS

In most cases, the analysis of all chemical components present in herb or herbal products is not viable, too costly, or simply unnecessary. A typical herbal preparation is manufactured according to specifications that call for a minimum amount of the active component(s)—i.e., "not less than." The level of other important constituents might also be listed, as well as any possible toxic components or compounds causing adverse effects. Applicable legislation may impose statements to be made regarding levels of environmental or processing pollutants, i.e., heavy metals, pesticide levels or herbicide levels.

Development of accurate quantitative methods of analysis for specific components (or marker) compounds in herbal manufacturing is critical for determining the quality and consistency of raw materials and finished products.

Typically, the first operation in the analysis of herbal products is to extract the compound(s) of interest with an appropriate solvent or solvent mixtures. The accuracy of analytical methods depends on the ability to completely and reproducibly extract the analyte of interest. Standard addition and matrix fortification (spiking) are the main methods used to determine the efficiency of extractions.

A frequent difficulty for analytical chemists working with herbal products is the lack of pure, well-characterized reference standards with known stability (Rodriguez et al., 1995; Jenke, 1996). Unfortunately, many herbal standards are either commercially unavailable or too costly to be acquired for routine analysis. Therefore, substantial investment of time and effort in isolation, purification and characterization of analytical standards is required.

There are some concurrent methodologies that allow one to evaluate and statistically validate the efficiency of extraction of specific marker compounds from crude herbal material without the use of pure reference standards (Anderson and Burney, 1998).

9.2. ANALYTICAL TECHNIQUES

Traditionally, analysis of commercial supplies of herbal products is done by visual inspection, taste and microscopic analysis. As the botanical industry grows and is challenged by clients and regulators, it will increasingly shift away from biological measurements toward chemical analyses. Most herbs, or more so their extracts, are not readily distinguishable by microscopic analysis. In addition, microscopic evaluation of plant material in finished products containing multiple botanicals is usually impossible.

Common chemical analyses include techniques such as thin-layer chromatography (TLC), high-performance chromatography (HPLC) and gas-liquid chromatography (GLC). Recently capillary electrophoresis (CE) is finding more applications in characterization of herbal products, especially testing for chiral purity and possible adulteration.

Capillary electrophoresis, developed as a technique in the late 1980s, has steadily grown in popularity. Essentially an analytical method, it has found application in the separation of small chemical molecules as well as biopolymers such as peptides, polysacharides, oligonucleotides, pharmaceuticals and natural products (Tomás-Barberán, 1995). Although the basic methodology involves the separation of molecules based on their charge-to-mass ratio, there are straightforward modifications to the procedure, borrowed from existing well-established techniques that allow separation based on size or isoelectric point. Another advantage is the possibility of separating noncharged molecules, while the degree of resolution can result in the separation of optical isomers. Such applications of CE have attracted a great deal of interest, as the separation of stereoisomers (enantiomers) is an important issue to both the pharmaceutical industry and the natural product chemistry. The ability to develop methods that are capable of resolving optical isomers is of particular relevance as the biological activity of a drug or a natural product is often restricted to one isomer and, even more importantly, there are instances where the "wrong" isomer can have deleterious effects. The best-known example is thalidomide and the devastating teratogenic effect of its isomer present in prescription drugs.

Enantiomers have identical chemical properties and classically are separated from each other by reaction with another chiral compound to give diastereoisomers with differing properties that are then exploited to achieve the separation. More recently, chiral HPLC has been developed where a stationary phase is derivatized with a chiral molecule, and separation is consequent upon the differing affinities between the enantiomers and the stationary phase. In CE, the principle is similar as the technique depends on the use of additives that, because of their own chirality, have differing affinities for the isomers and, hence, are usually termed enantioselective additives. These additives are not stationary in CE but move along the capillary. Cyclodextrins, bile salts and crown ethers are among the most commonly employed enantioselective agents (Ward, 1994; Soini et al., 1994).

The identification of synephrine enantiomers (Figure 13.1) is a good application of CE technology to the herbal industry. It is known that a common Chinese product known as Zhi-shi contains synephrine and is often used in herbal mixtures. The presence of both synephrine enantiomers is an indication

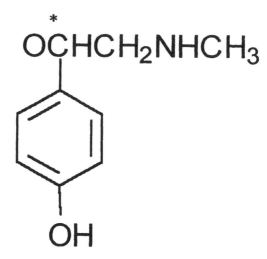

Figure 13.1 Synephrine structure. Asterisk indicates chiral carbon atom.

that the original plant material was adulterated with synthetic synephrine (Figure 13.2).

The application of CE to the area of natural products, biotechnology and the pharmaceutical industry has flourished in recent years, and over the next few years, it will likely further expand. The reason for this interest is the combination of low operating costs, short separation times, only very small amounts of material required for analysis and extremely high separation effi-

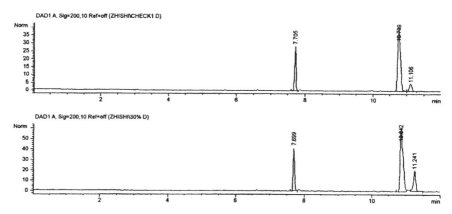

Figure 13.2 Application of capillary electrophoresis in the analysis of Zhi-shi (*Fructus Aurantii immaturus*). Upper frame: natural extract of Zhi-shi powder. Lower frame: extract of Zhi-shi adulterated with synthetic synephrine. The proportion of synephrine peaks corresponding to synephrine isomers is an indication that the original plant material was "improved" with a synthetic (racemic) compound.

ciency coupled with the option that CE can be combined with new mass spectrometry techniques (MALDI, FAB electrospray). Although the initial cost of the system is significant, the extensive range of very powerful methodologies that are applicable to the separation of a very wide range of molecules makes CE an important and versatile technique that has potential in the herbal industry for quality control.

Other useful techniques include nuclear magnetic resonance (NMR), mass spectrometry (MS) or a blend of the above methods. To the last category belong methods combining the chromatographic separation methods with the spectroscopic identification method. There are a wide array of possibilities for these so-called "hyphenated" methods. Examples include the following:

- gas chromatography-mass spectrometry (GC-MS),
- HPLC-MS
- CE-MS
- HPLC-UV often with photodiode array detector (PDA)

10. TECHNIQUE LIMITATIONS/CHALLENGES

No single analytical technique can uniquely identify a wide array of chemical compounds and confirm their structure with the possible exception of NMR, which lacks the sensitivity to address many biological problems. Mass spectrometry, on the other hand, is more sensitive but requires that the molecules be made volatile before they can be analyzed. Volatilization by conventional methods has, until recently, proven impossible for the larger molecules such as proteins, polysaccharides, glycoproteins and other macromolecular entities that exhibit biological activity and are often unique for a given medicinal plant. It is only with the advent of the desorption techniques such as fast atom bombardment (FAB), plasma desorption and laser desorption mass spectrometry that the intact molecules can be volatilized and ionized.

Matrix-assisted laser desorption/ionization (MALDI) mass spectrometry is a recent desorption technique that was originally introduced for the ionization of large peptides and proteins. However, it soon became apparent that many other types of compounds, including important bioactive macromolecules, could also be ionized. Sample consumption is minimal, with amounts only in the femtomole to low-picomole range currently required for analysis. It has been reported that MALDI is 10–100 times as sensitive as FAB mass spectrometry (Huberty et al., 1993) and that, unlike FAB, the technique is reasonably tolerant to the presence of buffer salts and other additives. Therefore, MALDI is a perfect "marriage" partner for the capillary electrophoresis (CE) technique and its application to herbs and their products, providing a powerful tool for bioactive component profiling and a rapid method for structural determination. In addition, unlike FAB mass spectrometry, no derivatization is required.

Recent developments in mass spectroscopy, especially MALDI and similar methods (electrospray, etc.), allow fast and cost-effective analyses of complex mixtures and identification of diagnostic ions. In combination with pattern recognition analysis or function-based neural network analysis (Zhao et al., 1998), it is possible to characterize complicated mixtures and analyze for the presence/ or absence of a given component (chemical entity) in an analyzed mixture.

The degree of sophistication in analytical methods depends on financial resources available to the testing or quality control laboratory, level of expertise residing with their staff and, finally, the price/value of the analyzed product. Obviously, in an industrial setting, there will either be no will or lack of economical justification for spending significant monies for analyses of very low-value product. The situation might be different if basic research or a regulatory agency becomes involved. In such situations, economics will not be the deciding factor. It has recently become obvious, through increased scrutiny and regulatory pressures, that insuring the proper identity and even quality of commercial supplies of herbal products is essential if the herbal industry is to continue development.

An example of GC-MS technique application is presented in Figure 13.3 for Kava (*Piper methysticum*), an herbal drink of South Pacific origin gaining popularity in North America and Europe. Kava's apparent activity is related to the reduction of anxiety, tension and excitedness. It is used as a traditional ceremonial drink in Polynesia and Oceania. There is no obvious chemical component responsible for physiological effects (Singh and Blumental, 1996). Despite this fact, lactones (two series of closely related compounds that are mainly either substituted alpha-pyrones or substituted 5,6-dihydro-alpha-pyrones) are being used as markers for quality evaluation in the trade of kava extracts. Usually, kava extracts are sold as a product containing 30% of the lactone mixture. There are several methods for evaluation of the lactones content in herbal preparations. The qualitative method might be based on TLC, however, for quantitative evaluation, HPLC or GC methods are preferred. In both methods, the exact content of lactones is determined by integrating the area under the peaks corresponding to given components. Because of possible overlapping peaks or the presence of compounds with similar chromatographic properties, positive identification of substances eluting at given times strengthens the results. In our laboratory, the GC-MS method is applied to analysis of kava lactones. Each major peak in the chromatogram is identified by its mass spectra, which includes the parent ion, corresponding to its molecular mass, as well as characteristic (unique) fragmentation patterns.

11. DEVELOPING PRECISE QUANTITATIVE METHODS FOR SPECIFIC MARKER COMPOUNDS IN HERBAL PRODUCTS

Over the last decade, transgenic or genetically engineered plants have entered the marketplace. Currently, a large percentage of important agriculture

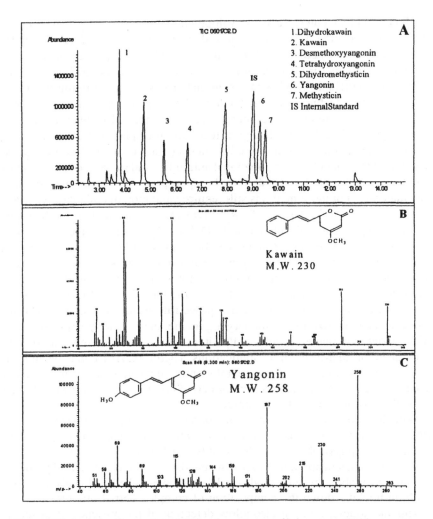

Figure 13.3 Analysis of kava extract. (A) Gas chromatogram of kava extract; (B) Mass spectrum of compound 2, kawain; (C) Mass spectrum of compound 6, yangonin. Conditions: HP-1GC column 0.32 mm × 25 m × 0.52 μm. Temperature gradient 180–295°C; rate 5°C/min. Detection: HP 5972 MS; scan range: 50–400 mu.

crops are produced using transgenic plants. In the U.S. over 40% of the soybeans, 20% of the corn and 5% of the cottonseed is transgenic. In Canada, about 50% of the canola crop is produced using transgenic seed. This process is slower within the European Community, where labeling of food products containing genetically modified organisms (GMO) is required (EC Novel Food Regulation 258/97 and Council Regulation EC 1138/98).

Existing technology in gene manipulation such as the above examples has opened doors to production in plants of new food additives, chemicals, antibodies and vaccines. Recent developments in production of chemicals via plant bioengineering have been reviewed (Kolodziejczyk and Shahidi, 1999). Recent and spectacular developments in reliable and quantitative analytical detection methods for analysis of protein and DNA may be applicable in the future for medicinal plant testing. Automated DNA fingerprinting has made a variety of new techniques available to medicinal plant research at reasonable prices. These techniques have been successfully applied to identification of genetic material in forensic analysis, recognition of plant varieties or testing for transgenic plant material in food products (Bates et al., 1996). Immunoassays based on monoclonal or polyclonal antibodies and ELISA (enzyme-linked immunosorbent assay) methods may be appropriate for testing unprocessed raw materials, but these methods still require validation in collaborative trials.

The specificity of qualitative PCR (polymerase chain reaction) methods can be highly dependent on the choice of the DNA and primers, and their sensitivity is encouraging. Amplified fragment polymorphism (AFLP) technology is a novel and powerful DNA fingerprinting technique (Zabeau and Vos, 1992; Vos et al., 1995; Perkin-Elmer Protocol, 1995). It is based on the selective amplification of restriction fragments from the total digest of genomic DNA.

The AFLP approach is particularly powerful because it requires no prior sequence characterization of the target genome and is readily applicable to a wide variety of plants. Additionally, it is easily standardized and readily automated for high-throughout applications. Therefore, AFLP may be an ideal tool for determining species or varietal identity, detecting admixtures of plant material or determining adulteration of herbal mixtures, but unfortunately, the technology is not applicable to extracts.

With the further development, automation and lowering of costs per analysis, DNA fingerprinting might become a versatile and complementary tool to chemical and physical methods for analysis of herbs.

When a sufficient amount of experimental data for important herbal products becomes available, these methods could supplement chemical analysis, especially in tracking down the composition of complex herbal mixtures.

This methodology might also allow detection of the presence of different plant species not using the chemical composition of plant metabolites, but identification through the analysis of their genetic profile.

As the evaluation of these methods becomes more sophisticated, this technology is expected to provide a high degree of precision, with the above method expected to provide an efficient tool to detect counterfeiting, mislabeling and erroneous use.

12. STANDARDIZATION DIFFICULTIES

Standardization of herbal products is not always an easy task, even for herbal products that are very popular and have been established on the market

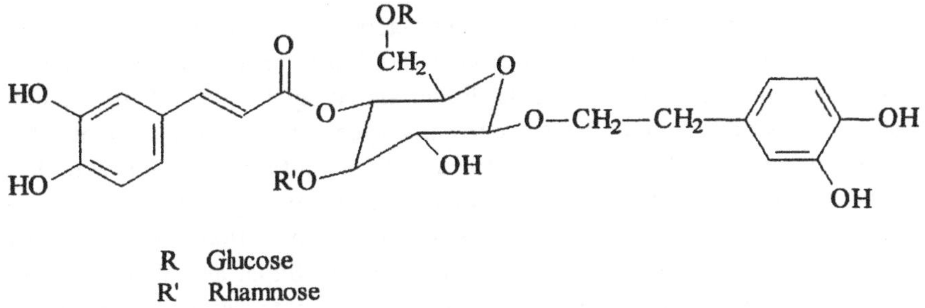

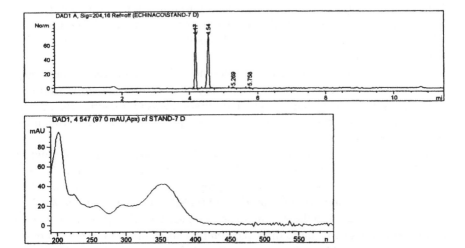

R Glucose
R' Rhamnose

Figure 13.4 Echinacoside structure, a commonly used marker for evaluation of *Echinacea* extract.

for many years. The reason for this difficulty is that often there is no single component responsible for the desired biological effect. There is evidence that a combination of bioactive components may be more effective than pure single chemical compound because of a synergistic effect (Foster, 1998).

A classic example of this controversy is the standardization of popular *Echinacea* preparations that are still the subject of many discussions. Some producers use echinacoside (Figures 13.4 and 13.5) as a reference standard. However, there are others (Schumacher and Friedberg, 1991) questioning the role of echinacoside as an active component and suggesting that other compounds such as alkylamides may be the more appropriate candidates for

Figure 13.5 CE Analysis of *Echinacea* extract. Upper frame: Electropherogram of *Echinacea* extract. Conditions: Glass uncoated capillary 50 μm × 56 cm, voltage 30 kV, temperature 30°C buffer pH 8 5,: 150 mM sodium borate, 75 mM SDS, UV detection at 204 nm. Lower frame. UV/VIS spectra of substance eluting at 4.54 min (Echinacoside), absorption spectrum and elution time confirm identity of analyte. Peak at 4.17—internal standard.

standardization. Of the proposed immuno-active components from *Echinacea* species, the polyacetylenes and chicoric acid are unstable and often are not detectable in commercial products (Bauer and Wagner, 1991). Only the polysaccharides and alkylamides are currently accepted as active components in commercial products. It was also shown that the content of major low-molecular weight components in *Echinacea* extracts is in the range of 0.1–1% (Wagner and Jurcic, 1991), and their profiles are distinctly different for *Echinacea pallida, E. angustifolia* and *E. purpurea* species (Bauer and Wagner, 1991).

The difficulty of determining an active component in a given herbal preparation is not so uncommon. Another example of a fashionable herb apart from *Echinacea*, is St. John's wort, or *Hypericum perforatum*. Recent popularity of this herb in North America came as an extension of its fame in Germany. Widely used as an alternative to synthetic antidepressants, St. John's wort is the main component of two popular medications against depression, accounting for almost 15% of the antidepressant market, or $55 million in 1996 (Briggs and Briggs, 1998). Initially, it was believed that antidepressant activity was related to the level of hypericin (Suzuki et al., 1984).

Neurotic depressive reactions are characterized by feelings of sadness, self-criticism, apathy and guilt, loss of initiative, and in more severe cases, suicidal tendencies and marked withdrawal. The neurochemical basis of these disorders is determined by the balance of activity between central monoamine-containing and acetylcholine-containing neurons. In depressive illness, the depressed behavioral activity appears to be due mainly to impairment of noradrenaline-containing systems coupled with uncompensated overactivity of acetylcholine-containing systems, and the depressed mood also involves dysfunction of seronitonine-containing neurons.

The mood-elevating effect of monoamine oxidase inhibitors is generally reckoned to be due to increases in availability of monoamine neurotransmitters at synapses in the central nervous system; brain levels of noradrenaline, dopamine and serotonin being increased as a consequence of their inactivation by monoamine oxidase.

Monoamine oxidase activity was believed to be due to hypericin (Figure 6.2), but subsequent studies suggest that other components in St. John's wort are responsible for its psychotherapeutic properties, possibly related to another neurotransmitter gamma aminobutyric acid (GABA) or to interleukin metabolism. Now, it is believed that the antidepressant activity of *Hypericum perforatum* is probably due to more than one component of the complex mixture of flavonoids, hyperoside and rutin, as well as biflavones (Hobbs, 1989).

Once again, similar to the case of *Echinacea*, there is no obvious indication that a marker constituent, hypericin or echinacoside, parallels the therapeutic effect. However, until better markers are identified, existing markers will be used for quality control during processing and trade. When the former is accomplished, some different approaches might be used and controlled and, to some extent, directed biosynthesis of natural products is possible.

Several commercial products derived from cell cultures are under development (Sicha et al., 1989). These cultures may have future importance in the production of large quantities of standardized "phytomedicines" for general use (easy to maintain pure line species).

13. SPECIAL PROBLEMS IN CONTROLLING QUALITY

13.1. ORGANICALLY GROWN STATUS

The recent onset of commercial harvesting may pose a threat to many wild species, for example, *Echinacea angustifolia*, a plant native to North America and growing wild on the plains. It has been used for generations by the First Nation people, but today, due to intensive harvesting, this plant has become almost extinct in its known habitat in the Souris River Valley in Saskatchewan (Harms, 1999). Harms goes on to say "the recent onset of commercial harvesting may now pose a threat to it and the wild harvesting of whole plants should not be allowed". Instead, a new set of practices stemming from Europe called Good Agriculture Practices (GAP) should be exercised for industrial production of high-purity herbal products. The so-called "organic farming," using none or strictly limited amounts of pesticides and herbicides and high-quality fertilizers free of heavy metals, is also recommended.

Herbal plants are not exempted from being attacked by insect pests and plant pathogens. Use of pest control agents can provide better yield and return for the herb producer. However, it might be a factor disqualifying the herbs from their application in medicinal use. International codes of pesticide residues in food plants are in place in most developed countries, but regulations on herbal products are rather scarce.

13.2. HEAVY METALS AND POLLUTANTS

Similar to any food product, possible contamination of herbal preparations with toxic contaminants should be reviewed. Standards for heavy metals in food are not really harmonized through different legislations/countries, but they are nonexistent for herbs.

As the demand for herbs increases and the price differential existing among countries influences trade, North America will face increased imports of less expensive herbal products from developing countries such as China, India and Brazil where pesticide and industrial contamination is common, and arsenic and mercury are intentionally added to the products (for rodent control).

13.3. ARE ENGINEERED PLANTS "NATURAL PRODUCTS"?

Recent advances in biotechnology allow genes to be expressed, leading to the production of desired chemicals in bioengineered (GMO) plants. Should these components still be considered "natural products"?

In 1998, the USDA asked for comments on whether food that has been irradiated, grown with sewage sludge or genetically engineered should be considered organic. These three options had already been rejected by the National Standards Board. In disclosed information (Hileman, 1998), the USDA received 18 comments in favor and 83,081 opposed for genetic engineering; 29 comments in favor of and 83,760 opposed to sewage sludge; and 18 comments supporting and 76,062 opposing the use of irradiation of organic food. Therefore, "organic, genetically modified products" continue to be a hard sell to the public.

14. CONCLUSION

Ensuring quality is much more than detection, specification and process control. It also includes awareness of every aspect of a manufacturing process from research to shipping. As the diversity of plant materials available for herbal applications increases, as the desire for more natural and herbal-based products becomes ingrained in our society, as the sale of plant materials having less well-defined qualitative markers (influencing sometimes not well-understood biological relationships) present situations for misinterpretation or misrepresentation, and as regulatory bodies strive to influence or define guidelines to ensure consumer health safety, there is an ever-increasing emphasis on the importance for the role of Quality Assurance/Quality Control units in overseeing GMP compliance and process validation (Wechsler, 1997).

15. REFERENCES

Akerele, O. 1993. "Summary of WHO guidelines for the assessment of herbal medicines," *HerbalGram*, 28:13–20.

Anderson M. L. and Burney, D. P. 1998. "Validation of sample preparation for botanical analysis," J. AOAC Int., 81:1005–1010.

Bates, S. R. E., Knorr, D. A., Weller, J. W. and Ziegle, J. S. 1996. "Instrumentation for automated molecular marker acquisition and analysis," in *The Impact of Plant Molecular Genetics*, ed., B. W. S. Sobral, Cambridge, MA: Birkhauser Boston, pp. 239–255.

Bauer, R. and Wagner, H. 1991. "*Echinacea* species as potential immunostimulatory drugs," in *Economic and Medicinal Plant Research*, vol. 5., eds., H. Wagner and N. R. Farnsworth, New York, NY· Academic Press, pp. 253–321.

Briggs, C. J. and Briggs, G. L. 1998. "Herbal Products in Depression Therapy," *CPJ*, Nov., 1998, pp. 40–44.

Cremer, C. J. 1996. "The evolution of the QC guy," *Quality Progress,* July, 1996, pp. 42–43.

DeCicco, J. 1997 "Importance and application of quality in the food, drug and cosmetic industry," *ASOC FDC Control,* 114 (Apr): 4–6.

De Sain, C. and Sutton, C. V. 1996. "Standard operating procedures: content, format, and management," *Pharmaceut. Technol.,* Oct., 1996, pp. 110–116.

Foster, S. 1998. "Standardized herbs: missing the mark," *Health Foods Business,* Dec., 1998, pp. 51–52.

Giese, J. H. 1996. "QA/QC, FAQ, IFT, WWW, HACCP, and ISO-IMHO," *Food Technol,* 50(3):42, 176.

Harms, V. 1999. "Prairie home to three distinct coneflowers," *Saskatoon Sun,* Feb. 21, 1999, p. A23.

Hileman, B. 1998. "Organic food proposal elicits flood of comments," *C&EN,* May 11, 1998, pp. 7–8.

Hobbs, C. 1989. "St. John's Wort (*Hypericum perforatum* L.)," *HerbalGram,* 18/19:24–33.

Huberty M. C., Vath J. E., Yu, W. and Martin, S. A. 1993 "Site specific carbohydrate identification in recombinant proteins using MALDI-TOF MS." *Anal. Chem.,* 65:2791–2800.

Immel, B. 1997. "30 Tips to bring you closer to compliance," *BioPharm,* Sept, 1997, pp. 66–71.

Jenke, D. R. 1996. "Chromatographic method validation· A review of current practices and procedures. Guidelines for primary validation parameters," *J. Liq. Chromatogr.,* 19:737–757.

Juran, J. M. and Gryna, F. M. 1988. *Juran's Quality Control Handbook.* 4th Ed. New York, NY: McGraw-Hill, Inc., 1,000 pp

Kieffer, R. G., Stoker, J. R. and Nally, J D 1998. "The knowledge and skills of the successful QA/QC manager," *Pharmaceut. Technol.,* Jan., 1998, pp. 70–78.

Kolodziejczyk, P P. and Fedec, P. 1999. "Recent progress in agriculture biotechnology and opportunities for contract research and development," in *Chemicals via Higher Plant Engineering,* eds., F. Shahidi, A Lopez-Munguia, P. P. Kolodziejczyk, G. Fuller and J. R. Whitaker, New York, NY: Klumer Academic/Plenum Publishers, pp 5–20.

Kolodziejczyk, P P and Shahidi, F. 1999. "Novel chemicals from plants via bioengineering," in *Chemicals via Higher Plant Engineering,* eds., F. Shahidi, A. Lopez-Munguia, P P. Kolodziejczyk, G. Fuller and J. R. Whitaker, New York, NY: Klumer Academic/Plenum Publishers, pp. 1–4.

Matsalla, R. J. 1992. "Canadian regulatory requirements for traditional herbal drugs," *C.H.F A. Newsletter,* Oct., 1992, pp. 8–9.

McCaleb, R. 1997. "FDA publishes proposed regulations for good manufacturing practices for dietary supplements," *HerbalGram,* 40·24–26.

Morrissey, P. 1993 "Specifications· An integral part of the QA program," *Frontiers,* Dec., 1993, pp 1–2.

Oliver, E. 1999. Personal communication. Columbia Pharma Consulting Services, Issaquah, WA, U S.

Perkin-Elmer Instruments. 1995. AFLP. Plant Mapping Kit Protocol (P/N 402083).

Pierson, M. D. and Corlett, Jr., D. A. 1992. *HACCP: Principles and Applications.* New York, NY: AVI: Van Nostrand Reinhold, 212 pp.

Rault, D. 1995. "Quality Assurance," *Grocer Today,* Dec , 1995, p. 46.

Rodriguez, L. C , Campana, A. M. C., Barrero, F. A., Linares, C. J. and Ceba, M. R. 1995. "Validation of an analytical instrumental method by standard addition methodology," *J. AOAC Int.,* 78·471–476.

Schumacher, A. and K. K Friedberg, 1991. "Analyses of the effect of *Echinacea angustifolia* on unspecified immunity of the mouse," *Arzneim -Forsch.* 41:141–147.

Sicha, J., Hubik, J. and Dusek, J. 1989. "Production of phenylpropanones by tissue cultures of the Genus *Echinacea*," *Ceskoslov. Farm.* 38 128–129.

Singh, Y N. and Blumental, M. 1996. "Kava an overview," *HerbalGram*, 39 34–56.

Soini, H., Stefansson, M., Riekkola, M. L. and Novotny, M. V. 1994. "Maltooligosaccharides as chiral selectors for the separation of pharmaceuticals by capillary electrophoresis," *Anal. Chem.*, 66·3477–3484.

Stoker, J. R 1997. "Top managers, top quality," *Pharmaceut. Exec.*, Feb., 1997, p. 64.

Suzuki, O , Katsumata, Y , Oya, M., Bladt, S. and Wagner, P. 1984. "Inhibition of monoamine oxidase by hypericin," *Planta Medica.*, 50·272–274.

Tomás-Barberán, F. A. 1995. "Capillary electrophoresis: A new technique in the analysis of plant secondary metabolites," *Phytochemical Analysis*, 6·177–192.

Townsend, P. I and Gebhardt, J. E. 1992. *Quality in Action, 93 Lessons in Leadership, Participation and Measurement.* New York, NY: John Wiley and Sons, Inc., 261 pp.

Turner, L. 1997. "Using custom manufacturers," *Supplement Industry Exec.*, Winter, pp. 28–32

Uys, P. 1994. *Good Manufacturing Practice. A Guide to Practical Quality Management.* Randburg, S. Africa: Knowledge Resources (Pty) Ltd , 164 pp

Vos, P , Hogers, R., Bleeker, M., Reljans, M., van de Lee, T., Hornes, M., Frijters, A., Pot, J , Peleman, J., Kuiper, M. and Zabeau, M. 1995. "AFLP· A new technique for DNA fingerprinting," *Nucleic Acids Res.*, 23:4407–4414.

Wagner, H. and K. Jurcic. 1991. "Immunological studies of plant extract combinations *in vitro* and *in vivo* on the stimulation of phagocytosis," *Arzn.-Forsch.*, 41:1072–1076.

Ward. T. J. 1994. "Chiral media for capillary electrophoresis," *Anal. Chem.*, 66:632–640.

Wechsler, J. 1997. "Regulatory reform boosts QC," *BioPharm*, Aug., 1997, pp. 51–52.

Yuan, R. 1998 "Herbal pharmaceutical industry," *Genetic Eng. News*, 18:1–3.

Zabeau, M. and Vos, P. 1992. "Selective restriction fragment amplification: A general method for DNA fingerprinting," European Patent Application 92402629.7.

Zhao, W., Chen, D. and Hu, S. 1998. "Potential function-based neural networks and its application to the classification of complex chemical patterns," *Comput. Chem.* (Oxford), 22:385–391.

Index

T - #0025 - 111024 - C0 - 229/152/25 - PB - 9780367398521 - Gloss Lamination